FLAVONOIDS - FROM BIOSYNTHESIS TO HUMAN HEALTH

Edited by **Gonçalo C. Justino**

Flavonoids - From Biosynthesis to Human Health
http://dx.doi.org/10.5772/65575
Edited by Gonçalo C. Justino

Contributors

Ana Rodrigues, Duraipandiyan Veeramuthu, William Raja T, Naif Abdullah Al-Dhabi, Ignacimuthu Savarimuthu, Beatriz Ramos Solano, Enrique Gutierrez, Ana García-Villaraco Velasco, Jose A. Lucas, F.Javier Gutierrez-Mañero, Weisheng Feng, Zhiyou Hao, Meng Li, Corina Danciu, Diana Antal, Florina Ardelean, Aimée Rodica Chis, Codruta Soica, Florina Andrica, Cristina Dehelean, Valentina Uivarosi, Alexandra Cristina Munteanu, Agustín G. Asuero, Julia Martín, Mª José Navas, Eugenia Marta Kuskoski, Maged Saad Abdel-Kader, Ahmed Foudah, Marisi Soares, João Luiz Baldim, Mário Ferreira Conceição Santos, Daniela Aparecida Chagas-Paula, João Henrique Ghilardi Lago, Welton Rosa, Maurice Ducret Awouafack, Marcos Soto-Hernandez, Graham Johnston, Jane Hanrahan, Tina Hinton, Raul Zamora-Ros, Romina Escobar-Cévoli, Carlota Castro-Espín, Virginie Béraud, Genevieve Buckland, Jose Luis Chavez-Servia, Araceli M. Vera-Guzmán, Elia N. Aquino-Bolaños, Elena Heredia-García, José C. Carrillo-Rodríguez, Sanjuana Hernández-Delgado, Hakan Cetinkaya, Muhammet Karaman, Halide Sedef Karaman, Ferudun Kocer, Muhittin Kulak, Chin Fei Chee, Noorsaadah Abd. Rahman, Ti Tee, Saul Ruiz-Cruz, Sarai Chaparro-Hernandez, Karen Lilian Hernández-Ruiz, Marco Antonio Lopez-Mata, Jesus Ornelas-Paz, Luis A. Cira-Chávez, María Isabel Estrada-Alvarado, Laura Elisa Gassos-Ortega, Rosa Isela GuzmánGerónimo, Mónica Herrera-Sotero, Francisco González-Cortés, Hugo García-Galindo, Enrique Juárez-Aguilar, Jose Chavez-Servia, Mauricio Rodríguez-Dorantes, Rosamaria Oliart, Erica Santos, Beatriz Helena Sales Maia, Aurea Ferriani, Sirlei Teixeira, Latifa Chebil, Manel Slimane, Mohamed Ghoul, Aude Cordin

Published by InTech
Janeza Trdine 9, 51000 Rijeka, Croatia

Publishing Process Manager Dajana Pemac
Technical Editor SPi Global
Cover InTech Design team

First published August, 2017
Printed in Croatia
Legal deposit, Croatia: National and University Library in Zagreb

Additional hard copies can be obtained from orders@intechopen.com

Flavonoids - From Biosynthesis to Human Health, Edited by Gonçalo C. Justino
p. cm.
Print ISBN 978-953-51-3423-7
Online ISBN 978-953-51-3424-4

For EU product safety concerns:
IN TECH d.o.o., Prolaz Marije Krucifikse Kozulić 3, 51000 Rijeka, Croatia,
info@intechopen.com or visit our website at intechopen.com.

Meet the editor

Gonçalo C. Justino is a postdoctoral research fellow at CQE/IST, Universidade de Lisboa (Portugal), and an assistant professor at ESTB, Instituto Politécnico de Setúbal. He received a PhD in Clinical and Pharmaceutical Biochemistry in 2007. His research interests have been focused on the mass spectrometry analysis of flavonoid derivatives and metabolites and more recently on the application of computational techniques in protein structure and drug design.

Contents

Preface

Flavonoids are abundant secondary metabolites found in plants and fungi that have various roles in these organisms, including pigmentation, cell signalling, plant defence and inter-organism communication. Due to their abundance in nature, flavonoids are also important components of the human diet, and the last four decades have seen an intense study focused on the structure characterization of flavonoids and on their roles in mammal metabolism.

Initially, flavonoids were identified as being potent antioxidants, capable of inhibiting many oxidative and nitrosative stress-induced lesions to biomolecules, which are considered to be at the onset of a number of human pathologies. They are also good chelating agents, and as they are partly planar, it has been shown that they are able to interfere with processes that involve DNA.

These antioxidant and chelating activities were present in both aglycones and glycosides but, most importantly, in the metabolites found in circulation in humans. This research area has evolved to include, first, the anti- and pro-inflammatory activities of flavonoids and, more recently, many other aspects of human biology, including modulation of neuronal signalling.

In this book, most of the well-established activities of flavonoids are reviewed, and recent research studies on the area of flavonoids are presented. Section 1 reviews the chemical aspects of structure characterization of flavonoids, also focusing on isoflavonoids and homoisoflavonoids. Section 2 addresses the biosynthesis of flavonoids in model plants as well as their role in abiotic stress situations and in agriculture and also presents a biomimetic approach to Diels-Alder flavonoid compounds. Section 3 deals with the role of flavonoids in metabolism and health, from their antioxidant and chelating action to their anticancer and neuronal modulation activities. Section 4 addresses the importance of flavonoids in foods, from consumption to their use as bioactive components.

Finally, a special thanks to Pedro Pinheiro for his invaluable contributions to this work.

Gonçalo C. Justino
Centro de Química Estrutural,
Instituto Superior Técnico,
Universidade de Lisboa,
Lisboa, Portugal

Section 1

The Chemistry of Flavonoids

Chapter 1

Flavonoids: Classification, Biosynthesis and Chemical Ecology

Erica L. Santos, Beatriz Helena L.N. Sales Maia,
Aurea P. Ferriani and Sirlei Dias Teixeira

Additional information is available at the end of the chapter

http://dx.doi.org/10.5772/67861

Abstract

Flavonoids are natural products widely distributed in the plant kingdom and form one of the main classes of secondary metabolites. They display a large range of structures and ecological significance (e.g., such as the colored pigments in many flower petals), serve as chemotaxonomic marker compounds and have a variety of biological activities. Therefore, they have been extensively investigated but the interest in them is still increasing. The topics that will be discussed in this chapter describe the regulation of flavonoid biosynthesis, the roles of flavonoids in flowers, fruits and roots and mechanisms involved in pollination and their specific functions in the plant.

Keywords: flavonoids, biosynthesis, pollination, allelochemicals, chemical ecology

1. Introduction

Flavonoids represent a highly diverse class of polyphenolic secondary metabolites, which are abundant in spermatophytes (seed-bearing vascular land plants: gymnosperms (cycades, conifers, ginkos and gnetophytes) and angiosperms) [1–3] but have also been reported from primitive taxa, such as bryophytes (nonvascular land plants, including liverworts, hornworts and mosses) [4, 5], pteridophytes (seedless vascular land plants, i.e., lycophytes, horsetails and all ferns) [6, 7] and algae [8, 9]. Overall, about 10,000 flavonoids have been recorded which represent the third largest group of natural products following the alkaloids (12,000) and terpenoids (30,000) [1, 10].

Flavonoids are essential constituents of the cells of all higher plants [11]. Plants have evolved to produce flavonoids to protect themselves against fungal parasites, herbivores, pathogens and ultraviolet (UV) radiation [10]. They resemble in their regulatory properties most of the lipid-soluble vitamins but serve, in addition, due to their color, as communicators with the environment. Flavonoids are recognized by pollinators, for example, insects, birds and animals, which contribute to the dispersion of seeds [11]. They act as symbionts, as allelochemicals, as antimicrobial and antiherbivory factors [10, 12]. Many studies have shown that flavonoids exhibit biological and pharmacological activities, including antioxidant, cytotoxic, anticancer, antiviral, antibacterial, anti-inflammatory, antiallergic, antithrombotic, cardioprotective, hepatoprotective, neuroprotective, antimalarial, antileishmanial, antitrypanosomal and antiamebial properties [13–15].

The topics that will be discussed in this chapter describe the regulation of flavonoid biosynthesis, the roles of flavonoids in flowers, fruits and roots and mechanisms involved in pollination and their specific functions in the plant.

2. The classification and biosynthesis of flavonoids

Flavonoids can be classified according to biosynthetic origin. Flavonoids are characterized by the presence of 15 carbon atoms in their basic skeleton, arranged in the form C6-C3-C6, which corresponds to two aromatic rings A and B linked by a unit of three carbon atoms, which may or may not give rise to a third ring. The rings are labeled A, B and C [15, 16]. The initial step in the biosynthesis of most flavonoids is the condensation of one *p*-coumaroyl-CoA molecule (shikimate derived, B ring) with three molecules of malonyl-CoA (polyketid origin, A ring) to give chalcone (2′, 4′, 6′, 4-tetrahydroxychalcone). This reaction is carried out by the enzyme chalcone synthase (CHS) [14–16]. Chalcone is subsequently isomerized by the enzyme chalcone flavanone isomerase (CHI) to flavanone. From these central intermediates, the pathway diverges into several side branches, each yielding a different class of flavonoids (**Figure 1**) [14, 16, 17].

Although the central pathway for flavonoid biosynthesis was conserved in plants, depending on the species, a group of enzymes, such as isomerases, reductases, hydroxylases, modifies the basic flavonoid skeleton, leading to the different flavonoid classes [1, 16], including chalcones and flavanones which are intermediary compounds in biosynthesis and final products present in various parts of the plant. Anthocyanins, proanthocyanidins, flavones and flavonols are other classes only known as end products of biosynthesis. The other important class is the isoflavonoids, which are formed by migration reaction of 2-aryl side chain to 3-position mediated by isoflavone synthase [1, 16, 18].

The retrochalcones are unusual flavonoids and have reversed A and B rings. The biosynthesis is not yet clearly defined but is likely to be derived from the common C_{15} intermediate of general flavonoid biosynthesis, more specifically from the reduction of dibenzoylmethanes [19–21] or by 2-hydroxylation of a flavanone [22, 23]. These compounds are restricted to relatively few plant species and have been isolated from some species of the families Leguminosae [24], Annonaceae and Basellaceae [25–28].

http://dx.doi.org/10.5772/67861

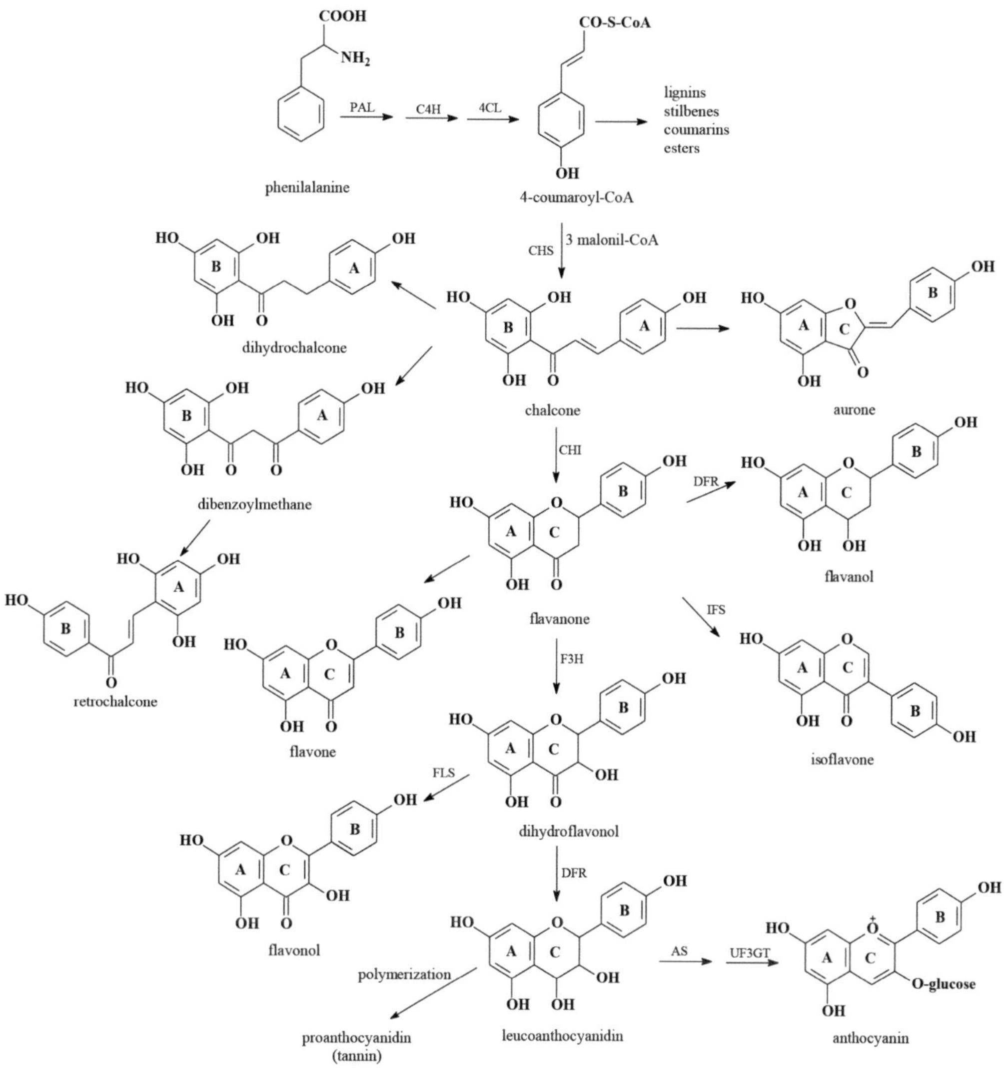

Figure 1. A diagram of the flavonoid biosynthetic pathway. Key enzymes catalyzing some reactions: PAL, phenylalanine amonialyase; C4H, cinnamate 4-hydroxylase; 4CL, 4-coumaroyl-coenzyme A ligase; CHS, chalcone synthase; CHI, chalcone flavanone synthase; F3H, flavanone 3β-hydroxylase; DFR, dihydroflavonol 4-reductase; FLS, flavonol synthase; IFS, isoflavonoid synthase; AS, anthocianin synthase and UF3GT, UDP glucose: flavonoid 3-O-glucosyltransferase. Adapted from Ref. [17].

Several species reported the presence of chalcone dimers bound by a cyclobutane (**Figure 2**) [29–35]. The phytochemical study of the roots of *Dahlstedtia grandiflora* was observed, and for the first time, the occurrence of dimerization in retrochalcones was noted [24]. The mechanisms of [2 + 2] cycloaddition involved in the formation of these compounds are suggested [30, 35]. In spite of the lack of biosynthetic studies of these natural products, much effort has been made in elucidating the biosynthetic pathways of flavonoids from a genetic perspective.

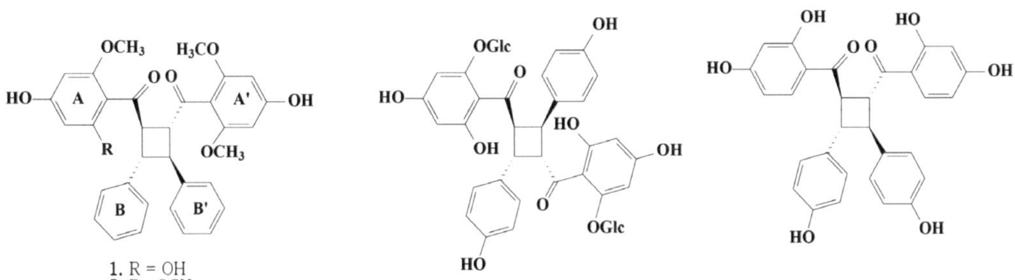

1. R = OH
2. R= OCH_3

Figure 2. Chalcone dimmers isolated from *Combretum albopunctatum* [31], *Helichrysum zivojinii* [29] and *Agapanthus africanus* [30].

Flavonoids occur naturally as compounds associated with sugar in conjugated forms (glycosides), without attached sugar as aglycones [1, 36]. They are often hydroxylated in positions 3, 5, 7, 3′, 4′ and 5′. Some of these hydroxyl groups are frequently methylated, acetylated or sulfated. Prenylation usually occurs directly at a carbon atom in the aromatic rings, but O-prenylation has also been found [11]. When glycosides are formed, the glycosidic linkage is normally located in position 3 or 7, and the carbohydrates are commonly L-rhamnose, D-glucose, glucose rhamnose, galactose or arabinose [1, 11]. These changes often alter their solubility, reactivity and stability. The majority of flavonoids are present in the form of glycosides under natural conditions [1].

2.1. The chemical structure of flavonoids

The chemical nature of flavonoids varies according to the hydroxylation pattern, conjugation between the aromatic rings, glycosidic moieties, methoxy groups and other substituents [37–39]. Flavonoids contain conjugated double bonds and groups (hydroxyl or other substituents) that can donate electrons through resonance to stabilize the free radicals, which originate in the electronic spectra of flavonoids [40].

Studies on flavonoids by UV spectroscopy have shown that most flavonoids consist of two major absorption maxima: band II (240–285 nm) which corresponds to the benzoyl system of the A ring, while band I (300–400 nm) represents the cinnamoyl system of the B ring (**Figure 3**) [36, 41].

Functional groups attached to the flavonoid skeleton may cause a shift in absorption. The application of standardized UV (or UV-Vis) spectroscopy has for years been used in analyses of flavonoids [11].

Flavonoids have the ability to sequester free radicals, are natural antioxidants derived from plants and are commonly found in foods and beverages [40]. The main structural features of flavonoids required for antioxidant activity can be determined by three fundamental factors: (1) a 3′,4′-dihydroxy (catechol) structure in the B ring favors the electron delocalization (**A**), (2) an unsaturated 2-3 bond in conjugation with a 4-keto group provides electron delocalization from the B ring (**B**) and (3) hydroxyl groups at positions 3 and 5 form intramolecular hydrogen bonding to the keto group (**C**) (**Figure 4**). These effects lead to the increases of the radical scavenging by delocalization of electrons or by donation of hydrogen [42].

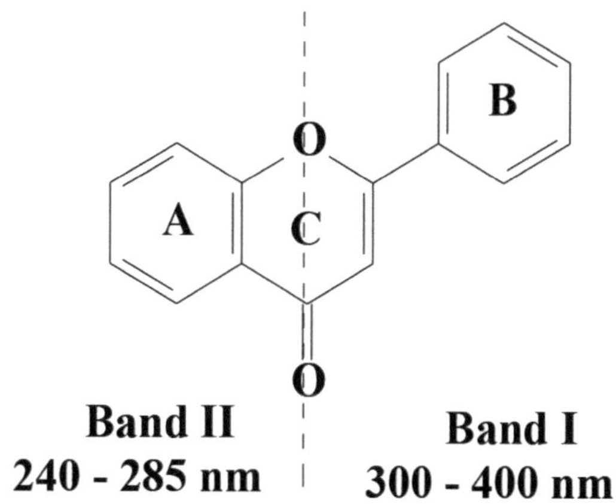

Figure 3. Band II absorption (originated from A-ring benzoyl system) and band I (from the B-ring cinnamoyl system). Adapted from Ref. [41].

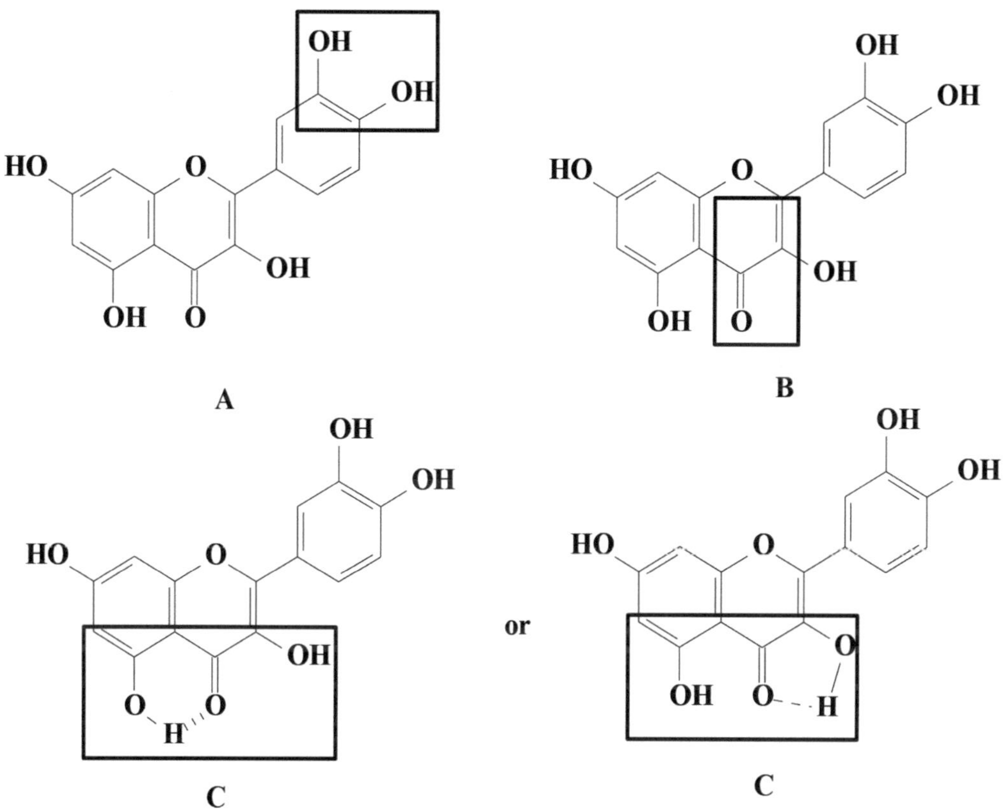

Figure 4. Structural groups for radical scavenging [42].

Flavonoids have different activity mechanisms such as free radical scavenging, inactivation of peroxides and other reactive oxygen species, chelation of metals and quenching of secondary lipid oxidation products [40]. The radical scavenging properties associated with the structure of flavonoids defend against oxidative stress and in doing so reduce heart disease, prevent cancer and slow down the aging processes in cells responsible for degenerative diseases [40, 42].

3. Ecology chemicals

3.1. Pigments and pollination

Plant compounds that are perceived by humans to have color are generally referred to as "pigments." The three main classes of pigments for coloration in plants are: betalains, carotenoids and flavonoids (anthocyanins) [43, 44]. All three classes of pigments act as visible signals to attract insects, birds and animals for pollination and seed dispersal [11, 43].

The pigments that color most flowers, fruits and seeds are flavonoids, which have the widest color range, from pale-yellow to blue [12]. Anthocyanins occur in all plant tissues and provide a wide range of colors ranging from orange/red to violet/blue [44]. They are formed by glycosides that may have several sugars in position 3; when there are no glycosides, the pigments are called anthocyanidins [12]. In addition to various modifications to their structures, their specific color also depends on pH, copigments and metal ions [11, 45].

The basic chromophore of anthocyanins is the flavylium ion [45, 46]. In acidic medium (pH below 2.5), anthocyanins show intensely reddish coloration or orange due the presence of flavylium cation form. When the pH increases from 2.5 to 4–6, the violet anhydrobase is formed first, but it decolorizes rapidly due the predominance of pseudobase carbinol formed by hydration (**Figure 5**) [11, 39, 47].

Since the flower cell sap is usually weakly acidic, in this pH region, most of the anthocyanins are in colorless form Ref. [47]. Hydration of the flavylium cation, which causes decoloration, may be prevented by formation of a complex between this ion and other substances. This phenomenon is called copigmenting [11, 47]. Such complexes are formed by intermolecular, intramolecular rearrangements and self-association, with organic molecules such as flavonoids, tannins, aromatic acids or metal for chelation [46, 47]. The copigmentation has a stabilizing effect as well as a bathochromic effect on anthocyanins [47]. Various flavonols and flavones act as copigments, with anthocyanins leading to an intensification of flower color [32, 45, 46]. So far, the main pigments targeted for flower and fruit color modifications are anthocyanins that contribute to a variety of colors such as red, pink and blue [44].

Humans recognize the color of a compound by perceiving reflected or transmitted light of wavelengths between 380 and 730 nm, while insects recognize light of shorter wavelengths [43]. Anthocyanins contribute to the UV patterns that are visible to insects and serve to signal flowers and fruits that are attractive to pollinators [45]. The light absorption of anthocyanins extends over most of the spectrum. Particularly, anthocyanins have an intense absorption in the 450–560 nm region (visible region), attributed to the hydroxyl cinnamoyl system of the B ring, while the absorption in the 240–280 nm region (UV region), characteristic of all flavonoids, corresponds to the A ring [32, 39].

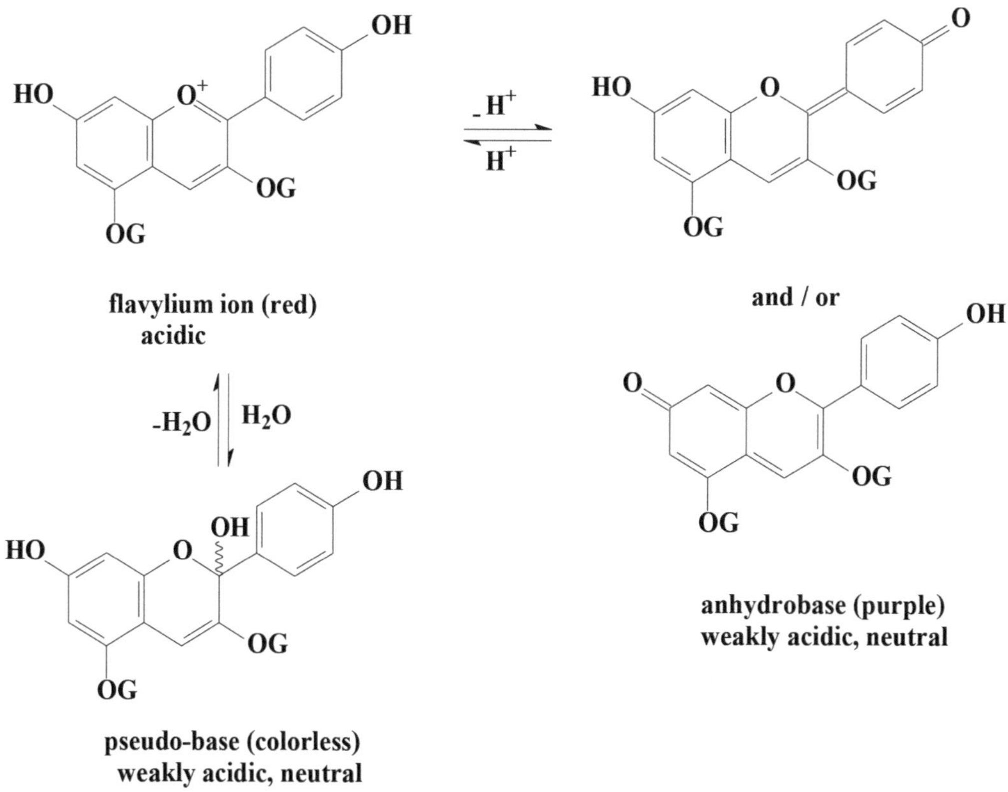

Figure 5. A change in the structure of an anthocyanin in aqueous solution as a function of pH [47].

The different colors produced by pigments are visible only to animals with the right photoreceptors, and many insects have limited color vision at the red end of the spectrum [40]. Due to the structural diversity of anthocyanins, the presence of one determined anthocyanin in the flower might affect the type of pollinators visiting the plant. The color preferences are different for different pollinators, and blue anthocyanins, for example, appear to attract bees more than red ones. Some butterfly and birds species visit red flowers, suggesting that both groups of animals are attracted to red anthocyanins [12, 40, 48].

A study of anthocyanins in two species of the genus *Schizanthus* Ruiz & Pav (Solanaceae) showed that the hummingbird-pollinated red flowers of *S. grahamii* contained a higher proportion of delphinidin 3-O-rutinoside (anthocyanin), whereas the bee-pollinated bluish-pink flowers of *S. hookeri* contained a higher proportion of petanin-derivatives (anthocyanin) [48].

Flavones and flavonols also contribute to flower color hue. Both groups of compounds comprise unpigmented or pale yellow flavonoids and are mostly invisible to the human eye [43, 44]. Studies on flavonoids by spectroscopy have revealed that most flavones and flavonols exhibit two major absorption bands: band I (320–385 nm) represents the B ring absorption, while band II (250–285 nm) corresponds to the A ring absorption [39]. As they absorb UV, which insects recognize, they give color and patterns to flowers to attract insects [43].

Chalcones and aurones, which provide yellow pigmentation in the flowers of several ornamental species, are relatively rare types of flavonoids [49]. The UV spectra of both compounds are characterized by an intense band I and diminished band II absorption [50]. The major absorption band in chalcones (band I) usually occurs in the range 340–390 nm. Band II is usually a minor peak in the 220–270 nm region, while the long wavelength absorption band in aurones is usually found in the 370–430 nm region. They produce the strongest yellow colors owing to their absorbance at longer wavelengths compared to the other types [50, 51]. Chalcones, flavonols, flavones or anthocyanins usually accumulate in sex organs of flowers, including the pollen. In contrast to man, some insects, especially bees, can perceive in the near ultraviolet (340–380 nm) as well as in the visible region. However, insects are possibly attracted to pollen whose color contrasts against petals due to UV reflective or absorptive flavonoids [46].

Pollination is an essential step in the reproductive process of the world's nearly 300,000 species of flowering plants because it is usually required for the production of seeds. Pollination can result from the action of abiotic forces such as wind and water, but 80% of the Angiosperms rely on animals, including bats, flies, butterflies, beetles and other insects [52]. Such diversity is acquired through evolutionary processes to ensure successful reproduction [44].

3.2. Allelopathy

The interactions between organisms are fundamental for the determination of plant abundance and distribution pattern in the community, of the productivity of several cultivated species and of the degree of interference on weeds [53]. Weeds are one of the most important factors that impose limitations on the development of agricultural activity in the world and are difficult to eradicate. The success of weeds in different cropping systems is associated, in part, with their ability to produce, store and release to the environment chemicals with allelopathic properties [54].

Allelopathy can be defined as a process by which compounds from the metabolism of a plant are released, preventing the germination, growth and development of other neighboring plants [55]. These compounds are involved in plant-plant interactions or allelopathy [56] and may influence, for example, in the vegetation of a local, in the succession of plants, in the germination of seeds and in the cultures productivity, among others [57]. Among the main groups of compounds with allelopathic potential are highlighted the benzoquinones, coumarins, flavonoids, terpenoids, glycosides, phenolic acids, alkaloids, rotenoids, catechins and tannins [58, 59].

Although flavonoids have many roles in plants, in relation to their role in allelopathy and the inhibition of seedling root growth [56], the activity of flavonoids in plant-plant interactions can be positive or negative [60]. The negative relations are mainly based on inhibiting germination and growth of other plants seedlings [56], as depicted in **Table 1**.

Some flavonoids present a level of phytotoxicity, indicating that allelopathy could be a beneficial function of the flavonoids to the producing plant [65]. Although the relative role of flavonoids in allelopathic interference has been less well-characterized than of some secondary metabolites, some examples of their involvement in autotoxicity and allelopathy are reported

Plant organism	Flavonoid	Function	References
Oryza sativa L.	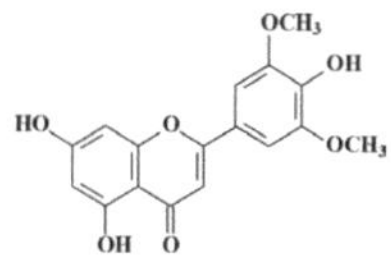	Allelopathic inhibitor of weeds and pathogens	[61]
-	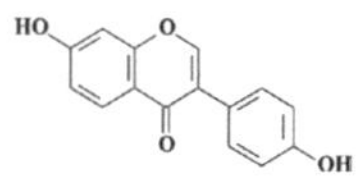	Allelopathic inhibitor of seed germination	[62]
Helianthus annuus	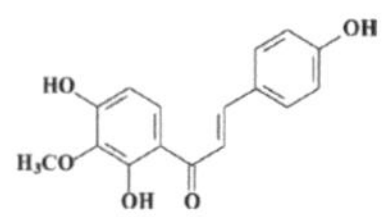	Inhibitor of seedling growth	[63]
-	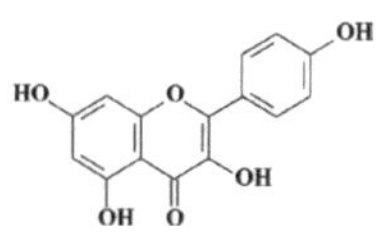	Allelopathic inhibitor of seedling growth	[64]

Table 1. Flavonoids of different classes with allelopathic potential.

[56]. In a previous study, see [65], presented flavonoids as are at least partly responsible for the strong phytotoxic effects of *Stellera chamaejasme* L. The potential allelopathic behavior may facilitate this weed to become a good competitor against other plant species in the environment.

Allied to the need for understanding the mechanism action of flavonoids, the importance of the study of allelopathy gains more and more attention in agriculture because these interactions could be employed for reducing weed growth.

Biopesticides based on flavonoids displaying allelopathic properties against weeds can potentially be an efficient natural defense against them [62]. In the study [63], the inhibiting activity against weeds of the species *Echinochloa crus-galli, Cyperus difformis* and *Cyperus iria* using the 5,7,4′-trihydroxy-3′,5′-dimethoxyflavone is shown.

4. Conclusion

Flavonoids are found in most plant tissues, provide a range of colors that attract pollinators, and, in fruit, they probably serve to attract frugivores that assist in seed dispersal. All of these pigments also function as antioxidants and sunscreens, absorbing wavelengths of ultraviolet. Their biosynthesis appears to be ubiquitous in plants and evolved early during land plant (from primitive green algae) evolution, aiding in plant protection and signaling. The precise mechanism by which flavonoids participate in allelopathy is still unknown, but the significance of allelopathy has gained more attention in agriculture, for example. Plant-plant interactions can influence or determine diversity, productivity and reproduction of a plant community beyond reduction or inhibition of weed growth.

Author details

Erica L. Santos[1]*, Beatriz Helena L.N. Sales Maia[1], Aurea P. Ferriani[1] and Sirlei Dias Teixeira[2]

*Address all correspondence to: ericaquimica@gmail.com

1 Federal University of Paraná, Curitiba, Paraná, Brazil

2 Tecnological Federal University of Paraná, Pato Branco, Paraná, Brazil

References

[1] Martens S, Preuss A, Matern U. Multifunctional flavonoid dioxygenases: flavonols and anthocyanin biosynthesis in *Arabidopsis thaliana* L. Phytochemistry. 2010;**71**:1040-1049.

[2] Jiang C, Schommer CK, Kim SY, Suh D-Y. Cloning and characterization of chalcone synthase from the moss, *Physcomitrella patens*. Phytochemistry. 2006;**67**:2531-2540.

[3] Oh H, Kim DH, Cho JH, Kim YC. Hepatoprotective and free radical scavenging activities of phenolic petrosins, flavonoids isolated from *Equisetum arvense*. Journal of Ethnopharmacology. 2004;**95**:421-424.

[4] Imperato, F. A new flavonoid glycoside from the fern *Dryopteris villarii*. Natural Product Communications. 2008;**3**:1709-1712.

[5] Vassao D, Kim KW, Davin LB, Lewis NG. Lignans (neolignans) and allyl/propenyl phenols: biogenesis, structural biology, and biological/human health considerations. In: Mander L, Liu H-WB, editors. Comprehensive natural products II (1st ed.). Oxford: Elsevier; 2010, pp. 815-928.

[6] Iwashina T. The structure and distribution of the flavonoids in plants. Journal of Plant Research. 2000;**113**:287-299.

[7] Markham KR. Distribution of flavonoids in the lower plants and its evolutionary significance. In: Harborne JB, editor. The flavonoids: advances in research since 1980 (1st ed.). London: Chapman & Hall; 1988, pp. 427-464.

[8] Markham KR, Porter LJ. Flavonoids in the green algae (chlorophyta). Phytochemistry. 1969;**8**:1777-1781.

[9] Feld H, Zapp J, Becker H. Secondary metabolites from liverwort *Tylimanthus renifolius*. Phytochemistry. 2003;**64**:1335-1340.

[10] Dixon RA, Pasinetti GM. Flavonoids and isoflavonoids: from plant biology to agriculture and neuroscience. Plant Physiology. 2010;**154**:453-457.

[11] Havsteen BH. The biochemistry and medical significance of the flavonoids. Pharmacology & Therapeutics. 2002;**96**:67-202.

[12] Harborne JB, Williams CA. Advances in flavonoid research since 1992. Phytochemistry. 2000;**55**:481-504.

[13] Tapas AR, Sakarkar DM, Kakde RB. Beydemir S. Flavonoids as nutraceuticals: a review. Tropical Journal of Pharmaceutical Research. 2008;**7**:1089-1099.

[14] Alzand KI, Mohamed M. Flavonoids: chemistry, biochemistry and antioxidant activity. Journal of Pharmacy Research. 2012;**5**:4013-4020.

[15] Ferreyra F, Rius SP, Casati P. Flavonoids: biosynthesis, biological functions, and biotechnological applications. Frontiers in Plant Science. 2012;**3**:1-15.

[16] Mierziak J, Kostyn K, Kulma A. Flavonoids as important molecules of plant interactions with the environment. Molecules. 2014;**19**:16240-16265.

[17] Koes RE, Quattrocchio F, Mol JNM. The flavonoid biosynthetic pathway in plants: function and evolution. BioEssays. 1994;**16**:123-132.

[18] Cushnie TPT, Lamb AJ. Antimicrobial activity of flavonoids. International Journal of Antimicrobial Agents. 2005;**26**:343-356.

[19] Shin-ichi A, Tsutomu F. Biosynthesis of a retrochalcone, echinatin: a feeding study with advanced precursors. Tetrahedron. 1981;**22**:2097-2098.

[20] Shin-ichi A, Tsutomu F. Studies on plant tissue cultures. Part 36. Biosynthesis of a retrochalcone, echinatin, and other flavonoids in the cultured cells of *Glycyrrhizaechinata*. A new route to a chalcone with transposed A- and B-rings. Journal of the Chemical Society, Perkin Transactions I. 1982:2725-2734.

[21] Shin-ichi A, Kumiko I, Tsutomu F. Stress-induced formation of echinatin and a metabolite, s-prenyl-licodione, in cultured *Glycyrrhiza echinata* cells. Phytochemistry. 1986;**25**:2803-2806.

[22] Nakamura K, Akashi T, Aoki T, Kawaguchi K, Ayabe S. Induction of isoflavonoid and retrochalcone branches of the flavonoid pathway in cultured *Glycyrrhiza echinata* cells treated with yeast extract. Bioscience Biotechnology and Biochemistry. 1999;**63:**1618-1620.

[23] Kajiyama K, Hiraga Y, Takahashi K, Hirata T, Kobayashi S, Sankawa U, Kinoshita T. Flavonoids and isoflavonoids of chemotaxonomic significance from *Glycyrrhiza pallidiflora* (Leguminosae). Biochemical Systematic and Ecology. 1993;**21**:785-793.

[24] Nepel A, Marques FA, Tozzi AMGA, Silva M, Maia BHLNS. Retrochalcones and its dimmers from roots of *Dahlstedtia grandiflora* – Fabaceae. In: 5th Brazilian Conference on Natural Products – BCNP and XXXI Annual Meeting on Micromolecular Evolution, Systematics and Ecology – RESEM, 26-29 October 2015, Atibaia. São Paulo; 2015.

[25] Furuya T, Matswnoto K, Hikichi M. Echinatin, a new chalcone from tissueculture of *Glycyrrhiza echinata*. Tetrahedron Letters. 1971;**27**:2567-2569.

[26] Calzada F, Mata R, Byes R, Linares E. A retrochalcone from *Anredera scandens*. Phytochemistry. 1990;**29**:2737-2738.

[27] Colegate LB, Ghisalbertit EL, Latiff A. Tepanone, a retrochalcone from *Ellipeza cuneifolia*. Phytochemistry. 1992;**31**:2123-2126.

[28] Haraguchi H, Ishikawa H, Mizutani K, Tamura Y, Kinoshita T. Antioxidative and superoxide scavenging activities of retrochalcones in *Glycyrrhiza inflate*. Bioorganic & Medicinal Chemistry. 1998;**6**:339-347.

[29] Aljancic IS, Vuckovic I, Jadranin M, Pesic M, Dordevic I, PodolskI-Renic A, Stojkovic S, Menkovic N, Vajs VE, Milosavljevic SM. Two structurally distinct chalcone dimers from *Helichrysum zivojinii* and their activities in cancer cell lines. Phytochemistry. 2014;**98**:190-196.

[30] Kamara BI, Manong DTL, Brandt EV. Isolation and synthesis of a dimeric dihydrochalcone from *Agapanthus africanus*. Phytochemistry. 2005;**66**:1126-1132.

[31] Katerere DR, Gray AI, Kennedy AR, Nash RJ, Waigh RD. Cyclobutanes from *Combretum albo punctatum*. Phytochemistry. 2004;**65**:433-438.

[32] Kumar M, Rawat P, Rahuja N, Srivastava AK. Antihyperglycemic activity of phenylpropanoyl esters of catechol glycoside and its dimers from *Dodecadenia grandiflora*. Phytochemistry.2009;**70**:1448-1455.

[33] Seidel V, Bailleul F, Waterman PG. (*Rel*)-1β,2α-di-(2,4-dihydroxy-6-methoxybenzoyl)-3β,4α-di-(4-methoxyphenyl)-cyclobutane and other flavonoids from the aerial parts of *Goniothalamus gardneri* and *Goniothalamus thwaitesii*. Phytochemistry. 2000;**55**:439-446.

[34] Zhang LIL, Wang S, Que S, Yang W, Zhang F, Gong N, Cheng W, Liang H, Ye M, Jia Y, Zhang Q. Oxyfadichalcones A–C: three chalcone dimers fused through a cyclobutane ring from Tibetan medicine *Oxytropis falcate* Bunge. Tetrahedron. 2013;**69**:11074-11079.

[35] Yang C, Wang X, Wang J, Luo J, Kong L. A [2 + 2] cycloaddition dimer and a diels-alder adduct from *Alpinia katsumadai*. Organic Letters. 2011;**13**:3380-3383.

[36] Fossen T, Andersen ØM. Spectroscopic techniques applied to flavonoids. In: Andersen OM, Markham KR, editors. Flavonoids chemistry, biochemistry and applications (1st ed.). New York: Taylor & Francis Group; 2006, pp. 100-118.

[37] Heim KE, Tagliaferro AR, Bobilya DJ. Flavonoid antioxidants: chemistry, metabolism and structure-activity relationships. Journal of Nutritional Biochemistry. 2002; **13**:572-584.

[38] Kumar S, Pandey AK. Chemistry and biological activities of flavonoids: an overview. The Scientific World Journal. 2013;**2013** 1-16.

[39] Eugster CH, Marki-Fischer E. The chemistry of rose pigments. Angewandte Chemie International Edition in English. 1991;**30**:654-672.

[40] Gupta J, Gupta A, Gupta AK. Flavonoids: its working mechanism and various protective roles. International Journal of Chemical Studies. 2016;**4**:190-198.

[41] Markham KR, Mabry TJ. Ultraviolet-visible and proton magnetic resonance spectroscopy of flavonoids. In: Harborne JB, Mabry TJ, editors. The flavonoids (1st ed.). London: Chapman & Hall; 1975, pp. 45-77.

[42] Croft KD. The chemistry and biological effects of flavonoids and phenolic acids. Annals of the New York Academy of Sciences. 1998;**854**:435-442.

[43] Tanaka Y, Sasaki N, Ohmiya A. Biosynthesis of plant pigments: anthocyanins, betalains and carotenoids. The Plant Journal. 2008;**54**:733-749.

[44] Nishihara M, Nakatsuka T. Biosynthesis of plant pigments: anthocyanins, betalains and carotenoids. Biotechnology Letters. 2011;**33**:433-441.

[45] Bueno JM, Sáez-Plaza P, Ramos-Escudero F, Jiménez AM, Fett R, Asuero AG. Analysis and antioxidant capacity of anthocyanin pigments. Part II: chemical structure, color, and intake of anthocyanins. Critical Reviews in Analytical Chemistry. 2012;**42**:126-151.

[46] Gould KS, Lister C. Flavonoid functions in plants. In: Andersen OM, Markham KR, editors. Flavonoids chemistry, biochemistry and applications (1st ed.). New York: Taylor & Francis Group; 2006, pp. 397-425.

[47] Goto T, Kondo T. Structure and molecular stacking of anthocyanins-flower color variation. Angewandte Chemie International Edition in English.1991;**30**:17-33.

[48] Sasaki K, Takahashi T. A flavonoid from *Brassica rapa* flower as the UV-absorbing nectar guide. Phytochemistry. 2002;**61**:339-343.

[49] Davies KM, Bloor SJ, Spiller GB, Deroles SC. Production of yellow colour in flowers: redirection of flavonoid biosynthesis in *Petunia*. The Plant Journal. 1998;**13**:259-266.

[50] Mabry T, Markham KR, Thomas MB. The ultraviolet spectra of chalcones and aurones spectroscopy of flavonoids. In: Mabry T, Markham KR, Thomas MB, editors. The systematic identification of flavonoids (1st ed.). New York: Springer-Verlag; 1970. pp. 227-230.

[51] Molitor C, Mauracher SG, Pargan S, Mayer RL, Halbwirth H, Rompel A. Latent and active aurone synthase from petals of *C. grandiflora*: a polyphenol oxidase with unique characteristics. Planta. 2015;**242**:519-537.

[52] Calderone NW. Insect pollinated crops, insect pollinators and US agriculture: trend analysis of aggregate data for the period 1992-2009. PLoS One. 2012;**7**:1-27.

[53] Trezzi MM. Avaliação do potencial alelopático de genótipos de sorgo[tese]. Porto Alegre: Universidade Federal do Rio Grande do Sul; 2002.

[54] Inoue MH, Santana DC, Souza Filho APS, Possamai ACS, Silva LE, Pereira MJB, Pereira KM. Potencial alelopático de *Annona crassiflora*: efeitos sobre plantas daninhas. Planta Daninha. 2010;**28**(3):489-498.

[55] Alves MCS, Medeiros Filho S, Innecco R, Torres SB. Alelopatia de extratos voláteis na germinação de sementes e no comprimento da raiz de alface. Pesquisa Agropecuária Brasileira. 2004;**39**:1083-1086.

[56] Weston LA, Mathesius U. Flavonoids: their structure, biosynthesis and role in the rhizosphere, including allelopathy. Journal of Chemical Ecology. 2013;**39**:283-297.

[57] Demuner AJ, Barbosa LCA, Chinelatto Junior LS, Reis C, Silva AA. Sorção e persistência da sorgoleona em um latossolo vermelho-amarelo. Química Nova. 2005;**28**:451-455.

[58] Vyvyan JR. Allelochemicals as leads for new herbicides and agrochemicals. Tetrahedron. 2002;**58**:1631-1646.

[59] Wang Q, Li Z, Ruan X, Pan C, Jiang D. Phenolics and plant allelopathy. Molecules. 2010;**15**:8933-8952.

[60] Almeida LFR. Composição química e atividade alelopática de extratos foliares de *Leonurussibiricus* L. (Lamiaceae). 2006, 105 p. Tese (Doutorado em Ciências Biológicas) – Instituto de Biociências, Universidade Estadual Paulista, Botucatu, 2006.

[61] Kong C, Xu X, Zhou B, Hu F, Zhang C, Zhang M. Two compounds from allelopathic rice accession and their inhibitory activity on weeds and fungal pathogens. Phytochemistry. 2004;**65**:1123-1128.

[62] Chang CF, Akinori S, Sumio K, Saburo T. Chemical studies on clover sickness. II. Biological functions of isoflavonoids and related compounds. From Agricultural and Biological Chemistry. 1969;**33**:398-408.

[63] Macías FA, Molinillo JMG, Torres A, Varela RM, Castellano D. Bioactive flavonoids from *Helianthus annuus* cultivars. Phytochemistry. 1997;**45**:683-687.

[64] Levizou E, Karageorgou P, Petropoulou Y, Grammatikopoulos G, Maneta SY. Induction of a geotropic response in lettuce radicle growth by epicuticular flavonoid aglycons of *Dittrichia viscosa*. Biologia Plantarum. 2004;**48**:305-307.

[65] Yan Z, Zeng L, Jin H, Qin B. Potential ecological roles of flavonoids from *Stellera chamaejasme*. Plant Signaling & Behavior, 2015;**10**:e1001225-1–e1001225-3.

Chapter 2

Isolation and Structure Identification of Flavonoids

Weisheng Feng, Zhiyou Hao and Meng Li

Additional information is available at the end of the chapter

http://dx.doi.org/10.5772/67810

Abstract

Flavonoids, which possess a basic C15 phenyl-benzopyrone skeleton, refer to a series of compounds in which two benzene rings (ring A and B) are connected to each other through three carbon atoms. Based on their core structure, flavonoids can be grouped into different flavonoid classes, such as flavonols, flavones, flavanones, flavanonols, anthocyanidins, isoflavones and chalcones. Flavonoids are often hydroxylated in positions 3, 5, 7, 3′, 4′ and/or 5′. Frequently, one or more of these hydroxyl groups are methylated, acetylated, prenylated or sulfated. In plants, flavonoids are often present as O- or C-glycosides. The O-glycosides have sugar substituents bound to a hydroxyl group of the aglycone, usually located at position 3 or 7, whereas the C-glycosides have sugar groups bound to a carbon of the aglycone, usually 6-C or 8-C. The most common carbohydrates are rhamnose, glucose, galactose and arabinose. This chapter mainly introduces the methods of isolation and structure identification of flavonoids.

Keywords: flavonoids, structures and classification, extraction, isolation, structure identification

1. Introduction

Flavonoids are important natural organic compounds of secondary metabolites that are produced during the long process of natural selection. They widely exist in the roots, stems, leaves, flowers and fruits of higher plants and ferns [1], which are of numerous categories and chemical structures. Because of the special chemical structures, flavonoids possess wide range of physiological and biochemical effects to the cells of mammal and other kinds of animals. Firstly, flavonoids possess strong chemical reactivity. For example, some flavonoids have antioxidant activity via scavenging free radicals in organism [2]. Furthermore, flavonoids possess various pharmacological activities of inhibiting the activity of enzymes, antitumor, antibiosis, antivirus, anti-inflammatory and so on [3–9]. The potential treatment and prevention effects have been shown in degenerative diseases such as tumors, aging and cardiovascular diseases [10–15]. Additionally, some compounds of flavonoids possess potential application prospects as weak hormones at treating menopausal syndrome of women [14–17].

2. Structures and classification

Flavonoids generally refer to the natural products of C_6-C_3-C_6 basic structure. Most of them are the chromone derivatives with the core structure of 2-phenylchromone and made up of three rings of A/B/C as shown in **Figure 1**.

According to connection mode of ring A with B, the connection position of ring B, the oxidation level of C_3 substructure and degree of polymerization, various type of flavonoids could be classified, as shown in **Table 1**.

The main factors of the structure diversity of flavonoids are as follows:

2.1. Change of ring system, degree of oxidation and number of core structure

Most of the flavonoids possess the core structure of C_6-C_3-C_6; few of them are C_6-C_1-C_6 (xanthone, for example). A few of them, such as homoisoflavones and rotenoids, possess C_6-C_4-C_6 structure skeleton. In most cases, C_3 part is formed to be a hexatomic or pentagon ring with C_6 part. It could also be aliphatic chain, such as chalcone and dihydrochalcone. Supposing that the double bond of ring C was hydrogenated, dihydro derivative was formed, such as flavanone and flavononol. Commonly, ring B is connected to C-2; it might be connected to C-3 or C-4 in a few cases, such as isoflavone and neoflavonoid. Most of the flavonoids have only one core structure; some of them possess two, however. We called them biflavonoids.

2.2. Various substituents at ring A and B

Generally, hydroxyl, methoxyl, methyl, isopentenyl, methylenedioxyl, benzyl, nitro groups and so on, could be substituted at ring A and/or B.

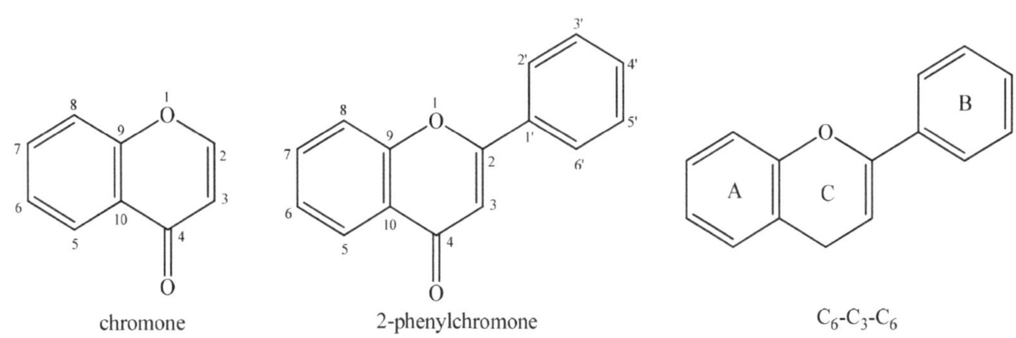

Figure 1. Basic structure of flavonoids.

http://dx.doi.org/10.5772/67810

Structure type	Basic structure	Structure type	Basic structure
Flavone	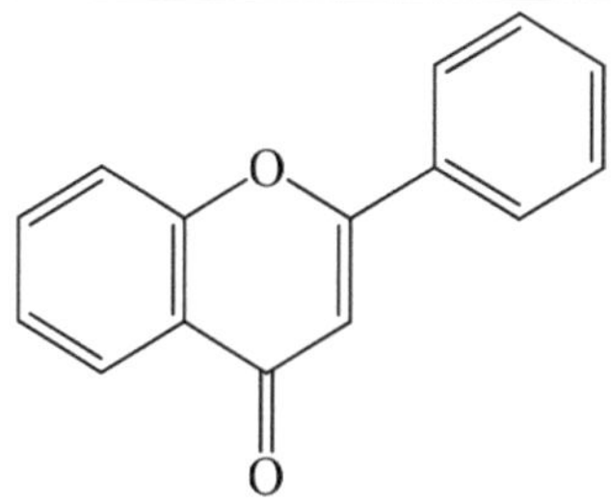	Chalcone	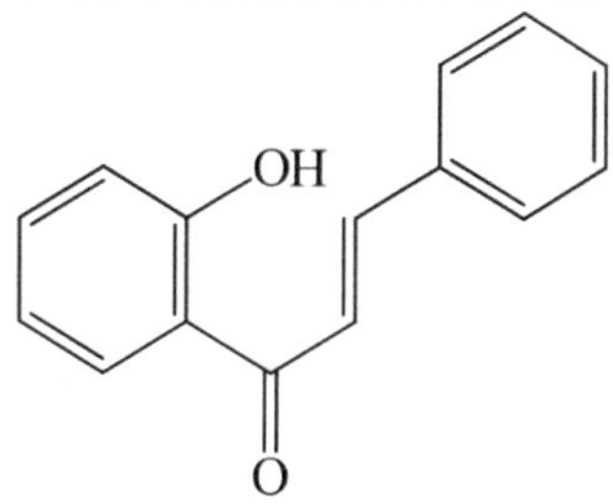
Flavonol	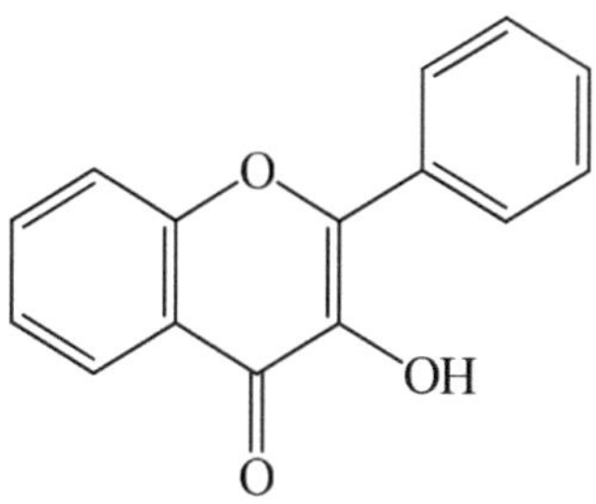	Dihydrochalcone	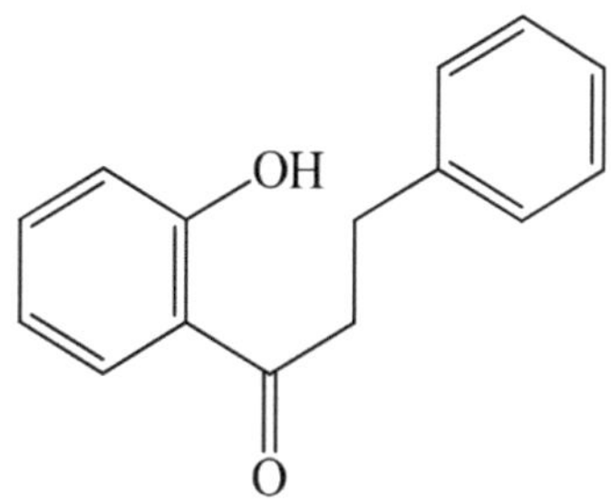
Flavanone	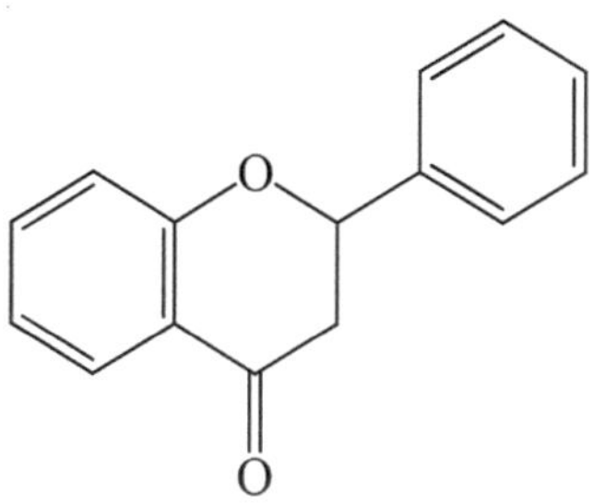	Anthocyanidin	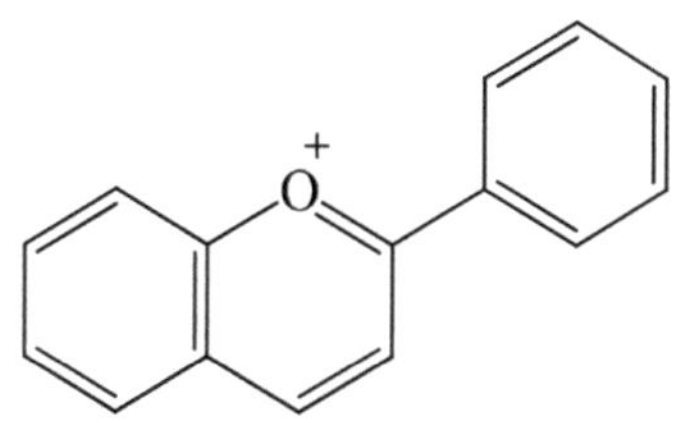
Flavanonol	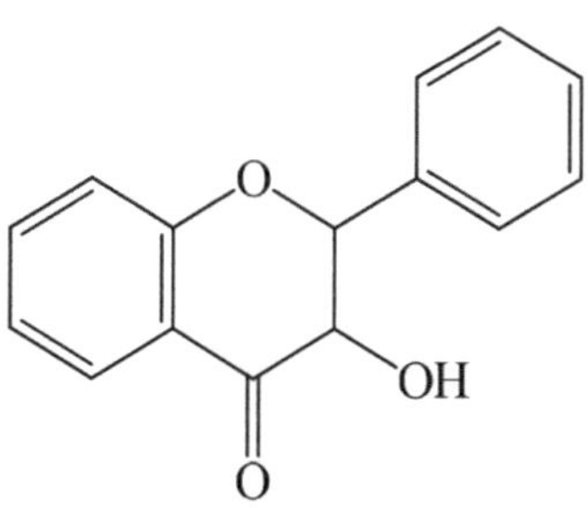	Flavan	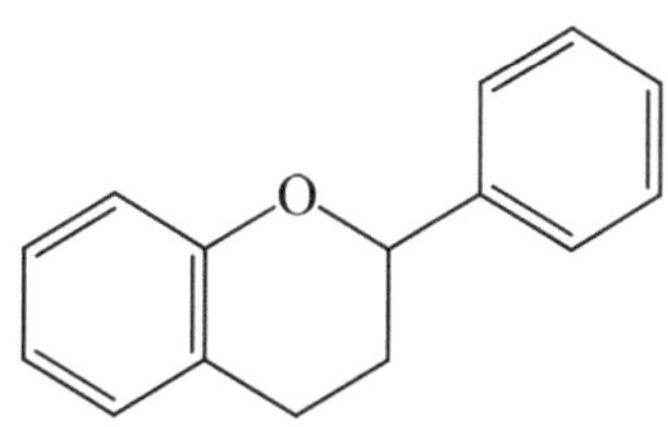
Isoflavone	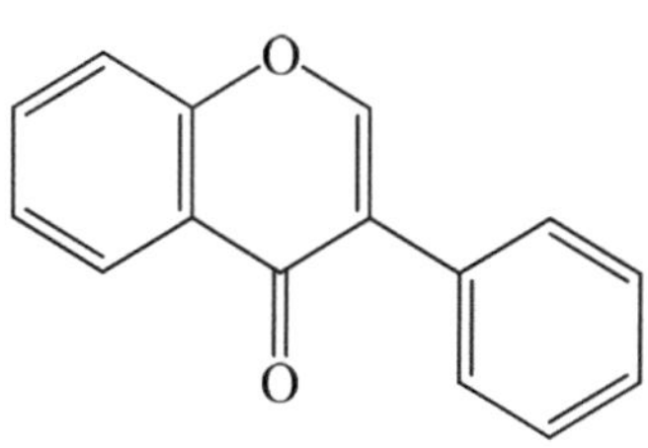	Aurone	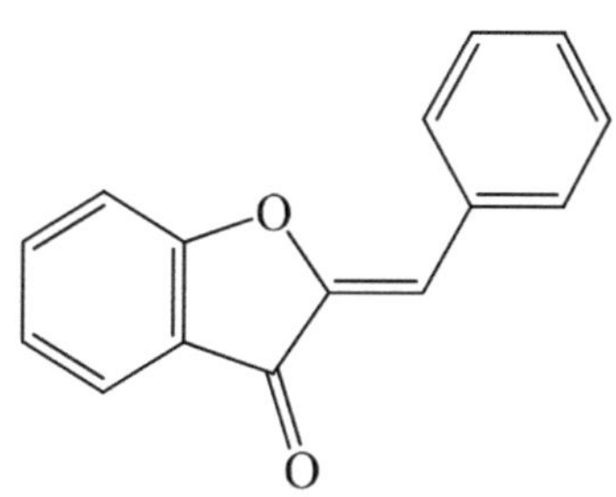

Structure type	Basic structure	Structure type	Basic structure
Isoflavanone	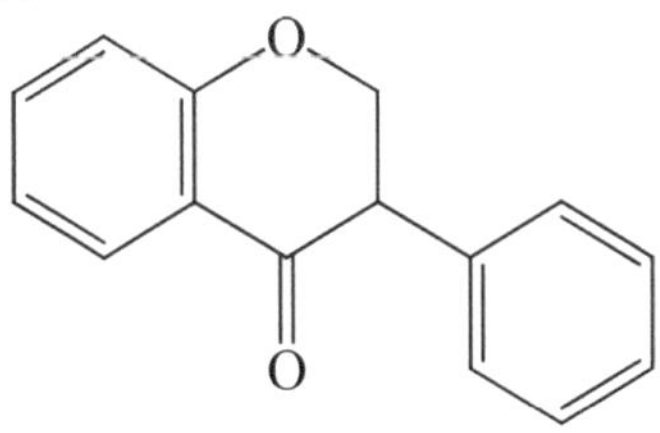	Homoisoflavanone	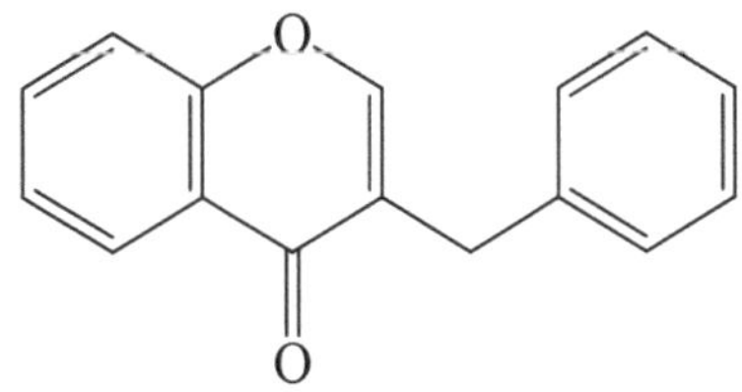
Rotenoid	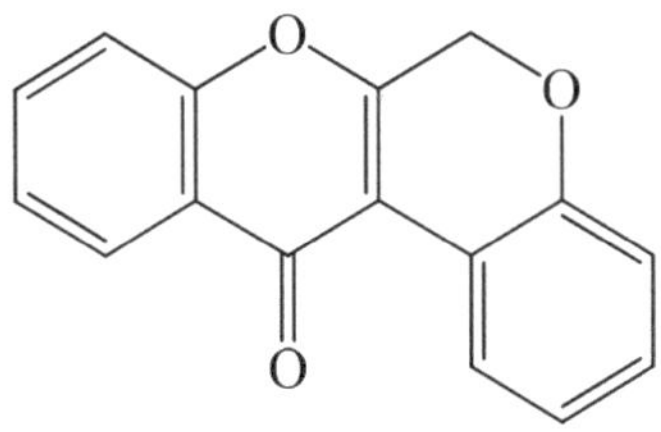	Xanthone	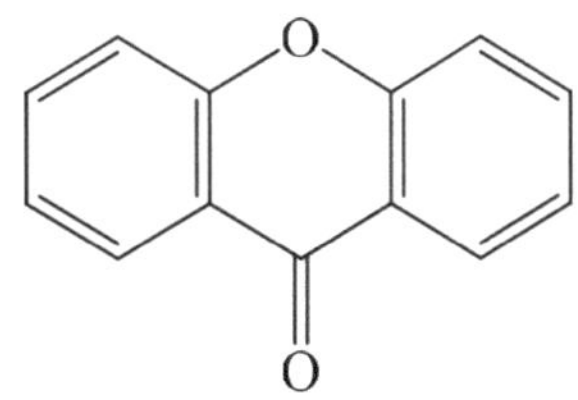
Pterocarpin	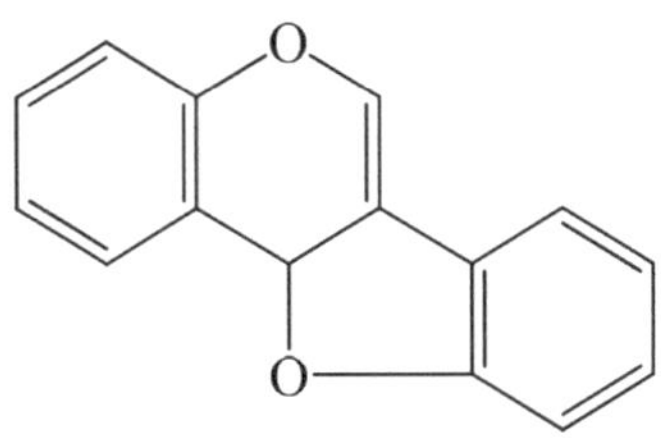	Neoflavonoid	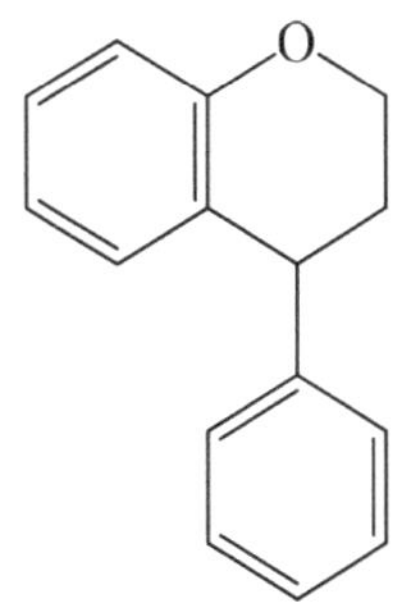

Table 1. Main structure types of flavonoids.

2.3. Glycosidation

Flavonoids are often glycosided to be O-glycosides or C-glycosides. During the glycosidation of flavonoids, almost every hydroxyl group could be the reaction position. However, the glycosidation reaction mostly occurred at 7-OH of flavone, flavanone and isoflavone, 3- and/or 7-OH of flavonol and flavanonol 3- and/or 5-OH of anthocyanidin. The glycosyl group of C-glycosides is often connected to C-6 and/or C-8.

2.4. Formation of complexes

Complexes of flavonoids could be formed with other types of compounds, such as phenylpropanoids, coumarins and alkaloids.

3. Extraction and isolation

3.1. Extraction

Traditional extraction methods of flavonoids often cause the problems of inefficiency, high energy consumption, more solvent consumption and so on. The new extraction methods and technologies occurred in recent years promoted the development of flavonoids. Because of the numerous types of flavonoids, the single extraction methods generally cannot meet the requirement. Traditional and modern methods should be applied together according to the extraction purpose.

3.1.1. *Traditional extraction methods*

Ethanol and methanol are frequently used to extract flavonoids. The common extraction methods include dipping, percolation, reflux, continuous reflux and so on. The alcohol of high concentration (90–95%) is applied to extract free flavonoids, and the alcohol at the concentration of about 60% is applied to extract flavonoid glycosides. For example, reflux method was applied to extract total flavonoids from leaves of *Ginkgo biloba* with 70% ethanol, and the product yield was significantly higher than the water decoction method [18].

Hot water extraction method is applied to flavonoid glycosides. It possesses the advantages of low cost, safety, simple equipment and could be applied in industrial production, but much water-soluble impurities, such as proteins and saccharides might be mixed into the product.

Most of the flavonoids are acidic because of hydroxyl or carboxyl groups, so they could be extracted with alkaline water or alkaline dilute alcohol. The commonly used solvents include dilute sodium hydroxide, lime water, 5% sodium hydroxide dilute ethanol solution and so on. Water-soluble impurities, such as tannins, pectins and mucilages, could be precipitated because of the formation of calcium salts during the extraction with lime water. It has often showed good results if 5% sodium hydroxide dilute ethanol solution was used. However, the product yield might be reduced because some flavonoids obtained after acidification might be dissolved in dilute ethanol solution. It should be noted that the concentration of alkali should not be excessive during the extraction, and the acidity of the solvent should not be excessive during acidification. Additionally, sodium borate should be used if adjacent phenolic hydroxyl groups are existed in the structures. Extraction of rutin from Flos Sophorae Immaturus is one example [19].

3.1.2. *New extraction technologies*

3.1.2.1. *Supercritical fluid extraction (SFE)*

The goal of selected extraction, isolation or purification might be achieved via controlling temperature, pressure and regulating the type and consumption of cosolvent during the supercritical fluid extraction. Cosolvent (e.g. ethanol) is usually added to induce product yield. For example, the product yield of supercritical CO_2 extraction of flavonoids from *Licorice* has been raised 2.2 times than the ordinary alcohol extraction [20].

3.1.2.2. *Ultrasonic extraction*

This method has been used in the quality analysis and small amount extraction of flavonoids. It's still seldom used in industrial production, however. For instance, ultrasonic extraction was used in the extraction of flavone from the bud of *Sophora japonica*, and the product yield was higher than reflux extraction method [21]. Ultrasonic extraction is superior to reflux method from the perspectives of energy saving, time saving and technology.

3.1.2.3. Microwave-assisted extraction

It has obtained good results in the extraction of flavonoids. However, it is confined to laboratories so far. It also can be applied combined with other methods to induce product yield. For example, refluxing extraction was used after treatment with microwave for a short time during the extraction of flavonoids from *Ophiopogon japonicus*. The product yield was induced significantly [22].

3.1.2.4. Enzyme method

The impurities, such as starches, pectins and proteins, could be removed after enzymolysis. Long extraction time is the limitation of this method. However, the mild operational conditions could overcome the shortcomings that some bioactive components may be decomposed under high temperature.

3.1.2.5. Macroporous adsorption resin

It has been used in the separation and enrichment of flavonoids. Suitable types should be chosen according to the nature of target constituents.

3.1.2.6. Ultrafiltration

The molecules of different molecular weight are separated depend on the pressure difference between both sides of ultrafiltration membrane. Proteins, polypeptide, polymeric pigments and starches could be removed largely. It possesses advantages of simple operations such as no need to heat and destroy the molecular structures. It could remove 69.4% pectins and 66% proteins during the preparation of soybean isoflavones [23].

3.1.2.7. Aqueous two phase extraction (ATPE)

Aqueous two phase system (ATPS) is formed when either two polymers and kosmotropic salt, or two salts (one chaotropic salt and the other a kosmotropic salt) are mixed at appropriate concentrations and at a particular temperature. The distribution coefficients are different in specific ATPS of different substances. The separation objective will be achieved via selective distribution between the two phases after adding substances into the system. It possesses advantages of time-saving, simple operation, mild condition, being easy to expand process, large treatment capacity and so on. The commonly used ATPS are high polymer system (e.g. PEG-Dextran system), high polymer-inorganic salt system and PEG-sulfate/phosphate system. The distribution characteristics of puerarin in the two-phase aqueous systems of PEG/$(NH_4)_2SO_4$ and acetone/K_2HPO_4 have been studied [24], and the best system has been determined.

3.2. Isolation

3.2.1. General methods

The isolation of flavonoids includes the separation of flavonoids and other kinds of compounds, and the obtaining of monomeric compounds. The choosing of isolation methods is

http://dx.doi.org/10.5772/67810

made primary according to polarity, acidity, molecular weight difference and special structure. Chromatography is still the first choice to isolate flavonoids.

3.2.1.1. Silica gel chromatography

Silica gel chromatography is the main method to isolate or identify flavonoids. It is applied to isolate low or medium polar constituents. Reversed phase silica gel (e.g. reversed phase C_{18} silica gel) is commonly used to isolate flavonoid glycosides.

3.2.1.2. Polyamide chromatography

Polyamide is a good adsorbent to isolate flavonoids. The adsorption strength hinges on hydrogen bonding associated between polyamide and flavonoids, which depends on the number and positions of hydroxyl groups in the molecules of flavonoids.

3.2.1.3. Polydextran gel chromatography

The most commonly used polydextran gel is sephadex LH-20 during the isolation of flavonoids. Adsorption is the main mechanism during the isolation of free flavonoids, and the adsorption strength is mainly based on the phenolic hydroxyl groups. However, molecular sieve effect plays the leading roles during the isolation of flavonoid glycosides.

3.2.2. Application of new isolation technologies

3.2.2.1. High-performance liquid chromatography (HPLC)

This technology has been widely used in the isolation and quality analysis of flavonoids and other kinds of natural products. The determination of chromatographic condition is the key to achieve separation purpose.

3.2.2.1.1. Choice of stationary phases

Silica gel and amino columns are mostly used during the operation of normal phase chromatography. In the reversed phase, HPLC (RP-HPLC), C_{18}, C_8, C_2, amino or phenyl columns could be applied, whereas C_{18} and C_8 columns are mostly used among them.

3.2.2.1.2. Choice of mobile phases

Methanol-water and acetonitrile-water system are commonly applied in RP-HPLC. In order to improve separation performance, minute quantity of acid (e.g. trifluoroacetic acid) could be added into mobile phase.

3.2.2.1.3. Detection

All of the flavonoids are able to absorb ultraviolet rays, so generally they could be detected by UV detectors. It is usually detected at 254–280 nm or 340–360 nm for flavones, flavonols and the corresponding glycosides, 520–540 nm for anthocyanidins and the corresponding glycosides, 250 nm for chromones.

3.2.2.2. High-speed counter current chromatography (HSCCC)

High-speed counter current chromatography (HSCCC) has been applied successfully to the isolation of flavonoids. The method is simple and quick to operate, and could get product with high purity. Furthermore, it is suitable to industrial production. For example, an HSCCC system has been employed to separate seven flavonoids from a methanolic extract of the leaves of *Oroxylum indicum* by a one-step isocratic elution using a chloroform-methanol-water (9.5:10.5) two-phase system [25].

3.2.2.3. Molecular imprinting technology (MIT)

Molecular imprinting technology (MIT) has been applied in recent years to isolation and active screening of flavonoids. As the study [26] of Pakade et al., molecularly imprinted polymers (MIPs) targeting quercetin were prepared from 4-Vinylpyridine and ethylene dimethacrylate (EDMA) under various solvent systems with the aim to form MIPs with high recognition for the quercetin molecule in aqueous systems at high temperature. The slopes for the effect of extraction time revealed that the mass transfer of the analytes was higher at 84°C than at 25°C. Also, the binding capacity for the most promising MIP and its corresponding NIP was higher at 84°C. The binding capacity for the MIP was similar to 30 μmol/g at 25°C and 120 μmol/g at 84°C, while for the corresponding NIP, it was similar to 15 and 90 μmol/g, at 25 and 84°C, respectively.

4. Structure identification of flavonoids

Generally, structure determination of flavonoids can be achieved easily because of the systematic research of their structures and the progress of spectroscopic technologies (Nuclear Magnetic Resonance spectroscopy, especially). Series of spectroscopic technologies, such as IR, UV, NMR, and MS, are often used during structure identification of flavonoids. In rare cases, total synthesis should be applied to verify the elucidated structures.

4.1. Ultraviolet spectrum (UV)

The positions, types and number of substituents in the conjugated systems could be speculated via means of UV spectrum. Most of the flavonoids in methanol possess two main absorption bands. Band I is at 300–400 nm, which is caused by electron transition of cinnamoyl group. Band II is at 240–280 nm, which is caused by electron transition of benzoyl group, as shown in **Figure 2**. The structure types and oxygen-bearing substituent types of flavonoids could be determined by the peak locations, shapes and strengths of band I and II, as shown in **Table 2** [27].

The locations and shapes of Band I and II will be affected by the substituents attached to rings A and B. Normally, red shift of band I increases accordingly when the number of hydroxyl groups located at ring B increases. Similarly, red shift of band II increases accordingly when the number of hydroxyl groups located at ring A increases, but it has trifling impact to band I, with the exception of 5-OH. The corresponding bands will be violet shifted 5–15 nm if the

http://dx.doi.org/10.5772/67810

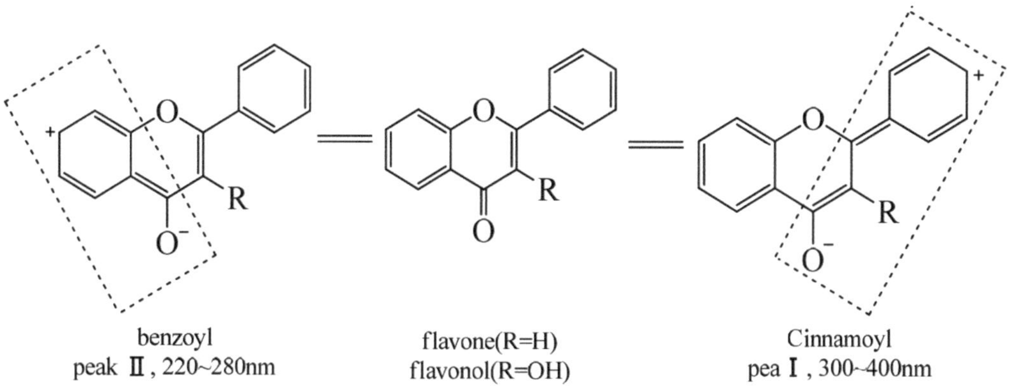

Figure 2. UV spectrum of flavonoids.

Structure type	Band II (nm)	Band I (nm)
Flavone	250–280	304–350
Flavonol (3-OH is substituted)	250–280	328–357
Flavonol (3-OH is free)	250–280	358–385
Isoflavone	245–270	310–330 (shoulder peak)
Flavanone and flavanonol	270–295	300–330 (shoulder peak)
Chalcone	220–270 (weak peak)	340–390
Aurone	230–270 (weak peak)	370–430
Anthocyanidin	270–280	465–560

Table 2. The spectral characteristics of UV-VIS spectrum of flavonoids.

particular hydroxyl is glycosided. Furthermore, the influence of the hydroxyl groups will almost disappear if they are acetylated.

4.2. Infrared spectrum (IR)

It is used mainly to determine the types of functional groups, substitution modes of aromatic rings and so on. The all functional groups, such as carbonyl, phenolic hydroxyl, phenyl and glycosyl, have possessed corresponding IR absorptions. The absorption band of hydroxyl groups are in the 3200–3650 cm^{-1} region, carbonyl groups are in 1660–1680 cm^{-1} region and the vibrations of benzene rings are at about 1500, 1580 and 1600 cm^{-1}.

4.3. Nuclear magnetic resonance spectrum (NMR)

Nuclear magnetic resonance spectrum (NMR) is the most powerful method to elucidate the structures of flavonoids. Kinds of solvents, such as $CDCl_3$, DMSO-d_6, C_5D_5N, $(CD_3)_2CO$ and CD_3OD, could be employed while performing NMR experiments. DMSO-d_6 is the optional solvent among them to perform NMR to flavonoids. Almost all kinds of flavonoids could be well dissolved in DMSO-d_6, and the resonance signals of flavonoids are rarely overlapped by solvent peaks (about δ2.5). Furthermore, NMR signals of phenolic hydroxyl groups could be displayed clearly with DMSO-d_6 as the solvent. The drawback of this solvent is high boiling point, which leads to difficulty in sample recovery.

4.3.1. ^{1}H-NMR spectrum

It provides information of chemical shifts, coupling constants and proton number. The types of flavonoids, substituted modes, number and configurations of glycosyls and so on, could be determined via ^{1}H-NMR spectrum.

4.3.1.1. Protons on ring C

^{1}H-NMR characteristics of protons on ring are shown in **Table 3** [28].

4.3.1.2. Protons on ring A

The ordinary substitution modes are 5,7-dioxygenation, 7-oxygenation, 5,6,7-trioxygenation and 5,7,8-trioxygenation, See **Figure 3**.

4.3.1.2.1. 5,7-Dihydroxyl substituted

5,7-Dihydroxyl flavonoids are most common. For this type of flavonoids, the signals of H-6 and H-8 are shown at δ5.7–6.9 as doublets, and the signal of H-6 is always at the higher field than H-8. The signals of both H-6 and H-8 shift to lower field after glycosidation of 7-OH.

Type	2-H	3-H	Note
Flavanone		δ6.3– 6.8 (s)	The signals maybe overlapped by H-6 or H-8.
Flavonol	None signal.		
Isoflavone	δ7.6–7.8 (s)		The signal is at rather low field because of influence of oxygen atom at position 1 and carbonyl at position 4.
Chalcone	α-H: δ6.7–7.4 (d, *J* = 17 Hz) β-H: δ7.3–7.7 (d, *J* = 17 Hz)		C-2 and C-3 form a *trans* double bond.
Flavanone	δ5.0–5.5 (dd, *J* = 11, 5 Hz)	δ2.3– 2.8 (2H) (dd, *J*=17, 11 Hz); (dd, *J* = 17, 5 Hz)	H-2 is coupled by two protons of position 3.
Flavanonol	4.8–5.0 (d, *J* = 11 Hz)	δ4.1–4.3 (d, *J* = 11 Hz)	Configurations of both C-2 and C-3 are R.
Flavanone-3-O-glycoside	δ5.0–5.6	δ4.3–4.6	After glycosidation of 3-OH, resonance signals of both H-2 and H-3 shift to low field.
Aurone	Exocyclic proton: δ6.5–6.7 (s)		

Table 3. Chemical shifts and coupling constants of ring C of common flavonoids.

http://dx.doi.org/10.5772/67810

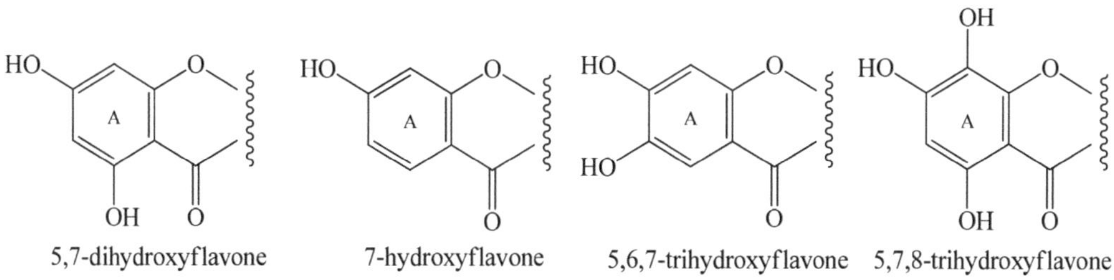

Figure 3. Substitution modes of ring A.

4.3.1.2.2. 7-Hydoxyl substituted

Signal of H-5 is shown to be a doublet since vicinal coupling exists between H-5 and H-6. Additionally, the chemical shift is at rather low field (about δ8.0) because of the shielding effect of carbonyl at position 4. H-6 is affected by H-5 and H-8, so it has showed a double-doublet (dd, J=2.0, 8.0 Hz). H-8 is showed to be a doublet (J=2.0Hz) because of the vicinal relationship with H-6. Signals of both, H-6 and H-8 are at δ6.3–7.1. The chemical shifts of protons on ring A are shown in **Table 4** [28].

4.3.1.3. Protons on ring B

There are a variety of substituted modes of ring B, such as non-substitution, 4′-oxygenation, 2′-oxygenation, 3′,4′-dioxygenation, 2′,4′-oxygenation, 3′,4′,5′-trioxygenation and 2′,4′,5′-trioxygenation, as shown in **Figure 4**. Generally, signals of protons on ring B are showed at slightly lower field, and the chemical shifts are usual at δ6.7-8.1. The substitution modes and structural information could be determined via the chemical shifts and coupling constants of ring B.

4.3.1.3.1. None substituent on ring B

For this mode, there are five protons on ring B. Signals of H-2′ and H-6′ are shown at lower field than H-3′, H-4′ and H-5′ because of the shielding effect of ring C. Furthermore, the peak shapes of all of the protons are complicated because of the coupling effects of the vicinal- and

	Type	**H-5**	**H-6**	**H-8**
5,7-Dihydroxyl	Flavone, flavonol, isoflavone		δ6.0–6.2 d	δ6.3–6.5 d
	7-O-Glucosides of above compounds		δ6.2–6.4 d	δ6.5–6.9 d
	Flavanone, flavanonol		δ5.75–5.95 d	δ5.9–6.1 d
	7-O-Glucosides of above compounds		δ5.9–6.1 d	δ6.1–6.4 d
7-Hydroxyl flavonoids	Flavone, flavonol, isoflavone	δ7.9–8.2 d	δ6.7–7.1 d	δ6.7–7.0 d
	Flavanone, flavanonol	δ7.7–7.9 d	5.7–6.0 d	δ5.9–6.1 d
5,6,7-Trihydroxyl flavonoids				δ6.95 s
5,7,8-Trihydroxyl flavonoids			δ6.3 s	

Table 4. Chemical shifts of protons on ring A.

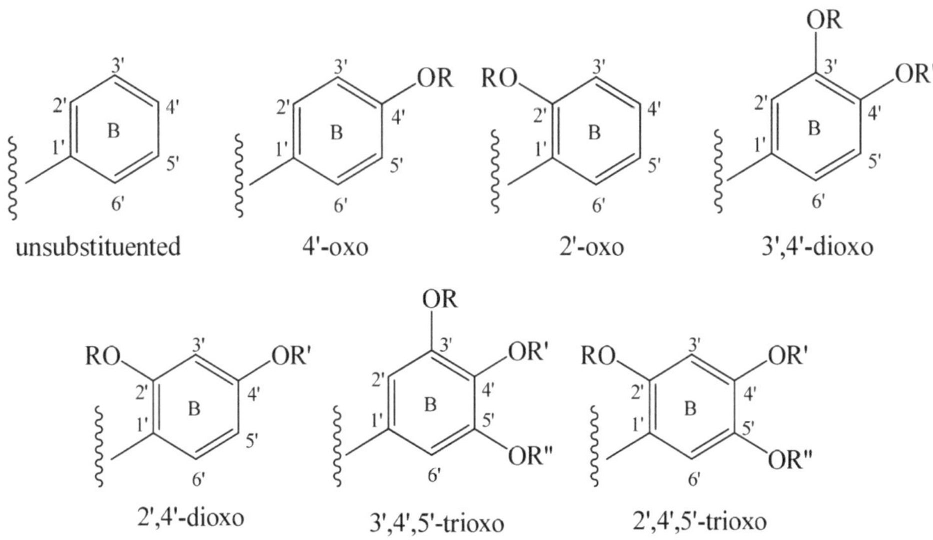

Figure 4. Substitution modes of ring B.

meta-coupling. The signals of H-2′ and H-6′ are usually at δ7.1–7.6 and of H-3′, H-4′ and H-5′ are at δ7.9–8.2.

4.3.1.3.2. 4′-Oxygenation

In this circumstance, ring B is a symmetrical substructure. One AA′BB′ coupling system is formed by four aromatic protons. The spectral characteristics are show in **Table 5** [28].

4.3.1.3.3. 3′,4′-Dioxygenation

In this circumstance, one ABX coupling system is formed by three aromatic protons, and three groups of signals are displayed as H-2′(1H, d, $J \approx 2.0$Hz), H-5′(1H, d, $J \approx 8.0$Hz) and H-6′(1H, dd, $J \approx 2.0, 8.0$Hz). The chemical shifts of protons on ring B are shown in **Table 6** [28].

Type	H-2′, H-6′	H-3′, H-5′
Flavanone	δ7.1–7.3	
Flavanonol	δ7.2–7.4	
Isoflavone	δ7.2–7.5	
Chalcone	δ7.4–7.6	δ6.5–7.1
Aurone	δ7.6–7.8	
Flavone	δ7.7–7.9	
Flavonol	δ7.9–8.1	

Table 5. Chemical shifts of protons on ring B of 4′-oxygenated flavonoids.

Type	Substituent mode	H-2′	H-5′	H-6′
Flavone	3′,4′-OH and 3′-OH, 4′-OCH_3	δ7.2–7.3	δ6.7–7.1	δ7.3–7.5
Flavonol	3′,4′-OH and 3′-OH, 4′-OCH_3	δ7.5–7.7		δ7.6–7.9
	3′-OCH_3, 4′-OH	δ7.6–7.8		δ7.4–7.6
	3′,4′-OH, 3-O-glc	δ7.2–7.5		δ7.3–7.7
Isoflavone, flavanone, flavanonol	3′,4′-OH	δ6.7–7.1		δ6.7–7.1

Table 6. Chemical shifts of protons on ring B of 3′,4′-dioxygenated flavonoids.

4.3.1.3.4. 2′-Oxygenation

ABCD coupling system is formed by the rest protons of ring B. The peak shapes are rather complicated. Signals of H-3′ and H-5′ are usually displayed at δ6.8–6.9, H-4′ at about δ7.2 and H-6′ at δ7.4–7.5.

4.3.1.3.5. 3′,4′,5′-Trioxygenation

If identical substituents are attached to C-3′ and C-5′, which allows the formation of a symmetrical substructure of ring B, H-2′ and H-6′ will display to be a singlet at δ6.5–7.5.

4.3.1.3.6. 2′,4′,5′-Trioxygenation

In the cases of this substituent mode, either of the two protons on ring B displays to be a singlet. Generally, signals of H-6′ in flavones and flavonols are showed at δ7.2–7.5, H-3′ at δ6.4–6.6. Signals of H-6′ are shown at slight higher field.

4.3.1.3.7. 2′,4′-Dioxygenation

In the cases of this mode, H-3′ will be showed at δ6.00–6.6 (d, $J \approx 2.0$ Hz), H-5′ at δ 6.6–6.5 (dd, $J \approx 2.0, 8.0$ Hz) and H-6′ at δ7.0–7.4 (d, $J \approx 8.0$ Hz). See **Table 7**.

4.3.1.4. Common substituents

The proton chemical shifts of common substituents of flavonoids are shown in **Table 8** [28].

4.3.2. ^{13}C-NMR spectrum

Strong regularities are also shown in ^{13}C-NMR spectra of flavonoids. The types of flavonoids, number and connection positions of glycosyls could be elucidated from ^{13}C-NMR spectra.

4.3.2.1. Identification of skeleton structures of flavonoids

The core structures are difficult to be elucidated by resonance signals of aromatic protons. However, the characteristic signals of carbons in ring C allowed the identification of different types of flavonoids, see **Table 9**.

Substituent mode	H-2′	H-3′	H-4′	H-5′	H-6′
None substituent on ring B	δ7.1–7.6 (m)	δ7.9–8.2 (m)	δ7.9–8.2 (m)	δ7.9–8.2 (m)	7.1-7.6 (m)
2′-Oxygenated		δ6.8–6.9 (m or dd)	δ7.2 (m or dd)	δ6.8–6.9 (m or dd)	7.4–7.5 (m or dd)
4′-Oxygenated	δ7.1–8.1 (2H, d, J≈8.0Hz)	δ6.5–7.1 (2H, d, J≈8.0Hz)		δ6.5–7.1 (2H, d, J≈8.0Hz)	7.1–8.1 (2H, d, J≈8.0Hz)
2′,4′-Dioxygenation		δ6.00–6.6(d, J≈2.0Hz)		δ6.30–6.50 (1H, dd, J≈8.0,2.0Hz)	7.0–7.4(d, J≈8.0Hz)
3′,4′-Dioxygenation	δ7.2–7.8(1H, d, J≈2.0Hz)			δ6.7–7.1(1H, d, J≈8.0Hz)	6.7–7.9(1H, dd, J≈2.0, 8.0Hz)
3′,4′,5′-Trioxygenation	δ6.5–7.5 (2H, s, H-2′,6′) as identical oxygen-bearing substituents are connected to position 3′,4′ and 5′.				
2′,4′,5′-Trioxygenation		δ6.4–6.6(s)			δ7.2–7.5(s)

Table 7. Chemical shifts of protons on ring B of various substituent modes [28].

Proton type	Chemical shift
Phenolic hydroxyl	5-OH (δ12.0–14.0),7-OH (δ10.8–11.0), 4′-OH (δ9.8–10.6), 3′-OH (δ9.2–10.4)
Methyl	C-6,8 (δ2.0–2.5); rha-CH_3 [δ9.2–10.4 (d, J≈6.5Hz)] CH_3CO [glc: δ1.65–2.10 (3H,s); aromatic -CH_3:δ2.3–2.5 (3H,s)
Methoxyl	δ3.5–4.1 (3H,s)
O-CH_2-O	δ6.0
Isopentenyl	δCH_2 (3–3.4); CH (5.2);CH_3 (δ1.7–1.8)
Terminal protons of glycosyl	δ4.5–5.5

Table 8. Chemical shifts of the protons on common substituents.

4.3.2.2. Determination of substituent modes of flavonoids

The substituent modes of core structures of flavonoids could be determined by the signals of aromatic carbons. The chemical shifts of carbons in ring A and B, if they are not substituent, are shown in **Table 10** [28].

4.3.2.2.1. Signal characteristics of ring A

Usually, the substituents, such as hydroxyl, methoxyl and isopentenyl groups, are attached at position 5 or/and 7 of ring A, which leads to the changes of chemical shifts of other carbons in ring A. It is shown in **Table 11** [28].

4.3.2.2.2. Signal characteristics of carbons on ring B

The signal characteristics of carbons on ring B are shown in **Table 12** [28].

http://dx.doi.org/10.5772/67810

Type	C-2	C-3	C-4
Flavone	δ160.0–165.5	δ104.0–112.0	δ175.0–184.0
Flavonol (3-OH is free)	δ145.0–149.5	δ135.6–139.0	δ172.0–178.0
Flavonol (3-OH is free)	δ155.7–157.0	δ133.2–134.3	δ172.0–178.0
Isoflavone	δ149.8–156.5	δ120.3–126.0	δ174.5–182.0
Flavanone	δ75.0–80.5	δ42.0–44.6	δ189.5–197.2
Flavanonol	δ82.7–83.5	δ71.2–73.0	δ188.0–197.0
Chalcone	δ136.9–145.4	δ116.8–128.1	δ188.0–194.6
Dihydrochalcone	δ39.2–45.5	δ28.5–31.2	δ199.6–205.5
Aurone	δ146.1–147.7	δ111.0–113.3	δ180.5–182.7
Flavan-3-ol	δ77.5–82.7	δ65.1–68.2	δ27.5–28.6
Flavan-3,4-diols	δ80.0–82.5	δ72.5–74.4	δ69.8–72.0

Table 9. Chemical shifts of carbons in ring C of flavonoids [28].

Ring A	Chemical shift	Ring B	Chemical shift
C-5	δ124–126	C-2′,6′	δ126.0–130.0
C-6	δ124–126	C-3′,5′	δ128–129
C-7	δ133–135	C-4′	δ128.5–132
C-8	δ116.5–119	C-1′ (flavone, flavonol, isoflavone)	δ130–133
C-9	δ154–156	C-1′ (flavanone, flavonol)	δ138–139
C-10	δ121–124		

Table 10. Chemical shifts of carbons in ring A and B if they are not substituent.

Substituent mode	Type	C-5	C-6	C-7	C-8	C-9	C-10
5-Oxygenated		≈δ160	δ109–111	δ135–137	δ107–108	δ157–160	δ111
7-Oxygenated	Flavone, flavonol	δ126	δ114	δ162–164	δ100–102	δ157	δ114–116
	Flavanone, flavanonol	δ128	δ109.5–110	δ165–166	δ100	δ162–164	δ114–116
	Isoflavone	δ127	δ115	δ161–163	δ102–104	δ156–158	δ117–119
5,7-Dioxygenation	Flavone, flavonol, isoflavone	δ160–162	δ97–100	δ162–165	δ93–95	δ155–158	δ103–109
	Flavanone, flavanonol	δ162–163	δ93–97	δ165–167	δ93–97	δ161–163	δ100–104

Table 11. Chemical shifts of carbons in ring A of flavonoids.

4.3.2.2.3. Signal characteristics of common substituents

The carbon chemical shifts of common substituents are shown in **Table 13** [28].

Substituent mode	Type	C-1′	C-2′	C-3′	C-4′	C-5′	C-6′
4′-Oxygenated	Flavone, flavonol, isoflavone	δ121–123	δ130	δ115	δ157–161	δ115	δ130
	Flavanone, flavanonol	δ128–130					
3′-OH,4′-OCH_3 (or 3′,4′-dihydroxyl)	Flavone, flavonol, isoflavone	δ121–125	δ113–114	δ145–147	δ149–151	δ112–116	δ118–122
3′-OCH3,4′-OH		δ121–125	δ110–111	δ150–152	δ148	δ115–117	δ118–122
3′-OH,4′-OCH_3 (or 3′,4′-dihydroxyl)	Flavanone, flavanonol	δ128–129	δ111–115	δ144–147	δ146–148	δ112–116	δ118–120
3′-OCH_3,4′-OH		δ131–132	δ111–115	δ147–149	δ146–147	δ112–116	δ118–120
2′,4′-Dioxygenation	Flavonoids	δ108–113	δ156–158	δ102–104	δ157–162	δ104–108	δ131–132
3′,4′,5′-Trioxygenation		δ120–126	δ106–109	δ146–153	δ93–97	δ136–142	δ106–109

Table 12. Chemicals shifts of carbons on ring B of flavonoids.

4.3.3. *Glycosides of flavonoids*

In plants, flavonoids are often present as O- or C-glycosides. The O-glycosides have sugar substituents bound to a hydroxyl group of the aglycone, usually located at position 3 or 7, whereas the C-glycosides have sugar groups bound to a carbon of the aglycone, usually 6-C or 8-C. The most common carbohydrates are rhamnose, glucose, galactose and arabinose.

Generally, the chemical shifts of terminal protons of glycosyls are at δ4.5–5.5 in ^{1}H-NMR. The terminal carbons of O-glycosides are at δ95–105 and at δ71–78 for C-glycosides. Furthermore, the number of glycosyls could be determined by combined analysis of ^{1}H and ^{13}C-NMR spectra. It is an effective method to determine the connection positions of glycosyls by glycosylation shifts, as shown in **Table 14** [28].

The configurations of glycosyls should be determined. The relative configurations of some glycosyl groups could be determined sometimes by coupling constants of terminal protons in ^{1}H-NMR spectra. The absolute configurations, however, should be determined by chemical methods and gas chromatography.

Substituent		Chemical shift
CH_3		6-CH_3(δ6–10);8-CH_3(δ20–30);$COCH_3$(δ17–22)
OCH_3		δ55–57
Isopentenyl [-$CH_2CH{=}CHCH_3$]		CH_2(δ21–22); CH(δ122–124; CH(δ129–131); CH_3(δ17–27)
O-CH_2-O		δ100–101
Terminal carbon of glycosyls	O-glycosides	δ95–105
	C-glycosides	δ71–80

Table 13. Chemical shifts of carbons of common substituents on flavonoids.

http://dx.doi.org/10.5772/67810

Glycosylation position	2	3	4	5	6	7	8	9	10	1′	2′	3′	4′	5′	6′
7-O-glu					+0.8	−1.4	+1.1		+1.7						
7-O-rha					+0.8	−2.4	+1.0		+1.7						
3-O-glu	+9.2	−2.1	+1.5	+0.4					+1.0	−0.8	+1.1	−0.3	+0.7		+1.5
3-O-rha	+10.3	−1.1	+2.0	+0.6					+1.1						
5-O-glu	−2.8	+2.2	−6.0	−2.7	+4.4	−3.0	+3.2	+1.4	+4.3	−1.3	−1.2	−0.4	−0.8	−1.0	−1.2
3′-O-glu	−0.5	+0.4									+1.6		+1.4	+0.4	+3.2
4′-O-glu	+0.1		+1.0							+3.7	+0.4	+2.0	−1.2	+1.4	

Table 14. Glycosylation shifts (average values) of flavonoids in ^{13}C-NMR spectrum.

As for the spectral method, the types and configurations could be speculated by the chemical shifts of glycosyl carbons in ^{13}C-NMR spectra, as shown in **Table 15** [28].

4.4. Mass spectral characteristics of flavonoids

ESI-MS and FAB-MS are widely applied in the studies of flavonoids. While the positive ion mode is employed, quasi-molecular ion peaks such as $[M+H]^+$, $[M+Na]^+$, $[M+K]^+$ and $[M+NH_4]^+$ will be displayed. $[2M+H]^+$, $[2M+Na]^+$ and so on will also be shown if the sample is concentrated. The MS fragmentation pathways of flavone and flavanone are shown in **Figures 5** and **6**.

4.5. Determination of absolute configuration

The absolute configuration should be determined if chiral atoms are existed in the structures. The main methods to elucidate absolute configuration include circular dichroism (CD), optical rotatory dispersion (ORD) and X-ray single crystal diffraction. Circular dichroism and ORD are mainly introduced here.

Type of glycosyl	C-1″	C-2″	C-3″	C-4″	C-5″	C-6″
β-D-glu	δ104.0	δ74.1	δ76.8	δ70.6	δ76.8	δ61.8
α-D-glu	δ100.0	δ72.2	δ74.1	δ70.6	δ72.5	δ61.6
β-D-gal	δ104.5	δ71.7	δ73.8	δ69.7	δ76.0	δ62.0
α- D- gal	δ100.1	δ69.2	δ70.5	δ70.2	δ71.6	δ62.2
β-D-man	δ102.3	δ71.7	δ74.5	δ68.4	δ77.6	δ62.6
α-D-man	δ102.2	δ71.4	δ72.1	δ68.3	δ73.9	δ62.5
β-D-rha	δ102.4	δ71.8	δ74.1	δ73.4	δ73.4	δ17.9
α-D-rha	δ102.1	δ71.2	δ71.5	δ73.3	δ69.5	δ17.9

Table 15. Carbon chemical shifts of common glycosyls.

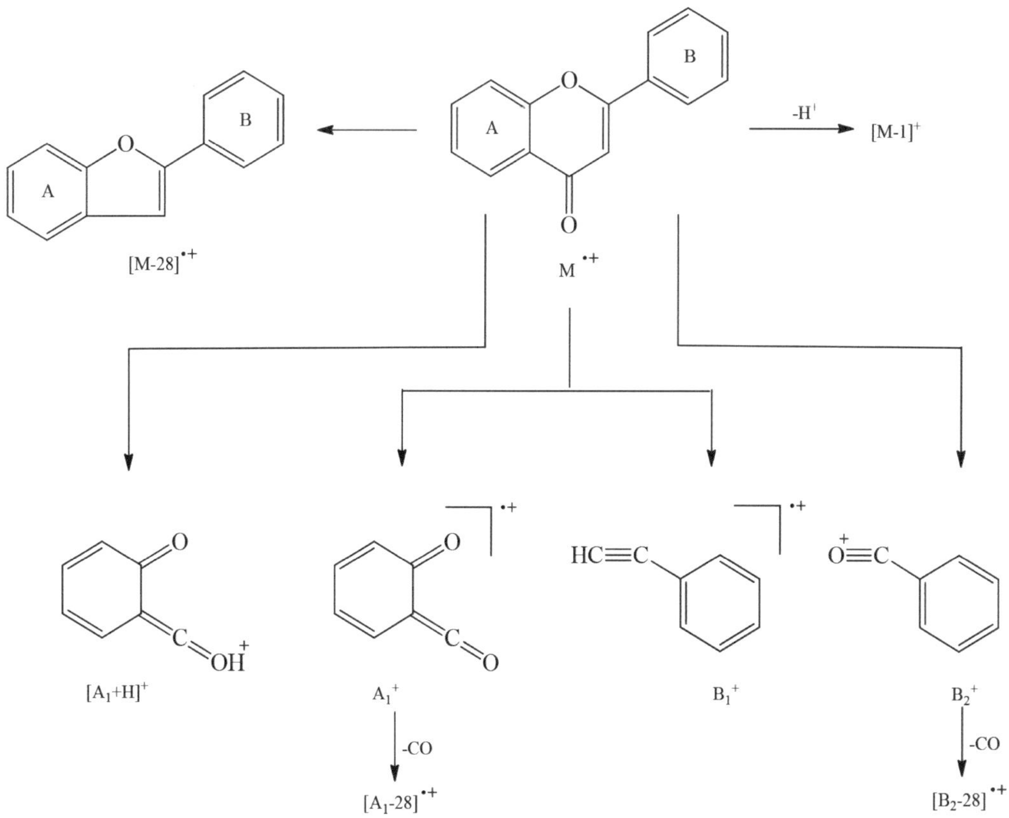

Figure 5. MS fragmentation pathway of flavone.

4.5.1. Optical rotatory dispersion (ORD)

For the flavonoids possess chiral centers, their optical activities (589.0 nm, Na-D light source) are correlative with spatial configurations, as shown in **Table 16** [27].

4.5.2. Circular dichroism (CD)

It is the most used method to elucidate the absolute configurations of flavonoids via cotton effect (CE) of CD spectra.

4.5.2.1. Flavanone

Most of the protons of flavanones at position 2 are axial ($J \approx 11.0$Hz). The characteristics of CE are shown in **Table 17**.

As reported in literature [29], the absolute configurations of the enantiomeric flavanone pair (2S)-6-formyl-5,7-dihydroxyflavanone (1a) and (2R)-6-formyl-5,7-dihydroxyflavanone (1b) were assessed via their chiroptical data. The ECD curves of compound (1a) showed sequential positive and negative cotton effects near 310 and 280 nm for the $n \rightarrow \pi^*$ and $\pi \rightarrow \pi^*$ electronic transitions, respectively. These cotton effects are reminiscent of flavanones exhibiting

http://dx.doi.org/10.5772/67810

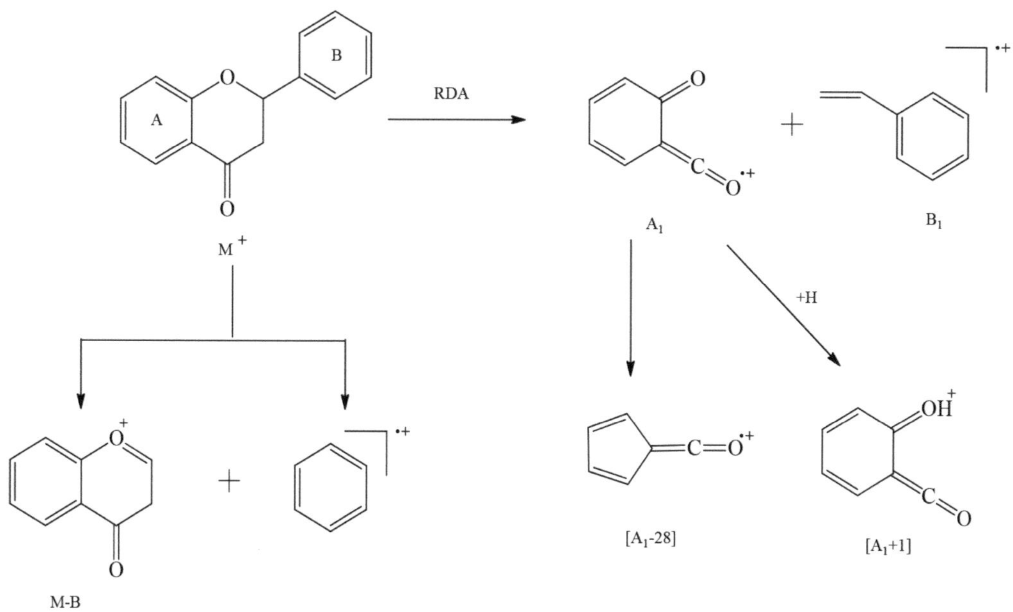

Figure 6. MS fragmentation pathway of flavanone.

Type	Optical activity	Spatial structure
Flavan	–	2S
Flavanonol (*trans*-form)	+	2R,3R
Epicatechin (*cis*-form)	-	2R,3R
Epicatechin (*cis*-form)	+	2S,3S
Catechin (*trans*-form)	+	2R,3S
Pterocarpin (*cis*-form)	–	6aR, 11aR
6a-Hydroxyl pterocarpin (*cis*-form)	–	6aS, 11aS
6a-Hydroxyl pterocarpin (*trans*-form)	+	6aR, 11aR

Table 16. Optical activities of flavonoids.

CE		Absolute configuration of C-2
270–290 nm	320–330 nm	
–	+	S
+	–	R

Table 17. Relationship between CE and absolute configurations of flavanones.

P-helicity of the conformational flexible heterocycle with a C-2 equatorial B ring and, hence, (2S) absolute configuration. The mirror image related to ECD spectrum of **1b** accordingly confirmed its (2R) absolute configuration. It is shown in **Figure 7**.

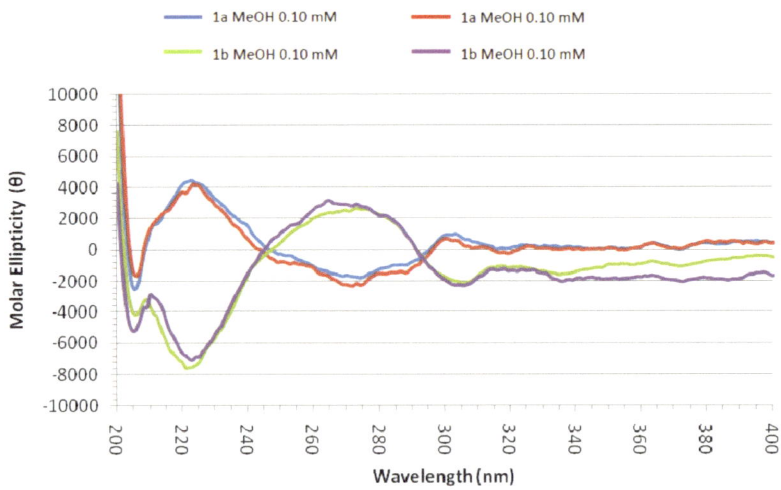

Figure 7. ECD spectra for compounds **1a** and **1b**.

4.5.2.2. Flavanonol

Four possible structures are existed in nature because of the existence of two chiral centers (C-2 and C-3), while 2R, 3R configurations are commonest. The relative configuration could be determined by coupling constant between H-2 and H-3 and then CD spectrum is employed to elucidate the absolute configuration, as shown in **Table 18** [30].

4.5.2.3. 3-Hydroxyl flavans

Similarly, C-2 and C-3 are also the chiral centers of 3-hydroxyl flavans. The characteristics are shown in **Table 19** [31].

4.5.2.4. 4-Hydroxyl flavans

The relative configuration could be determined by coupling constant of H-2 and H-4 combined with NOE spectra and then CD spectrum could be employed to elucidate the absolute configuration, as shown in **Table 20** [32].

Relative configuration	Cotton effect (300–340)	Absolute configuration
trans-	+	2R, 3R
	–	2S, 3S
cis-	+	2R, 3S
	–	2S, 3R

Table 18. Relationship between absolute configurations of flavanonol with CE.

http://dx.doi.org/10.5772/67810

Relative configuration	CE		Absolute configuration
	240 nm	280 nm	
Trans-	+	–	2R, 3S
	–	+	2R, 3R
Cis-	–	+	2S, 3R
	+	+	2S, 3S

Table 19. Relationship between absolute configurations of 3-hydroxyl flavans with CE.

Relative configuration	CE		Absolute configuration
	240 nm	280 nm	
Trans-	–	–	2S, 4S
	+	+	2S, 4R
Cis-	–	–	2S, 4S
	+	+	2R, 4R

Table 20. Relationship between absolute configurations of 4-hydroxyl flavans with CE.

Relative configuration between C-2 and C-3	Relative configuration between C-3 and C-4	Cotton effect (280 nm)	Absolute configuration
Cis-	*Cis-*	+	2S,3S,4S
	Trans-	+	2S,3S,4R
Trans-	*Cis-*	+	2S,3R,4R
	trans-	+	2S,3R,4S
Cis-	*Cis-*	–	2R,3R,4R
	Trans-	–	2R,3R,4S
Trans-	*Cis-*	–	2R,3S,4S
	Trans-	–	2R,3S,4R

Table 21. Relationship of absolute configurations of 3,4-dihydroxyl flavans with CE.

4.5.2.5. 3,4-Dihydroxyl flavans

More absolute configurations are existed because of three chiral centers, as shown in **Table 21** [33].

4.5.2.6. Flavans

The cotton effects of flavans are show in **Table 22** [34].

4.5.2.7. Isoflavans

The CE characteristics are shown in **Table 23**.

Cotton effect (280 nm)	Absolute configuration
+	2R
–	2S

Table 22. Cotton effects of flavans.

CE		Absolute configuration
260–320 nm	220–260 nm	
+	–	3R
–	+	3S

Table 23. Cotton effects of isoflavans.

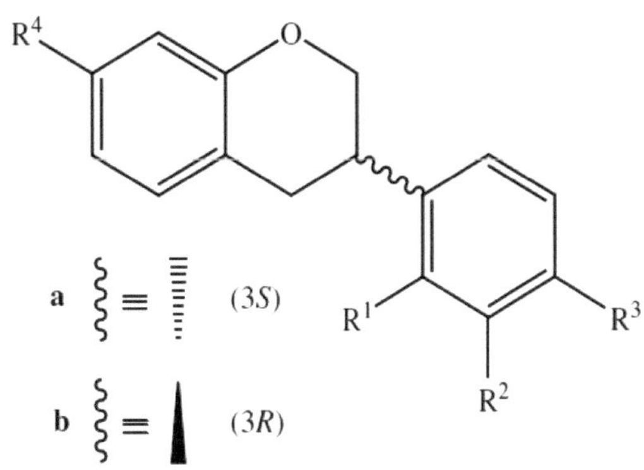

Compound	R1	R2	R3	R4
30a/b	OH	H	OMe	OH
31a/b	H	H	OMe	H
32a/b	H	OMe	OMe	H
33a/b	OMe	H	OMe	H
34a/b	H	H	OMe	OMe
35a/b	OMe	H	H	OMe
36a/b	OMe	H	OMe	OMe

Figure 8. Synthetic isoflavans (31a/b–36a/b).

Versteeg et al. [35] synthesized six isoflavans and their enantiomers (31a/b-36a/b), and used authentic 3S- and 3R-vestitol (30a and 30b) derivatives to establish the absolute configuration at C3 of the synthetic isoflavans (**Figure 8**). (3S)-Isoflavans with oxygenation at both the A- and B-rings (34a, 35a and 36a) display positive and negative CEs in the 240 (1La) and 270–280 nm (1Lb) regions, respectively, and conversely for the 3R-enantiomers (34b, 35b and 36b) (**Figure 9**).

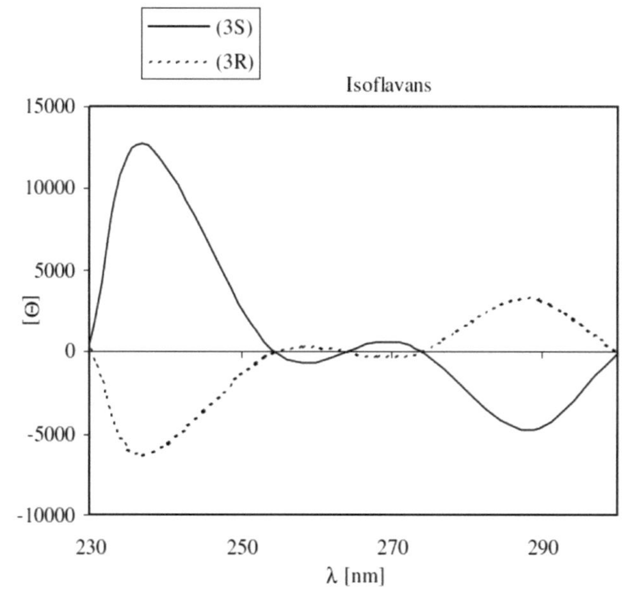

Figure 9. CD spectra of isoflavans with oxygenation at both the A and B rings.

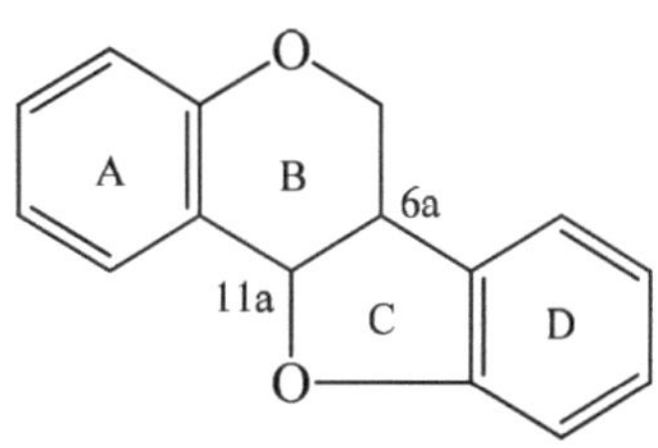

Relative configuration of between C-6a and C-11a	CE		Absolute configuration
	260–310 nm	220–250 nm	
Trans-	+		6aS, 11aR
	–		6aR, 11aS
Cis-	+	–	6aR, 11aR
	–	+	6aS, 11aS

Table 24. Relationship of absolute configurations of pterocarpins with CE.

4.5.2.8. Pterocarpins

The spectral characteristics are shown in **Table 24** [36].

The relationships between the CE and absolute configurations will change after a hydroxyl group is attached to position 6a, as shown in **Table 25** [37].

Relative configuration of between C-6a and C-11a	CE		Absolute configuration
	260–310 nm	220–250 nm	
Trans-	−	−	6aS, 11aR
	+	+	6aR, 11aS
Cis-	+	−	6aS, 11aS
	−	+	6aR, 11aR

Table 25. Relationship of absolute configurations of 6a-hydroxyl pterocarpins with CE.

Author details

Weisheng Feng*, Zhiyou Hao and Meng Li

*Address all correspondence to: fwsh@hactcm.edu.cn

School of Pharmacy, Henan University of Chinese Medicine, Zhengzhou, China

References

[1] Bandele, O.J. & Osheroff, N. (2007). Bioflavonoids as poisons of human topoisomerase IIα and IIβ. Biochemistry, Vol. 46, pp. 6097–6108.

[2] Saija, A.; Scalese, M.; Lanza, M.; Marzullo, D.; Bonina, F. & Castelli, F. (1995). Flavonoids as antioxidant agents: importance of their interaction with biomembranes. Free Radical Biology and Medicine, Vol. 19, pp. 481–486.

[3] El-Subbagh, H.I.; Abu-Zaid, S.M.; Mahran, M.A.; Badria, F.A. & Al-Obaid, A.M. (2000). Synthesis and biological evaluation of certain α, β-unsaturated ketones and their corresponding fused pyridines as antiviral and cytotoxic agents. Journal of Medicinal Chemistry, Vol. 43, pp. 2915–2921.

[4] Miyoshi, N.; Naniwa, K.; Yamada, T.; Osawa, T. & Nakamura, Y. (2007). Dietary flavonoid apigenin is a potential inducer of intracellular oxidative stress: the role in the interruptive apoptotic signal. Archives of Biochemistry and Biophysics, Vol. 466, pp. 274–282.

[5] Shukla, S.; MacLennan, G.T.; Flask, C.A.; Fu, P.; Mishra, A.; Resnick, M.I. & Gupta, S. (2007). Blockade of β-catenin signaling by plant flavonoid apigenin suppresses prostate carcinogenesis in TRAMP mice. Cancer Research, Vol. 67, pp. 6925–6935.

[6] Walle, T. & Walle, U.K. (2007). Novel methoxylated flavone inhibitors of cytochrome P450 1B1 in SCC-9 human oral cancer cells. Journal of Pharmacy and Pharmacology, Vol. 59, pp. 857–862.

[7] Patanasethanont, D.; Nagai, J.; Matsuura, C.; Fukui, K.; Sutthanut, K.; Sripanidkulchai, B. O.; Yumoto, R. & Takano, M. (2007). Modulation of function of multidrug resistance associated-proteins by Kaempferia parviflora extracts and their components. European Journal of Pharmacology, Vol. 566, pp. 67–74.

[8] Walle, T.; Ta, N.; Kawamori, T.; Wen, X.; Tsuji, P.A. & Walle, U.K. (2007). Cancer chemopreventive properties of orally bioavailable flavonoids—methylated versus unmethylated flavones. Biochemical Pharmacology, Vol. 73, pp. 1288–1296.

[9] Verschoyle, R.D.; Greaves, P.; Cai, H.; Borkhardt, A.; Broggini, M.; D'Incalci, M.; Riccio, E.; Doppalapudi, R.; Kapetanovic, I.M.; Steward, W.P. & Gescher, A.J. (2006). Preliminary safety evaluation of the putative cancer chemopreventive agent tricin, a naturally occurring flavone. Cancer Chemotherapy and Pharmacology, Vol. 57, pp. 1–6.

[10] Graf, B.A.; Milbury, P.E. & Blumberg, J.B. (2005). Flavonols, flavones, flavanones, and human health: epidemiological evidence. Journal of Medicinal Food, Vol. 8, pp. 281–90.

[11] Woodman, O.L.; Meeker, W.F. & Boujaoude, M. (2005). Vasorelaxant and antioxidant activity of flavonols and flavones: structure-activity relationships. Journal of Cardiovascular Pharmacology, Vol. 46, pp. 302–309.

[12] Chen, B.; Cai, J.; Song, L.S.; Wang, X. & Chen, Z. (2005). Effects of ginkgo biloba extract on cation currents in rat ventricular myocytes. Life Sciences, Vol. 76, pp. 1111–1121.

[13] Jones, D.J.; Lamb, J.H.; Verschoyle, R.D.; Howells, LM.; Butterworth, M.; Lim, C.K.; Ferry, D.; Farmer, P.B. & Gescher, A.J. (2004). Characterisation of metabolites of the putative cancer chemopreventive agent quercetin and their effect on cyclo-oxygenase activity. British Journal of Cancer, Vol. 91, pp. 1213–1219.

[14] Liu, Y.; Teng, Y.P. & Zhang, Y.M. (2000). Research progress of dietary flanones. Journal of Preventive Medicine of Chinese People's Liberation Army, Vol. 18, pp. 384–386.

[15] Akama, T.; Shida, Y.; Sugaya, T.; Ishida, H.; Gomi, K. & Kasai, M. (1996). Novel 5-aminoflavone derivatives as specific antitumor agents in breast cancer. Journal of Medicinal Chemistry, Vol. 39, pp. 3461–3469.

[16] Ullmannova, V. & Popescu, N.C. (2007). Inhibition of cell proliferation, induction of apoptosis, reactivation of DLC1, and modulation of other gene expression by dietary flavone in breast cancer cell lines. Cancer Detection and Prevention, Vol. 31, pp. 110–118.

[17] Ahmed, N.; Dubuc, C.; Rousseau, J.; Bénard, F. & van Lier, J.E. (2007). Synthesis, characterization, and estrogen receptor binding affinity of flavone-, indole-, and furan-estradiol conjugates. Bioorganic & Medicinal Chemistry Letters, Vol. 17, pp. 3212–3216.

[18] Li, C. & Fan, J. (1998). Extraction and determination of total flavone from foliage of Gingk. Anhui Agricultural Science Bulletin, Vol. 4, pp. 20–22.

[19] Ma, G.G. & Wang, J.Z. (2007). Review on the extraction method of rutin from *Sophora japonica L.* Chinese Wild Plant Resources, Vol. 26, pp. 5–8.

[20] Fu. Y.J.; Zu, Y.G.; Zhao, C.J. & Li, C.Y. (2003). Supercritical CO_2 extraction of flavonoids from licorice. Chinese Journal of Applied Chemistry, Vol. 20, pp. 1217–1219.

[21] Lan, C.Y.; Zhou, C.S.; Fan, B.W.; Zhou & Y.L. (2005). Research on ultrasoic ware extraction of flavone from bud of *Sophora japonica*. Natural Product Research and Development, Vol. 17, pp. 55–58.

[22] Lv, G.Y.; Ye, L. & Yu, J.W. (2009). Study on microwave-assisted extraction technique of flavonoid from *Ophiopogon japonicus*. China Journal of Traditional Chinese Medicine and Pharmacy, Vol. 24, pp. 216–218.

[23] Huang, C.H.; Wang, Y.; Ou, S.Y.; Duan, H.Y. & Yang, A.H. (2006). Preparation of soybean isoflavone by inorganic UF membrane. China Oils and Fats, Vol. 31, pp. 53–55.

[24] Huo, Q. (2006). Measurement and correlation of partition coefficients of puerarin in aqueous two-phase PEG-$(NH_4)_2SO_4$ and acetone-K_2HPO_4 systems. Journal of Shenyang Pharmaceutical University, Vol. 23, pp. 139–144.

[25] Yuan, Y.; Hou, W.l.; Tang, M.H.; Luo, H.D.; Chen, L.J.; Guan, Y.H. & Sutherland, I.A. (2008). Separation of flavonoids from the leaves of *Oroxylum indicum* by HSCCC. Chromatographia, Vol. 68, pp. 885–892.

[26] Pakade, V.; Lindahl, S.; Chimuka, L. & Turner, C. (2012). Molecularly imprinted polymers targeting quercetin in high-temperature aqueous solutions. Journal of Chromatography A, Vol. 1230, pp. 15–23.

[27] Markham K.R. (1990). Techniques of Flavonoid Identification. Science Press, Beijing, China.

[28] Fang Q.C. (2006). Research of Natural Pharmaceutical Chemistry. Pecking Union Medical College Press, Beijing, China.

[29] Zaki, M.A.; Nanayakkara, N.P.; Hetta, M.H.; Jacob, M.R.; Khan, S.I.; Mohammed, R.; Ibrahim, M.A.; Samoylenko, V.; Coleman, C.; Fronczek, F.R.; Ferreira, D. & Muhammad, I. (2016). Bioactive formylated flavonoids from *Eugenia rigida*: isolation, synthesis, and X-ray crystallography. Journal of Natural Products, Vol. 79, pp. 2341–2349.

[30] Gaffield, W. (1970). Circular dichroism, optical rotatory dispersion and absolute configuration of 3-hydroxyflavanones and their glycosides. Determination of aglycone chirality in flavanone glycosides. Tetrahedron, Vol. 26, pp. 4093–4108.

[31] Rensburg, H.V.; Steynberg, P.J.; Burger, J.F.W.; Heerden, P.S.V. & Ferreira, D. (1999). Circular dichroic properties of flavan-3-ols. Journal of Chemistry Research, Synopses, Vol. 7, pp. 450–451.

[32] Wu, L.J. (2007). Applied Natural Organic Chemistry. People's Medical Publishing House, Beijing, China.

http://dx.doi.org/10.5772/67810

[33] Slade, D.; Ferreira, D. & Marais, J.P. (2005). Circular dichroism, a powerful tool for the assessment of absolute configuration of flavonoids. Phytochemistry, Vol. 66, pp. 2177–2215.

[34] Antus, S.; Kurtán, T.; Juhász, L.; Kiss, L.; Hollósi, M. & Májer, Z. (2001). Chiroptical properties of 2,3-dihydrobenzo[b]furan and chromane chromophores in naturally occurring O-heterocycles. Chirality, Vol. 13, pp. 493–506.

[35] Versteeg, M.; Bezuidenhoudt, B.C.B. & Ferreira, D. (1999). Stereoselective synthesis of isoflavonoids. (R)- and (S)-isoflavans. Tetrahedron, Vol. 55, pp. 3365–3376.

[36] Kiss, L.; Kurtán, T.; Antus, S. & Bényei, A. (2003). Chiroptical properties and synthesis of enantiopure *cis* and *trans* pterocarpan skeleton. Chirality, Vol. 15, pp. 558–563.

[37] Harborne, J.B. (1994). The Flavonoids Advances in Research Since 1986. Chapman and Hall Ltd., London.

Chapter 3

Isolation and Structure Characterization of Flavonoids

Maurice D. Awouafack, Pierre Tane and
Hiroyuki Morita

Additional information is available at the end of the chapter

http://dx.doi.org/10.5772/67881

Abstract

Flavonoids are one of the most important classes of secondary metabolites from natural products due to their several applications in medicine, foods, diet industries, and so on. Even though a huge number has been reported from natural and synthetic sources, scientists are still interested in flavonoids and derivatives. The biggest challenge for working on secondary metabolites is related to the use of the predicted theoretical method to isolate the expected compound and finally analyse the spectroscopic data to elucidate and fully characterize the structure. This chapter was designed to document useful techniques for isolation and structure characterization of flavonoids. Besides the well-known methods that have been used so far, we would also put together updated information about novel challenge techniques published in recent articles on isolation and characterization of flavonoids. Our data were obtained mainly from academic library and from reported data online by using research links such as Google Scholar, Scopus, SciFinder, Scirus, PubMed, and so on. Our field experience on phytochemistry of isolation and characterization of flavonoids was also used in this chapter.

Keywords: natural products, flavonoids, techniques, isolation, characterization

1. Introduction

Flavonoids are a large group of C-15 (C_6–C_3–C_6) secondary metabolites widespread in higher plants and are also detected in some lower plants such as algae. An important number has been reported from natural and synthetic sources due to their several applications in pharmaceutical and diet industries. Flavonoids occur in natural products specially blooming plant species, and colours of flowers could be indicative for the class of compounds. Flavonoids are mostly obtained as yellow pale, white, red, purple, blue, and so on from species of several plant families but are known to be widespread in the Fabaceae family. Flavonoids could be detected in natural products by using some analytical methods such as the Shinoda [1], sodium hydroxide [1] and *p*-dimethylaminocinnamaldehyde tests [2].

2. Classification and basic skeletons of flavonoids

Flavonoids have a diversity of chemical structures constituted of 15 carbon atoms in their basic skeletons with a C_6–C_3–C_6 framework made by two aromatic rings (A and B) linked by a three-carbon unit that may or may not form a third ring (C). Generally, carbons are referred to by a numbering system, which utilizes ordinary numerals for the A- and C-rings and 'primed' numerals for the B-ring (**1–3**), but this is not respected when referring to chalcones (**3**) [3, 4]. The B-ring could be linked to C-ring at position C-2, C-3 or C-4 to form most classes of this secondary metabolite known as the flavonoids (flavone, flavonol, flavonone, flavan), isoflavonoids (isoflavone, isofavonone, isoflavonol, isoflavan, rotenoid, coumestane, pterocarpan, isoflavene) and neoflavonoids (**4**) (arylcoumarin, neoflavene, etc.), respectively. Some minor flavonoids are also known such as aurone, chalcone, and dihydrochalcone which is the first class of flavonoids obtained by biosynthesis and therefore the precursor of other classes. The chemical structure diversity of flavonoids is particularly obtained from glycosylation, methoxylation, prenylation, hydroxylation that usually took place with some specific positions in these different classes [3, 5–7] (**Figure 1**).

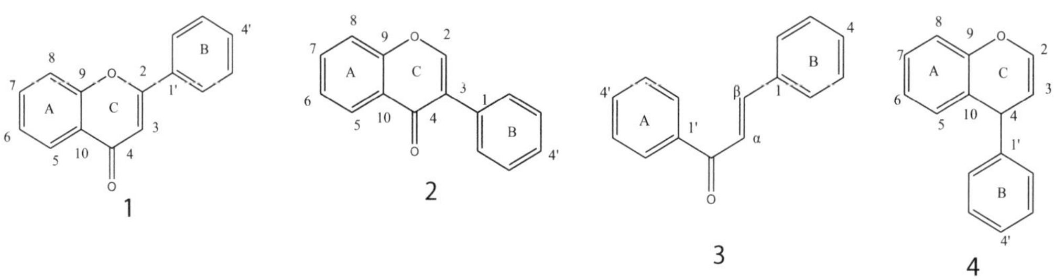

Figure 1. Predominant basic skeletons of flavonoids and carbon atoms numbering pattern.

3. Isolation techniques of flavonoids

3.1. Sample preparation

Flavonoids especially those in plants could be extracted from several parts such as roots, barks, leaves, fruits, woods and flowers. Samples are more often dried and ground before the extraction process. This initial treatment of samples helps in facilitating the extraction yields as well as preserving constituents. In some cases, the extraction is carried out on fresh plant materials. The dried plant materials have been reported in several investigations to contain most flavonoids than fresh samples [8–10]. Obviously, the ground samples always gave higher yields of extraction, and this could be justified by the fact that the solvent has contact with surface constituents when the powder has smaller particles. The extraction yields of flavonoids from natural products are also affected by some factors such as temperature, time and ratio of water in case of aqueous mixing solvents [11].

Several methods have been used for extracting flavonoids in plant materials. These include maceration, infusion, decoction, percolation, hot continuous extraction (soxhlet), ultrasound-assisted extraction and microwave-assisted extraction, using solvents as water, ethanol, methanol, n-butanol, acetone, ethyl acetate, chloroform, and so on. Polar solvents are used to obtain flavonoid glycosides, whereas non-polar solvents extracted mostly their aglycones. Most of the investigations conducted in the extraction of flavonoids in plant materials have been done by maceration and infusion [10, 12]. A herbal tea from *Viscum album* L. was prepared using maceration and infusion to yield 31 and 43% flavonoid-like substances, respectively [13]. Ethanol, methanol and acetone are among the best solvents for extracting flavonoids [14, 15]. Acetone was reported to be the best solvent to extract flavonoids from a bitter melon—*Momordica charantia*—and *Tagetes patula* while the ethanol extract from *Trigonella foenum-graecum* had the highest flavonoid contents [11, 12, 16].

Decoction process is presented as a simple, cheap and convenient extraction method that may be useful in poor-equipped laboratories. A Thai medicinal plant called Siamese neem tree (*Azadirachta indica* A. Juss. var. *siamensis Valeton*) is well known to have flavonoids (rutin and quercetin) as main bioactive constituents. The decoction provided an extract with the highest amount of total flavonoids (17.54 mgRE/g extract) when using six different extraction techniques such as maceration, percolation, decoction, soxhlet extraction, ultrasonic extraction and microwave-assisted extraction in dried young flowers [17]. However, it may also depend on the plant material including the part used, number of constituents present and some conditions mentioned above that influence the extraction process. This was the case for the whole plant of *Senecio anteuphorbium* collected from Sidi Ifni, Southern Anti-Atlas of Morocco that was extracted using soxhlet extraction, decoction and maceration, and the methanolic extract from the soxhlet extraction showed the highest total flavonoid content (26.59 ± 0.24 mg QE/gE or 39.47 ± 1.01 mg RE/gE) while the aqueous maceration had the lowest (6.52 ± 0.09 mg QE/gE or 9.68 ± 0.22 mg RE/gE) [18].

The extraction of powered seeds of *Ziziphus mauritiana* using different methods such as maceration, decoction, soxhlet extraction and sonication with 50 and 80% ethanol, and water (decoction) as solvents, was reported and the high total flavonoid contents was obtained from the sonication technique [19].

In the basic mechanism of the extraction techniques, the microwave-assisted extraction follows several steps when comparing to conventional extractions. These include the penetration of the solvent into the solid matrix, the solubilization and/or breakdown of constituents, the transportation of the solute outside of the solid matrix, the migration of the solute from the external surface of the solid into the solution, the movement of the extract with respect to the solid, and the separation and discharge of the extract and solid [20–22]. The main difference between the microwave-assisted extraction and conventional extractions being the directions of heat and mass gradients during the extraction: for the first process, both move from inside to outside while in the second case, the mass transfer goes from inside to outside when heat occurs from outside to inside of the subtract [20–22].

Following the traditional Indian medicinal preparations, Krishnan and Rajan recently reported a suitable extraction of flavonoids from *Terminalia bellerica* Roxb., by the microwave-assisted solid-liquid method, an investigation conducted in view to study the influence of solvent-to-feed ratio and temperature on kinetics and thermodynamics of aqueous extraction [23]. Total flavonoids with good yield (1.13%) obtained under optimum conditions (ultrasonic power 500 W, extraction time 20 min, material solvent ratio 1:20, and ethanol concentration 30%) using ultrasound-assisted extraction were reported from the corn silk (*Zea mays* L.), a Chinese medicinal herb, with a recommendation for this plant to be developed as food natural antioxidant reagents [24]. Ultrasonic extract of flower from Lythrum *salicaria* L. was reported to possess good scavenging of hydrogen peroxide owing to the higher phenolic and flavonoid contents when using three methods of extraction such as percolation, ultrasonic-assisted extraction and polyphenol fraction [25].

All these techniques allow to have flavonoids in the crude extract with good yield before the application of different fractionation and purification procedures for their isolation.

3.2. Chromatography as a main tool for isolation of flavonoids

The isolation of flavonoids from natural sources is conducted by repeated and successive chromatography techniques such as open column chromatography (CC), preparative thin-layer chromatography (prep. TLC), centrifugal preparative thin-layer chromatography (CPTLC), high-speed counter-current chromatography (HSCCC), medium-pressure liquid chromatography (MPLC), high-pressure preparative liquid chromatography (prep. HPLC), and so on.

In column chromatography method, stationary phases could be normal or reverse phase silica gels, Sephadex (LH-20, G-10, G-25 and G-50). In view to have flavonoids-rich fractions, it is recommended to use some preliminary liquid-liquid extraction methods or polymeric resins such as Diaion HP-20, Amberlites (XAD-2, XAD-7) from the crude extract. These polymeric resins are very useful when the absorption of extracts is eluted in the open column chromatography with an increasing gradient of methanol in water.

The open column chromatography (CC) is still the most useful and easy isolation technique for natural products isolation and by means that of purification of flavonoids. The choice of the good solvents system for mobile phase is important and should be made from the check-up TLC on the crude or the flavonoids-rich fraction. Combination and polarity of solvents should be used depending on the class of flavonoids targeted. After the flash column, some major fractions could directly be subjected to Sephadex LH-20 or prep. TLC if they do not contain complex mixture of flavonoids. During the separation process, constituents from the flavonoid-rich fractions could have closer retention factors (Rf) based on their polarities. The change in phase of the adsorbent in some cases is useful to have good separation in either small open CC purification or prep. TLC. Several investigations reported the isolation of new flavonoids using CC. This included two dihydrochalcones, rare natural resources secondary metabolites, from *Eriosema glaumerata* [26], two polyhydroxylated flavones having antioxidant activity from *E. robustum* [3], one isoflavanol from *Kotschya strigosa* [27], two

http://dx.doi.org/10.5772/67881

glucoside isoflavones from *Iris kashmiriana* [28], four dimeric chalcone derivatives from *Uvaria siamensis* [29], five flavonoids from *Millettia griffithii* [6], one pterocarpan, three isoflavones from the root, stem bark and leaves of *Erythrina schliebenii* [30], four flavonoid C-glycosides with anti-inflammatory properties from the leaves of *Piper aduncum* [31]. The number of recent published articles using CC is indicative for the useful and convenience of this method. Nevertheless, the prep. HPLC technique has been widely used for isolating commonly polyphenols and more specifically flavonoids. The suitability of this method for this class of secondary metabolites is associated with its high absorption in UV that is used as detector during the isolation.

The advantage of this technique is also associated with its analytical version that could help in qualitative characterization of flavonoids in the analysing sample. The diode array detector (DAD) and photodiode array detector (PDA) are commonly used. Further detectors such as mass spectrometry (SM) and nuclear magnetic resonance (NMR) could be combined with UV for more characterization of each flavonoid detected [32–35].

Several works on isolation of flavonoids from natural products using prep. HPLC have been published so far, and some of these compounds, recently reported, are documented in **Table 1** as well as their sources, column characteristics and mobile phases used (**Table 1** and **Figure 2**).

The application of other chromatography techniques, such as circular liquid chromatography (CLC), centrifugal preparative thin layer chromatography (CPTLC), high speed counter current chromatography (HSCCC), medium pressure liquid chromatography (MPLC), and so on, has also led to the isolation of numerous structures of flavonoids [39, 47–49]. Most flavonoids were isolated with combination of these techniques with prep. HPLC: four flavonoids (4′,5-dihydroxy-3′,7-dimethoxyflavanone, 5-hydroxy-7,3′,4′-trimethoxyflavanone, 5,4′-dihydroxy-3,7,3′-trimethoxyflavone, and 5-hydroxy-3,7,4′-tetramethoxyflvone) were isolated from *Pogostemon cablin* (Blanco) Benth. using the HSCCC technique with two phase solvent system made of n-hexane–ethyl acetate–methanol–water (11:5:11:5, v/v/v/v) followed by further purification on prep. HPLC [39]. The combination of HSCCC and semi-prep. HPLC was used to isolate three flavonoid glycosides (orientin, vitexin, quercetin-3-*O*-neohesperidoside) from *Trollius ledebouri* Reichb. [48]. Two new flavonoids (*rac*-6-formyl-5,7-dihydroxyflavanone and 2′,6′-dihydroxy-4′-methoxy-3′-methylchalcone) were recently reported from *Eugenia rigida* using CPTLC and prep. HPLC [50]. Flavoalkaloids and flavonol glucosides were reported from *Astragalus monspessulanus* using the combination of CC, low-pressure liquid chromatography (LPLC) and prep. HPLC [51].

Flavonoids could also be isolated as enantiomers from natural products. Lachnoisoflavones A (**5**) and B (**6**) were isolated from *Crotalaria lachnophora* using prep. HPLC as two enantiomer isoflavones as preliminary indicated by their $[\alpha]_D$ value [0.002 (*c* 0.1, MeOH)] [36]. The presence of the racemic mixture of **5** was successfully confirmed by a chiral HPLC-MS^2 separation that exhibited, on the chromatogram, two signals having the same peak area (**Figure 3**) [36]. This indicates the advantage of HPLC techniques for the isolation and structure characterization of flavonoids.

Names and sources	Classes	Mobile phases	Column characteristics
Lachnoisoflavones A (**5**) and B (**6**), *Crotalaria lachnophora* [36]	Isoflavone	H_2O (0.1% FA) – MeOH (0.1% FA), gr.	Nucleodur C_{18} 5 μm (250 × 16 mm)
Mansoins A (**7**) and B (**8**), *Mansoa hirsute* [37, 38]	Flavanone	H_2O – MeCN, gr.	Luna OSD C_{18} 5 μm (250 x 21.2 mm) & (250 x 10 mm)
4′,5-Dihydroxy-3′,7-dimethoxyflavanone (**9**), 4′,5-dihydroxy-3,7,3′-trimethoxyflavone (**10**), *Pogostemon cablin* (Blanco) Benth [39]	Flavanone Flavonol	MeOH – AcOH (0.1% aq.), is. (7:3) (75:25)	YMC C_{18} 5 μm (250 × 10 mm)
Brutieridin (**11**), Melitidin (**12**), *Citrus bergamia* [40]	Flavanone	H_2O (0.1% FA) – MeCN, gr.	ONYX C_{18} (100 × 3 mm)
Cyanidin 3-[3″-(*O*-β-D-glucopyranosyl)-6″-(*O*-α-L-rhamnopyranosyl)-*O*-β-D-glucopyranoside] (**13**), Cyanidin 3-rutinoside (**14**), *Asparagus officinalis* [41]	Anthocyanin	[AcOH:MeCN:H_2O (1:4:5)] – [AcOH:H_2O (1:9)], gr.	Cosmosil 5C_{18} AR II (250 × 20 mm)
*O*3--(6-*E*-Feruloyl)-β-D-glucopyranosyl-(1→2)-[β-D-xylopyranosyl-(1→2)-]α-L-rhamnopyranosyl-quercetin, Gallocatechin, *Alphitonia neocaledonica* [42]	Flavone Flavanol	MeCN – H_2O (3:7 and 6:4), is. MeCN – H_2O (0.025% TFA), gr.	Luna C_{18} 5 μm (250 × 10 mm)
Pelargonidin 3-(6″-*p*-coumarylglucoside)-5-(4‴-malonylglucoside), Pelargonidin 3-(6″-malonylglucosid, *Ficus padana* Burm. L. [43]	Anthocyanin	[H_2O (2% FA) – MeCN:H_2O:FA (49:49:2)], gr.	Shimpack PRC-ODS 5 μm (250 × 20 mm)
Isoschaftoside (**15**), Orientin (**16**), Isoorientin (**17**), *Mauritia flexuosa* [44]	Flavone	MeCN – H_2O, gr. [MeOH (0.1% TFA) – H_2O (0.1% TFA)], gr.	Shimpack C_{18} 5 μm (250 × 20 mm)
Trilobatin (**18**), Phloretin (**19**), 3-Hydroxyphloretin (**20**), Phlorizin (**21**), *Malus* crabapples "*Radiant*" [45]	Dihydrochalcone	MeOH – 0.01% TFA (3:2)	Agilent Extend C_{18} (250 × 9.4 mm)
Diplotrin A (**22**), Diplotasin (**23**), *Mimosa diplotricha* [46]	Flavone	H_2O – MeOH (2:3), is.	Cosmosil 5C_{18}-AR-II 5 μm (250 × 20 mm)

FA, formic acid; TFA, trifluoracetic acid; gr, gradient polarity, is, isocratic.

Table 1. Some recent flavonoids isolated from natural products using HPLC as well as columns and mobile phases used.

http://dx.doi.org/10.5772/67881

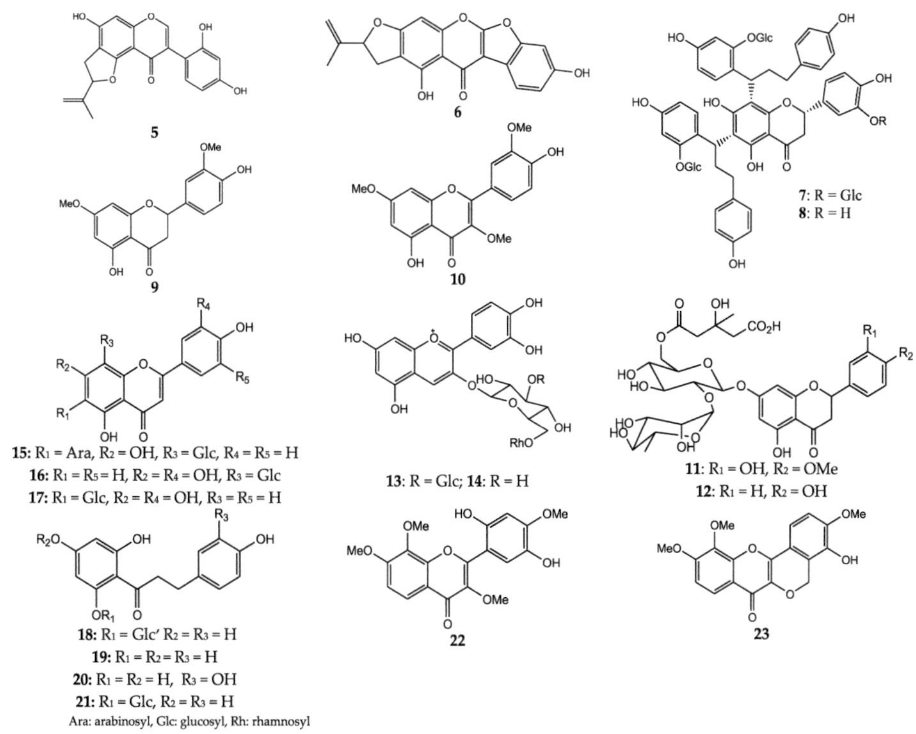

Figure 2. Some flavonoids recently isolated by Prep. HPLC from natural resources.

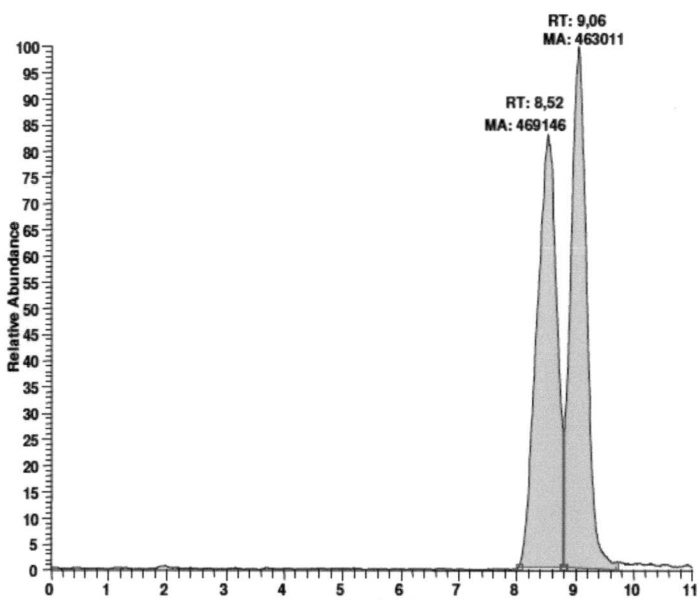

Figure 3. Chromatogram of chiral separation by LC-MS^2 of **5**.

4. Structure characterization of flavonoids

The structure characterization of flavonoids is related to the elucidation of their spectroscopic spectra obtained by techniques such as nuclear magnetic resonance (NMR), mass spectrometry (MS), spectrophotometric ultra-violet (UV) and infrared (IR). Physical properties of the flavonoids as melting point (mp), circular dichroism (CD), optical rotatory power ($[\alpha]_D$) are also useful for full characterization of the isolated flavonoid specially when its contains stereocenter for CD and $[\alpha]_D$. Some chapters in this book would provide more details about the use of spectroscopic analysis to characterize flavonoids. Nevertheless, the NMR spectroscopy is mainly divided into 1D and 2D analyses. The 1D NMR analysis includes the proton (^1H), carbon-13 (^{13}C) and distortionless enhancement by polarization transfer experiment (DEPT) that provide information about the signals of protons, carbons and type of carbons (C, CH, CH_2 or CH_3) in the structure of flavonoid under elucidation. The ^1H NMR spectrum is very useful as it provides the number (integration value) and the type of proton involved. The chemical shift (δ_H) values are usually exhibited within 0 (reference standard value for TMS) and 14 ppm in ^1H NMR of flavonoids while in ^{13}C NMR, they (δ_C) appeared between 0 and 220 ppm. Characteristic proton and carbon chemical shift values for some flavonoid classes were summarized [52, 53] (**Table 2**).

The 2D NMR is composed mainly with the proton-proton correlated spectroscopy ($^1H\,^1H$ COSY), the heteronuclear multiple quantum coherence (HMQC)/heteronuclear single quan-tum coherence (HSQC), the heteronuclear multiple bond connectivity (HMBC), the nuclear overhauser spectroscopy (NOESY), the rotative-frame overhauser spectroscopy (ROESY) and the

Chemical shifts (ppm)	**^1H**
2–3	H-3 (Flavanone), CH_3 aromatic
4–6	H-2 (Flavanone, dihydroflavonol)
6–8	A- and B-ring protons
8–8.5	H-2 isoflavone
12–14	5-OH when C=O at C-4 (usually observed in DMSO-d_6)
Chemical shifts (ppm)	**^{13}C**
210–170	C=O
165–155 (no ortho/para oxygenation)	Oxygenated aromatic carbons
150–130 (with ortho/para oxygenation)	Oxygenated aromatic carbons
135–125 (para substitution)	Non-oxygenated aromatic carbons
125–90 (with ortho/para oxygenation)	Non-oxygenated aromatic carbons
80–40	Non-oxygenated (C-2, C-3 flavanone/flavanol)
28–35	C-4, flavanol

Table 2. Characteristic proton and carbon chemical shifts for some flavonoids.

total correlated spectroscopy (TOCSY) experiments. The ^{13}C data of flavonoids in several cases could also be assigned from HMQC and HMBC spectra.

The infrared spectroscopy compared to other spectroscopic techniques exhibits little but useful information in the structure characterization of flavonoids. Most of hydroxylated flavones, isoflavones and chalcones or dihydrochalcones showed maxima large band absorptions around 3300–3600 cm^{-1} due to hydroxyl groups. Additionally, intense band absorption characteristic for flavonoid carbonyl groups (C=O) is observed around 1680 cm^{-1} and is shifted approximately to 1620 cm^{-1} when the hydroxyl is chelated with a C=O. From the IR spectrum of flavonoids, a sharp and intense absorption band is also observed between 1600 and 1500 cm^{-1} due to aromatic double bonds (aromatic rings).

Ultra-violet (UV) absorption spectroscopy of flavonoids has two maxima absorptions around 300–350 and 240–285 nm corresponding to bands I and II from A- and B-rings, respectively. This technique is used for identification of the flavonoid type and its oxygenation pattern. UV-shift reagents ($AlCl_3$, NaOMe, NaOAc, NaOAc + H_3BO_3, $AlCl_3$ + HCl) are mostly used in the sample solution to confirm the presence and the substitution pattern of hydroxyl groups in flavonoids. The presence of ortho-dihydroxylated groups could be detected by the bathochromic shift of band I after addition of NaOAc/H_3BO_3 while the addition of $AlCl_3$ led to the bathochromic effect of band I when the flavonoid with a carbonyl at C-4 had hydroxyl group at positions C-3 or C-5. The bathochromic shift of band II occurs especially when NaOAc is added to a solution of flavonoids having a free hydroxyl group at C-7 [3, 54, 55]. Characteristic UV absorption bands I and II due to different classes of flavonoids have been reported elsewhere (**Table 3**) [54].

The mass spectrometry technique is very helpful in the structure elucidation of flavonoids. It is used in the determination of the molecular weight for establishing the distribution of substituents between the A- and B-rings and in the determination of the nature and site of attachment of the sugar(s) in flavonoid C- and *O*-glucosides. The molecular weight of the basic

Band II (nm)	Band I (nm)	Flavonoid class
250–280	310–350	Flavone
250–280	330–360	Flavonols (3-OH substituted)
250–280	350–385	Flavonols (3-OH free)
245–275	310–330 shoulder	Isoflavone
	C. 320 peak	Isoflavones (5-deoxy-6,7-dioxygenated)
275–295	300–330 shoulder	Flavonones and dihydroflavonols
230–270	340–390	Chalcones
230–270	380–430	Aurones
270–280	465–560	Anthocyanidins and anthocyanins

Table 3. Ultra-violet absorption ranges for flavonoids.

flavonoid nucleus is 222 *a.m.u.* for flavones, isoflavone and aurone; 224 *a.m.u.* for flavanones and chalcones; 238 *a.m.u.* for flavonols; and 240 *a.m.u.* for the dihydroflavonols. The molecular weight of the unknown flavonoid could be deduced by addition of atomic mass units of all its substituents [16 a.m.u (-OH), 30 a.m.u. (-OCH_3), and so on] to one of the basic molecular weights above. The loss of some ion-fragments from the molecular or pseudo-molecular ion is very characteristic in the mass spectra of flavonoids. Peaks obtained during this fragmentation process represent accurately the corresponding ion-fragments that are expressed as mass-to-charge ratio (m/z). The exact molecular weight for each fragment may be measured to the nearest 0.0001 mass unit if the mass spectrometer is operating in high resolution. This information enables calculation of precise molecular formula from the molecular ion peak and ion-fragments [54]. A prerequisite for successful mass spectrometry is that the flavonoid should be sufficiently volatile in the high vacuum within the mass spectrometer. Most aglycones are sufficiently volatile at probe temperature of 100–230°C, higher temperatures being required for the more polar polyhydroxyflavones and flavonols. Glycosides, anthocyanidins and biflavonoids, however, are not sufficiently volatile and should therefore be derivatized to improve their volatility. Some standard methods used for derivatization of compounds are permethylation or perdeuteromethylation and trimethylsilylation [54].

Natural products in general or flavonoids in particular remain an important source for drug discovery. Determination of their absolute configurations is one of the most challenging tasks in the structure elucidation of chiral flavonoids. It has been proven that the change in absolute configuration of secondary metabolites consequently affected the difference in pharmacological activity of both stereo-compounds. Methods such as chiroptical approaches, chemical synthesis, analytical chemistry, chiral derivatization and X-ray crystallography could be used to determine the absolute configuration of flavonoids. An important investigation was reported on the determination of absolute configuration of natural products and some flavonoids using experimental and calculated electronic circular dichroism (ECD) data [56].

5. Conclusion

The extraction, isolation and characterization of flavonoids from natural products have been carried out successfully by natural product chemists and phytochemists using relevant techniques and new methods. Some of these techniques and methods have been documented in this chapter with illustrations owing to some flavonoids recently reported. It is clear that the HPLC and its combination with other available techniques of isolation are being often used to obtain flavonoids from natural sources especially from plant species. The characterization of flavonoids remains basically focused on the analysis of their spectroscopic, mass and UV data and some chemical investigations depending on the nature of the structure under elucidation. The need of flavonoids in agriculture, food and drug industries still one of the worldwide up-to-date research interests. Natural resources and especially medicinal plants are still available to discover novel or efficient antioxidant flavonoids that could be used as drugs to fight against degenerative diseases one of the issues the global health is facing today.

http://dx.doi.org/10.5772/67881

Author details

Maurice D. Awouafack[1,2*], Pierre Tane[1] and Hiroyuki Morita[2]

*Address all correspondence to: amauduc2@yahoo.com

1 Laboratory of Natural Products Chemistry, Department of Chemistry, Faculty of Science, University of Dschang, Dschang, Cameroon

2 Institute of Natural Medicine, University of Toyama, Toyama, Japan

References

[1] Bello IA, Ndukwe GI, Audu OT, Habila JD. A bioactive flavonoid from *Pavetta crassipes* K. Schum. Organic and Medicinal Chemistry Letters. 2011;1(1):14.

[2] Delcour JA, Varebeke DJD. A new colourimetric assay for flavanoids in pilsner beers. Journal of the Institute of Brewing. 1985;91(1):37-40.

[3] Awouafack MD, Tane P, Eloff JN. Two new antioxidant flavones from the twigs of *Eriosema robustum* (Fabaceae). Phytochemistry Letters. 2013;6(1):62-66.

[4] DeTommasi N, Pizza C, Aquino R, Cumanda J, Mahmood N. Flavonol and chalcone ester glycosides from *Bidens leucantha*. Journal of Natural Products. 1997;60(3):270-273.

[5] Tagousop CN, Ngnokam D, Harakat D, Voutquenne-Nazabadioko L. Three new flavonoid glycosides from the aerial parts of *Graptophyllum grandulosum* Turril (Acanthaceae). Phytochemistry Letters. 2017;19:172-175.

[6] Tang H, Pei HY, Wang TJ, Chen K, Wu B, Yang QN, Zhang Q, Yang JH, Wang XY, Tang MH, Peng AH, Ye HY, Chen LJ. Flavonoids and biphenylneolignans with anti-inflammatory activity from the stems of *Millettia griffithii*. Bioorganic & Medicinal Chemistry Letters. 2016;26(18):4417-4422.

[7] Al-Aamri KK, Hossain MA. New prenylated flavonoids from the leaves of *Dodonea viscosa* native to the sultanate of Oman. Pacific Science Review A: Natural Science and Engineering. 2016;18(1):53-61.

[8] Rabeta MS, Lin SP. Effects of different drying methods on the antioxidant activities of leaves and berries of *Cayratia trifolia*. Sains Malaysiana. 2015;44(2):275-280.

[9] Vaidya BN, Brearley TA, Joshee N. Antioxidant capacity of fresh and dry leaf extracts of sixteen *Scutellaria* species. Journal of Medicinally Active Plants. 2014;2(3):42-49.

[10] Vongsak B, Sithisarn P, Mangmool S, Thongpraditchote S, Wongkrajang Y, Gritsanapan W. Maximizing total phenolics, total flavonoids contents and antioxidant activity of *Moringa oleifera* leaf extract by the appropriate extraction method. Industrial Crops and Products. 2013;44:566-571.

[11] Tan SP, Parks SE, Stathopoulos CE, Roach PD. Extraction of flavonoids from bitter melon. Food and Nutrition Sciences. 2014;5(5):458-465.

[12] Munhoz VM, Longhini R, Souza JRP, Zequi JAC, Mello E, Lopes GC, Mello JCP. Extraction of flavonoids from *Tagetes patula*: process optimization and screening for biological activity. Revista Brasileira De Farmacognosia-Brazilian Journal of Pharmacognosy. 2014;24(5):576-583.

[13] Jäger S, Beffert M, Hoppe K, Nadberezny D, Frank B, Scheffler A. Preparation of herbal tea as infusion or by maceration at room temperature using mistletoe tea as an example. Scientia Pharmaceutica. 2010;79(1):145-156.

[14] Iloki-Assanga SB, Lewis-Lujan LM, Lara-Espinoza CL, Gil-Salido AA, Fernandez-Angulo D, Rubio-Pino JL, Haines DD. Solvent effects on phytochemical constituent profiles and antioxidant activities, using four different extraction formulations for analysis of *Bucida buceras* L. and *Phoradendron californicum*. BMC Research Notes. 2015;8:396.

[15] Dai J, Mumper RJ. Plant phenolics: extraction, analysis and their antioxidant and anticancer properties. Molecules. 2010;15(10):7313-7352.

[16] Sambandam B, Thiyagarajan D, Ayyaswamy A, Raman P. Extraction and isolation of flavonoid quercetin from the leaves of *Trigonella foenum-graecum* and their antioxidant activity. International Journal of Pharmacy and Pharmaceutical Sciences. 2016;8(6):120-124.

[17] Chaisawangwong W, Gritsanapan W. Extraction method for high free radical scavenging activity of Siamese neem tree flowers. Sonklanakarin Journal of Science and Technology. 2009;31(4):419-423.

[18] Lahlou FA, Hmimid F, Loutfi M, Bourhim N. Antioxidant activity phenolics flavonoids and proanthocyanidins content of *Senecio anteuphorbium*. International Journal of Biochemistry Research & Review. 2014;4(6):550-558.

[19] San A, Sithisarn P, Gritsanapan W. Appropriate extraction method for high contents of total phenolics, total flavonoids and free radical scavenging properties of *Ziziphus mauritiana* seed extracts. Planta Medica. 2013;79(13):1272.

[20] Veggi PC, Martinez J, Meireles MAA. Fundamentals of microwave extraction. In: Chemat F, Cravotto G, editors. Microwave-assisted extraction for bioactive compounds: theory and practice, food engineering series 4. Springer Science + Business Media, New York; 2013, pp. 15-52.

[21] Chemat F, Abert-Vian M, Zill-e-Huma Y-J. Microwave assisted separations: green chemistry in action. In: Pearlman JT, editor. Green chemistry research trends. Nova Science Publisher, New York; 2009, pp. 33-62.

[22] Aguilera JM. Solid–liquid extraction. In: Tzia C, Liadakis G, editors. Extraction optimization in food engineering. Marcel Dekker, New York; 2003, pp. 35-55.

[23] Krishnan RY, Rajan KS. Microwave assisted extraction of flavonoids from *Terminalia bellerica*: study of kinetics and thermodynamics. Separation and Purification Technology. 2016;157:169-178.

[24] Zheng LL, Wen G, Yuan MY, Gao F. Ultrasound-assisted extraction of total flavonoids from corn silk and their antioxidant activity. Journal of Chemistry. 2016;2016:1-5.

[25] Jamshidi M, Shabani E, Hashemi Z, Ebrahimzadeh MA. Evaluation of three methods for the extraction of antioxidants from leaf and aerial parts of *Lythrum salicaria* L. (Lythraceae). International Food Research Journal. 2014;21(2):783-788.

[26] Awouafack MD, Kouam SF, Hussain H, Ngamga D, Tane P, Schulz B, Ivan R. Green, Karsten Krohn. Antimicrobial prenylated dihydrochalcones from *Eriosema glomerata*. Planta Medica. 2008;74(1):50-54.

[27] Awouafack MD, Tchuenguem RT, Ito T, Dzoyem JP, Tane P, Morita H. A new isoflavanol from the fruits of *Kotschya strigosa* (Fabaceae). Helvetica Chimica Acta. 2016;99(4):321-324.

[28] Alam A, Jaiswal V, Akhtar S, Jayashree BS, Dhar KL. Isolation of isoflavones from *Iris kashmiriana* Baker as potential anti proliferative agents targeting NF-kappaB. Phytochemistry. 2017;136:70-80.

[29] Salae A-W, Chairerk O, Sukkoet P, Chairat T, Prawat U, Tuntiwachwuttikul P, Chalermglin P, Ruchirawat S. Antiplasmodial dimeric chalcone derivatives from the roots of *Uvaria siamensis*. Phytochemistry. 2017;135:135-143.

[30] Nyandoro SS, Munissi JJ, Kombo M, Mgina CA, Pan F, Gruhonjic A, Fitzpatrick P, Lu Y, Wang B, Rissanen K, Erdélyi M. Flavonoids from *Erythrina schliebenii*. Journal of Natural Products. 2017;80(2):377-383.

[31] Thao NP, Luyen BTT, Widowati W, Fauziah N, Maesaroh M, Herlina T, Manzoor Z, Ali I, Koh YS, Kim YH. Anti-inflammatory flavonoid C-glycosides from *Piper aduncum* leaves. Planta Medica. 2016;82(17):1475-1481.

[32] Simirgiotis MJ. Antioxidant capacity and HPLC-DAD-MS profiling of Chilean Peumo (*Cryptocarya alba*) fruits and comparison with German Peumo (*Crataegus monogyna*) from Southern Chile. Molecules. 2013;18(2):2061-2080.

[33] Lin LZ, Harnly JM. LC-PDA-ESI/MS identification of the phenolic components of three compositae spices: Chamomile, Tarragon, and *Mexican Arnica*. Natural Product Communications. 2012;7(6):749-752.

[34] Braunberger C, Zehl M, Conrad J, Fischer S, Adhami H-R, Beifuss U, Krenn L. LC–NMR, NMR, and LC–MS identification and LC–DAD quantification of flavonoids and ellagic acid derivatives in *Drosera peltata*. Journal of Chromatography B. 2013;932:111-116.

[35] Liu EH, Qi LW, Cao J, Li P, Li CY, Peng YB. Advances of modern chromatographic and electrophoretic methods in separation and analysis of flavonoids. Molecules. 2008;13(10):2521-2544.

[36] Awouafack MD, Spiteller P, Lamshoft M, Kusari S, Ivanova B, Tane P, Spiteller M. Antimicrobial isopropenyl-dihydrofuranoisoflavones from *Crotalaria lachnophora*. Journal of Natural Products. 2011;74(2):272-278.

[37] Campana PRV, Coleman CM, Teixeira MM, Ferreira D, Braga FC. TNF-alpha inhibition elicited by mansoins A and B, heterotrimeric flavonoids isolated from *Mansoa hirsuta*. Journal of Natural Products. 2014;77(4):824-830.

[38] Campana PRV, Coleman CM, Sousa LP, Teixeira MM, Ferreira D, Braga FC. Mansoins C-F, oligomeric flavonoid glucosides isolated from *Mansoa hirsuta* fruits with potential anti-inflammatory activity. Journal of Natural Products. 2016;79(9):2279-2286.

[39] Li K, Zhang H, Xie H, Liang Y, Wang X, Ito Y. Preparative isolation and purification of five flavonoids from *Pogostemon cablin* Benth by high-speed countercurrent chromatography and preparative high-performance liquid chromatography. Journal of Liquid Chromatography & Related Technologies. 2011;34(15):1617-1629.

[40] Di Donna L, De Luca G, Mazzotti F, Napoli A, Salerno R, Taverna D, Sindona G. Statin-like principles of bergamot fruit (*Citrus bergamia*): isolation of 3-hydroxymethylglutaryl flavonoid glycosides. Journal of Natural Products. 2009;72(7):1352-1354.

[41] Sakaguchi Y, Ozaki Y, Miyajima I, Yamaguchi M, Fukui Y, Iwasa K, Motoki S, Suzuki T, Okubo H. Major anthocyanins from purple asparagus (*Asparagus officinalis*). Phytochemistry. 2008;69(8):1763-1766.

[42] Muhammad D, Hubert J, Lalun N, Renault JH, Bobichon H, Nour M, Voutquenne-Nazabadioko L. Isolation of flavonoids and triterpenoids from the fruits of *Alphitonia neocaledonica* and evaluation of their anti-oxidant, anti-tyrosinase and cytotoxic activities. Phytochemical Analysis. 2015;26(2):137-144.

[43] Syukri D, Darwis D, Santoni A. Preparative HPLC for the purification of major anthocyanins from *Ficus padana* Burm L. Research Journal of Chemical Sciences. 2013;3(12):60-64.

[44] de Oliveira DM, Siqueira EP, Nunes YRF, Cota BB. Flavonoids from leaves of *Mauritia flexuosa*. Revista Brasileira De Farmacognosia-Brazilian Journal of Pharmacognosy. 2013;23(4):614-620.

[45] Qin XX, Xing YF, Zhou ZQ, Yao YC. Dihydrochalcone compounds isolated from crabapple leaves showed anticancer effects on human cancer cell lines. Molecules. 2015;20(12):21193-21203.

[46] Lin LC, Chiou CT, Cheng JJ. 5-deoxyflavones with cytotoxic activity from *Mimosa diplotricha*. Journal of Natural Products. 2011;74(9):2001-2004.

[47] Peng JY, Yang GJ, Fan GR, Wu YT. Preparative isolation and separation of a novel and two known flavonoids from *Patrinia villosa* Juss by high-speed counter-current chromatography. Journal of Chromatography A. 2005;1092(2):235-240.

[48] Zhou X, Peng JY, Fan GR, Wu YT. Isolation and purification of flavonoid glycosides from *Trollius ledebouri* using high-speed counter-current chromatography by stepwise increasing the flow-rate of the mobile phase. Journal of Chromatography A. 2005;1092(2):216-221.

[49] Marston A, Hostettmann K. Separation and quantification of flavonoids. In: Andersen ØM, Markham KR, editors. Flavonoids: chemistry, biochemistry, and applications. CRC Press and Taylor & Francis Group, New York; 2006, pp. 1-21.

[50] Zaki MA, Nanayakkara NPD, Hetta MH, Jacob MR, Khan SI, Mohammed R, Ibrahim MA, Samoylenko V, Coleman C, Fronczek FR, Ferreira D, Muhammad I. Bioactive formylated flavonoids from *Eugenia rigida*: isolation, synthesis, and X-ray crystallography. Journal of Natural Products. 2016;79(9):2341-2349.

[51] Krasteva I, Bratkov V, Bucar F, Kunert O, Kollroser M, Kondeva-Burdina M, Ionkova I. Flavoalkaloids and flavonoids from *Astragalus monspessulanus*. Journal of Natural Products. 2015;78(11):2565-2571.

[52] Markham KR, Mabry TJ. Ultraviolet–visible and proton magnetic resonance spectroscopy of flavonoids. In: Harborne JB, Mabry TJ, Mabry H, editions. The Flavonoids. Chapman and Hall, London; 1975, pp. 45-77.

[53] Levy GC, Nelson GL. Carbone-13 nuclear magnetic resonance for organic chemists. Wiley, New York; 1972.

[54] Markham KR. Techniques of flavonoids identification. Academic Press, London; 1982, p. 144.

[55] Fathiazad F, Delazar A, Amiri R, Sarker S. Extraction of flavonoids and quantification of rutin from waste tobacco leaves. Iranian Journal of Pharmaceutical Research. 2010;5(3):222-227.

[56] Li XC, Ferreira D, Ding YQ. Determination of absolute configuration of natural products: theoretical calculation of electronic circular dichroism as a tool. Current Organic Chemistry. 2010;14(16):1678-1697.

Chapter 4

Isoflavonoids

Ahmed I. Foudah and Maged Saad Abdel-Kader

Additional information is available at the end of the chapter

http://dx.doi.org/10.5772/intechopen.68701

Abstract

Isoflavonoids are interesting class of natural products due to their positive effects on human health. Isoflavonoids include isoflavones, isoflavanones, isoflavans, rotenoids and pterocarpans. Although they are reported from many plant families, most isoflavonoids are produced by the subfamily *Papilionaceae* of the Fabaceae. Various chromatographic methods have been applied for the purification of isoflavonoids. Simple Ultra Violet (UV) absorption spectra as well as both One and two dimensional NMR (1D- and 2D-NMR) are critical for the identification of isoflavonoids. Each class of isoflavonoids has its unique feature in both ^{1}H- and ^{13}C-NMR that enable their proper characterization. High Resolution Mass Spectrometry (HRMS) is a substantial tool in such challenge. *In vitro* experiments indicated that isoflavonoids possess antioxidant, antimutagenic, antiproliferative as well as cancer preventive effects. Epidemiological studies provide support for some of these effects on human. Members of this class also are reported to have antimicrobial activity. In this chapter, isoflavones, isoflavanones, isoflavans, homoisoflavonoids and isoflavenes will be discussed in relation to their occurrence, methods of purification, spectral characters helpful in structure elucidation as well as their biological importance.

Keywords: isoflavones, isoflavanones, isoflavans, homoisoflavonoids, isoflavenes

1. Introduction

Genstin (**1**) was the first isolated isoflavone from *Genista tinctoria* known as Dyer's Brrom in 1899 [1]. Later in 1926 [2], the structure was identified. Genstin (**1**) was isolated from Soybeans in 1941 [3]. Although the main source of isoflavonoids is member of the Fabaceae [4], some were reported from other families such as Amaranthaceae [5, 6], Rosacease [7] and Poaceae [8]. Isoflavonoids were also reported from fungi [9] and Propolis [10]. The dietary consumption of isoflavonoid-rich sources is linked with health advantages toward osteoporosis, postmenopausal symptoms, cardiovascular diseases and chemo-prevention [11]. People from SE

Asia have much lower risk of developing prostate cancer compared to Americans due to high consumption of soy rich in isoflavonoids. Upon immigration to the USA and changing the dietary components, this difference rapidly disappears [12]. Isoflavonoids are also classified as dietary antioxidants [13]. These facts were the driving force behind the use of isoflavonoid-rich sources as nutraceutical and dietary supplements [14].

Isoflavonoids are a large subclass of the most common plant polyphenols containing 15 carbon atoms known as flavonoids [15]. In isoflavonoids (3-phenylchromans), the phenyl ring B is attached to heterocyclic ring C at position 3 rather than 2 in flavonoids [16]. Generally, flavonoids are biosynthesised via Shikimic acid pathway. Shikimic acid is also a precursor for the biosynthesis of phenylpropanoids and aromatic acids. At certain stages, the activity of the key enzyme chalcone isomerase (CHI) resulted in the formation of flavanones that converted to isoflavonoids under the influence of isoflavone synthase [17]. The biosynthesis of isoflavonoids, consequently, is considered as an offshoot from the flavonoids biosynthetic pathway [18]. Highest level of isoflavonoids occurs usually in roots, seedlings and seeds [18, 19].

Isoflavonoids are sub-classified into many subclasses based on the oxidation status of ring C as well as the formation of a forth ring 'D' by coupling between rings B and C. Subclasses free from ring D include isoflavones, isoflavanones, isoflavan-4-ol, homoisoflavonoids, isoflavans and isoflav-3-ene. Rotenoids, pterocarpans, coumaronochromones and coumaronochromene represent the subclasses with additional ring D formation [11].

This chapter will deal with the different aspects of the isoflavonoid subclasses keeping the original three-ring skeleton (**Figure 1**). Occurrence, isolation, key spectroscopic characters and biological activities will be covered starting from 2000 to date.

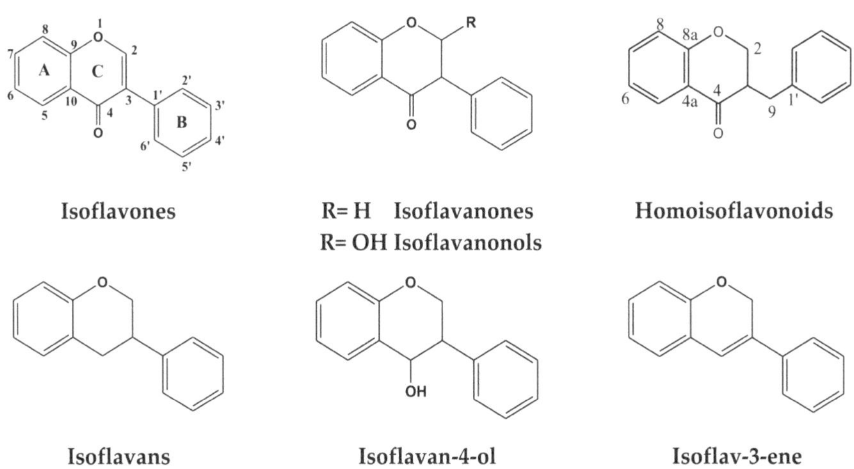

Figure 1. The skeletons of the isoflavonoids with three-ring structures.

2. Extraction and purification

The most popular method used for extraction of isoflavonoids is maceration with either MeOH or EtOH containing various percentages of H_2O at room temperature followed by liquid-liquid

http://dx.doi.org/10.5772/intechopen.68701

fractionation using solvents with different polarities [6, 10, 19–32]. Another method of extraction used MeOH or EtOH under reflux or in soxhlet apparatus [5, 33–36]. Mixture of MeOH and $CHCl_3$ or CH_2Cl_2 (1:1) was also applied for extraction [37–41]. Other research groups extracted the plant materials with acetone [42–44], $CHCl_3$ [45, 46], CH_2Cl_2 [47–50] or diethyl ether [51] at room temperature. Successive extraction starting with petroleum ether or hexane, $CHCl_3$, EtOAc and MeOH using soxhelt apparatus [52–56] was also reported. The isoflavone contents of soybeans were extract using supercritical fluid extraction [57].

The majority of purification and isolation steps utilized silica gel in the form of column, Preparative Thin Layer Chromatography (PTLC) or Centrifugal Preparative Thin Layer Chromatography (CPTLC) [19, 21, 45]. Combination of silica gel and Sephadex LH-20 was also applied for isoflavonoid purification [6, 10, 54, 55]. In addition to silica gel, semi-preparative C_{18} High Performance Liquid Chromatography (HPLC) columns were used for final purification of isoflavonoids [23, 30, 31, 38, 48]. The polar *n*-butanol fraction of *Ononis serrata* was fractionated on C_{18} silica gel applying the Vacuum Liquid Chromatography (VLC) technique followed by normal silica gel column for purification of isoflavonoid glucosides [27]. Two isoflavenes were isolated from *Lespedeza homoloba* after chromatography on porous polymer gel Diaion followed by silica gel column. Final purification step was performed on preparative C_{18} HPLC column [36]. Isoflavonoids from *Iris germanica* were purified by silica gel VLC and CC, and final purification was achieved via LiChrolut EN/RP-18 solid phase extraction tubes [26]. High-speed counter-current chromatography (HSCCC) was applied for the purification of flavan glycoside and isoflavones from *Astragalus membranaceus*, the seeds of *Millettia pachycarpa* and soy flour [20, 58, 59]. Isolation and identification of isoflavanones, biflavanones and bisdihydrocoumarins were achieved using Liquid Chromatography- Mass Spectrometry (LC-MS), Liquid Chromatography- Solid Phase Extraction- Nuclear Magnetic Resonance (LC-SPE-NMR) and Electronic Circular Dichroism (ECD). In this method, MS of target compounds was measured directly in the LC effluent. For NMR analyses, the peaks were collected from 20 LC runs, loaded on SPE cartilages, dried with nitrogen gas and finally eluted with CD_3OD [32].

3. Spectroscopic identification

3.1. Infrared (IR) transmission spectra

Both phenolic hydroxyls and carbonyl groups are present in most of the isoflavonoid classes. However, the most characteristic feature of isoflavans and isoflavenes is the lack of carbonyl function bands. The absorption bands for the C-4 carbonyl in isoflavones and isoflavanones present in the range 1606–1694 cm^{-1} [9, 23–26]. Differentiation between isoflavones and isoflavanones from the position of C-4 carbonyl bands in the IR spectra is not achievable.

3.2. Ultra Violet (UV) absorption spectra

In spite of the tremendous advances in 2D-NMR and MS, the UV absorption spectra in MeOH and MeOH with shift reagent still can provide useful information for flavonoids identification. In all isoflavonoids except isoflavenes, ring B has no or little conjugation with the main chromophore composed of rings A and C. This fact is expressed as intense band II and diminished band I [60].

For isoflavones, band II shows absorption at λ_{max} 245–275 nm. Shift reagents can be used to detect hydroxylation at ring A. NaOAc induces 6–20 nm bathochromic shift as an indication of free 7-hydroxyl group. The 10–14 nm shift with $AlCl_3$/HCl is diagnostic for free 5-OH group. The absence of any shift with NaOMe is an evidence for the absence of free hydroxyls in ring A [19, 27, 28, 50, 60].

The UV spectra of about 28 published isoflavanone were reviewed. Band II absorption was found in the range 270–295 nm [5, 9, 23, 25, 29, 33, 39, 41, 43, 44, 47–50, 55, 61, 62]. Among these publications, only three used shift reagents with five isolated isoflavanones. Analysis of the obtained results revealed that $AlCl_3$ induced 17–23 nm bathochromic shift in band II due to the complex formed between C-4 carbonyl and C-OH groups. All the entitled compounds contain C-7 free hydroxyl groups, and NaOAc produced 34–37 nm bathochromic shift in band II [39, 47, 50]. However, more data are required to draw a solid conclusion.

The few available UV data of homoisoflavonoids showed band II absorption in the same range reported for isoflavanones [63].

Isoflavans UV spectra show one prominent maxima representing band II between 270 and 295 nm [21, 37, 38, 45]. The available UV data of isoflavenes indicated the presence of two bands at 235–245 and 320–337 nm along with a shoulder 287–300 nm [29, 30, 31, 35, 36].

3.3. Circular Dichroism (CD) Spectroscopy

Saturation of the double bond between C-2 and C-3 creates a new asymmetric center in the molecules. The orientation at these centers is in most cases determined from the CD spectra.

Isoflavanones show three absorption maxima at 200–240, 260–300 and 320–352 nm. Determination of the absolute configuration at C-3 is based on the n→π* carbonyl transition between 320 and 352 nm. The positive sign at this region is diagnostic for (3*R*) orientation with ring B having equatorial position. The coupling constant between the *trans*-diaxial $H2_{\beta}$ and H3 can confirm the equatorial orientation of ring B [64]. Optical inactivity of isoflavanones most probably is a result of racemization that can occur during extraction and purification [64]. The isolation of two racemic mixtures, 3*S*- and 3*R*-7-O-glucosyldiphysolones (**2, 3**) and (3*S*)- and (3*R*)-7,4′-di-O-glucosyldiphysolones (**4, 5**), from *Ormocarpum kirkii* was explained as result of isomerization in aqueous solution [32]. The same observation was reported in three isolated isoflavanones from *Platycelphium voënse* and *Desmodium canum* [41, 47]. Due to the positive cotton effect at 337 nm, the (3*R*) orientation was assigned to eryzerin B (**6**). However, eryzerin A (**7**) was reported in the same publication with undetermined absolute stereochemistry [44]. The (3*R*) orientation was also assigned to 2,3-dihydro-7-demethylrobustigenin (**8**) and saclenone (**9**) isolated from *Erythrina sacleuxii* based on the positive cotton effect at 320 and 334 nm, respectively [49].

Isoflavans configuration is much more complicated. The heterocyclic ring C is expected to have the half-chair form a fact that can be diagnosed from the vicinal coupling constants between H-2, H-3 and H-4 protons. Such *J* values along with the CD curves can then lead to determination of the absolute configuration [64]. (3*S*)-isoflavans with oxygenation at both the A and B rings display positive and negative cotton effects at 240 and 270–280 nm regions, respectively. The opposite was observed for the (3*R*)-enantiomers. The 7-deoxy (3*S*)-isoflavans

http://dx.doi.org/10.5772/intechopen.68701

with mono- and di-oxygenation at ring B displayed negative cotton effects in both the 230–240 and 270–290 nm regions, and the opposite was observed for the (3*R*)-enantiomers [64]. The difficulty in assigning the absolute configuration of isofalvans was reflected by Bedane et al. [37]. The authors isolated two new isoflavans, erylivingstone J (**10**) and erylivingstone K (**11**). The measured CD spectrum showed negative cotton effect near 306 nm and a positive cotton effect near 240 nm supporting (*S*)-configuration. Three known compounds, 2′-methoxyphaseollinisoflavan (**12**), 7,4′-dihydroxy-2′,5-dimethoxy isoflavan (**13**) and 7,4′-dihydroxy-2′-methoxy-3′-(3-methylbut-2-enyl) isoflavan (**14**), with (*R*)-absolute configuration were isolated from the same source in this study. Suspicions about the purity of the new compounds and isolation of compounds with (*R*)-absolute configuration led the authors to report the new compounds without absolute configuration [37]. The enantiomer (3*S*) (+) 2′-*O*-methylphaseollidinisoflavan (**15**) was isolated from *Erythrina caffra* along with the (3*R*) (−) erythbidin A (**16**). The configuration was assigned based on ^{1}H-NMR *J* values, optical rotation and CD spectra. However, the reported CD data did not cover the lower range of the spectrum near 240 nm [45]. The absolute configuration of abruquinone L (**17**) was successfully assigned by combination of ^{1}H-NMR analyses of the *J* values between ring C protons and the CD spectrum which showed a strong positive cotton effect at 202 nm and two negative cotton effects at 212 and 233 nm [38]. Due to the positive cotton effect at 337 nm, the (3*R*) orientation was assigned to eryzerin C (**18**). However, eryzerin D (**19**) was reported with undetermined absolute stereochemistry [44].

In case of isoflavan-4-ol, C-4 becomes a new chiral center and 4 isomers could exist. Out of the possible isomers, two are *cis*- and two are *trans*-. Hata et al. synthesized and compared the CD spectra of four stereoisomers. The 3*R*, 4*S*-*trans*-isoflavan-4-ol stereoisomer showed negative cotton effect between 250 and 300 nm and positive cotton effect between 220 and 240 nm. The other 3*S*, 4*R*-*trans*-isoflavan-4-ol stereoisomer showed CD spectrum having cotton effect at the same ranges but with opposite sign. The 3*S*, 4*S* *cis*-isoflavan-4-ol stereoisomer expressed positive cotton effect between 245 and 300 nm, while the other enantiomer 3*R*, 4*R*-isoflavan-4-ol has a negative cotton effect at the same region [65].

3.4. Nuclear Magnetic Resonance (NMR) Spectroscopy

3.4.1. ^{1}H- and ^{13}C-NMR

^{1}H- and ^{13}C-NMR spectra provide key information for the identification of the isoflavonoids skeleton. The proton and carbon signals for positions 2–4 in ring C (**Table 1**) provide a unique feature for each class.

The simplest ring C spectrum is that of isoflavones as it shows only one downfield proton singlet for H-2. The oxygenated C-2 chemical shift is also characteristic for isoflavones. The wide range for C-4 carbonyl resulted from the effect of C-5 substitutions. The lack of C5 free hydroxyl resulted in the upfield shift of the C-4 carbonyl chemical shift to a value less than 175.0 ppm in most cases [27, 34]. With the presence of C-5 free hydroxyl and formation of hydrogen bond C-4 carbonyl, the carbonyl chemical shift value is usually above 180.0 ppm [19, 24, 28].

Saturation of the double bond between C-2 and C-3 of isoflavones leads to the formation of the isoflavanone skeleton. Such array contains a CH_2-O and CH-aryl and renders the

	Position 2		Position 3		Position 4	
	^{1}H	^{13}C	^{1}H	^{13}C	^{1}H	^{13}C
Isoflavones	7.82–8.45 s	150.9–155.0	–	121.5–125.5	–	173.9–181.5
Isoflavanones	4.46–4.76 (dd, ax) 4.34–4.63 (dd, eq)	69.6–72.3	3.93–4.32 (dd)	45.3–51.1	–	193.0–198.8
Homoisoflavonoids	4.06–4.32 (dd)	68.8–69.3	2.65–2.80 (m)	46.8–48.7	–	192.7–198.3
Isoflavans	4.33–3.83 (t, ddd, tdd, dt, dd)	69.2–71.2	3.36–3.55 (tdd, dd, dddd, m)	30.79–33.6	2.64–2.98 (dd, ddd)	26.1–31.9
Isoflavan-4-ol	4.21–3.60 (dd, t)	66.8–66.9	3.52–3.49 (ddd)	40.5–40.6	5.47–5.49 (d)	79.0–79.6
Isoflavenes	4.83–5.25 (s, d)	67.6–68.8	–	127.5–129.6	6.47–6.74 (s, d)	118.3–121.9

Table 1. Key ^{1}H- and ^{13}C-NMR spectral data for identification of isoflavonoid classes.

^{1}H-NMR signals of ring C more complex making an AMX spin system. The three protons appear as dd with different *J* values due to *ax-ax, ax-eq* and/or *eq-eq* splitting. In some cases, some signals may appear as *t* or interfere with other signals in the molecule [23, 41, 55, 61, 62]. Absolute configuration of isoflavanones was determined by a simple ^{1}H-NMR experiment in the presence of (*R*)- and (*S*)-binol as chiral solvating agent. The presence of (*R*)- or (*S*)-binol produces variable changes in the chemical shifts of the most downfield H-2 proton. Comparing these chemical shift changes enables the assignment of the absolute configuration [66].

No significant difference can be observed when the chemical shifts of positions 2–4 are compared in the ^{1}H-and ^{13}C-NMR spectra of isoflavanones and homoisoflavonoids. The splitting pattern of H-3 is expected to be much more complex. However, the additional C-9 in homoisoflavonoids provides the key evidence for their identification. The H-9 protons appear in the range of δ_H 2.62–3.13 (dd) as a result of coupling with H-3 proton. The C-9 methylene appears at δ_C 31.9–32.2 ppm [63, 67].

Isoflavans lacks the C-4 carbonyl present in isoflavanones with expected two more proton signals from ring C to form an ABMXZ spin system. Although the H-4 proton signals are more upfield compared to H-2 and H-3, the splitting pattern is more complex than the corresponding isoflavanones. This pattern along with the ^{13}C-NMR chemical shifts of C-2, C-3 and C-4 is the diagnostic feature for the isoflavan nucleus [20–22]. Isoflavan-4-ol is characterized by two oxygenated methines in both ^{1}H- and ^{13}C-NMR spectra.

Formation of double bond between C-3 and C-4 in isoflavans led to the emerging of the isoflav-3-ene class. The ring C ^{1}H-NMR signals of isoflavenes is simplified to two singlet for the 2H of C-2 and 1H of C-4. In some reports, a long-range coupling with small *J* value (1–2 Hz) was observed between H-2 and H-4 protons [35, 36, 43, 56].

3.4.2. 2D-NMR

^{1}H-NMR and different ^{13}C-NMR experiments like Distortionless Enhancement by Polarization Transfer (DEPT 45, DEPT 90 and DEPT 135) in most cases enable the identification of the main skeleton of the isoflavonoids as well as the substitution pattern. Heteronuclear Single-Quantum Correlation (HSQC) experiment is applied to correlate protons and carbons through one bond. So, assignment of protons and carbons as CH_3, CH_2 and CH can be confirmed undoubtfully. ^1H-^1H-Correlation Spectroscopy (COSY) or similar experiments are applied to identify the spin systems in the compounds. These experiments identified protons separated by 3 bonds as well as different arrays present in the aromatic systems. The obtained COSY data allow the identification of the adjacent groups in the compounds and substitution pattern in the aromatic systems. Heteronuclear Multiple-Bond Correlation (HMBC) experiment acquired at different *J* values can identify correlation between protons and carbons through 2, 3 or sometimes 4 bonds especially in the aromatic systems. HMBC data play a key role in the determination of substituents location on the main skeleton. For example, the location of the furan ring in 4′-O-methylerythrinin C (**20**) at C-6 was assigned from HMBC correlations [28]. The location of the prenyl group at C-8 in erysubin F (**21**) was also assigned from correlations obtained from HMBC experiment [42].

Nuclear Overhauser Effect (NOE) is an effect observed between protons close to each other in space regardless to the number of bonds separating them [68]. The NOE effect can be clarified via One dimensional Nuclear Overhauser effect (1D-NOESY), Gradient-Enhanced Nuclear Overhauser Effect (GOESY) experiments or the now more favorable 2D-NOESY or Rotating Frame Nuclear Overhauser Effect (ROESY) experiments. The NOE effect is sometimes crucial for correct assignments of substitutions especially in the absence of significant UV data with shift reagents that can give information about OH group positions. The NOE effect in some situations is more decisive than HMBC due to the few number of correlations that can be observed and the fact that correlations are dependent on distance in space rather than direct bond correlations.

The positions of ring B substituents in lysisteisoflavanone (**22**) were assigned utilizing GOESY experiment where irradiation of the OCH_3 and H-1″ of the prenyl group resulted in enhancement in their neighboring protons [50]. The NOE enhancement experiment was utilized to determine the position of OCH_3 in olibergin B (**23**) [24]. Position of OCH_3 in platyisoflavanone B (**24**) [41], vestitol (**25**), lotisoflavan (**26**) [21], erypoegin D (**27**) [43] and eryzerin B (**6**) [44] was assigned based on NOESY experiment results. The NOESY experiment was also employed to determine the position of glucose in ormosinoside A (**28**) [25].

NOESY data were also utilized to analyse the relative stereochemistry of the isoflavanol pumilanol (**29**) ring C protons [46].

3.5. Mass Spectroscopy (MS)

Mass spectroscopy with different techniques and the great advances in instrumentation can provide accurately the molecular weight and the exact molecular formula. In addition, some common routes of fragmentation can provide additional evidences about the substitution pattern on both rings A and B. The mass fragments derived from a *retro*-Diels Alder (RDA) type cleavage give an idea about the substituent's on ring A and ring B as well (**Figure 2**). These MS fragments were used for the confirmation of ring A and ring B substitution pattern in the structure elucidation. Observation of MS ion fragments at *m*/*z* 177 and 153 as a result of *RDA* type cleavage followed by a hydrogen transfer indicated the location of two

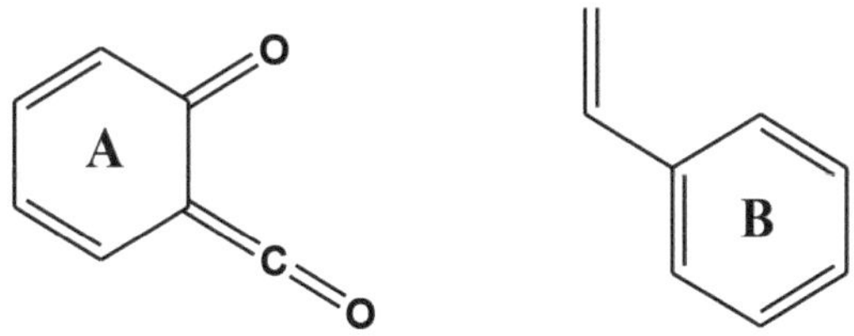

Figure 2. Main fragments of retro-diels–alder (RDA) type cleavage.

methoxyls and a hydroxyl group on the B ring of the isoflavone olibergin A (**30**) [24]. The placement of two hydroxyl group at ring A and methylenedioxy and one methoxyl at ring B in the structure of (±)5,7-dihydroxy-2′-methoxy-3′,4′-methylenedioxyisoflavanone (**31**) was confirmed by MS fragments [33]. (S)-Platyisoflavanone A (**32**) mass spectrum showed fragment at *m*/*z* 232 indicating two methoxyls and 3-methylbut-2-enyl group at ring B [41]. The base peak in the MS spectrum of uncinanone D (**33**) at *m*/*z* 194 [$C_{11}H_{14}O_3$] resulted from *retro*-Diels Alder (RDA) cleavage of ring C supported the presence of 3 methoxyl groups at ring B [48]. Similarly, the location of three methoxyl groups on ring B and two hydroxyl groups on ring A in the structure of the isoflavanone (±)5,7-Dihydroxy-2′,3′,4′-trimethoxy-isoflavanone (**34**) was supported by MS fragmentation [33]. The fragmentation of 5,7-Dihydroxy-2′,4′,5′-trimethoxyisoflavanone (**35**) generated mass fragments at *m*/*z* 153 corresponding with ring A with two hydroxyls and at *m*/*z* 194 for ring B with three methoxyls [39]. The location of the methyl group in desmodianone A (**36**), desmodianone B (**37**), desmodianone D (**38**), desmodianone E (**39**) and 6-methyltetrapterol A (**40**) at C-6 was confirmed from the MS fragment at *m*/*z* 167 for A-ring [47]. The MS fragments at *m*/*z* 346 $[508{-}163{+}H]^+$ and 194 indicated the presence of a sugar moiety in the A ring and three methoxyl groups in the B ring in the structure of 5,7-dihydroxy-2′,3′,4′-trimethoxy-isoflavanone 7-O-β-glucopyranoside (**41**) [33]. With a fragment 30 mass units less at *m*/*z* 164 in the spectrum of 5,7-Dihydroxy-2′,4′-dimethoxy-isoflavanone 7-O-β-glucopyranoside (**42**), only two methoxyls were assigned to ring B and sugar was placed on ring A [33].

In addition to providing the M+ at 328 *m*/*z* of 2-methoxyjudaicin (**43**) the fragment at *m*/*z* 297 due to loss of the two methoxyls was very supportive for the structure since the MS spectrum of judaicin (**44**) show only fragment due to loss of one methoxyl group at C-2′. The MS data of judaicin 7-*O*-glucoside (**45**) and judaicin 7-*O*-(6″-*O*-malonylglucoside) (**46**) showed common ion at *m*/*z* 298 corresponding to the aglycone part after the loss of the glycosyl moieties at C-7 [30, 31].

4. Isolated compounds update

The isolated isoflavonoids from natural sources are presented in **Tables 2–6**, and their structures are provided in **Figures 3–7**. Isoflavones, isoflavanones and isoflavans from 2000 to date are arranged according to publication date in **Tables 2–4**, respectively. Due to the limited number of isoflavenes, the current survey includes all isolated members available in the literature (**Table 5**). Synthetic compounds are not included in this chapter.

Name	Source	Ref.
2,3-Dehydrokievitone (**47**)	*Erythrina sacleuxii*	[49]
5′-Prenylpratensein (**48**)	*Erythrina latissima*	[39]
Erysubin F (**21**)	*Erythrina suberosa*	[42]
6″-*O*-Malonylgenistin (**49**)	*Glycine max*	[59]
Irisolone (**50**)	*Polygala stenopetala*	[69]
Isoerysenegalensein E (**51**), Alpinumisoflavone (**52**), Wighteone (**53**)	*Erythrina lysistemon*	[50]
2″,6″-O-Diacetyloninin (**54**)	*Glycine max*	[70]
Isoprunetin 7,4′-di-O-*β*-D-glucopyranoside (**55**) Genistein 7,4′-di-O-*β*-D-glucopyranoside (**56**)	*Genista morisii*	[54]
Genistein (**57**)	*Desmodium uncinatum*	[62]
Olibergin A (**30**), Olibergin B (**23**), Genistein (**57**), Formononetin (**58**) Biochanin A (**59**)	*Dalbergia oliveri*	[24]
Rothindin(**60**)	*Ononis serrata*	[27]
4′-*O*-Methylerythrinin C (**20**), 4′-*O*-Methylalpinumisoflavone (**61**) 4′-*O*-Methyl-2″-hydroxydihydroalpinumisoflavone (**62**) 7-O-Methylbiochanin A (**63**)	*Lotus polyphyllos*	[28]
Genistin (**1**), Genistein (**57**), Daidzein (**64**), Daidzin (**65**) Glycitein (**66**), Glycitin (**67**)	*Semen sojae praeparatum*	[71]
7-O-Geranylbiochanin A (**68**)	*Tephrosia tinctoria*	[72]
Olibergin B (**23**), Biochanin A (**59**), 8-C-Geranyl-7-O-methylbiochanin A (**69**)	*Dalbergia paniculata*	[73]
Biochanin A (**59**), 6-Hydroxy-7,4′-dimethoxyflavone (**70**) 6,7,4′-Trimethoxyflavone (**71**)	*Gynerium sagittatum*	[8]
4′-O-Methylderrone (**72**)	*Lotus polyphyllos*	[19]
4′,5′-Dimethoxy-6,6-dimethylpyranoisoflavone (**73**)	*Millettia pachycarpa*	[58]
Erypoegin D (**27**), Alpinumisoflavone (**52**), Wighteone (**53**) 5,4′-Dihydroxy-7-methoxy-3′-(3-methylbuten-2-yl)isoflavone (**74**) 5,2′,4′-Trihydroxy-7-methoxy-5′-(3-methylbuten-2-yl)isoflavone (**75**) 5,4′-Dihydroxy-7-methoxy-3′-(3-methyl-2-hydroxybuten-3-yl) isoflavone (**76**) 3′-Formyl-5,4′-dihydroxy-7-methoxyisoflavone (**77**) 5-Hydroxy-3″-hydroxy-2″,2″dimethyldihydropyrano[5″,6″:3′,4′] isoflavone (**78**) 3′-Isoprenylgenistein (**79**), Isolupabigenin (**80**)	*Erythrina poeppigiana*	[74]
Genistein (**57**), Formononetin (**58**), Biochanin A (**59**), Calycosin (**81**) Ononin (**82**), Sissotrin (**83**)	*Cicer arietinum*	[75]
Tlatlancuayin (2′,5-dimethoxy-6,7-methylenedioxyisoflavone) (**84**)	*Iresine herbstii*	[5]
2′-Hydroxygenistein (**85**), 3′-Omethylorobol (**86**) 7-O-Methyltectorigenin (**87**), Prunetin (**88**), Licoagroisoflavone (**89**) Cajanin (**90**), Lachnoisoflavone A (**91**)	*Crotalaria lachnophora*	[76]

Name	Source	Ref.
Pierreione A (**92**), Pierreione B (**93**), Pierreione C (**94**), Pierreione D(**95**)	*Antheroporum pierrei*	[77]
Genistein 5-*O*-*β*-glucopyranoside (**96**), Prunetin 5-*O*-*β*-glucopyranoside (**97**)	*Potentilla astracanica*	[7]
Erysubin F (**21**), Erythraddison I (**98**), Erythraddison II (**99**) Echrenone b10 (**100**)	*Erythrina addisoniae*	[23]
Ormosinosides A (**28**), Genistein (**57**), Biochanin A (**59**), Daidzein (**64**) Daidzin (**65**), Sissotrin (**83**), 7-O-Methylbiochanin A (**63**) Isoformononetin (**101**), 4′,7-Di-O-methyldaidzein (**102**), Isoprunetin (**103**) Sophoricoside (**104**), Isoprunetin-7-O-*β*-D-glucoside (**105**) 6″-*β*-D-Xylose-genistin (**106**)	*Ormosia henryi*	[25]
Genistein (**57**),Biochanin A (**59**), Daidzein (**64**) 3′-Hydroxydaidzein-7-O-glucopyranoside (**107**) Calycosin-7-O-glucopyranoside (**108**)	*Trifolium scabrum*	[78]
5,6-Dihydroxy-7,8,3′,5′-tetramethoxyisoflavone (**109**)	*Iris pseudacorus*	[79]
Formononetin (**58**), Ononin (**82**), Calycosin (**81**) Calycosin-7-O-glucopyranoside (**108**)	*Astragalus mongholicus*	[80]
Formononetin (**58**)	*Dalbergia oliveri*	[53]
Genistein (**57**), Biochanin A (**59**), Calycosin-7-O-glucopyranoside (**108**)	*Dalbergia odorifera*	[81]
Neobavaisoflavone (**110**)	*Erythrina excels, Erythrina senegalensis*	[40]
Biochanin A (**59**)	*Dothideomycetes fungus CMU-99*	[9]
Neoraudiol (**111**)	*Neorautanenia mitis*	[52]
Genistin (**1**), Daidzein (**64**), Daidzin (**65**), Puerarin (**112**)	*Pueraria lobata*	[34]
Formononetin (**58**), Ononin (**82**), 3-(4-(Glucopyranosyloxy)-5-hydroxy-2-methoxyphenyl)-7-hydroxy-4H-chromen-4-one (**113**)	*Ononis angustissima*	[82]
7,2′,5′-Trimethoxy-3′,4′-methylenedioxyisoflavone (**114**) 6,7-Dimethoxy-3′,4′-methylenedioxyisoflavone (**115**) 5,4′-Dihydroxy-7,2′,5′-trimethoxyisoflavone (**116**)	*Piscidia carthagenensis*	[83]
Isosideroxylin (**117**)	*Leiophyllum buxifolium*	[84]
Achyranthoside A (**118**), Achyranthoside B (**119**)	*Achyranthes bidentata*	[6]
Genistein (**57**), Biochanin A (**59**), Prunetin (**88**), Tectorigenin (**120**)	*Dalbergia odorifera*	[85]
8-Hydroxyirilone 5-methyl ether (**121**), 8-Hydroxyirilone (**122**) Irilone 4′-methyl ether (**123**), Irilone (**124**), Irisolidone (**125**) Irigenin S (**126**), Irigenin (**127**), Iridin S (**128**), Iridin (**129**) 4′-*O*-*β*-ᴅ-glucopyranoside (**130**)	*Iris germanica*	[26]

Table 2. Isolated isoflavones from natural sources since 2000 to date.

http://dx.doi.org/10.5772/intechopen.68701

Name	Source	Ref.
(*R*)-2,3-Dihydro-7-demethylrobustigenin (**8**), (*R*)-saclenone (**9**)	*Erythrina sacleuxii*	[49]
5,7-Dihydroxy-2′,4′,5′-trimethoxyisoflavanone (**35**)	*Erythrina latissima*	[39]
Bolusanthol B (**131**), Bolusanthol C (**132**) 5,7,3′-Trihydroxy-4′-methoxy-5′-γ,γ-dimethylallylisoflavanone (**133**) 5,7,2′-Trihydroxy-4′-methoxy-6,5′-di(γ,γ-dimethylallyl)isoflavanone (**134**) 5,7,2′,4′-Tetrahydroxy-8,3′-di(γ,γ-dimethylallyl)-isoflavanone (**135**)	*Bolusanthus speciosus*	[86]
Lysisteisoflavanone (**22**)	*Erythrina lysistemon*	[50]
Seputheisoflavone (**136**)	*Ptycholobium contortum*	[87]
Dihydrodaidzin (**137**), Dihydrogenistin (**138**)	*Glycine max*	[70]
Erypoegin C (**139**), Erypoegin D (**140**)	*Erythrina poeppigiana*	[43]
Eryzerin B (**6**), Eryzerin A (**7**)	*Erythrina zeyheri*	[44]
Erypoegin G (**141**)	*Erythrina poeppigiana*	[61]
Cajanol (**142**)	*Crotalaria lachnophora*	[76]
7,4′-Dihydroxy-2′-methoxy-6-geranylisoflavanone (**143**) 2′,4′-Dihydroxy-6″-methyl-6″-(4‴-methylpent-3-enyl) pyrano(3″,2″:6,7)-isoflavanone (**144**)	*Lespedeza bicolor*	[88]
Desmodianone A (**36**), Desmodianone B (**37**), Desmodianone D (**38**) Desmodianone E(**39**), 6-Methyltetrapterol A (**40**)	*Desmodium canum*	[47]
Uncinanone A (**145**), Uncinanone B (**146**), Uncinanone C (**147**)	*Desmodium uncinatum*	[62]
(±)5,7-Dihydroxy-2′-methoxy-3′,4′-methylenedioxyisoflavanone (**31**) (±)5,7-Dihydroxy-2′,3′,4′-trimethoxy-isoflavanone (**34**) 5,7-Dihydroxy-2′,3′,4′-trimethoxy-isoflavanone 7-O-*β*-glucopyranoside (**41**) 5,7-Dihydroxy-2′-methoxy-3′,4′-methylenedioxyisoflavanone 7-O-*β*-glucopyranoside (**148**) 5,7-Dihydroxy-2′,4′-dimethoxy-isoflavanone 7-O-*β*-glucopyranoside (**42**) 5,7,4′-Trihydroxy-2′,3′-dimethoxy-isoflavanone 7-O-*β*-glucopyranoside (**149**)	*Desmodium styracifolium*	[33]
Uncinanone D (**33**), Uncinanone E (**150**)	*Desmodium uncinatum*	[48]
Ferreirin (**151**), Dihydrocajanin (**152**), Dalbergioidin (**153**) Dihydrobiochanin A (**154**)	*Gynerium sagittatum*	[8]
5,7-Dihydroxy-2′-methoxy-3′,4′-methylenedioxyisoflavanone (**31**) Uncinanone A (**37**), Dalbergioidin (**153**) 4′,5-Dihydroxy-2′,3′-dimethoxy-7-(5-hydroxyoxychromen-7yl)-isoflavanone (**155**), Parvisoflavanone (**156**), Isoferreirin (**157**)	*Uraria picta*	[55]
Dalhorridin (**158**), Dalhorridinin (**159**)	*Dalbergia horrida*	[89]
5,3′-Dihydroxy-4′-methoxy-5′-(3-methyl-1,3-butadienyl)-2″,2″-dimethylpyrano[5, 6:6,7]isoflavanone (**160**) 5,3′-Dihydroxy-5′-(3-hydroxy-3-methyl-1-butenyl)-4′-methoxy-2″,2″-dimethylpyrano[5, 6:6,7]isoflavanone (**161**)	*Erythrina costaricensis*	[90]
Sophoronol A (**162**), Sophoronol B (**163**), Sophoronol C (**164**) Sophoronol D (**165**), Sophoronol E (**166**), Sophoronol F (**167**)	*Sophora mollis*	[91]

Name	Source	Ref.
3-Hydroxy-kenusanone B (**168**), Sophoraisoflavanone A (**169**) Kenusanone H (**170**)	*Echinosophora koreensis*	[92]
Desmodianone F (**171**), Desmodianone G (**172**)	*Desmodium canum*	[93]
5,7,3′-Trihydroxy-4′-methoxy-6,5′-di(γ, γ-dimethylallyl)-isoflavanone (**173**) 5,3′-Dihydroxy-4′-methoxy-5′-γ,γ-dimethylallyl-2″,2″-dimethylpyrano[5, 6: 6,7]isoflavanone (**174**) 5,3′-Dihydroxy-2″,2″-dimethylpyrano[5, 6: 6,7]-2‴,2‴-dimethylpyrano[5, 6: 5,4]isoflavanone (**175**)	*Erythrina costaricensis*	[94]
Glabraisoflavanone A (**176**), Glabraisoflavanone B **(177)**	*Glycyrrhiza glabra*	[95]
Isodarparvinol B (**178**), Dalparvin (**179**), (3*S*)-Sativanone (**180**)	*Dalbergia parviflora*	[96]
2′,2,5-Trimethoxy-6,7-methylenedioxyisoflavanone (**181**)	*Iresine herbstii*	[5]
Erythraddison III (**182**), Erythraddison IV (**183**)	*Erythrina addisoniae*	[23]
Dalbergioidin (**153**)	*Lespedeza cyrtobotrya*	[29]
3(*R*)-2′-Methoxyl-5,7,4′-trihydroxy-6-(3-methylbut-2-enyl)-isoflavanone (**184**) 3′-Geranyl-3,5,7,2′,4′-pentahydroxyflavonol (**185**)	*Campylotropis hirtella*	[97]
Triquetrumone E (**186**), Triquetrumone F (**187**)	*Tadehagi triquetrum*	[98]
Hirtellanine H (**188**), Hirtellanine I (**189**), Hirtellanine J (**190**)	*Campylotropis hirtella*	[99]
Ormosinol (**191**)	*Ormosia henryi*	[25]
7-O-Glucosyldiphysolone (**2, 3**), (3*R*)-7,4′-Di-O-glucosyldiphysolone (**4**) (3*S*)-7,4′-Di-O-glucosyldiphysolone (**5**), 4″-hydroxydiphysolone (**192**)	*Ormocarpum kirkii*	[32]
Platyisoflavanone B (**24**), Platyisoflavanone A) (**32**) Platyisoflavanone C (**193**), Platyisoflavanone D (**113**) Sophoraisoflavanone A (**169**), Glyasperin F (**194**)	*Platycelphium voënse*	[41]
(+)-Violanone (**195**)	*Dalbergia oliveri*	[53]
(3*S*)-2′,4′-Dimethoxy-3,7-dihydroxyisoflavanone (**196**) (3*S*)-2′,4′,5′-Trimethoxy-7-hydroxyisoflavanone (**197**) (3*R*)-4′-Methoxy-2′,3,7-trihydroxyisoflavanone (**198**) (3*R*)-Violanone (**199**), (3R)-3′-O-methylviolanone (**200**) (3*R*)-Sativanone (**201**)	*Dalbergia odorifera*	[100]
Dalbergioidin (**153**) (3*R*) 5,7,3′,4′-Tetrahydroxy-2′-methoxyisoflavanone (**202**) (3*R*) 5′,8-Di-(γ,γ-dimethylallyl)-2′,5-dihydroxyl-4′,7-dimethoxyl-isoflavanone (**203**) 5,7-Dihydroxy-2′,4′-dimethoxyisoflavanone (**204**)	*Uraria clarkei*	[101]
Uncinanone E (**150**) 5,7-dihydroxy-2′-methoxy-3′,4′-methylenedioxy isoavanone (**155**) (3*R*) 7,2′,4′-Trihydroxy-3′-methoxy-5-methoxycarbonylisoflavanone (**205**) (3*R*) 7,2′-Dihydroxy-3′,4′-dimethoxy-5-methoxycarbonylisoflavanone (**206**)	*Cassia siamea*	[102]
Sigmoidin H (**207**)	*Erythrina excels*, *Erythrina senegalensis*	[40]
6,3′-di(3-hydroxy-3-methylbutyl)-5,7,2′, 4′-tetrahydroxyisoflavanone (**208**) 3(R)-6,3′-di(3-hydroxy-3-methylbutyl)-2′-methoxyl-5,7,4′-trihydroxyisoflavanone (**209**)	*Campylotropis hirtella*	[103]

Name	Source	Ref.
Uncinanone D (**33**), Desmodianone E (**144**), Desmodianone F (**171**) Grabraisoflavanone A (**176**) (3*R*)-7-Hydroxy-4′-methoxy-5-methoxycarbonyl-isoflavanone (**210**) (3*R*)-8-Hydroxy-4′-methoxy-7-methoxycarbonyl-isoflavanone (**211**) (3*R*)-7,2′,4′-Trihydroxy-3′-methoxy-5-methoxycarbonyl-isoflavanone (**205**)	*Desmodium oxyphyllum*	[104]
Glycitein (**66**), Dihydrodaidzein (**133**), Dihydrogenistein (**134**) Dothideoisoflavanone (**212**), (3*S*)-3,4′,7-trihydroxyisoflavanone (**213**)	Dothideomycetes fungus CMU-99	[9]
Neotenone (**214**)	*Neorautanenia mitis*	[52]
Eryvarins Y (**215**), Eryvarins Z (**216**), Orientanol E (**217**) 2,3-Dihydroauriculatin (**218**)	*Erythrina variegata*	[105]

Table 3. Isolated isoflavonones from natural sources since 2000 to date.

Name	Source	Ref.
5,7-Dimethoxy-3-(4-hydroxybenzyl)-4-chromanone (**219**) 5,6-Dihydroxy-7-methoxy-3-(4-hydroxybenzyl)-4-chromanone (**220**) 7-*O*-Methyl-3,9-dihydropunctatin (221) 5,7-Dihydroxy-3-(4-hydroxybenzyl)-4-chromanone (222)	*Drimiopsis burkei* *Drimiopsis maculata*	[63]

Table 4. Isolated homoisoflavonoids from natural sources since 2000 to date.

Name	Source	Ref.
Bolusanthol A (**223**)	*Bolusanthus speciosus*	[86]
Neocandenatone (**224**)	*Dalbergia congestiflora*	[22]
(3*R*)-(−)-7,2′-Dihydroxy-3′,4′-dimethylisoflavan-7-O-*β*-D-glucopyranoside (**225**)	*Astragalus membranaceus*	[20]
Eryzerin C (**18**), Eryzerin D (**19**)	*Erythrina zeyheri*	[44]
6-Desmethyldesmodian A (**226**), Desmodian A (**227**) Desmodian B (**228**), 6-Desmethylesmodian B (**229**) Desmodian C (**230**), 3′-Hydroxydesmodian B (**231**)	*Desmodium canum*	[106]
Pumilanol (**29**)	*Tephrosia pumila*	[46]
Salisoflavan (**232**)	*Salsola imbricata*	[107]
Desmodian A (**227**), Desmodian D (**233**)	*Desmodium canum*	[93]
3*S* (+) 2′-O-Methylphaseollidinisoflavan (**15**) 3*R*(-)Erythbidin A (**16**)	*Erythrina caffra*	[45]
Vestitol (**25**), Neovestitol (**234**)	Brazilian propolis	[10]
(3*S*,4*R*)-4′-Hydroxy-6,3′-dimethoxyisoflavan-4-ol (**235**)	*Taxus yunnanensis*	[108]
Cordifoliflavanes A (**236**), Cordifoliflavanes B (**237**)	*Codonopsis cordifolioidea*	[109]
Vestitol (**25**), Lotisoflavan (**26**)	*Lotus lalambensis*	[21]
Abruquinone A (**238**), Abruquinone D (**239**), Abruquinone J (**240**) Abruquinone K (**241**), Abruquinone L (**17**)	*Abrus precatorius*	[38]

Name	Source	Ref.
Erylivingstone J (**10**), Erylivingstone K (**11**) 2′-Methoxyphaseollinisoflavan (**12**) 7, 4′-Dihydroxy-2′,5′-dimethoxy isoflavan (**13**) 7,4′-Dihydroxy-2′-methoxy-3′-(3-methylbut-2-enyl) isoflavan (**14**)	*Erythrina livingstoniana*	[37]
Kotstrigoisoflavanol (**242**)	*Kotschya strigosa*	[110]

Table 5. Isolated isoflavans from natural sources since 2000 to date.

Name	Source	Ref.
Neorauflavene (**243**)	*Neorautanenia edulis*	[51]
Sepiol (**244**), 2′-O-Methylsepiol (**245**)	*Gliricidia speium*	[111]
Dimethoxytrihydroxyisoflavene (**246**)	*Baphia nitida*	[56]
Haginin A (**247**), Haginin B (**248**)	*Lespedeza cyrtobotrya*	[35]
7,3′,4′-Triacetoxy-6′-methoxyisoflav-3-ene (**249**) 7, 2′-Diacetoxy-4′-methoxyisoflav-3-ene (**250**)	*Millettia* sp.	[112]
2-Methoxyjudaicin (**43**)	*Cicer bijugum*	[30]
Judaicin (**44**), Judaicin 7-*O*-glucoside (**45**) Judaicin 7-*O*-(6″-*O*-malonylglucoside) (**46**)	*Cicer judaicum*	[31]
Haginin C (**251**), Haginin D (**252**)	*Lespedeza cyrtobotrya*	[113]
Haginin D (**253**), Haginin E (Phenoxodiol) (**254**)	*Lespedeza homoloba*	[36]
Erypoegin A (**255**), Erypoegin B (**256**)	*Erythrina poeppigiana*	[43]
Glabrene (**257**)	*Glycyrrhiza glabra*	[114]
Haginin A (**247**)	*Lespedeza cyrtobotrya*	[29]
Haginin E (Phenoxodiol) (**254**)	*Dothideomycetes* fungus CMU-99	[9]

Table 6. Isolated isoflavenes from natural sources.

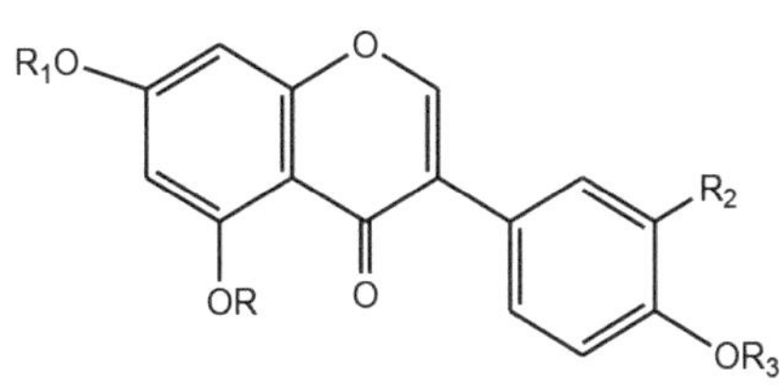

(**1**) R= R_2= R_3= H, R_1= Glucose
(**28**) R= Me, R_1= 6"-(β-D-Xylosyl)-glucopyranoside, R_2= R_3= H
(**49**) R= R_2= R_3= H, R_1= 6"-O-(Malonyl)-glucopyranoside
(**55**) R= Me, R_1= R_3= Glucose, R_2= H
(**56**) R= R_2= H, R_1= R_3= Glucose
(**57**) R= R_1= R_2= R_3= H
(**59**) R= R_1= R_2= H, R_3= Me
(**63**) R= R_2= H, R_1= R_3= Me
(**68**) R= R_2= H, R_1= Geranyl, R_3= Me
(**74**) R= R_3= H, R_1= Me, R_2 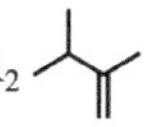
(**77**) R= R_3= H, R_1= Me, R_2= Formyl
(**79**) R= R_1= R_3= H, R_2= Prenyl
(**83**) R= R_2= H, R_1= Glucose, R_3= Me
(**86**) R= R_1= R_3= H, R_2= OMe
(**88**) R= R_2= R_3= H, R_1= Me
(**96**) R= Glucose, R_1= R_2= R_3= H
(**97**) R= Glucose, R_1= Me, R_2= R_3= H
(**103**) R= Me, R_1= R_2= R_3= OH
(**104**) R= R_1= R_2= H, R_3= Glucose
(**105**) R= Me. R_1= Glucose, R_2= R_3= H
(**106**) R= R_2= R_3= H, R_1= 6"-(β-D-Xylosyl)-glucopyranoside

http://dx.doi.org/10.5772/intechopen.68701

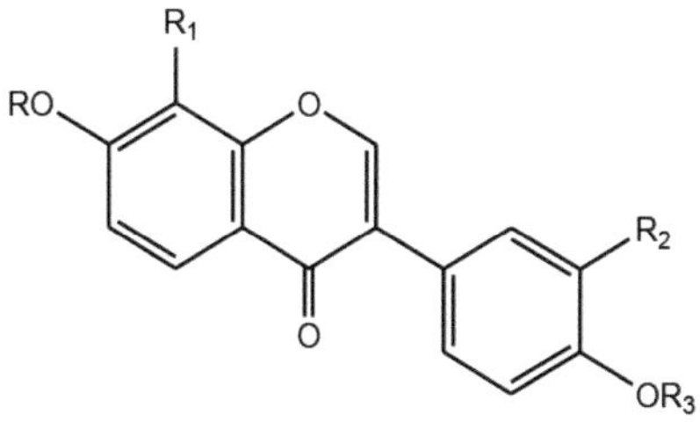

(21) $R = R_3 = H$, $R_1 = R_2 =$ Prenyl
(69) $R = R_3 = H$, $R_1 =$ Geranyl, $R_2 = H$
(112) $R = R_2 = R_3 = H$, $R_1 =$ Glucose

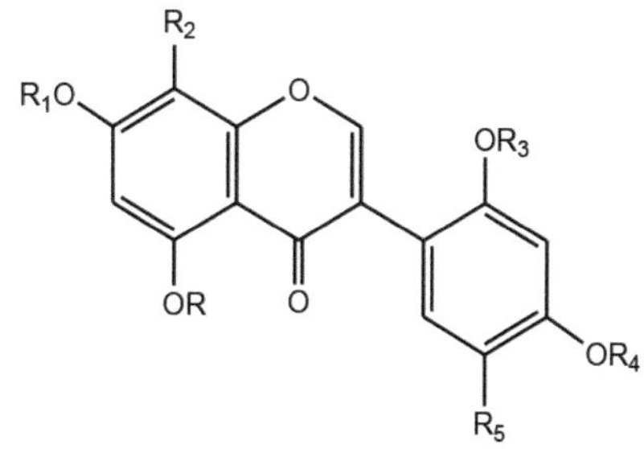

(27) $R = R_2 = R_4 = H$, $R_1 = R_3 =$ Me, $R_5 =$ Prenyl
(30) $R = R_1 = R_2 = R_4 = H$, $R_3 =$ Me, $R_5 =$ OMe
(47) $R = R_1 = R_3 = R_4 = R_5 = H$, $R_2 =$ Prenyl
(75) $R = R_2 = R_3 = R_4 = H$, $R_1 =$ Me, , R_5 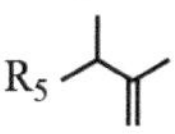
(85) $R = R_1 = R_2 = R_3 = R_4 = R_5 = H$
(109) $R = R_2 = R_3 = R_4 = R_5 = H$ $R_1 =$ Me
(116) $R = R_2 = R_4 = H$, $R_1 = R_3 =$ Me, $R_5 =$ OMe

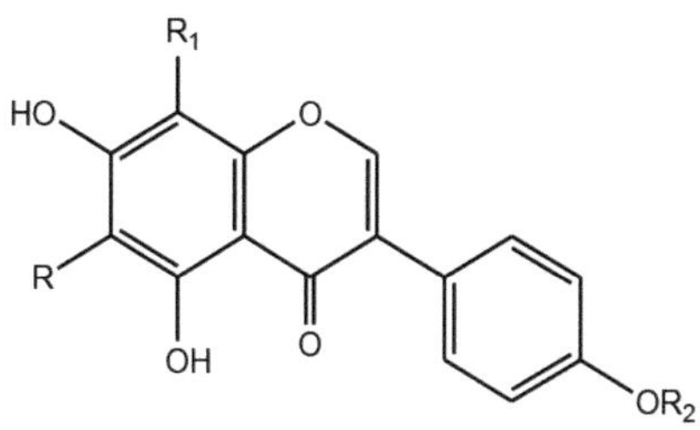

(23) $R = H$, $R_1 =$ Geranyl, $R_2 =$ Me
(51) 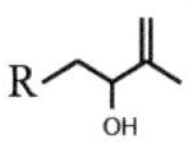 , $R_1 =$ Prenyl, $R_2 = H$

(53) $R =$ Prenyl, $R_1 = R_2 = H$
(120) $R =$ OMe, $R_1 = R_2 = H$

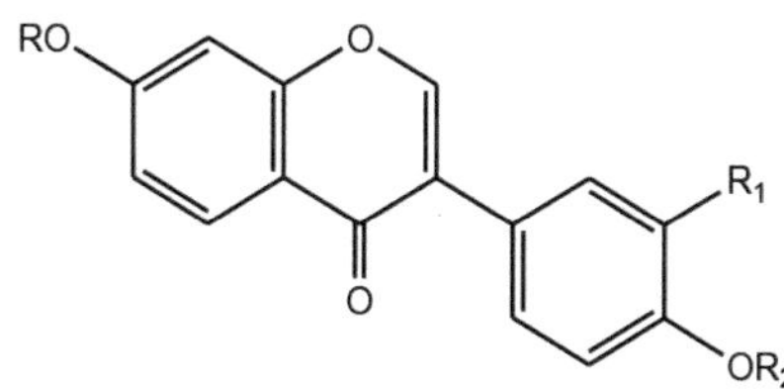

(54) R= 2", 6"-Diacetylglucoside, $R_1 = H$, $R_2 =$ Me
(58) $R = R_1 = H$, $R_2 =$ Me
(64) $R = R_1 = R_2 = H$,
(65) $R =$ Glucose, $R_1 = R_2 = H$
(81) $R = H$, $R_1 =$ OH, $R_2 =$ Me
(82) $R =$ Glucose, $R_1 = H$, $R_2 =$ Me,
(101) $R =$ Me, $R_1 = R_2 = H$
(102) $R = R_2 =$ Me, $R_1 = H$
(110) $R = R_2 = H$, 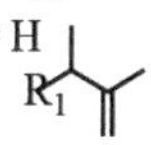

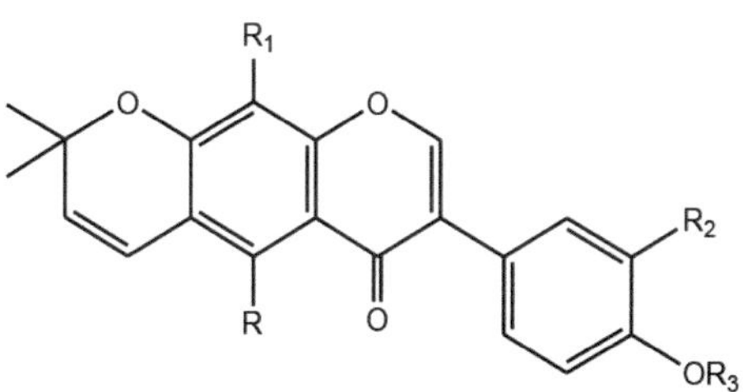

(52) $R =$ OH, $R_1 = R_2 = R_3 = H$
(61) $R =$ OH, $R_1 = R_2 = H$, $R_3 =$ Me
(92) $R = R_2 =$ OMe, $R_1 = H$, 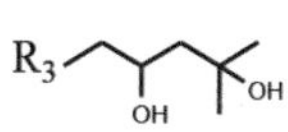

(93) $R = R_1 = H$, $R_2 =$ OMe, 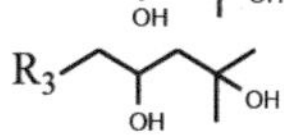

(94) $R = R_2 =$ OMe, $R_1 = H$, 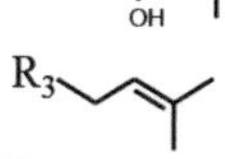

(95) $R = R_1 = H$, $R_2 =$ OH, 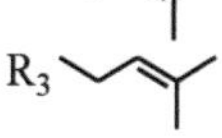

(98) R=OH, 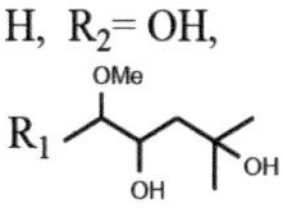 , $R_2 = R_3$=H

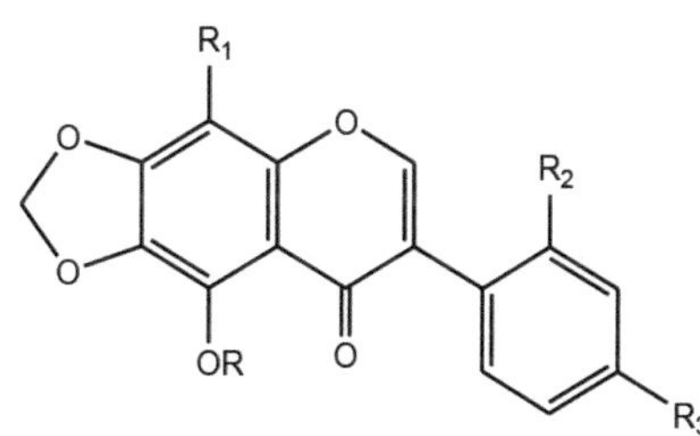

(50) $R =$ Me, $R_1 = R_2 = H$, $R_3 =$ OH
(84) $R =$ Me, $R_1 = R_3 = H$, $R_2 =$ OMe
(121) $R =$ Me, $R_1 = R_3 =$ OH, $R_2 = H$
(122) $R = R_2 = H$, $R_1 = R_3 =$ OH,
(123) $R = R_1 = R_2 = H$, $R_3 =$ OMe
(124) $R = R_1 = R_2 = H$, $R_3 =$ OH
(130) $R = R_1 = R_2 = H$, $R_3 =$ O-Glucose

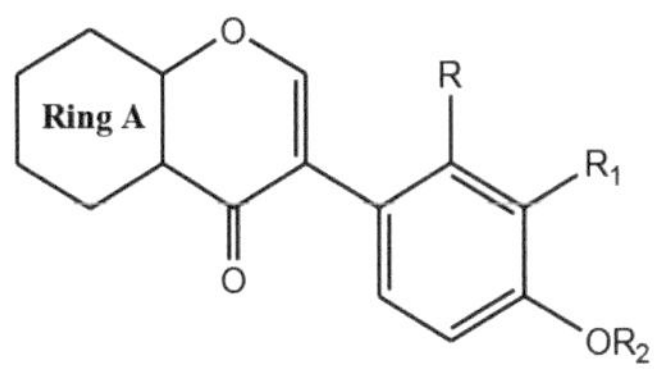

(**62**) $R= R_1= H$, $R_2= Me$
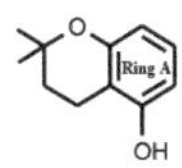

(**72**) $R= R_1= H$, $R_2= Me$
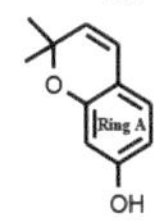

(**73**) $R= H$, $R_1= OMe$, $R_2= Me$
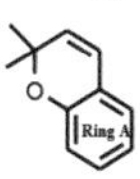

(**89**) $R= R_1= R_2= H$
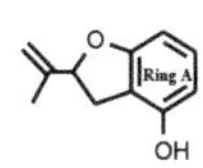

(**91**) $R= OH$, $R_1= R_2= H$
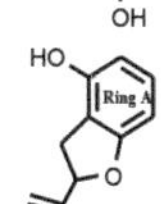

(**99**) $R= OH$, $R_1= R_2= H$
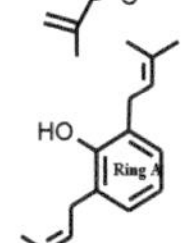

(**111**) $R= R_2= H$, $R_1= OH$
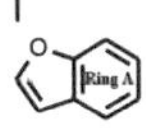

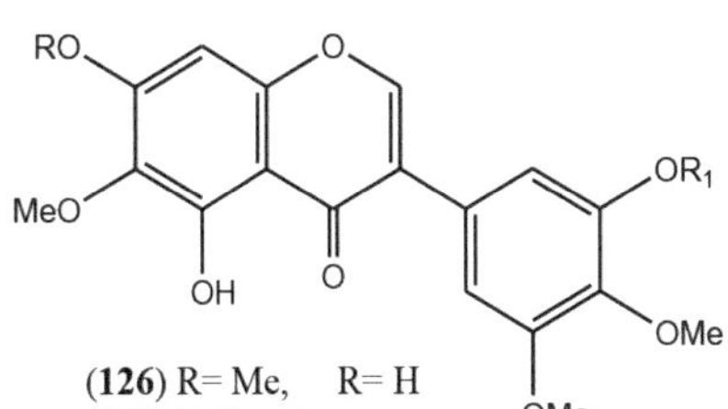

(**126**) $R= Me$, $R= H$
(**127**) $R, R_1= H$
(**128**) $R= Glucose$, $R_1= Me$
(**129**) $R= Glucose$, $R_1= H$

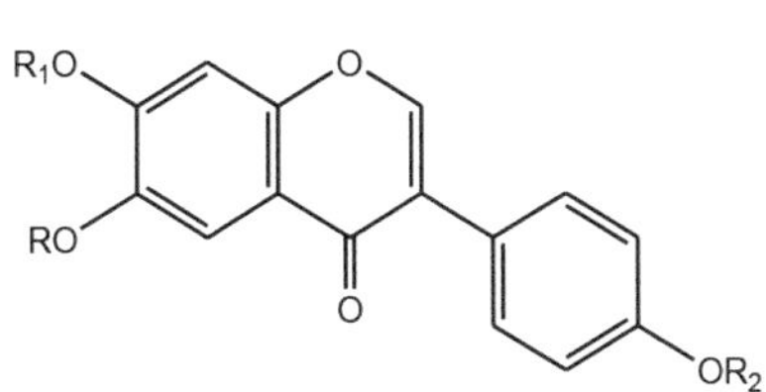

(**66**) $R= Me$, $R_1= R_2= H$
(**67**) $R= Me$, $R_1= Glucose$, $R_2= H$
(**70**) $R= H$, $R_1= R_2= Me$
(**71**) $R= R_1= R_2= Me$

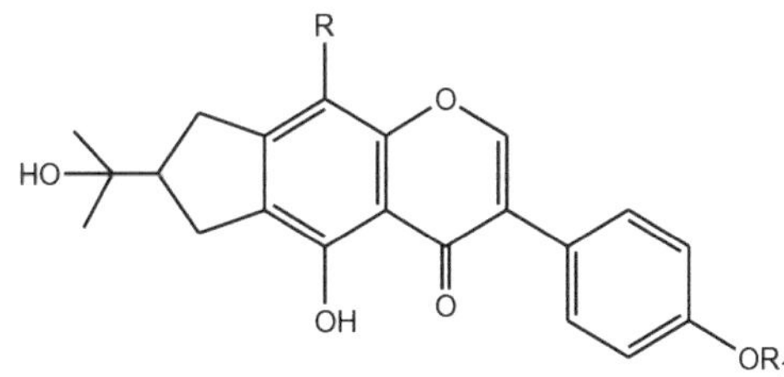

(**20**) $R= H$, $R_1= Me$
(**100**) $R= Prenyl$, $R_1= H$

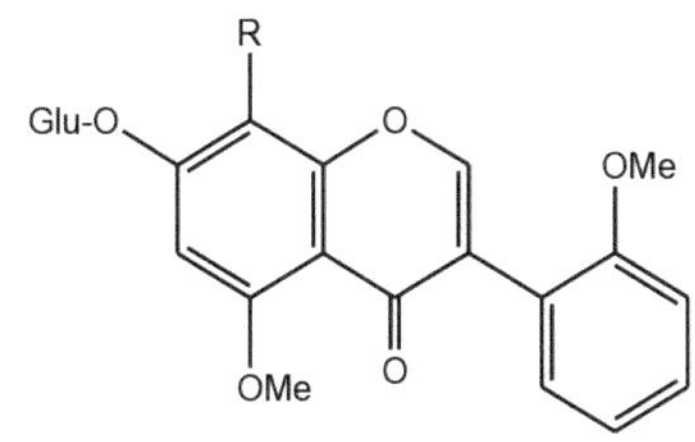

(**118**) $R= CH_2OMe$
(**119**) $R= OMe$

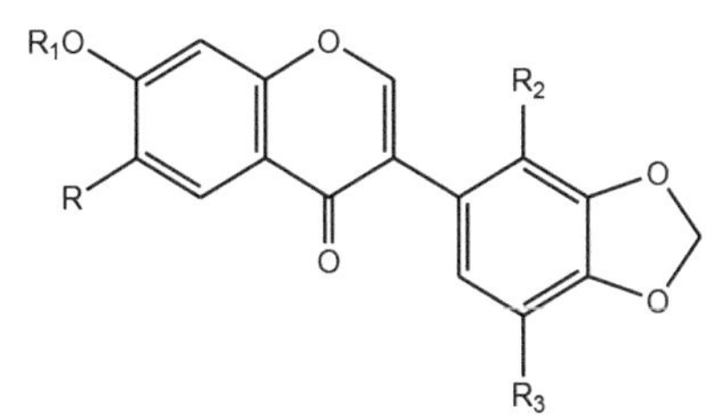

(**60**) $R= R_2= R_3 = H$, $R_1= Glucose$
(**114**) $R= H$, $R_1= Me$, $R_2= R_3 = OMe$
(**115**) $R= OMe$, $R_1= Me$, $R_2= R_3 = H$

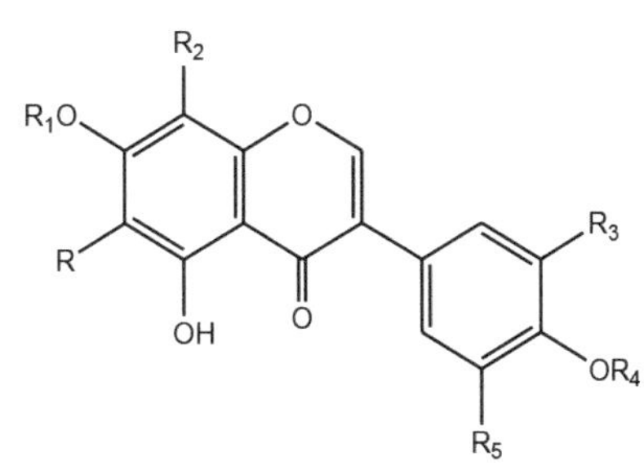

(**48**) $R= R_1= R_2= R_4= H$, $R_3= Prenyl$, $R_5= OMe$
(**80**) $R= R_1= R_4= R_5= H$, $R_2= R_3= Prenyl$
(**87**) $R= OMe$, $R_1= Me$, $R_2= R_3= R_4= R_5= H$
(**109**) $R= OH$, $R_1= R_4= Me$, $R_2= R_3= OMe$, $R_5= H$
(**117**) $R= R_1= R_2= Me$, $R_3= R_4= R_5= H$
(**125**) $R= OMe$, $R_1= R_2= R_3= R_5= H$, $R_4= Me$

Figure 3. Isolated isoflavones from natural sources since 2000 to date.

http://dx.doi.org/10.5772/intechopen.68701

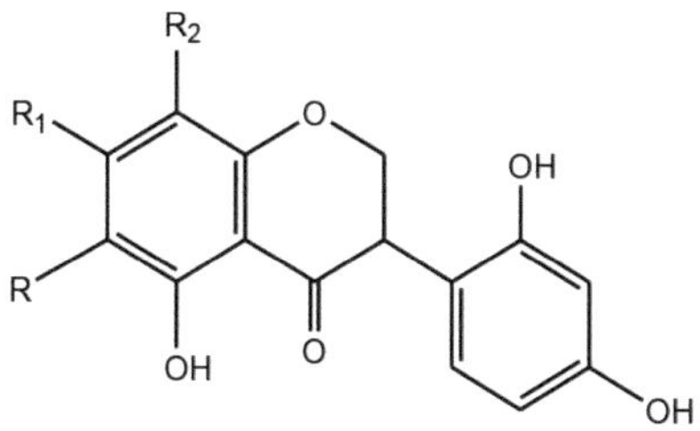

(2,3) R= Prenyl, R_1= O-Glucose, R_2 = H
(145) R= Prenyl, R_1= OH, R_2 = H
(152) R= R_2= H, R_1 = OMe
(153) R= R_2= H, R_1= OH
(170) R= H, R_1= OH, R_2 = Geranyl
(184) R= Prenyl, R_1= OH, R_2= H
(192) R= 4-Hydroxyprenyl, R_1= OH, R_2= H
(217) R= R_2= Prenyl, R_1= OH

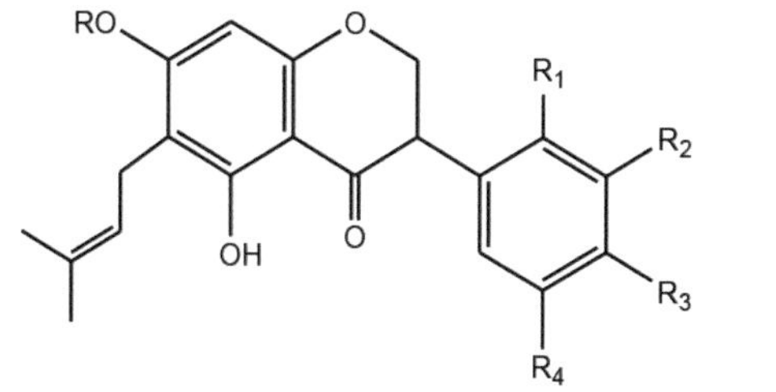

(5) R= Glucose, R_1= OH, R_2= R_4= H, R_3= O-Glucose
(33) R= R_4= H, R_1= R_2= R_3= OMe
(132) R= R_1=R_4= H, R_2= Prenyl, R_3= OH
(134) R= R_2= H, R_1= OH, R_3= OMe, R_4= Prenyl
(135) R= R_1= H, R_2= R_3=OH, R_4= Prenyl
(173) R= R_1= H, R_2= OH, R_3=OMe, R_4= Prenyl

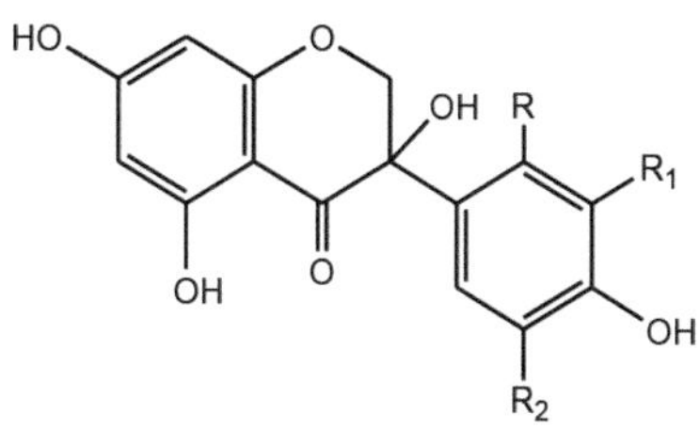

(165) R= OH, R_1= Prenyl, R_2= OH
(185) R= OH, R_1= Geranyl, R_2= H
(212) R= R_1= R_2= H

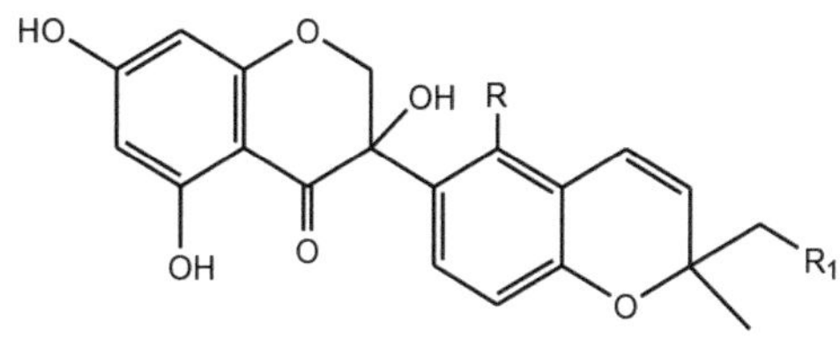

(162) R= R_1= H
(163) R= OOH, R_1= OH

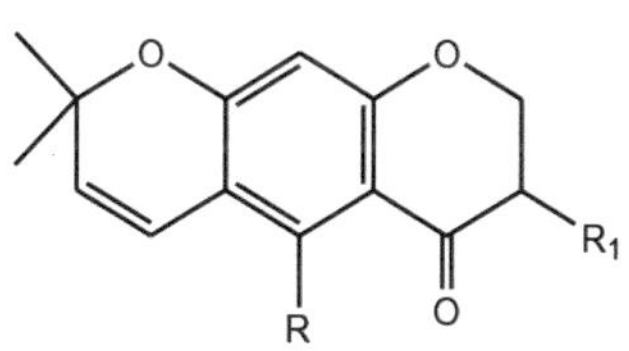

(9) R= OH, R_1

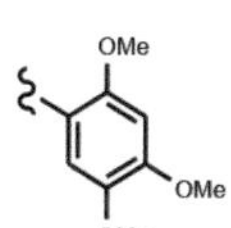

(174) R= H, R_1

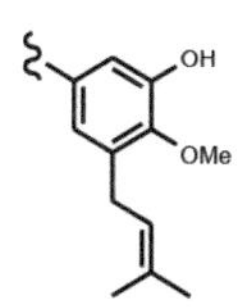

(175) R= H, R_1

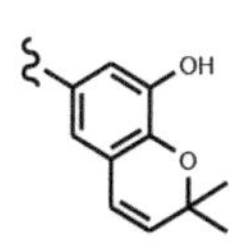

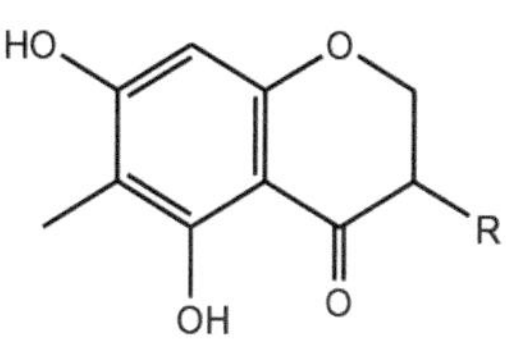

(36) R

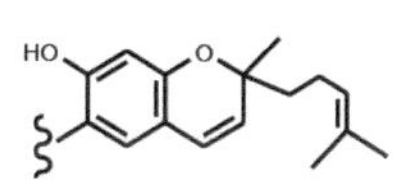

(37) R

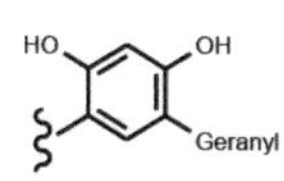

(38) R

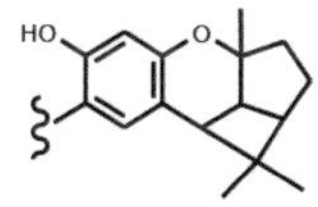

(39) R

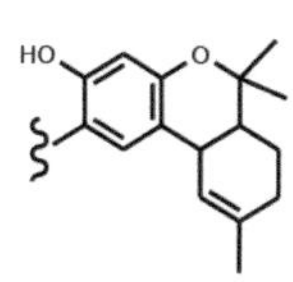

(40) R

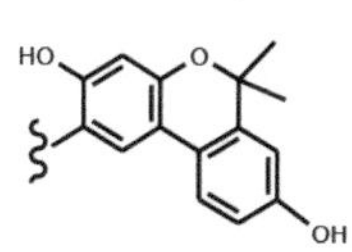

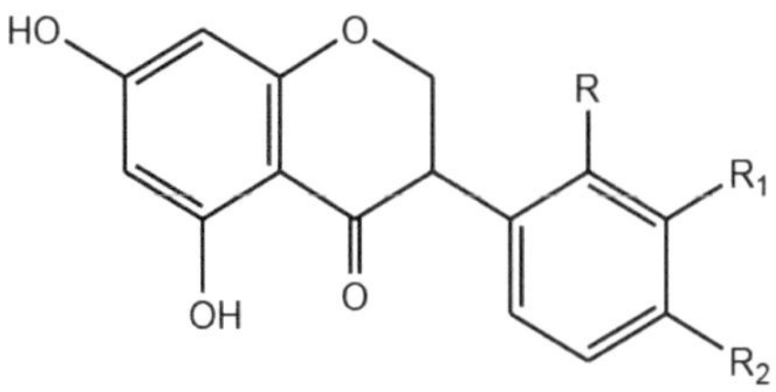

(**31**) R= OMe, R_1, R_2 = Methylenedioxy
(**32**) R= OMe, R_1= Prenyl, R_2= OMe
(**24**) R= OH, R_1= Prenyl, R_2= OMe
(**34**) R= R_1= R_2 = OMe
(**137**) R= R_1= H, R_2= OH
(**151**) R= OH, R_1= H, R_2= OMe
(**154**) R= R_1= H, R_2= OMe
(**156**) R= R_1= OMe, R_2= OH
(**157**) R= OMe, R_1= H, R_2= OH
(**169**) R= OMe, R_1= Prenyl, R_2= OH
(**182**) R= H, R_1= Prenyl, R_2= OMe
(**202**) R= OH, R_1= OMe, R_2= OH
(**204**) R= R_2= OMe, R_1= H

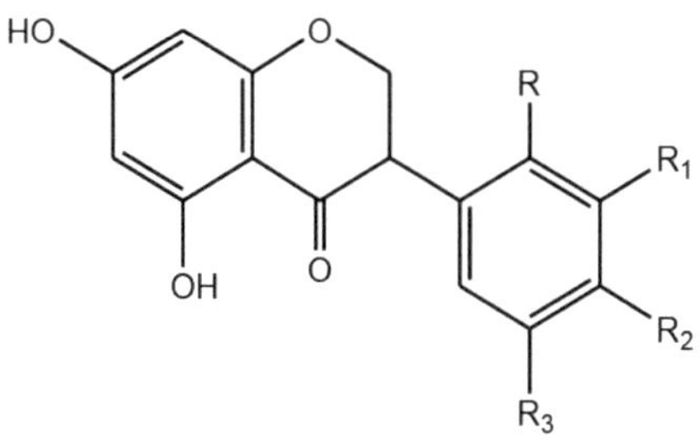

(**8**) R= R_2= R_3= OMe, R_1= H
(**22**) R= OH, R_1= H, R_2= OMe, R_3 = Prenyl
(**123**) R= H, R_1= Prenyl, R_2 = R_3= OH
(**125**) R= H, R_1= OH, R_2= OMe, R_3 = Prenyl
(**131**) R= OMe, R_1= H, R_2= OH, R_3 = Prenyl
(**179**) R= OMe, R_1= H, R_2= OMe, R_3= OH

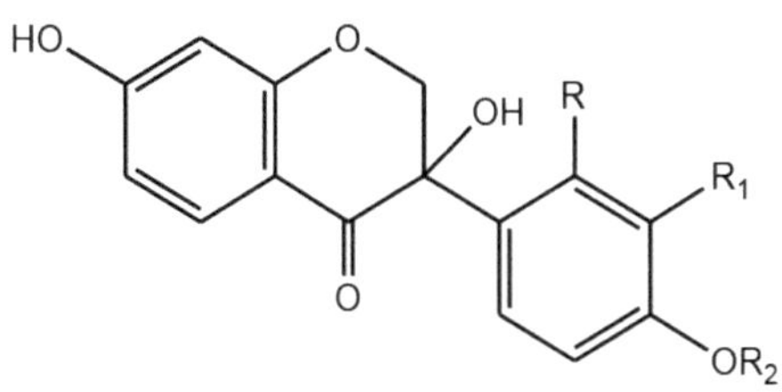

(**178**) R, R_1= OH, R_2= Me
(**196**) R= OMe, R_1= H, R_2= Me
(**198**) R= OH, R_1= H, R_2= Me
(**213**) R= R_1= R_2= H

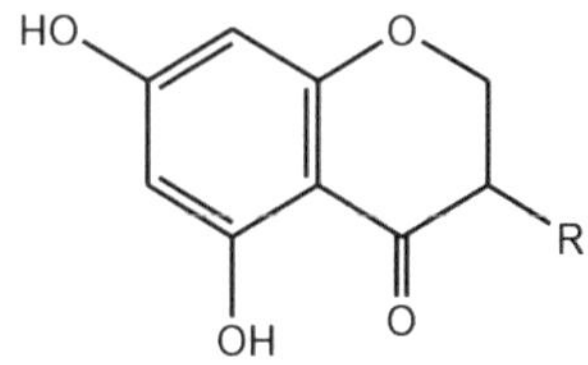

(**113**) R 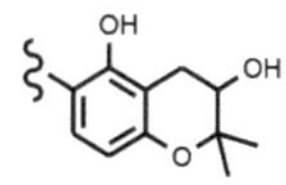

(**136**) R 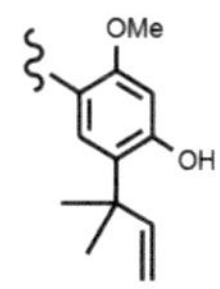

(**183**) R 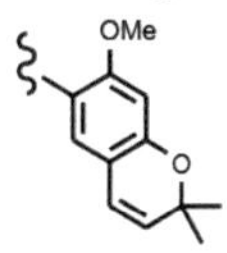

(**188**) R 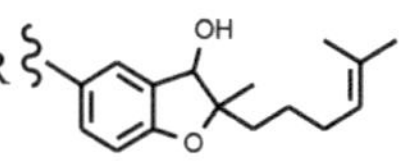

(**190**) R 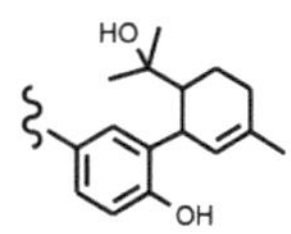

(**193**) R 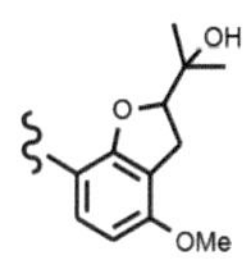

(**194**) R 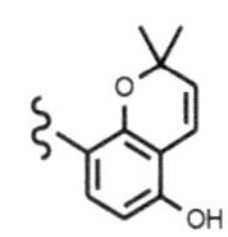

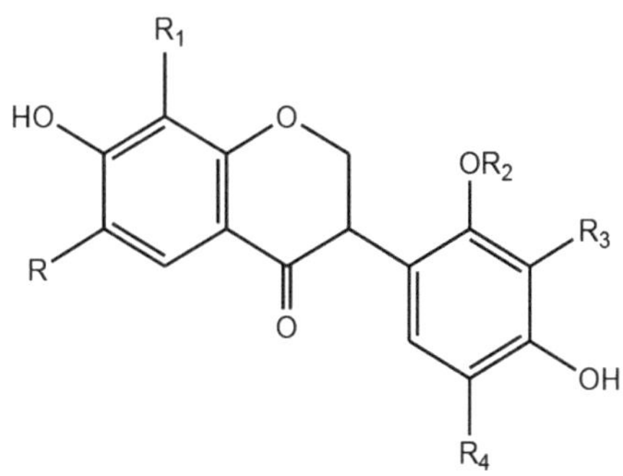

(**6**) R= R_1= Prenyl, R_2= R_3= R_4= H
(**7**) R= R_2= R_4= H, R_1= R_3= Prenyl
(**191**) R= R_1= R_4= Prenyl, R_2= R_3= H
(**215**) R= Prenyl, R_1= 2-Hydroxyprenyl, R_2= R_3= R_4= H
(**216**) R= 2-Hydroxyprenyl, R_1= Prenyl, R_2= R_3= R_4= H

http://dx.doi.org/10.5772/intechopen.68701

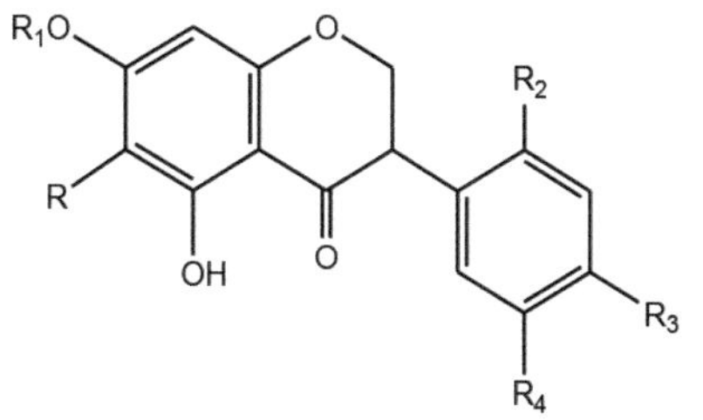

(**35**) R= R_1= H, R_2= R_3=R_4= OMe
(**138**) R= R_2= R_4= H, R_1= Glucose, R_3= OH
(**142**) R= R_4= H, R_1=Me, R_2= R_3= OMe
(**143**) R= Geranyl, R_1= R_4= H, R_2= OMe, R_3= OH
(**150**) R= R_1=Me, R_2= OMe, R_3=OH, R_4= H

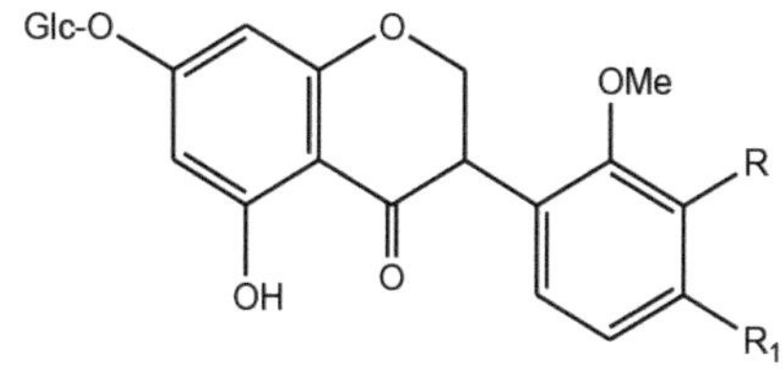

(**41**) R= R_1= OMe
(**42**) R= H, R_1= OMe
(**148**) R, R_1= Methylenedioxy
(**149**) R= OMe, R_1= OH

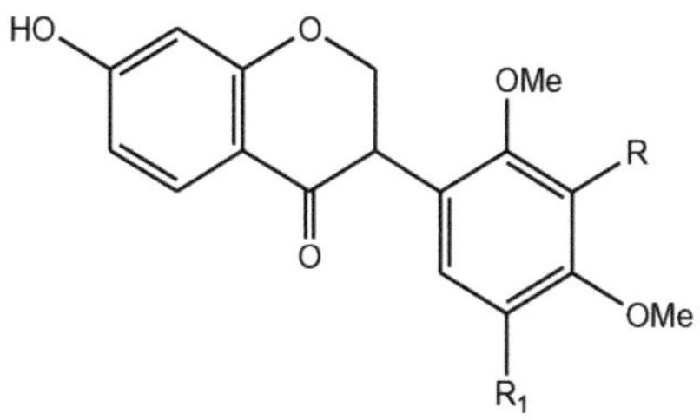

(**180, 201**) R= R_1= H
(**195, 199**) R= OH, R_1= OH
(**197**) R= H, R_1= OMe
(**200**) R= OMe, R_1= H

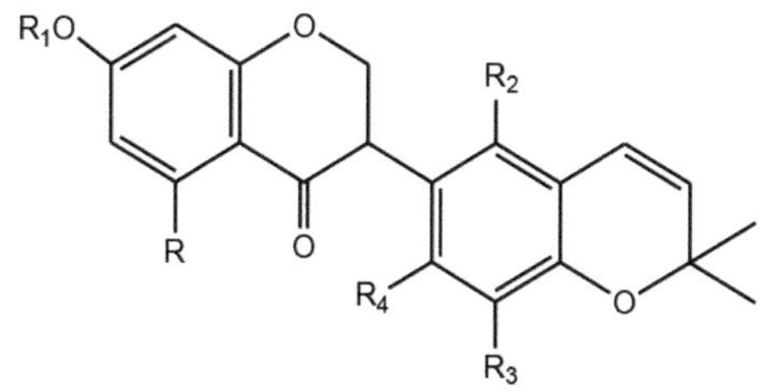

(**141**) R= OH, R_1= Me, R_2= R_3=H, R_4= OMe
(**189**) R= R_1= R_3= H, R_2= OMe, R_4= OH
(**207**) R= R_1=R_2=R_3= H, R_4= OMe

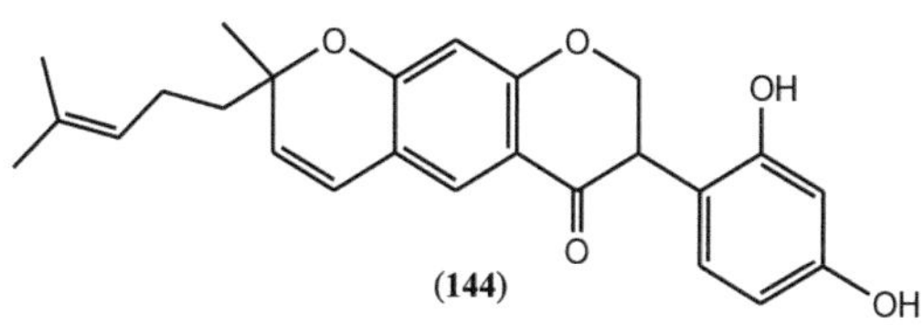

(**144**)

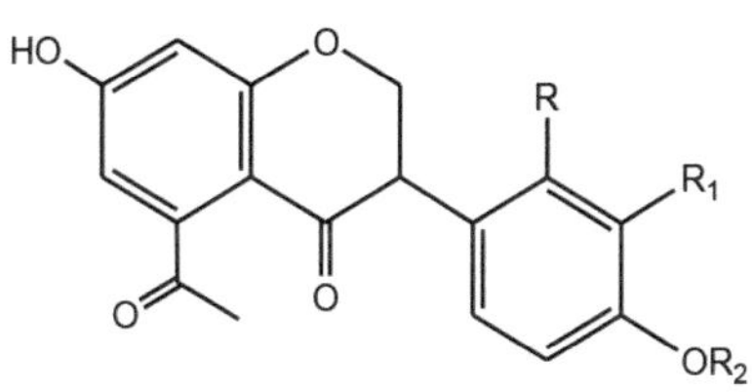

(**205**) R= OH, R_1= OMe, R_2= H
(**206**) R= OH, R_1= OMe, R_2= Me
(**210**) R= R_1= H, R_2= Me

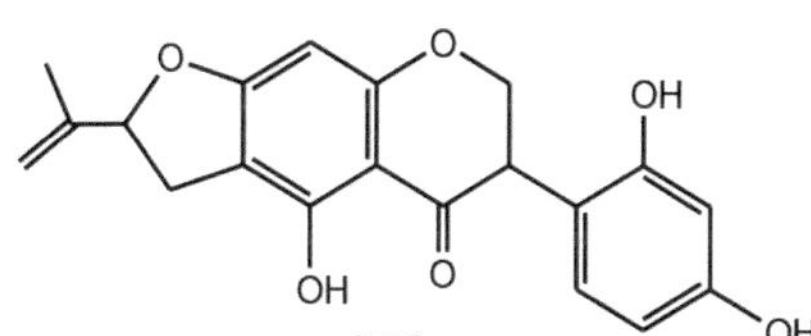

(**146**)

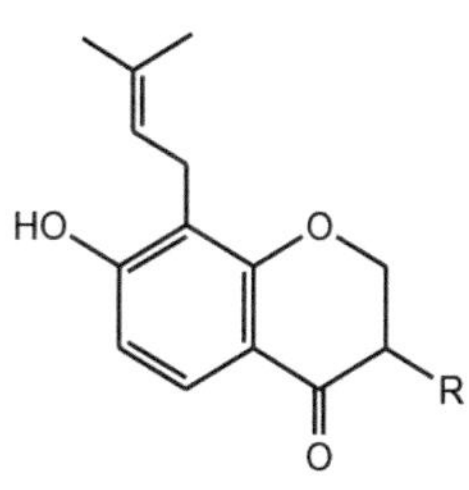

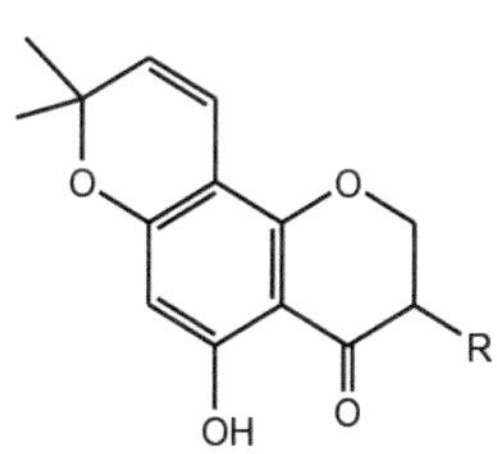

(**176**) R
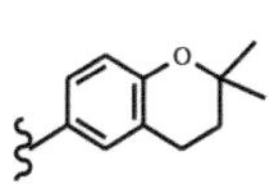

(**186**) R
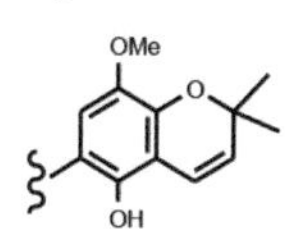

(**177**) R
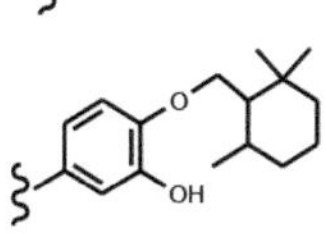

(**187**) R
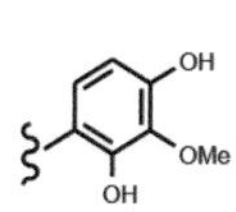

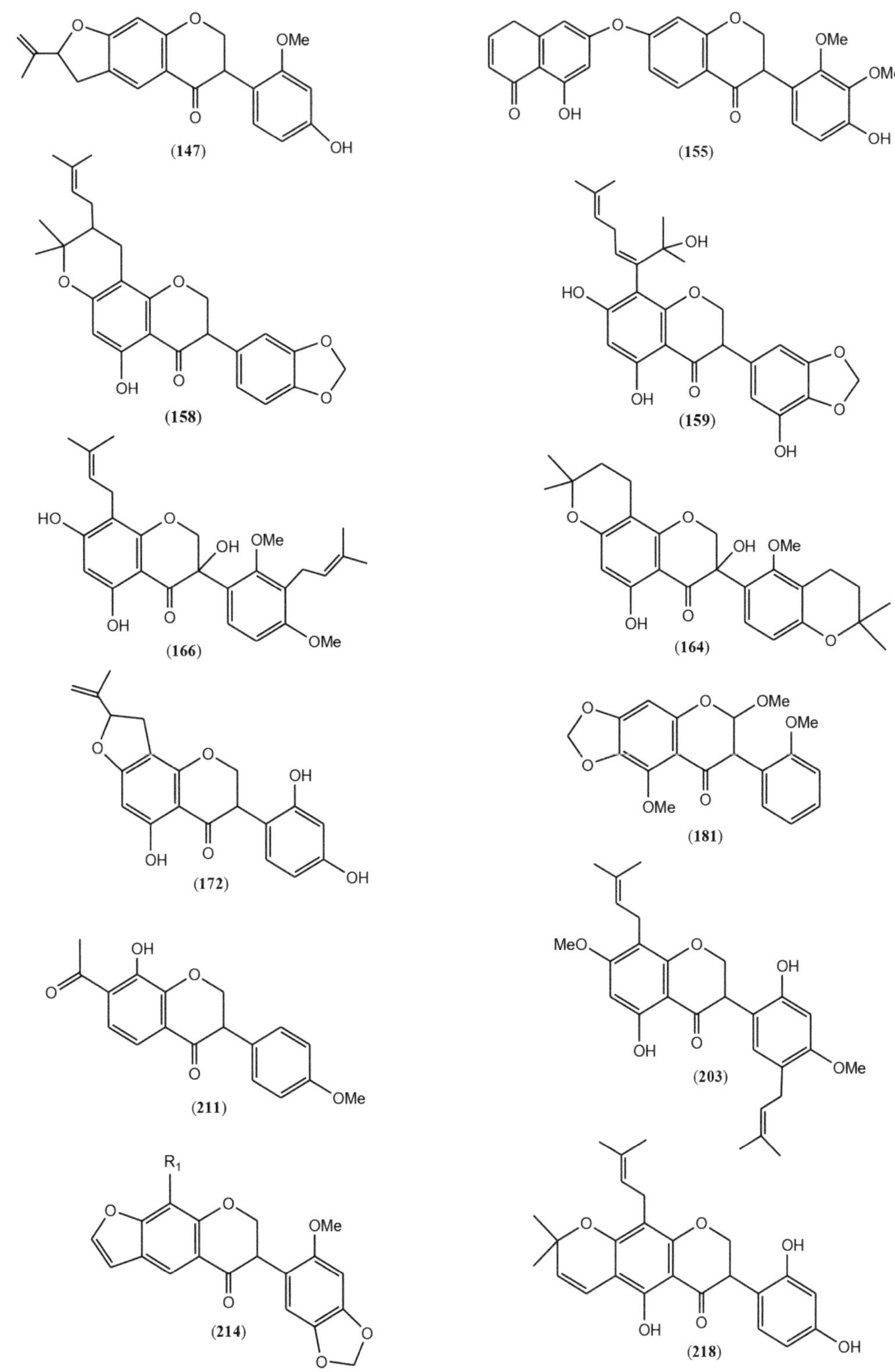

Figure 4. Isolated isoflavanones from natural sources since 2000 to date.

http://dx.doi.org/10.5772/intechopen.68701

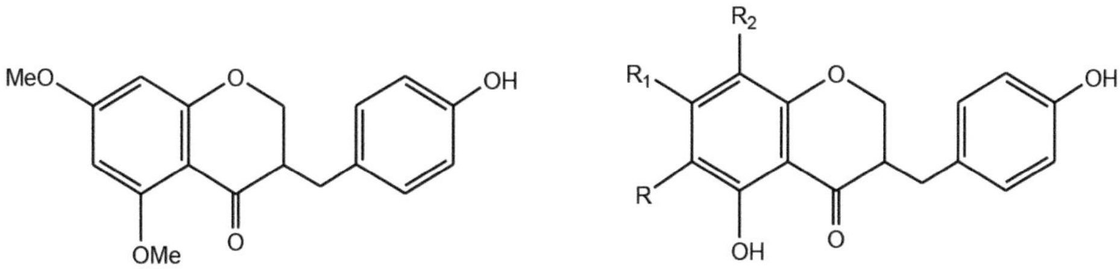

(219)

(220) R= OH, R_1= OMe, R_2= H
(221) R= H, R_1= R_2= OMe
(222) R= R_2=H, R_1= OH

Figure 5. Isolated homoisoflavonoids from natural sources since 2000 to date.

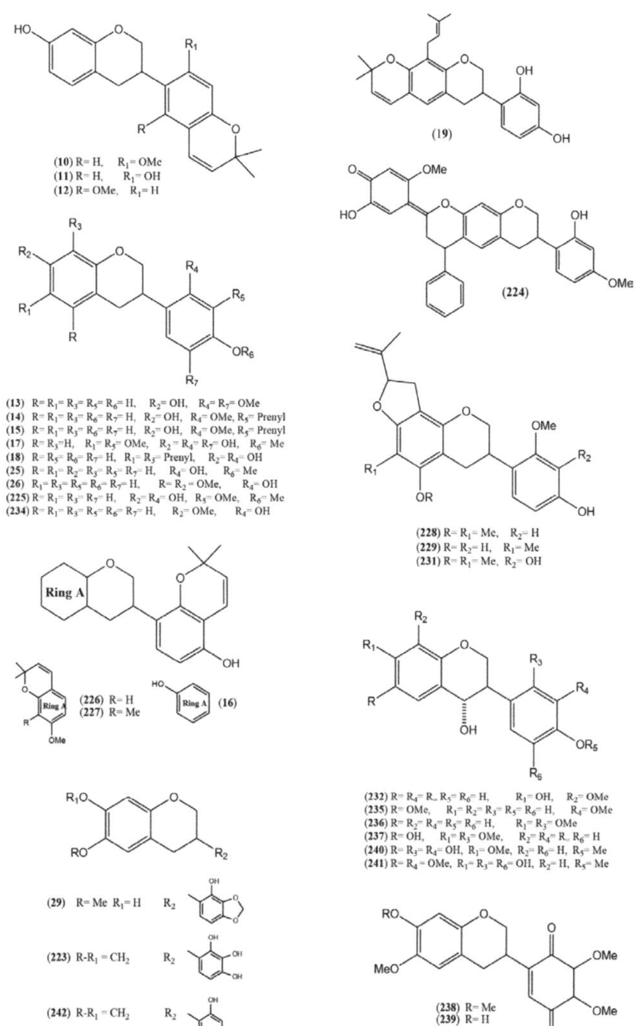

Figure 6. Isolated isoflavans from natural sources since 2000 to date.

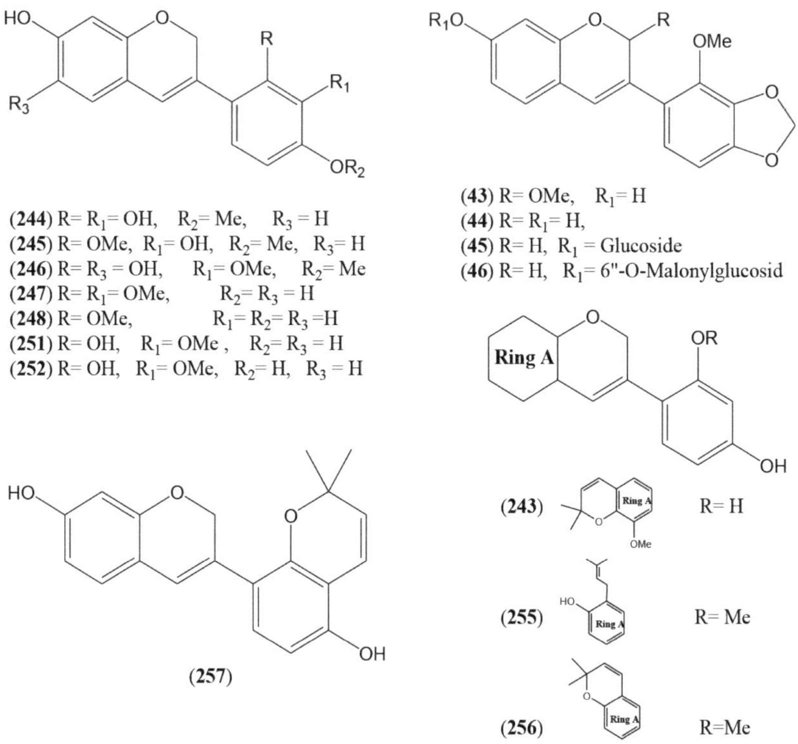

Figure 7. Isolated isoflavenes from natural sources since 2000 to date.

5. Biological activities

Isoflavonoids are reported to have a variety of bioprotective effects, including antioxidant, antimutagenic, anticarcinogenic and antiproliferative activities. Isoflavonoids may protect the body from hormone-related cancers, like breast, endometrial (uterine) and prostatic [115–119]. Isoflavonoids have gained a lot of public interest due to the possible correlation between their dietary consumption and health beneficial effects toward osteoporosis and post-menopausal symptoms [120, 121].

Among the isoflavonoids isolated from dothideomycetes fungus CMU-99, Biochanin A (**59**) showed weak cytotoxic activity against lung cancer cells (NCI-H137) and noncancerous Vero cells. Dothideoisoflavanone (**212**) exhibited cytotoxic effect against oral human carcinoma (KB) but was non-toxic against noncancerous Vero cells [9]. Among the isoflavonoids isolated from *Erythrina addisoniae*, Echrenone b10 (**100**) was found to be more than three times as potent as tamoxifen against MCF7/ADR and MDA-MB-231. Erythraddison III (**182**) was twice as potent as tamoxifen [23]. The isoflavanone Ormosinol (191) significantly inhibited adenocarcinomic human alveolar basal epithelial cells (A549) and human hepatic cell line (HepG2) [25]. Neobavaisoflavone (**110**) and Sigmoidin H (**207**) were selectively active *in vitro* against the resistant cancer cells 6/9, 4/9, CCRF-CEM, HCT116 (p53+/+),

MDA-MB-231-BCRP and U87MG [40]. Platyisoflavanone A (**32**) showed cytotoxic effect against noncancerous Vero cells [41]. 2″,6″-O-diacetyloninin (**54**) was active against human stomach carcinoma (Hs 740.T, Hs 756 T), breast adenocarcinoma (Hs 578 T, Hs 742.T) and prostate carcinoma (DU 145, LNCaP-FGC) cell lines [70]. Pierreione A (**79**) and Pierreione B (**93**) demonstrated selective toxicity to solid tumor cell lines with minimal cytotoxicity [77]. Isosideroxylin (**117**) was selectively active against the against ER^- MDA-MB-231 breast cancer cell line [84]. (3*R*) 5′,8-Di-(γ,γ-dimethylallyl)-2′,5-dihydroxyl-4′,7-dimethoxyl-isoflavanone (**203**) isolated from *Uraria clarkei* possessed good activity against the tested Hela, K562 and HL60 cell lines [101].

Haginin E (Phenoxodiol) (**254**) inhibits cell proliferation of a wide range of human cancer cell lines including leukemia, breast and prostate carcinomas, and is 5–20 times more potent than genistein [122]. Primary ovarian cancer cells resistant to conventional chemotherapy undergo apoptosis following Haginin E (Phenoxodiol) (**254**) treatment. Haginin E (Phenoxodiol) (**254**) is an efficient inducer of cell death in ovarian cancer cells and sensitizes the cancer cells to Fas-mediated apoptosis [123]. Haginin E (Phenoxodiol) (**254**) also exhibits significant ability to induce cell death in the prostate cancer cell lines LNCaP, DU145 and PC3 that utilize different signaling pathways than those reported in ovarian cancer studies [124]. Haginin E (Phenoxodiol) (**254**) development as an antitumor drug was based to a large extent on its low toxicity in normal tissues, but potent topoisomerase-II inhibitory effects in rapidly dividing tumor cells. This advantage led to its fast-track FDA approval for Phase II/III clinical trials [125].

Platyisoflavanone A (**32**) showed antibacterial activity against *Mycobacterium tuberculosis* (TB) in the microplate alamar blus assay (MABA) [41]. Isoflavonoids isolated from roots of *Erythrina zeyheri* were tested against methicillin-resistant *Staphylococcus aureus* (MRSA). Anti-MRSA potency of the isoflavan Eryzerin C (**18**) was the highest followed by Eryzerin D (**19**) [44]. 5,7,3′-Trihydroxy-4′-methoxy-6,5′-di(γ, γ-dimethylallyl)-isoflavanone (**173**) isolated from *Erythrina costaricensis* was also active on MRSA [94]. The two isoflavans 3*S* (+) 2′-O-Methylphaseollidinisoflavan (**15**) and 3*R*(-)Erythbidin A (**16**) isolated from *E. caffra* as well as the two isoflavanones 5,7-Dihydroxy-2′-methoxy-3′,4′-methylenedioxyisoflavanone (**31**) and 4′,5-Dihydroxy-2′,3′-dimethoxy-7-(5-hydroxyoxychromen-7yl)-isoflavanone (**155**) isolated from *Uraria picta* were active against *S. aureus* [45, 55]. The isoflavone Neoraudiol (**111**) displayed antimicrobial activity on *Bacillus subtilis, Salmonella typhii* and *Candida albicans* [52]. Lachnoisoflavone A (**91**) from *Crotalaria lachnophora* showed moderate inhibitory activities against *Escherichia coli* and *Klebsiella pneumonia* [76].

Isoflavanones from the Stem of *Cassia siamea* were evaluated for their anti-tobacco mosaic virus (Anti-TMV) activities [102]. (3*R*)7,2′,4′-Trihydroxy-3′-methoxy-5-methoxycarbonylisoflavanone (**205**) was the most active among the tested compounds [102]. In addition to anti-TMV, cordifoliflavanes A (**236**) cordifoliflavanes B (**237**) expressed anti-HIV-1 activities [109].

As a part of plant phenolics, isoflavonoids are expected to have antioxidant activities. Ormosinol (**191**) showed significant antioxidant activity against DPPH radicals [25]. The isoflavene Haginin A (**247**) and the isoflavonones dalbergioidin (**153**) showed antioxidant properties in both 2,2′-Azino-bis(3-ethylbenzthiazoline-6-sulfonic acid) (ABTS) and 1,1-Diphenyl-2-picryl-hydrazyl (DPPH) assays [29]. Seputheisoflavone (**132**) from *Ptycholobium contortum* was active in the ABTS assay [87]. Isoflavones from the *Astragalus mongholicus* were examined for antioxidant potential in DPPH assay. Results indicated that Calycosin (**81**) and Calycosin-7-O-glucopyranoside (**108**) are more active than Formononetin (**58**) [80].

The *in vitro* antiprotozoal activity of isoflavan quinines from *Abrus precatorius* was tested against *Plasmodium falciparum* (K1 strain), *Trypanosoma brucei rhodesiense* (STIB 900 strain), *Trypanosoma cruzi* (Tulahuen strain C2C4 w/LacZ) and *Leishmania donovani* (strain MHOM/ET/67/L82). Abruquinone D (**239**) and abruquinone K (**241**) were the most active against *T. brucei rhodesiense* [38]. Pumilanol (**29**), an isoflavan from *Tephrosia pumila*, exhibited significant antiprotozoal activities against *T. brucei rhodesiense, T. cruzi* and *L. donovani* [46]. Sophoronol C (**164**) and Sophoronol E (**166**) exhibited moderate anitplasmodial activity against the CQS D10 strain of *P. falciparum* [91].

Daidzein (**64**) and Daidzin (**65**) possess a vasorelaxant action through opening of K^+ channels and inhibition of Ca^{2+} influx in the vascular smooth muscle cells. This cerebral vasodilator activity may be beneficial to patients with obstructive cerebrovascular diseases [126].

Other studies reported on the effects of isoflavonoids on specific enzymes are presented in **Table 7**.

Compound name	Activity	Significance	Ref.
Tlatlancuayin (**84**) 2′,2,5-Trimethoxy-6,7-methylenedioxyisoflavanone (**181**)	a-glucosidase inhibitory	Weak	[5]
Achyranthoside A (**118**) Achyranthoside B (**119**)	Lipopolysaccharide (LPS)-induced nitric oxide (NO) production	Significant inhibition	[6]
Erysubin F (**21**), Erythraddison II (**99**) Echrenone b10 (**100**), Erythraddison III (**182**) Erythraddison IV (**183**)	Protein tyrosine phosphatase 1B (PTP1B)	Significant inhibition	[23]
Sophoraisoflavanone A (**169**) Kenusanone H (**170**)	Alcohol dehydrogenase (ADH) Aldehyde dehydrogenase (ALDH)	Significant activation	[92]
Glabrene (**257**)	Tyrosinase inhibition	Significant inhibition	[114]

Table 7. Effect of isoflavonoids on specific enzyme activities.

Dedication

Professor Dr. Ahmed A. Seif El-Dien, Department of Pharmacognosy, College of Pharmacy, Alexandria University, Alexandria 21215, Egypt. (Born in: 9 September 1948–passed away in: 19 December 2016).

Author details

Ahmed I. Foudah[1] and Maged Saad Abdel-Kader[1,2]*

*Address all correspondence to: mpharm101@hotmail.com

1 Department of Pharmacognosy, College of Pharmacy, Prince Sattam Bin Abdulaziz University, Al-Kharj, Saudi Arabia

2 Department of Pharmacognosy, College of Pharmacy, Alexandria University, Alexandria, Egypt

References

[1] Perkin AG, Newbury FG. LXXIX.—The colouring matters contained in dyer's broom (*Genista tinctoria*) and heather (*Calluna vulgaris*). Journal of the Chemical Society, Transactions. 1899;**75**:830-839. DOI: 10.1039/CT8997500830

[2] Baker W, Robinson R. CCCLIX.—Synthetical experiments in the isoflavone group. Part II. A synthesis of methylgenistein (methylprunetol) dimethyl ether and the constitution of prunetol (genistein). Journal of the Chemical Society (Resumed). 1926;**129**:2713-2720. DOI: 10.1039/JR9262902713

[3] Walter ED. Genistin (an isoflavone glucoside) and its aglucone, genistein, from soybeans. Journal of the American Chemical Society. 1941;**63**(12):3273-3276. DOI: 10.1021/ja01857a013

[4] Veitch NC. Isoflavonoids of the leguminosae. Natural Product Reports. 2009;**26**(6):776-802. DOI: 10.1039/c3np70024k

[5] Valentová M, Marek R, Švajdlenka E, Kubínová R, Suchý V. A new isoflavanone from *Iresine herbstii*. Fitoterapia. 2011;**82**(2):272-275. DOI: 10.1016/j.fitote.2010.10.010

[6] Tan Y, An N, Li Y, Cheng S, Zhang J, Zhang X, Li Y. Two new isoflavonoid glucosides from the roots of *Achyranthes bidentata* and their activities against nitric oxide production. Phytochemistry Letters. 2016;**17**:187-189. DOI: 10.1016/j.phytol.2016.07.028

[7] Şöhretoğlu D, Sterner O. Isoflavonoids, flavonoids and flavans from *Potentilla astracanica*. Biochemical Systematics and Ecology. 2011;**39**(4):666-668. DOI: 10.1016/j.bse.2011.05.020

[8] Benavides A, Bassarello C, Montoro P, Vilegas W, Piacente S, Pizza C. Flavonoids and isoflavonoids from *Gynerium sagittatum*. Phytochemistry. 2007;**68**(9):1277-1284. DOI: 10.1016/j.phytochem.2007.03.007

[9] Boonyaketgoson S, Trisuwan K, Bussaban B, Rukachaisirikul V, Phongpaichit S. Isoflavanone and xanthone derivatives from Dothideomycetes fungus CMU-99. Tetrahedron Letters. 2015;**56**(9):1057-1059. DOI: 10.1016/j.tetlet.2015.01.088

[10] Oldoni TL, Cabral IS, d'Arce MA, Rosalen PL, Ikegaki M, Nascimento AM, Alencar SM. Isolation and analysis of bioactive isoflavonoids and chalcone from a new type of Brazilian propolis. Separation and Purification Technology. 2011;**77**(2):208-213. DOI: 10.1016/j.seppur.2010.12.007

[11] Simons R, Gruppen H, Bovee TF, Verbruggen MA, Vincken JP. Prenylated isoflavonoids from plants as selective estrogen receptor modulators (phytoSERMs). Food & Function. 2012;**3**(8):810-827. DOI: 10.1039/c2fo10290k

[12] Hirayama T. Epidemiology of prostate cancer with special reference to the role of diet. Journal of the National Cancer Institute Monographs. 1979;**53**:149-155. PMID:537622

[13] Reiter R, Paredes S, Korkmaz A, Jou MJ, Tan DX. Melatonin combats molecular terrorism at the mitochondrial level. Interdisciplinary Toxicology. 2008;**1**(2):137-149. DOI:10.2478/v10102-010-0030-2

[14] Boue SM, Cleveland TE, Carter-Wientjes C, Shih BY, Bhatnagar D, McLachlan JM, Burow ME. Phytoalexin-enriched functional foods. Journal of Agricultural and Food Chemistry. 2009;**57**(7):2614-2622. **DOI:** 10.1021/jf8040403

[15] Crozier A, Jaganath IB, Clifford MN. Dietary phenolics: Chemistry, bioavailability and effects on health. Natural Product Reports. 2009;**26**(8):1001-1043. DOI: 10.1039/b802662a

[16] Han RM, Tian YX, Liu Y, Chen CH, Ai XC, Zhang JP, Skibsted LH. Comparison of flavonoids and isoflavonoids as antioxidants. Journal of Agricultural and Food Chemistry. 2009;**57**(9):3780-3785. DOI: 10.1021/jf803850p

[17] Dixon RA, Pasinetti GM. Flavonoids and isoflavonoids: From plant biology to agriculture and neuroscience. Plant Physiology. 2010;**154**(2):453-457. DOI: 10.1104/pp.110.161430

[18] Matsumura E, Matsuda M, Sato F, Minami H, Ramawat KG, Mérillon JM. Isoflavonoids. In: Natural Products: Phytochemistry, Botany and Metabolism of Alkaloids, Phenolics and Terpenes. Springer-Verlag Berlin Heidelberg; 2013. pp. 1849-1865. ISBN: 978-3-642-22143-9 (Print) 978-3-642-22144-6 (Online).ch60

[19] Abdel-Kader MS, Basudan OA, Parveen M, Amer ME. A new 3-arylcoumarin from the roots of an Egyptian collection of *Lotus polyphyllos*. Natural Product Research. 2008;**22**(5):448-452. DOI: 10.1080/14786410701591812

[20] Ma X, Tu P, Chen Y, Zhang T, Wei Y, Ito Y. Preparative isolation and purification of isoflavan and pterocarpan glycosides from *Astragalus membranaceus* Bge. var. mongholicus (Bge.) Hsiao by high-speed counter-current chromatography. Journal of Chromatography A. 2004;**1023**(2):311-315. PMID:14753698

[21] Alqasoumi SI, Al-Rehaily AJ, Abdel-Kader MS. Two isoflavans and a 3-arylcoumarin from the roots of Lotus lalambensis *growing in Saudi Arabia*. Revista Latinoamericana de Química. 2013;**41**(1):61-67. ISSN 0370-5943

[22] Barragan-Huerta BE, Peralta-Cruz J, González-Laredo RF, Karchesy J. Neocandenatone, an isoflavan-cinnamylphenol quinone methide pigment from *Dalbergia congestiflora*. Phytochemistry. 2004;**65**(7):925-928. DOI: 10.1016/j.phytochem.2003.11.011

[23] Nguyen PH, Sharma G, Dao TT, Uddin MN, Kang KW, Ndinteh DT, Mbafor JT, Oh WK. New prenylated isoflavonoids as protein tyrosine phosphatase 1B (PTP1B) inhibitors from *Erythrina addisoniae*. Bioorganic & Medicinal Chemistry. 2012;**20**(21):6459-6464. DOI: 10.1016/j.bmc.2012.08.024

[24] Ito C, Itoigawa M, Kanematsu T, Ruangrungsi N, Mukainaka T, Tokuda H, Nishino H, Furukawa H. Isoflavonoids from *Dalbergia olivari*. Phytochemistry. 2003;**64**(7):1265-1268. PMID:14599524

[25] Feng S, Hao J, Xu Z, Chen T, Qiu SX. Polyprenylated isoflavanone and isoflavonoids from *Ormosia henryi* and their cytotoxicity and anti-oxidation activity. Fitoterapia. 2012;**83**(1):161-165. DOI: 10.1016/j.fitote.2011.10.007

[26] Ibrahim SR, Mohamed GA, Zayed MF, Ross SA. 8-Hydroxyirilone 5-methyl ether and 8-hydroxyirilone, new antioxidant and α-amylase inhibitors isoflavonoids from *Iris germanica* rhizomes. Bioorganic Chemistry. 2017;**70**:192-198. DOI: 10.1016/j.bioorg.2016. 12.010

[27] Abdel-Kader MS. Two isoflavonoid glucoside derivatives from *Ononis serrata* growing in Egypt. Natural Product Sciences. 2004;**10**(6):321-324. ISSN: 1226-3907

[28] Abdel-Kader MS, Amer ME, Tang S, Kingston DG. Two new isoflavone derivatives from the roots of an Egyptian collection of *Lotus polyphyllos*. Natural Product Sciences. 2006;**20**(10):922-926. DOI: 10.1080/14786410500364627

[29] Lee MY, Kim JH, Choi JN, Kim J, Hwang GS, Lee C. The melanin synthesis inhibition and radical scavenging activities of compounds isolated from the aerial part of *Lespedeza cyrtobotrya*. Journal of Microbiology and Biotechnology. 2010;**20**(6):988-994. PMID:20622497

[30] Veitch NC, Stevenson PC. 2-Methoxyjudaicin, an isoflavene from the roots of *Cicer bijugum*. Phytochemistry. 1997;**44**(8):1587-1589. DOI: 10.1016/S0031-9422(96)00757-1

[31] Stevenson PC, Veitch NC. Isoflavenes from the roots of *Cicer judaicum*. Phytochemistry. 1996;**43**(3):695-700. DOI: 10.1016/0031-9422(96)00346-9

[32] Xu YJ, Foubert K, Dhooghe L, Lemière F, Maregesi S, Coleman CM, Zou Y, Ferreira D, Apers S, Pieters L. Rapid isolation and identification of minor natural products by LC–MS, LC–SPE–NMR and ECD: Isoflavanones, biflavanones and bisdihydrocoumarins from *Ormocarpum kirkii*. Phytochemistry. 2012;**79**:121-128. DOI: 10.1016/j.phytochem.2012.04.004

[33] Zhao M, Duan JA, Che CT. Isoflavanones and their O-glycosides from *Desmodium styracifolium*. Phytochemistry. 2007;**68**(10):1471-1479. DOI: 10.1016/j.phytochem.2007.02.01

[34] Mun SC, Mun GS. Dynamics of phytoestrogen, isoflavonoids, and its isolation from stems of *Pueraria lobata* (Willd.) Ohwi growing in Democratic People's Republic of Korea. Journal of Food and Drug Analysis. 2015;**23**(3):538-544. DOI: 10.1016/j.jfda.2015.04.003

[35] Miyase T, Ueno A, Noro T, Fukushima S. Studies on the constituentes of *Lespedeza cyrtobotrya* Miq. I. The structures of a new chalcone and two new isoflav-3-ens. Chemical and Pharmaceutical Bulletin. 1980;**28**(4):1172-1177. DOI: 10.1248/cpb.28.1172

[36] Miyase T, Sano M, Nakai H, Muraoka M, Nakazawa M, Suzuki M, Yoshino K, Nishihara Y, Tanai J. Antioxidants from *Lespedeza homoloba*.(I). Phytochemistry. 1999;**52**(2):303-310. DOI: 10.1016/S0031-9422(99)00195-8

[37] Bedane KG, Masesane IB, Majinda RR. New isoflavans from the root bark of *Erythrina livingstoniana*. Phytochemistry Letters. 2016;**17**:55-58. DOI: 10.1016/j.phytol.2016.07.023

[38] Hata Y, Ebrahimi SN, De Mieri M, Zimmermann S, Mokoka T, Naidoo D, Fouche G, Maharaj V, Kaiser M, Brun R, Potterat O. Antitrypanosomal isoflavan quinones from *Abrus precatorius*. Fitoterapia. 2014;**93**:81-87. DOI: 10.1016/j.fitote.2013.12.015

[39] Wanjala CC, Majinda RR. A new isoflavanone from the stem bark of *Erythrina latissima*. Fitoterapia. 2000;**71**(4):400-405. PMID:10925011

[40] Kuete V, Sandjo LP, Kwamou GM, Wiench B, Nkengfack AE, Efferth T. Activity of three cytotoxic isoflavonoids from *Erythrina excelsa* and *Erythrina senegalensis* (neobavaisoflavone, sigmoidin H and isoneorautenol) toward multi-factorial drug resistant cancer cells. Phytomedicine. 2014;**21**(5):682-688. DOI: 10.1016/j.phymed.2013.10.017

[41] Gumula I, Heydenreich M, Derese S, Ndiege IO, Yenesew A. Four isoflavanones from the stem bark of *Platycelphium voënse*. Phytochemistry Letters. 2012;**5**(1):150-154. DOI: 10.1016/j.phytol.2011.11.012

[42] Tanaka H, Etoh H, Watanabe N, Shimizu H, Ahmad M, Rizwani GH. Erysubins C–F, four isoflavonoids from *Erythrina suberosa* var. glabrescences. Phytochemistry. 2001;**56**(7):769-773. DOI: 10.1016/S0031-9422(00)00441-6

[43] Tanaka H, Oh-Uchi T, Etoh H, Shimizu H, Tateishi Y. Isoflavonoids from the roots of *Erythrina poeppigiana*. Phytochemistry. 2002;**60**(8):789-794. DOI: 10.1016/S0031-94 22(02) 00164-4

[44] Tanaka H, Oh-Uchi T, Etoh H, Sako M, Asai F, Fukai T, Sato M, Murata J, Tateishi Y. Isoflavonoids from roots of *Erythrina zeyheri*. Phytochemistry. 2003;**64**(3):753-758. DOI: 10.1016/S0031-9422(03)00390-X

[45] El-Masry S, Hamoda HM, Zaatout HH, Abdel-Kader MS. Constituents of *Erythrina caffra* stem bark grown in Egypt. Natural Product Reports. 2010;**16**(4):211-216

[46] Ganapaty S, Pannakal ST, Srilakshmi GV, Lakshmi P, Waterman PG, Brun R. Pumilanol, an antiprotozoal isoflavanol from *Tephrosia pumila*. Phytochemistry Letters. 2008;**1**(4):175-178. DOI: 10.1016/j.phytol.2008.09.006

[47] Botta B, Gacs-Baitz E, Vinciguerra V, Delle Monache G. Three isoflavanones with cannabinoid-like moieties from *Desmodium canum*. Phytochemistry. 2003;**64**(2):599-602. DOI:10.1016/S0031-9422(03)00201-2

[48] Guchu SM, Yenesew A, Tsanuo MK, Gikonyo NK, Pickett JA, Hooper AM, Hassanali A. C-methylated and C-prenylated isoflavonoids from root extract of *Desmodium uncinatum*. Phytochemistry. 2007;**68**(5):646-651. DOI: 10.1016/j.phytochem.2006.11.035

[49] Yenesew A, Midiwo JO, Heydenreich M, Schanzenbach D, Peter MG. Two isoflavanones from the stem bark of *Erythrina sacleuxii*. Phytochemistry. 2000;**55**(5):457-459. DOI: 10.1016/S0031-9422(00)00349-6

[50] El-Masry S, Amer ME, Abdel-Kader MS, Zaatout HH. Prenylated flavonoids of *Erythrina lysistemon* grown in Egypt. Phytochemistry. 2002;**60**(8):783-787. DOI: 10.1016/S0031-9422(02)00202-9

[51] Brink AJ, Rall GJ, Engelbrecht JP. Phenolic neorautanenia isoflavanoids: The isolation and structures of neorauflavene,(−)-neorauflavane and neoraufurane, three novel isoflavanoids from *Neorautanenia edulis*. Tetrahedron. 1974;**30**(2):311-314. DOI: 10.1016/S0040-4020(01)91463-2

[52] Lasisi AA, Adesomoju A. Neoraudiol, a new isoflavonoid and other antimicrobial constituents from the tuberous root of *Neorautanenia mitis* (A. Rich) Verdcourt. Journal of Saudi Chemical Society. 2015;**19**(4):404-409. DOI: 10.1016/j.jscs.2012.04.011

[53] Pluempanupat S, Kumrungsee N, Pluempanupat W, Ngamkitpinyo K, Chavasiri W, Bullangpoti V, Koul O. Laboratory evaluation of *Dalbergia oliveri* (Fabaceae: Fabales) extracts and isolated isoflavonoids on *Aedes aegypti* (Diptera: Culicidae) mosquitoes. Industrial Crops and Products. 2013;**44**:653-658. DOI: 10.1016/j.indcrop.2012.09.006

[54] Giachi I, Manunta A, Morelli I, Pistelli L. Flavonoids and isoflavonoids from *Genista morisii*. Biochemical Systematics and Ecology. 2002;**30**(8):801-803. DOI: 10.1016/S0305-1978(01)00116-8

[55] Rahman MM, Gibbons S, Gray AI. Isoflavanones from *Uraria picta* and their antimicrobial activity. Phytochemistry. 2007;**68**(12):1692-1697. DOI: 10.1016/j.phytochem.2007.04.015

[56] Arnone A, Camarda L, Merlini L, Nasini G, Taylor DA. Isoflavonoid constituents of the West African red wood *Baphia nitida*. Phytochemistry. 1981;**20**(4):799-801. DOI: 10.1016/0031-9422(81)85178-3

[57] Rostagno MA, Araujo JMA, Sandi D. Supercritical fluid extraction of isoflavones from soybean flour. Food Chemistry. 2002; 78: 111-117. DOI: 10.1016/S0308-8146(02)00106-1

[58] Ye H, Chen L, Li Y, Peng A, Fu A, Song H, Tang M, Luo H, Luo Y, Xu Y, Shi J. Preparative isolation and purification of three rotenoids and one isoflavone from the seeds of *Millettia pachycarpa* Benth by high-speed counter-current chromatography. Journal of Chromatography A. 2008;**1178**(1):101-107. DOI: 10.1016/j.chroma.2007.11.060

[59] Degenhardt A, Winterhalter P. Isolation and purification of isoflavones from soy flour by high-speed countercurrent chromatography. European Food Research and Technology. 2001;**213**(4):277-280.DOI: 10.1007/s002170100386

[60] Mabry TJ, Markham KR, Thomas MB. Reagents and procedures for the ultraviolet spectral analysis of flavonoids. In: The Systematic Identification of Flavonoids. Springer-Verlag Berlin Heidelberg New York; 1970. pp. 35-40. ISBN: 978-3-642-88460-3 (Print) 978-3-642-88458-0 (Online)

[61] Tanaka H, Oh-Uchi T, Etoh H, Sako M, Sato M, Fukai T, Tateishi Y. An arylbenzofuran and four isoflavonoids from the roots of *Erythrina poeppigiana*. Phytochemistry. 2003;**63**(5):597-602. DOI: 10.1016/S0031-9422(03)00184-5

[62] Tsanuo MK, Hassanali A, Hooper AM, Khan Z, Kaberia F, Pickett JA, Wadhams LJ. Isoflavanones from the allelopathic aqueous root exudate of *Desmodium uncinatum*. Phytochemistry. 2003;**64**(1):265-273. DOI: 10.1016/S0031-9422(03)00324-8

[63] Koorbanally C, Mulholland DA, Crouch NR. Norlignans and homoisoflavanones from two South African *Drimiopsis* species (Hyacinthaceae: Hyacinthoideae). Biochemical Systematics and Ecology. 2006;**34**(7):588-592. DOI: 10.1016/j.bse.2005.12.011

[64] Slade D, Ferreira D, Marais JPJ. Circular dichroism, a powerful tool for the assessment of absolute configuration of flavonoids. Phytochemistry. 2005;**66**(18):2177-2215. DOI: 10.1016/j.phytochem.2005.02.002

[65] Won D, Shin B-K, Kang S, Hur H-G, Kim M, Han J. Absolute configurations of isoflavan-4-ol stereoisomers. Bioorganic & Medicinal Chemistry Letters. 2008;**18**(6):1952-1957. DOI: 10.1016/j.bmcl.2008.01.116

[66] Yi J, Du G, Yang Y, Li Y, Li Y, Guo F. Chiral discrimination of natural isoflavanones using (*R*)- and (*S*)-BINOL as the NMR chiral solvating agents. Tetrahedron Asymmetry. 2016;**27**:1153-1159. DOI: 10.1016/j.tetasy.2016.09.002

[67] Adinolfi M, Lanzetta R, Laonigro G, Parrilli M, Breitmaier E. ^{1}H and ^{13}C chemical shift assignments of homoisoflavanones. Magnetic Resonance in Chemistry. 1986;**24**(8):663-666. DOI: 10.1002/mrc.1260240806

[68] Friebolin H. Basic One- and Two-Dimensional NMR Spectroscopy. 5th ed.Weinhiem: Wiley-VCH; 2010. ISBN: 978-3-527-32782-9

[69] Chaudhary BA, Irshad A. Isolation of an isoflavone from *Polygala stenopetala*. Journal of Pure and Applied Sciences. 2001;**20**(1):71-74

[70] Hosny M, Rosazza JPN. New isoflavone and triterpene glycosides from soybeans. Journal of Natural Products. 2002;**65**(6):805-813. PMID: 12088420

[71] Qu L-P, Fan G-R, Peng J-Y, Mi H-M. Isolation of six isoflavones from *Semen sojae praeparatum* by preparative HPLC. Fitoterapia. 2007;**78**(3):200-204. DOI: 10.1016/j.fitote.2006.11.002

[72] Reddy BAK, Khalivulla SI, Gunasekar D. A new prenylated isoflavone from *Tephrosia tinctoria*. Indian Journal of Chemistry. 2007;**46B**:366-369

[73] Khalivulla SI, Reddy BAK, Gunasekar D, Murthy MM, Rao TP, Blond A, Bodo B. A new C-geranylated isoflavone from *Dalbergia paniculata*. Natural Product Communications. 2007;**2**(11):1109-1111

[74] Djiogue S, Halabalaki M, Alexi X, Njamen D, Fomum ZT, Alexis MN, Skaltsounis AL. Isoflavonoids from *Erythrina poeppigiana*: Evaluation of their binding affinity for the estrogen receptor. Journal of Natural Products. 2009;**72**(9):1603-1607. DOI: 10.1021/np900271m

[75] Zhao S, Zhang L, Gao P, Shao Z. Isolation and characterisation of the isoflavones from sprouted chickpea seeds. Food Chemistry. 2009;**114**(3):869-873. DOI: 10.1016/j.foodchem.2008.10.026

[76] Awouafack MD, Spiteller P, Lamshöft M, Kusari S, Ivanova B, Tane P, Spiteller M. Antimicrobial Isopropenyl-dihydrofuranoisoflavones from *Crotalaria lachnophora*. Journal of Natural Products. 2011;**74**(2):272-278. DOI: 10.1021/np1005218

[77] Gao S, Xu Y-M, Valerioteand FA, Gunatilaka LAA. Pierreiones A–D, solid tumor selective pyranoisoflavones and other cytotoxic constituents from *Antheroporum pierrei*. Journal of Natural Products. 2011;**74**(4):852-856. DOI: 10.1021/np100763p

[78] Kowalska I, Jedrejek D, Ciesla L, Pecio L, Masullo M, Piacente S, Oleszek W, Stochmal A. Isolation, chemical and free radical scavenging characterization of phenolics from *Trifolium scabrum* L. Aerial Parts. Journal of Agricultural and Food Chemistry. 2013;**61**(18):4417-4423. DOI: 10.1021/jf400153c

[79] Kovalev VN, Zatyl'nikova OA, Kovalev SV. A new isoflavone from *Iris pseudacorus*. Chemistry of Natural Compounds. 2013;**49**(1):34-35. DOI: 10.1007/s10600-013-0499-y

[80] Yu D, Duan Y, Bao Y, Wei C, An L. Isoflavonoids from *Astragalus mongholicus* protect PC12 cells from toxicity induced by L-glutamate. Journal of Ethnopharmacology. 2005;**98**(1-2):89-94. DOI: 10.1016/j.jep.2004.12.027

[81] Zhang DY, Zu YG, Fu YJ, Luo M, Gu CB, Wang W, Yao XH. Negative pressure cavitation extraction and antioxidant activity of biochanin A and genistein from the leaves of *Dalbergiaodorifera* T. Chen. Separation and Purification Technology.2011;83:91-99. DOI: 10.1016/j.seppur.2011.09.017

[82] Ghribi L, Waffo-Téguo P, Cluzet S, Marchal A, Marques J, Mérillon J-M, Ben Jannet H. Isolation and structure elucidation of bioactive compounds from the roots of the Tunisian *Ononis angustissima* L. Bioorganic & Medicinal Chemistry Letters. 2015;**25**(18):3825-3830. DOI: 10.1016/j.bmcl.2015.07.076

[83] Ortega AR, Toscano RA, Hernández-Barragán A, Alvarez-Cisneros C, Joseph-Natha P. Structure elucidation of a new isoflavone by exclusive use of ^{1}H NMR measurements. Magnetic Resonance in Chemistry. 2015;**53**(10):860-865. DOI: 10.1002/mrc.4278

[84] Tian D, Porter JR. An isoflavone from *Leiophyllum buxifolium* and its antiproliferative effect. Journal of Natural Product. 2015;**78**(7):1748-1751. DOI: 10.1021/acs.jnatprod.5b00100

[85] Li L, Liu J-Z, Luo M, Wang W, Huang Y-Y, Efferth T, Wang H-M, Fu Y-J. Efficient extraction and preparative separation of four main isoflavonoids from *Dalbergia odorifera* T. Chen leaves by deep eutectic solvents-based negative pressure cavitation extraction followed by macroporous resin column chromatography. Journal of Chromatography B. 2016;**1033-1034**:40-48. DOI: 10.1016/j.jchromb.2016.08.005

[86] Bojase G, Wanjala CCW, Majinda RRT. Flavonoids from the stem bark of *Bolusanthus speciosus*. Phytochemistry. 2001;**56**(8):837-841. PMID:11324914

[87] Fotso GW, Maher FA, Ngnintedo D, Ango PY, Kapche DGFW, Ngameni B, Ngwenya B, Yeboah SO, Ngadjui BT, Andrae-Marobela K. Three new isoflavonoids with antioxidant properties from *Ptycholobium contortum* (N.E.Br.) Brummitt (Leguminosae). Phytochemistry Letters. 2015;**14**:254-259. DOI: 10.1016/j.phytol.2015.10.015

[88] Maximov OB, Kulesh NI, Stepanenko LS, Dmitrenok PS. New prenylated isoflavanones and other constituents of *Lespedeza bicolor*. Fitoterapia. 2004;**75**(1):96-98. DOI: 10.1016/j.fitote.2003.07.012

[89] Narayanan MC, Rao PR, Shanmugam NN, Gopalakrishnan SM, Devi K. Isolation and characterisation of bioactive isoflavonoids from the roots of *Dalbergia horrida*. Natural Product Research. 2007;**21**(10):903-909. DOI: 10.1080/14786410601130091

[90] Tanaka H, Hattori H, Oh-Uchi T, Sato M, Yamaguchi R, Sako M, Tateishi Y. Two new isoflavanones from *Erythrina costaricensis*. Journal of Asian Natural Products Research. 2008;**10**(10):983-987. DOI: 10.1080/10286020802217598

[91] Zhang GP, Xiao ZY, Rafique J, Arfan M, Smith PJ, Lategan CA, Hu LH. Antiplasmodial Isoflavanones from the Roots of Sophora mollis. Journal of Natural Product. 2009;**72**(7):1265-1268. DOI: 10.1021/np900144c

[92] Choi EJ, Kwon HC, Sohn YC, Nam CW, Park HB, Kim CY, Yang HO. Four flavonoids from *Echinosophora koreensis* and their effects on alcohol metabolizing enzymes. Archives of Pharmacal Research. 2009;**32**(6):851-855. DOI: 10.1007/s12272-009-1606-2

[93] Zappia G, Menendez MP, Sampaio de Andrade Lima C, Botta B. A *p*-quinol isoflavan and two new isoflavanones from *Desmodium canum*. Natural Product Research. 2009;**23**(7):665-671. DOI: 10.1080/14786410802362717

[94] Tanaka H, Hattori H, Oh-Uchi T, Sato M, Sako M, Tateishi Y, Rizwani GH. Three new isoflavanones from *Erythrina costaricensis*. Natural Product Research. 2009;**23**(12):1089-1094. DOI: 10.1080/14786410802292286

[95] Suman A, Ali M, Alam P. New prenylated isoflavanones from the roots of *Glycyrrhiza glabra*. Chemistry of Natural Compounds. 2009;**45**:487-491. DOI: 10.1007/s10600-009-9403-1

[96] Umehara K, Nemoto K, Matsushita A, Terada E, Monthakantirat O, De-Eknamkul W, Miyase T, Warashina T, Degawa M, Noguchi H. Flavonoids from the Heartwood of the Thai Medicinal Plant *Dalbergia parviflora* and their effects on estrogenic-responsive human breast cancer cells. Journal of Natural Products. 2009;**72**(12):2163-2168. DOI: 10.1021/np900676y

[97] Sheng Z, Qing T, Qingyao S, Guowei Q, Zhengwu S. Two new isoflavonones from radix *Campylotropis hirtella*. Huaxue xuebao (Acta Chimica Sinica). 2010;**68**(21):2227-2230

[98] Zhang R-T, Cheng G-G, Feng T, Cai X-H, Luo X-D. Four new isoflavanones from *Tadehagi triquetrum*. Natural Products and Bioprospecting. 2011;**1**:121-123. DOI: 10.1007/s13659-011-0033-5

[99] Tan Q, Zhang S, Shen Z. Flavonoids from the roots of *Campylotropis hirtella*. Planta Medica. 2011;**77**(16):1811-1817. DOI: 10.1055/s-0030-1271185

[100] Lee C, Lee JW, Jin Q, Jang DS, Lee S-J, Lee D, Hong JT, Kim Y, Lee MK, Hwang BY. Inhibitory constituents of the heartwood of *Dalbergia odorifera* on nitric oxide production in RAW 264.7 macrophages. Bioorganic & Medicinal Chemistry Letters. 2013;**23**(14):4263-4266. DOI: 10.1016/j.bmcl.2013.04.032

[101] Huang X-Z, Bai X-S, Liang H, Wang C, Li W-J, Guo J-M, Jiang Z-Y. Cytotoxic Isoflavanones from *Uraria clarkei*. Bulletin of the Korean Chemical Society. 2013;**34**(5):1421-1424. DOI: 10.5012/bkcs.2013.34.5.1421

[102] Hu Q-F, Niu D-Y, Zhou B, Ye Y-Q, Du G, Meng C-Y, Gao X-M. Isoflavanones from the stem of *Cassia siamea* and their Anti-tobacco mosaic virus activities. Bulletin of the Korean Chemical Society. 2013;**34**(10):3013-3016. DOI: .org/10.5012/bkcs.2013.34.10.3013

[103] Li X, Xuan B, Shou Q, Shen Z. New flavonoids from *Campylotropis hirtella* with immunosuppressive activity. Fitoterapia. 2014;**95**:220-228. DOI: 10.1016/j.fitote.2014.03.028

[104] Li Y-P, Li Y-K, Du G, Yang H-Y, Gao X-M, Hu Q-F. Isoflavanones from *Desmodium oxyphyllum* and their cytotoxicity. Journal of Asian Natural Products Research. 2014;**16**(7):735-740. DOI: 10.1080/10286020.2014.906406

[105] Tanaka H, Atsumi I, Hasegawa M, Hirata M, Sakai T, Sato M, Yamaguchi R, Tateishi Y, Tanaka T, Fukai T. Two new isoflavanones from the roots of *Erythrina variegate*. Natural Product Communications. 2015;**10**(3):499-501. PMID:25924538

[106] Lima CSdeA, Zappia G, Monache GD, Botta B. Uncommon 5-Methoxyisoflavans from *Desmodium canum*. European Journal of Organic Chemistry. 2006;**2006**(24):5445-5448. DOI: 10.1002/ejoc.200600742

[107] Saleem M, Akhter N, Ali MS, Nazir M, Riaz N, Moazzam M, Arshad M, Jabbar A. Structure determination of salisomide and salisoflavan, two new secondary metabolites from *Salsola imbricata*, by 1D and 2D NMR spectroscopy. Magnetic Resonance in Chemistry. 2009;**47**:263-265. DOI: 10.1002/mrc.2361

[108] Tezuka Y, Morikawa K, Li F, Auw L, Awale S, Nobukawa T, Kadota S. Cytochrome P450 3A4 inhibitory constituents of the wood of *Taxus yunnanensis*. Journal of Natural Products. 2011;**74**(1):102-105. DOI:10.1021/np100665j

[109] Gao X, Mu H, Li X, Yang G, Li G, Hu Q. Two new 4-hydroxyisoflavanes from the root of codonopsis cordifolioidea and their Anti-virus activities. Journal of the Chinese Chemical Society. 2012;**59**(4):540-543. DOI: 10.1002/jccs.201100525

[110] Awouafack MD, Tchuenguem RT, Ito T, Dzoyem JP, Tane P, Morita H. A new isoflavanol from the fruits of *Kotschya strigosa* (Fabaceae). Helvetica Chimica Acta. 2016;**99**(4):321-324. DOI: 10.1002/hlca.201500508

[111] Jurd L, Manners GD. Isoflavene, isoflavan, and flavonoid constituents of *Gliricidia speium*. Journal of Agricultural and Food Chemistry. 1977;**25**(4):723-726. DOI: 10.1021/jf60212a034

[112] Mitsuanga T, Kondo R, Imamura H. The chemistry of the color of wood. III. The isoflavonoids of the heartwood of murasakitagayasan (*Millettia* sp.). Mokuzai Gakkaishi Journal of the Japan Wood Research Society. 1987;**33**(3):234-238

[113] Miyase T, Ueno A, Noro T, Fukushima S. Studies on the Constituents of *Lespedeza cyrtobotrya* Miq. II. The structures of Haginin C, Haginin D and Lespedeol C. Chemical and Pharmaceutical Bulletin. 1981;**29**(8):2205-2209

[114] Nerya O, Vaya J, Musa R, Izrael S, Ben-Arie R, Tamir S. Glabrene and isoliquiritigenin as tyrosinase inhibitors from licorice roots. Journal of Agricultural and Food Chemistry. 2003;**51**(5):1201-1207. DOI: 10.1021/jf020935u

[115] Birt DF, Hendrich S, Wang W. Dietary agents in cancer prevention: flavonoids and isoflavonoids. Pharmacology & Therapeutics. 2001;**90**(2-3):157-177. DOI: 10.1016/S0163-7258(01)00137-1

[116] Miadokova E, Masterova I, Vlckova V, Duhova V, Toth J. Antimutagenic potential of homoisofl avonoids from *Muscari racemosum*. Journal of Ethnopharmacology. 2002;**18**:381-386. DOI: 10.1016/S0378-8741(02)00135-6

[117] Ryan-Borchers TA, Park JS, Chew BP, McGuire MK, Fournier LR, Beerman KA. Soy isoflavones modulate immune function in healthy postmenopausal women. American Journal of Clinical Nutrition. 2006;**83**:1118-1125

[118] Iwasaki M, Inoue M, Otani T, Sasazuki S, Kurashi N, Miura T, Yamoto S, Tsugane S. Plasma isoflavone level and subsequent risk of breast cancer among Japanese women: A nested case-control study from Japan Public Health Center-base prospective study group. Journal of Clinical Oncology. 2008;**26**(10):1677-1683. DOI: 10.1200/JCO.2007.13.9964

[119] Scarpato R, Paganucci L, Bertoli A, Fiore L, Pistelli L, Federico G. Licoflavone C attenuates the genotoxicity of cancer drugs in human peripheral lymphocytes. Phytotherapy Research. 2008;**22**(12):1650-1654. PMID: 16685055

[120] Messina M. Soy foods, isoflavones, and the health of postmenopausal women. .American Journal of Clinical Nutrition. 2014;**100**:1S-8S. DOI: 10.3945/ajcn.113.071464

[121] Wei P, Liu M, Chen Y, Chen DC. Systematic review of soy isoflavone supplements on osteoporosis in women. Asian Pacific Journal of Tropical Medicine. 2012;**5**(3):243-248. DOI: 10.1016/S1995-7645(12)60033-9

[122] Constantinou A, Husband A. Phenoxodiol (2H-1-benzopyran-7-0,1,3-(4-hydroxyphenyl)), a novel isoflavone derivative, inhibits DNA topoisomerase II by stabilizing the cleavable complex. Anticancer Research. 2002;**22**:2581-2585. PMID: 12529967

[123] Kamsteeg M, Rutherford T, Sapi E, Hanczaruk B, Shahabi S, Flick M, Brown D, Mor G. Phenoxodiol – an isoflavone analog – induces apoptosis in chemoresistant ovarian cancer cells. Oncogene. 2003;**22**:2611-2620. DOI: 10.1038/sj.onc.1206422

[124] Mahoney S, Arfuso F, Rogers P, Hisheh S, Brown D, Millward M, Dharmarajan A. Cytotoxic effects of the novel isoflavone, phenoxodiol, on prostate cancer cell lines. Journal of Biosciences. 2012;**37**:73-84. PMID:22357205

[125] Georgaki S, Skopeliti M, Tsiatas M, Nicolaou KA, Ioannou K, Husband A, Bamias A, Dimopoulos MA, Constantinou AI, Tsitsilonis OE. Phenoxodiol, an anticancer isoflavene, induces immunomodulatory effects *in vitro* and *in vivo*. Journal of Cellular and Molecular Medicine. 2009;**13**(9b):3929-3938. DOI: 10.1111/j.1582-4934.2009.00695.x

[126] Deng Y, Ng ESK, Yeung JHK, Kwan YW, Lau CBS, Koon JCM, Zhou L, Zuo Z, Leung PC, Fung KP, Lam FFY. Mechanisms of the cerebral vasodilator actions of isoflavonoids of Gegen on rat isolated basilar artery. Journal of Ethnopharmacology. 2012;**139**(1):294-304. DOI: 10.1016/j.jep.2011.11.021

Chapter 5

Homoisoflavonoids from *Caesalpinia* spp.: A Closer Look at Chemical and Biological Aspects

João Luiz Baldim, Welton Rosa,
Mário Ferreira Conceição Santos,
Daniela Aparecida Chagas-Paula,
João Henrique Ghilardi Lago and
Marisi Gomes Soares

Additional information is available at the end of the chapter

http://dx.doi.org/10.5772/67723

Abstract

Homoisoflavonoids are rare compounds distributed within a few families of plants including species from Fabaceae. The genus *Caesalpinia,* the main focus of this chapter, is a prolific source of these unique natural products. Homoisoflavonoids from *Caesalpinia* spp. are associated to ethnopharmacological uses for diverse purposes. In this sense, the following chapter sheds light on the occurrence, biosynthesis, isolation, synthesis, and structural analysis of these compounds from species of the genus *Caesalpinia* and their biological potential.

Keywords: *Caesalpinia,* Homoisoflavonoids, natural products, biological activities

1. Introduction

The genus *Caesalpinia* comprises more than 500 species around the world, existing essentially in tropical and subtropical zones. These species are correlated to ethnopharmacological uses due to their biological properties, which include analgesic, adaptogenic, antiangiogenic, antiulcer, anthelmintic, antibacterial, insecticidal, antifungal, anti-inflammatory, antipyretic, antioxidant, antiproliferative, antiviral, antimalarial, immunomodulatory, and immunosuppressive activities, as well as glutathione S-transferase (GST) inhibition, xanthine oxidase (XO) inhibition suppression of melanin synthesis, inhibition of viral neuraminidases, and other properties which will be further discussed ahead [1–5].

Associated to these biological properties, these plants are chemically composed by different classes of metabolites including steroids, organic acids, chromenes, diterpenes, triterpenes, polyphenols, tannins, anthraquinones, alkaloids, and flavonoids, which comprise the natural product diversity of this genus.

Besides these compounds, species from the genus *Caesalpinia* interestingly produce unusual compounds such as uncommon biflavonoids and a rare subclass of flavonoids, named homoisoflavonoids. The first group is more distributed within plants, while homoisoflavonoids are restrict only to some vegetal species including those from Fabaceae and Asparagaceae [6]. These compounds are also encountered, although less common, in other families as Gentianaceae, Polygonaceae, Portulacaceae, and Orchidaceae. There are two different work concerning about homoisoflavonoids, which relate the existence of approximately 240 naturally occurring compounds [6, 7].

In this sense, it is important to define the general characteristics of flavonoids, once they are the core subunits of biflavonoids and cover the rare class of homoisoflavonoids. In general, flavonoids are low molecular weight polyphenols, brightly colored due to their absorptions of UV light, and the most common structures are associated to antioxidant properties [8–10]. Flavonoids, classified as phytoalexins, are produced as a response to microbial infection in plants. They have a notorious participation into the scientific scenario due to the beneficial association to the humans' daily basis intake of nutrients as functional foods improving human health [8, 10].

The consumption of functional foods, or nutraceuticals, is strongly associated to these compounds. In addition, the ingestion of flavonoids from functional foods implicates in lowering blood triglycerides and homocysteine, decreasing blood pressure, acting against inflammatory, platelet antiaggregation processes, and the improvement of endothelial function [11]. These compounds are also associated to another range of biological properties lowering the incidence of cancer, including prostate, stomach, breast, and lung cancers [12]. In addition, various protective effects of flavonoids have demonstrated them as important multi-target agents [13, 14].

In that regard, the genus *Caesalpinia* is considered a rich source of common flavonoids. However, this genus is also associated to unique biflavonoids constituted by homoisoflavonoids subunits and a considerable amount of representatives from the class of naturally occurring homoisoflavonoids. Up to date, there are reports pointing to the existence of about 240 naturally occurring homoisoflavonoids [6, 7].

An interesting point is that homoisoflavonoids can also be found as dimers. Biflavonoids compounds are dimers of flavonoids assembled in diverse manners by different species. The number of possibilities for these structures (involving all classes of flavonoids) points to more than 20,000 different molecules. However, not all these have been encountered in nature so far, summing to 500 representatives [15]. From these, less than 10 are constituted by homoisoflavonoids subunits.

As homoisoflavonoids and their dimers from the genus *Caesalpinia* are unique compounds, this chapter proposes to gather the available data from the literature in a systematic overview associating them to biological properties aiming to demonstrate these compounds as notable representatives composing the chemical space associated to natural products.

http://dx.doi.org/10.5772/67723

2. General classification and biosynthesis of flavonoids

The classification of flavonoids consists in two main groups, the 2-phenylchromans and the 3-phenylchromans. Compounds presenting the 2-phenylchroman core, in which the aromatic ring B is connected to C-2 atom, include flavonols, flavanones, flavan-3-ols, flavones, anthocyanins, and proanthocyanidins. On the other hand, compounds with the 3-phenylchroman group, in which the aromatic ring B is connected to C-3 atom, include isoflavonoids named isoflavones, isoflavans, and pterocarpans. Another group, named neoflavonoids, in which the benzene ring B is connected to C-4 atom, is less common. There are cases in which the ring C occurs as an isomeric form presenting a five-membered ring, which is associated to the formation of aurones. Another class of phenolic compounds, named chalcones, is not considered true flavonoids due to their lack on the aromatic C ring but still considered members of the flavonoids family. In the same way, a closely related group compounds, the stilbenes, are important due to their biological potential [16]. A brief representation of each class of flavonoids and their sources is demonstrated in **Figure 1**.

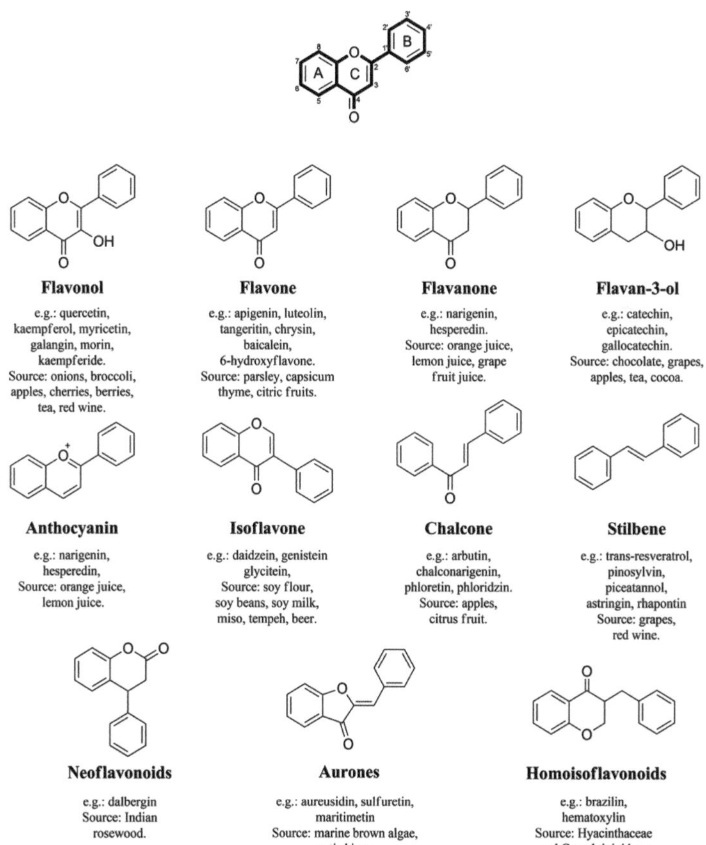

Figure 1. Classification of flavonoids, general structures, examples, and biological sources.

These structures are important for the recognition and classification of biflavonoids moieties, once they could exist as complex structures presenting aurones, isoflavonoids, neoflavonoids, chalcones, and other moieties as well as dimers of homoisoflavonoids.

Flavonoids are products from the phenylpropanoid building block cinnamoyl-CoA, in which chain extension is provided by three units of malonyl-CoA [17]. Cinnamoyl-CoA is derived from the amino acids phenylalanine and tyrosine which are converted by phenylalanine and tyrosine ammonia lyases to cinnamic acid and *para*-hydroxycinamic acid, respectively [18]. The aromatic polyketide formed from the union of cinnamoyl-CoA and three units of malonyl-CoA might form the benzo-γ-pyrone nucleus containing aromatic rings A, B, and a heterocyclic ring C, substituted or not. This nucleus is precursor of a great number of flavonoids. In this sense, flavonoids are characterized by the classic flavan nucleus presenting a C_6-C_3-C_6 skeleton. In addition, chalcones might undergo different cyclization with the addition of a single carbon, provided by S-methyl moiety of methionine, which lead to the formation of the homoisoflavonoid nucleus, which can be converted to the other classes of homoisoflavonoids (**Figure 2**).

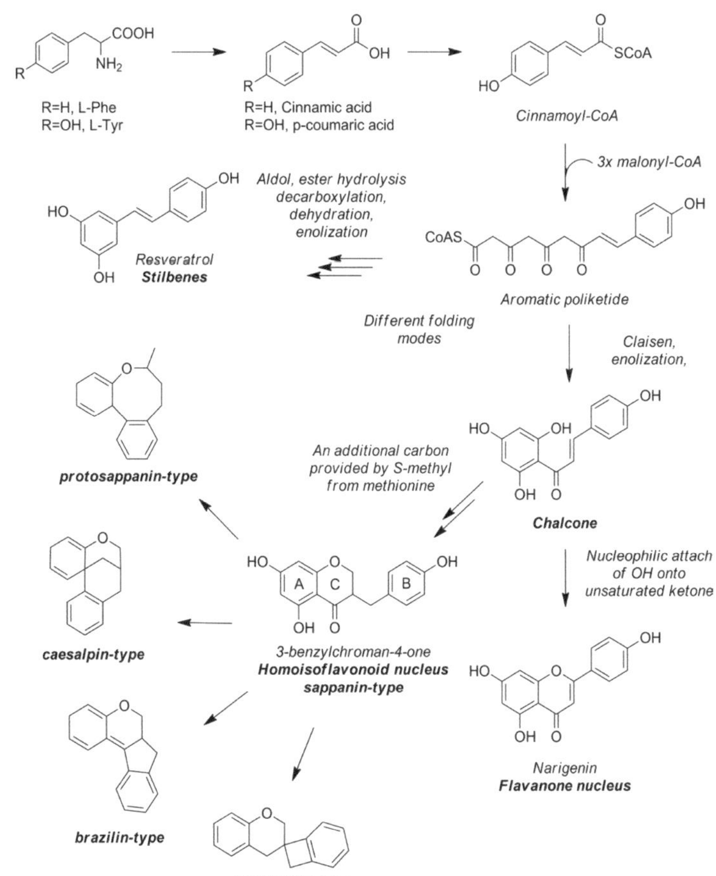

Figure 2. Biosynthetic scheme for the formation of a flavonoid nucleus (monomeric structure of biflavonoids) and the formation of the existing types of homoisoflavonoid nucleus.

3. Occurrence of homoisoflavonoids in *Caesalpinia* spp.

Homoisoflavonoids have a general structure of 16 carbons containing two phenyl rings and one heterocyclic ring. Homoisoflavonoids are biosynthesized from cinnamic acid derivatives along with malonyl-CoA subunits. The resulting compound, an aromatic polyketide, is the precursor of chalcones. In the following step, the aromatic polyketide undergoes a Claisen and enolization reactions, which lead to the formation of the chalcone backbone. An additional carbon is added to the chalcone, provided by S-methyl moiety from methionine, creating the homoisoflavonoid skeleton containing 16 carbons. Thus, there is the formation of 3′-hydroxyl-chalcone as a precursor, which is transformed to 3-benzylchroman-4-one. Subsequently, different cyclization leads to the formation of other types of homoisoflavonoids (**Figure 2**).

The existence of these compounds is associated to the genus *Caesalpinia* involving species as *C. pulcherrima* [19, 20], *C. echinata* [1, 21, 22], *C. bonduc* [3], *C. sappan* [4, 23–28], *C. japonica* [29], and *C. milletti* [30]. However, the diversity of compounds (in number and structurally) is associated to *C. sappan,* a prolific source of homoisoflavonoids with important ethnopharmacological applications. The crude extract of *C. sappan,* named Sappan lignum, is widely studied and used for the treatment of diverse diseases [28].

The classification of homoisoflavonoids comprises five main groups named scillascillin, brazilin, caesalpin, protosappanin, and sappanins. Homoisoflavonoids from the class scillascillins exhibit a spiro ring with four members between rings C and D. However, species from the genus *Caesalpinia* do not produce scillascillins. These compounds are encountered only in the family Asparagaceae [7].

The most common class of homoisoflavonoids in the genus *Caesalpinia* is the sappanin-type. This class presents a 3-benzyl chromanone unit. The diversity of these compounds is associated to a wide variation of substituents, such as hydroxyl, methoxyl, formyl, methyl groups, among others, which confer to sappanin-type the position of the most abundant. In this chapter, the sappanin-type homoisoflavonoids corresponded approximately to 70% of the compounds.

In this aspect, the species *C. pulcherrima,* which is a perennial large shrub, widely distributed in the tropical and subtropical areas of Americas, South India, Taiwan and South-East Asian countries [20, 31]. It is used in the folk medicine due to its medicinal properties for the treatment of skin diseases, tumors, and fevers, and association to antibacterial, antidiarrheal, cytotoxic, and antiulcer properties [32]. *C. pulcherrima* produces a large variety of sappanin-type homoisoflavonoids exhibiting a diverse pattern of substitution such as bonducellin **(1)**, isobonducellin **(2)**, 7-*O*-methylbonducellin **(3)**, 2′-*O*-methylbonducellin **(4)**, sappanone A **(5)**, (3*E*)-3-(1,3-benzodioxol-5-ylmethylene)-2,3-dihydro-7-hydroxy-4H-1-benzopyran-4-one **(6)**, (3*E*)-3-(1,3-benzodioxol-5-ylmethylene)-2,3-dihydro-7-methoxy-4H-1-benzopyran-4-one **(7)**, (*E*)-3-(3-hydroxy-4-methoxybenzylidene)-6,7-dimethoxychroman-4-one **(8)**, (3*E*)-2,3-dihydro-7-hydroxy-3-[(3-hydroxy-4-methoxyphenyl)-methylene]-4H-1-benzopyran-4-one **(9)**, (3*E*)-2,3-dihydro-3-[(3,4-dimethoxyphenyl)methylene]-7-methoxy-4H-1-benzopyran-4-one **(10)**, (*E*)-7-methoxy-3-(4-methoxybenzylidene)chroman-4-one **(11)**, (*E*)-7-hydroxy-3-(3,4,5-trimethoxybenzylidene)chroman-4-one **(12)** [19, 20, 31]. Some of these compounds were tested against Gram-positive microorganisms such as *Bacillus subtilis, Bacillus sphaericus,* and

Staphylococcus aureus exhibiting moderate antimicrobial activity. However, they were inactive or weakly active against Gram-negative microorganisms such as *Pseudomonas aeruginosa, Klebsiella aerogenes,* and *Chromobacterium violaceum*. Concerning the antifungal activity, these compounds presented moderate activity against *Aspergillus niger* and *Candida albicans* in comparison with standard compounds Clotrimazole (antifungal), Streptomycin (antibacterial), and Penicillin G (antibacterial) [19]. Compounds **5** and **6** presented moderate activity against *Staphylococcus aureus* (inhibition zone of 11–15 cm) at 100 μg/mL, while **6** and **10** presented moderate activity against *Klebsiella aerogenes* (inhibition zone of 11–15 cm) at 100 μg/mL. Streptomycin presented an inhibition zone of 21–25 cm at 100 μg/mL. Compounds **5**, **8**, and **9** exhibited moderated activity against *Aspergillus niger* (inhibition zone of 5–10 cm at 150 μg/mL). On the other hand, the compounds **4**, **8**, **7**, **9**, and **10** were moderately active against *Candida albicans* at 150 μg/mL (inhibition zone of 5–10 cm). To comparison, positive control clotrimazole was active against all strains at 100 μg/mL (inhibition zone 21–25 cm) [19].

Rao and collaborators tested the compound **2** against the inflammatory process and described that **2** inhibits the production of NO, TNF-α, and IL-12. In fact, **2** was the most active compound in the experiments at the concentration of 40 μM, reducing 92% of the NO production (IC_{50} = 20 μM) in mouse peritoneal macrophages induced by LPS + IFN-γ. The authors suggested that the mode of action of **2** probably affects the production of NO by the induction of LPS + IFN-γ in mouse peritoneal macrophages [20].

The species *C. echinata*, commonly known as *Pau-brasil* (brazilwood), is endemic from Brazil and played an important historical role in the country [1]. This species has been reported to contain a large range of polyphenols including the homoisolflavonoids brazilin **(13)** and brazilin **(14)**. The compound **15** is a natural dye and is also abundant in the species *C. sappan* (from 8 to 22%). The species *C. echinata*, considered the first source of brazilin, is used for diverse purposes such as healing agent, oral analgesic, and tonics. The species *C. echinata* has also demonstrated antitumor effect *in vivo* against cells strains of Ehrlich Carcinoma and Sarcoma 180. In addition, an interesting antiangiogenic effect was noticed [21]. On the other hand, compound **14**, an oxidation product of brazilin, was considered effective against the inflammatory and cytotoxic processes [22]. Compound **14** displayed cytotoxic effects against human cancer cell lines, such as HepG2 and Hep3B (liver), MDA-MB-231 and MCF-7 (breast), A549 (pulmonary), and CA9-22 (gingival) [22].

Phytochemical studies on ethanolic extracts of *C. bonduc* yielded two sappanin-type homoisoflavonoids identified as caesalpinianone **(15)** and 6-*O*-methylcaesalpinianone **(16)**, which exhibited different levels of GST inhibition and antifungal activities [3]. The IC_{50} values of compounds **15** and **16** were determined as 16.5 and 17.1 μM, respectively for GST inhibition. Ethacrynic acid, a standard substrate GST inhibitor, exhibited a IC_{50} = 17.6 μM, suggesting that homoisoflavonoids have significant inhibition of GST activity [3].

The species *C. sappan* is the most prolific source of homoisoflavonoids with many representatives involving brazilin-, caesalpin-, protosappanin- and sappanin-types. Extracts of *C. sappan*, known as sappan lignum, have been used as emmenagogue, hemostatic, anti-inflammatory and for treatment of thrombosis. There are also relates about its antimicrobial activity against *Staphylococcus, Diplococcus, Corynebacterium,* and *Shigella baydii* [24].The species *C. sappan* afforded brazilin- and sappanin-types homoisoflavonoids such as compounds **13**, **14**, caesalpin P **(17)**, 3′-*O*-methylbrazilin **(18)**, brazilide A **(19)**, 3′-deoxy-4-*O*-methylsappanol **(20)**, sappanol **(21)**, 4-*O*-methylsappanol (**22**, **Figure 3**), in which compounds **13**, **14**, and **20**, and were active to the suppression of melanin synthesis [4]. Melanin is important to the protection of the skin from UV radiation, and its excessive synthesis could lead to melasma and lentigo. Compound **13** exhibited strong suppression of melanogenesis (EC_{50} = 3.0 μM) and cell viability around 95%. Furthermore, compound **20** also exhibited expressive activity (EC_{50} = 4.6 μM) with nonsignificant toxicity (cell viability around 92%). The other compounds displayed high cytotoxicity against HMV-II cells [4].

Species *C. sappan* constitute a source of sappanin-type homoisoflavonoids. Related compounds such as 4-(7-hydroxy-2,2-dimethyl-9βH-1,3,5-trioxa-cyclopenta[α]naphthalene-3-lymethyl)-benzene-1,2-diol **(23)**, 7,3′,4′-trihydroxy-3-benzyl-2H-chromene **(24)** exhibited moderate activity as inhibitors of viral neuraminidases. Viral neuraminidases are considered essential to viral replication cycle and a valid therapeutic target for antiviral drugs. Compound **5** presented the best activity against H1N1 (IC_{50} = 0.7 μM); H3N2 (IC_{50} = 1.1 μM); and H9N2 (IC_{50} = 1.0 μM) [23].

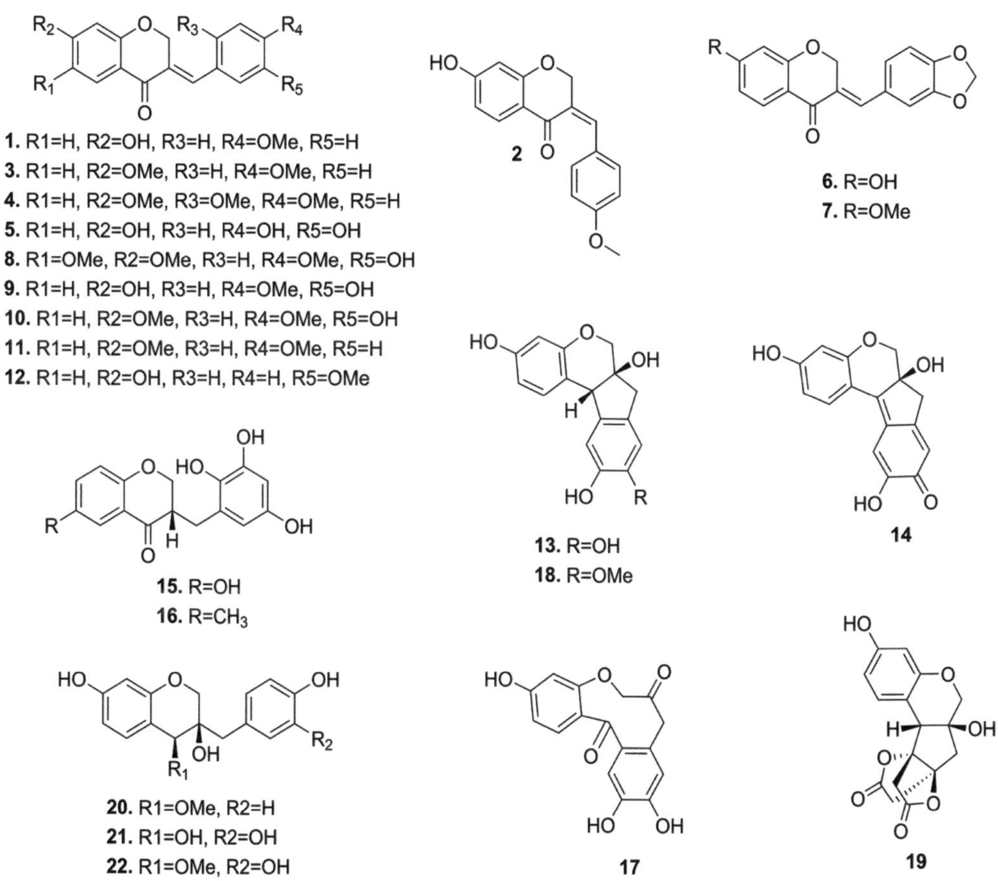

Figure 3. Sappanin-type and brazilin-type homoisoflavonoids from *Caesalpinia* spp.

Other sappanin-type compounds such as (3*R*,4*S*)-3-(4′-hydroxybenzyl)-3,4-dihydro-2″,3″-dimethyl-3H-[1,3]dioxolo[4,5-c]chromen-7-ol **(25)**, and (3a*R*,9b*S*)-3a-(4-hydroxy-3-methoxy-benzyl)-2,2-dimethyl-4,9b-dihydro-3aH-[1,3]dioxolo[4,5-c]chromen-7-ol **(26)** are associated to the inhibition of NO production [27]; **21** and **22** associated to the inhibition of melanin synthesis [4]. Sappanol derivatives were also identified from *C. sappan* as in the case of 3′-*O*-methylsappanol **(27)**, 3′-*O*-methylepisappanol **(28)**, and a unique lactone-based homoisoflavonoid named caesalpiniaphenol B **(29)** [6]. In addition, the compound caesalpin J **(30)**, one of the only seven caesalpin-type homoisoflavonoids reported in the literature, was isolated from *C. sappan*. Caesalpin J exhibited weak to moderate antimicrobial effects [25].

The species *C. japonica* is considered another source of biologically active homoisoflavonoids in which diverse homoisoflavonoids including **5**, **13**, **20**, **21**, **22**, protosappanin A-C **(31–33)**, 4-*O*-methylepisappanol **(34)**, episappanol **(35)**, and sappanone B (**36, Figures 4–6**) have been isolated and characterized [29].

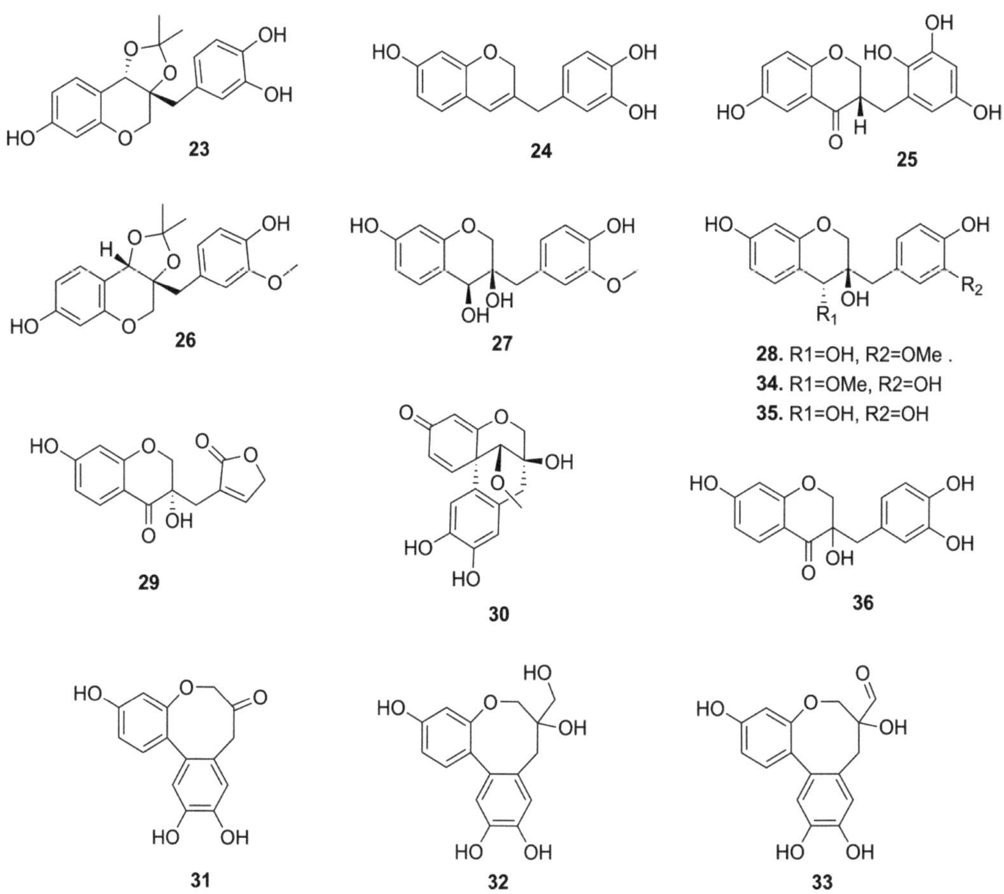

Figure 4. Sappanin-, casealpin-, and protosappanin-type structures of homoisoflavonoids from *Caesalpinia* spp.

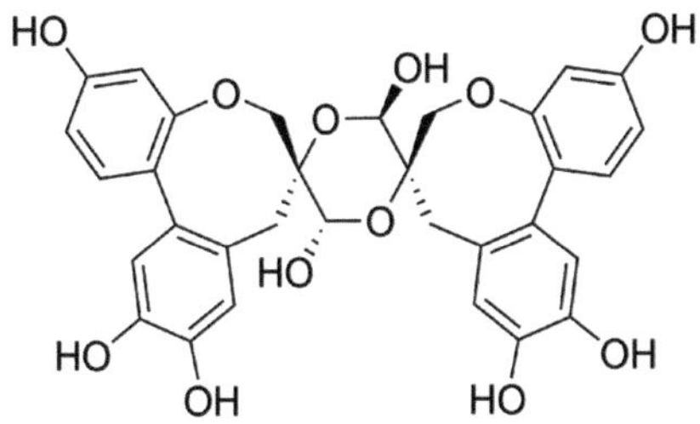

37

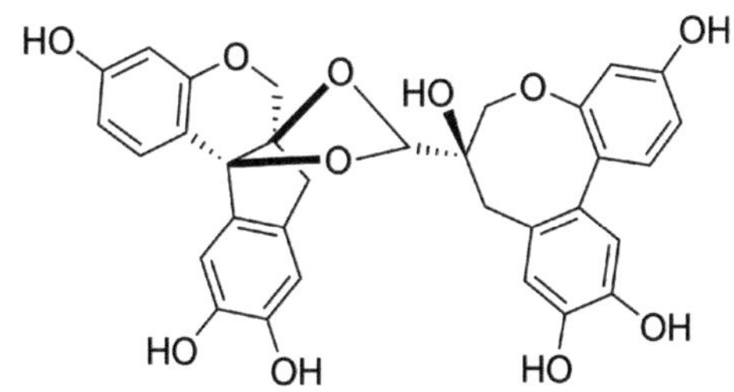

38

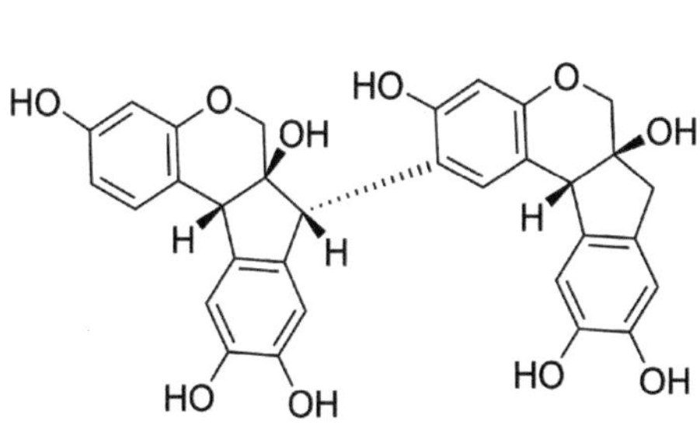

39

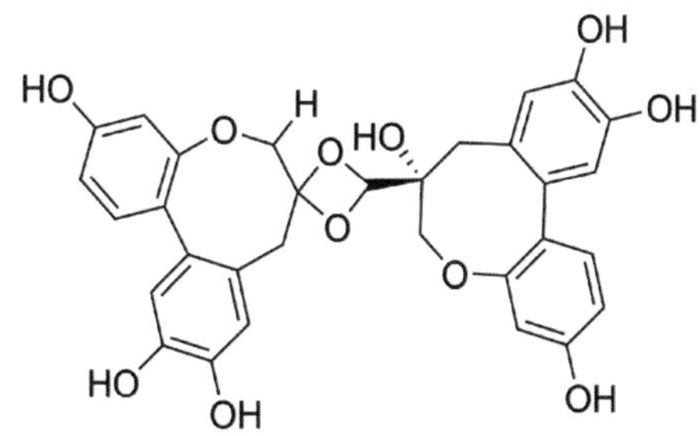

40

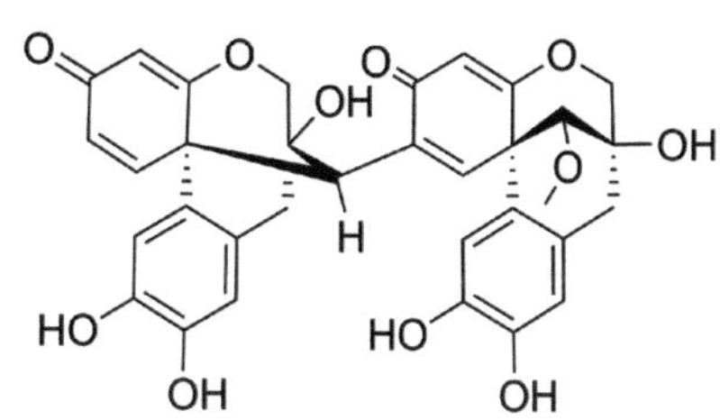

41

Figure 5. Bihomoisoflavonoids from species of the genus *Caesalpinia*.

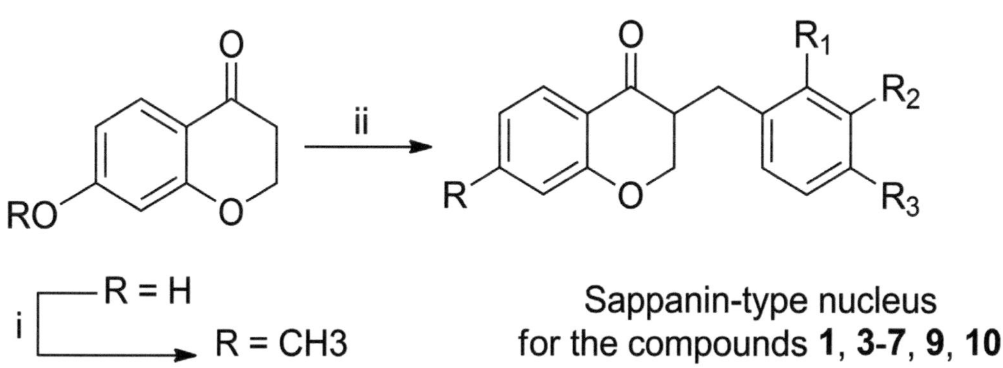

Figure 6. Scheme of synthesis of sappanin-type homoisoflavonoids obtained from *C. pulcherrima*. Reagent and conditions: (i) CH_3I, K_2CO_3, acetone, 2h, reflux, (91%); (ii) substituted benzaldehyde, piperidine, 2h (58–69%). Adapted from Ref. [19].

Homoisoflavonoids classified as protosappanins are commonly associated to the species *C. sappan* and *C. japonica*. These compounds are resulting from the connection of C-4 and C-4a atoms forming an eight-membered ring. There are only eleven protosappanins reported so far [26]. Compounds **31–33** did not show significant cytotoxicity against MCF7, A549, LN229 cell lines. In addition, compounds **32** and **33** were also tested against the inflammatory process exhibiting weak to moderate activity [27].

4. Biflavonoids containing homoisoflavonoids subunits in *Caesalpinia* spp.

Flavonoids can also exist as dimers, named biflavonoids, which represents flavonoids linked by C–C or C–O–C bond in order to form a flavonoid-flavonoid structure. The connection can occur in several modes in the three rings of the flavan nucleus. The ring A could be linked to the ring A′, indicated as A-A. This could also occur between the rings A-C, B-B, C-B, among other possibilities that are enlarged by functional groups as OH, MeO, C=O, C=C. The occurrence of common biflavonoids in the genus *Caesalpinia* is known only to some species, such as *C. ferrea* [33], *C. pyramidalis* [34], *C. pluviosa* [35].

Furthermore, certain species from the *Caesalpinia*, mostly *C. sappan*, are associated to the production of biflavonoids containing homoisoflavonoids subunits such as protosappanin D **(37)**, a biflavonoid which exhibit two subunits of **33**, and protosappanin E **(38)**, which display **13** and **33** as subunits. Compounds **37** and **38** were tested against the inflammatory process associated to the inhibition of iNOS and PGE2 production, as well as the suppression of TNF-α and COX-2. Washiyama and collaborators suggested that the protosappanin skeleton and the functional group at C7 would be important to the activity of **37** and **38** [27]. The investigation of sappan lignum as a possible XO inhibitor leads to the isolation of diverse compounds including neoprotosappanin **(39)** and protosappanin E-2 **(40, Figure 5)**. These two compounds presented IC_{50} of 38.3 and 18.9 μM, respectively, exhibiting a concentration-dependent behavior. In what extent their mode of inhibition, the biflavonoid **39** was considered a noncompetitive XO inhibitor while **40**, a competitive inhibitor [2]. The inhibition of XO is associated to improvements in cardiovascular health as well as, the reduction of ROS, and the amelioration of gout cases [36].

The presence of rare caesalpins is correlated to *C. sappan*. A dimer named neosappanone **(41)** was isolated from this species and evaluated against XO. The IC_{50} of **41** was determined as 29.7 μM and associated to a competitive inhibition of XO. Therefore, these results showed that the traditional use of *C. sappan* for rheumatism and inflammatory diseases could be attributed to its phenolic composition, specifically to dimers of homoisoflavonoids [2].

5. Isolation, synthesis, and structural analysis of homoisoflavonoids

Due to the intrinsic interest in homoisoflavonoids and their biological activities, several works have been discussing different structural aspects of homoisoflavan nucleus-bearing organic compounds [6, 7]. The structural uniqueness of these compounds and their potent biological activities makes them a target of choice for studies in natural products research on the determination of absolute configurations, organic synthesis, isolation, and structural determination [7].

The isolation of homoisoflavonoids involves different chromatographic techniques. Homoisoflavonoids are generally separated after treatment of the organic extract (MeOH, $CHCl_3$) with several chromatographic phases. The use of column chromatography steps (using silica gel and/or Sephadex LH-20), preparative thin layer chromatography, as well as high performance liquid chromatography (HPLC) methods (semi-preparative and preparative) have been used to purification [7]. In addition, there are other methods used to the isolation of flavonoids, such as counter current chromatography [37] can be adopted for the isolation of homoisoflavonoids and flavonoids.

Besides the isolation of naturally occurring homoisoflavonoids from the species *C. pulcherrima*, a synthetic approach of the isolated homoisoflavonoids **1**, **3**, **4**, **5**, **6**, **7**, **9**, and **10** employed the piperidine catalyzed condensation as key steps in this synthesis of these structures (**Figure 6**). This procedure afforded products reaching around 60% yielding following conditions exhibited in **Figure 3** [19].

The structures of homoisoflavonoids have been unambiguously established by analysis of spectroscopic NMR data supported by analysis of UV and MS spectra. These analyses confirm the presence of the 15-carbon backbone related to classic flavonoids, and the 16-carbon skeleton with two phenyl rings (A and C) and one heterocyclic ring (B) separated by an additional carbon, forming the homoisoflavonoids skeleton [24, 29, 31].

Analysis of the ^{1}H and ^{13}C NMR spectra indicates the presence of carbonyl groups at δ_C 170.7–220.0 as well as those assigned to carbons/hydrogens of aromatic ring at δ_C 100.0–170.0/δ_H 6.00–8.50 and hydroxyl derivate group as characteristic signs. The homoisoflavonoids, when existing as dimers, exhibited their ^{13}C and ^{1}H NMR spectra typically duplicate and superposed when presenting the same subunits.

The extra carbon existing in homoisoflavonoids compared to ordinary flavonoids can be aliphatic displaying ^{13}C and ^{1}H NMR signs at δ_C 30.0–35.0/δ_H 2.60–3.00; or olefinic at δ_C 100.0–140.0/δ_H 5.30–6.00, respectively. Correlations in the HMBC, HSQC and COSY spectrum resolve all ambiguities to the structure of these compounds.

The compounds **17**, **19**, **29**, **37**, **38** are homoisoflavonoids derived from the auto-oxidation of precursors or present differentiated biosynthesis. These compounds present uncommon chemical structures with ^{1}H, and ^{13}C NMR spectra relativity complex, in some cases, exhibiting signs that indicate the presence of lactone and others characteristic group.

Homoisoflavonoids present signals of absorptions in the UV spectrum at λ_{max} 222–230, 270–280, and 300–310 nm, depending on the presence of conjugated double bonds on their structures. The MS spectrum present $[M+H]^+$ ions between m/z 250–350 to homoisoflavonoids and $[M+H]^+$ ions between m/z 500–600 to dimers of homoisofavonoids with some minor compounds with $m/z > 600$ [24, 29, 31].

The absolute stereochemistry can be established by circular dichroism analysis by comparison to models with known stereochemistry. The compound **28**, isolated from *C. sappan*, had its absolute stereochemistry determined by this method compared with data from 3-deoxysappanol. The result suggested that **28** has the absolute stereochemistry at the C-3 and C-4 positions to be (3*R*,4*S*) [24].

6. Conclusion

The chemical space related to natural products is associated to important scientific findings in what to extent the discovery of important new chemical entities. In this sense, the genus *Caesalpinia* is a prolific source of secondary metabolites which also biosynthesize homoisoflavonoids. In addition, only a few species are reported to produce dimers of homoisoflavonoids as secondary metabolites. Homoisoflavonoids and its dimers are classified as unusual natural products with strict occurrence in nature present the most diverse types of biological activities. These compounds are isolated through HPLC methods and identified by different techniques as NMR, UV, MS based on the biosynthesis of chalcones with an additional carbon provided by *S*-methyl from methionine. It is important to highlight that these phenolic compounds are part of ethnopharmacological applications by people whose access and use the biodiversity for the improvement of health conditions are made for centuries. Furthermore, the investigation of this class of phenolic compounds provides chemosystematics data for classification and discovery of pharmacologically efficient compounds from species of *Caesalpinia* spp.

Acknowledgements

The authors acknowledge the CAPES, CNPq, FAPEMIG, FAPESP, and FINEP for subsidies and funding.

Conflict of interests

The authors declare no conflict of interests.

Author details

João Luiz Baldim[1,2], Welton Rosa[1], Mário Ferreira Conceição Santos[1], Daniela Aparecida Chagas-Paula[1], João Henrique Ghilardi Lago[2] and Marisi Gomes Soares[1]*

*Address all correspondence to: marisigs@gmail.com

1 Institute of Chemistry, Federal University of Alfenas, Unifal-MG, Alfenas-MG, Brazil

2 Centre of Human and Natural Sciences, Federal University of ABC, UFABC, Santo André-SP, Brazil

References

[1] Baldim Zanin JL, De Carvalho BA., Martineli PS, et al. The genus *Caesalpinia* L. (Caesalpiniaceae): Phytochemical and pharmacological characteristics. Molecules. 2012;**17**(7):7887-7902. doi:10.3390/molecules17077887

[2] Nguyen MTT, Awale S, Tezuka Y, Tran Q Le, Kadota S. Xanthine oxidase inhibitors from the heartwood of Vietnamese *Caesalpinia sappan*. Chem Pharm Bull (Tokyo). 2005;**53**(8):984-988. doi:10.1248/cpb.53.984

[3] Ata A, Gale EM, Samarasekera R. Bioactive chemical constituents of *Caesalpinia bonduc* (Fabaceae). Phytochem Lett. 2009;**2**(3):106-109. doi:10.1016/j.phytol.2009.02.002

[4] Mitani K, Takano F, Kawabata T, et al. Suppression of melanin synthesis by the phenolic constituents of sappanwood (*Caesalpinia sappan*). Planta Med. 2013;**79**(1):37-44. doi:10.1055/s-0032-1327897

[5] Jeong HJ, Kim YM, Kim JH, et al. Homoisoflavonoids from *Caesalpinia sappan* displaying viral neuraminidases inhibition. Biol Pharm Bull. 2012;**35**(5):786-790. doi:10.1248/bpb.35.786

[6] Lin LG, Liu QY, Ye Y. Naturally Occurring Homoisoflavonoids and Their Pharmacological Activities. Planta Med. 2014;**80**(13):1053-1066. doi:10.1055/s-0034-1383026

[7] Abegaz BM, Mutanyatta-Comar J, Nindi M. Naturally occurring homoisoflavonoids: Phytochemistry, biological activities and synthesis. Nat Prod Commun. 2007;**2**(4):475-498.

[8] Havsteen BH. The Biochemistry and Medical Significance of the Flavonoids. Pharmacol. Ther. 2002;**96**(2):67-202. doi:10.1016/S0163-7258(02)00298-X.

[9] Hertog M, Feskens E, Hollman P, Katan M, Kromhout D. Dietary antioxidants flavonoids and risk of coronary heart disease: the Zutphen Elderly Study. Lancet. 1993; **342**(8878):1995:1007-1011. doi: 10.1016/0140-6736(93)92876-U

[10] Cook NC, Samman S. Dietary sources. J Nutr Biochem. 2006;2863(Ldl):66-76.

[11] Alissa EM, Ferns GA. Functional foods and nutraceuticals in the primary prevention of cardiovascular diseases. J Nutr Metab. 2012;2012. doi:10.1155/2012/569486

[12] Shashank K, Abhay K. Chemistry and biological activities of flavonoids: An overview. Sci World J. 2013;**4**(2):32-48.

[13] Pietta PG. Flavonoids as antioxidants. J Nat Prod. 2000;**63**(7):1035-1042. doi:10.1021/np9904509

[14] Heim KE, Tagliaferro AR, Bobilya DJ. Flavonoid antioxidants: chemistry, metabolism and structure-activity relationships. J Nutr Biochem. 2002;**13**(10):572-584. doi:10.1016/S0955-2863(02)00208-5

[15] Mercader AG, Pomilio AB. Biflavonoids: Occurrence, Structural Features and Bioactivity. 1st ed. (Mercander AG, Pomilio AB, eds.). New York: Novinka; 2012.

[16] Tsao R. Chemistry and biochemistry of dietary polyphenols. Nutrients. 2010;**2**(12):1231-1246. doi:10.3390/nu2121231

[17] Dewick PM. Medicinal Natural Products: A Biosynthetic Approach. 3rd ed. Chichester, UK: John Wiley & Sons, Inc.; 2009. doi:10.1016/j.jep.2009.05.024

[18] Batra P, Sharma AK. Anti-cancer potential of flavonoids: Recent trends and future perspectives. Biotech. 2013;**3**(6):439-459. doi:10.1007/s13205-013-0117-5

[19] Das B, Thirupathi P, Ravikanth B, Aravind Kumar R, Sarma AVS, Basha SJ. Isolation, synthesis, and bioactivity of homoisoflavonoids from *Caesalpinia pulcherrima*. Chem Pharm Bull (Tokyo). 2009;**57**(10):1139-1141. doi:10.1248/cpb.57.1139

[20] Rao YK, Fang SH, Tzeng YM. Anti-inflammatory activities of flavonoids isolated from *Caesalpinia pulcherrima*. J Ethnopharmacol. 2005;**100**(3):249-253. doi:10.1016/j.jep.2005.02.039

[21] da Silva Gomes ECB, Jimenez GC, da Silva LCN, et al. Evaluation of antioxidant and antiangiogenic properties of *Caesalpinia echinata* extracts. J Cancer. 2014;**5**(2):143-150. doi:10.7150/jca.7439

[22] Yen CT, Nakagawa-Goto K, Hwang TL, et al. Antitumor agents. 271: Total synthesis and evaluation of brazilein and analogs as anti-inflammatory and cytotoxic agents. Bioorganic Med Chem Lett. 2010;**20**(3):1037-1039. doi:10.1016/j.bmcl.2009.12.041

[23] Jeong HJ, Kim YM, Kim JH, et al. Homoisoflavonoids from *Caesalpinia sappan* displaying viral neuraminidases inhibition. Biol Pharm Bull. 2012;**35**(5):786-790. doi:10.1248/bpb.35.786

[24] Namikoshi M, Nakata H, Yamada H, Nagai M, Saitoh T. Homoisoflavonoids and related compounds. II. Isolation and absolute configurations of 3,4-dihydroxylated homoisoflavans and brazilins from *Caesalpinia sappan* L. Chem Pharm Bull (Tokyo). 1987;**35**(7):2761-2773. doi:10.1248/cpb.35.2761

[25] Niranjan Reddy VL, Ravikanth V, Jansi Lakshmi VVNS, Suryanarayan Murty U, Venkateswarlu Y. Inhibitory activity of homoisoflavonoids *from Caesalpinia sappan* against *Beauveria bassiana*. Fitoterapia. 2003;**74**(6):600-602. doi:10.1016/S0367-326X(03)00153-9

[26] Wang Z, Sun JB, Qu W, Guan FQ, Li LZ, Liang JY. Caesappin A and B, two novel protosappanins from *Caesalpinia sappan* L. Fitoterapia. 2014;**92**:280-284. doi:10.1016/j.fitote.2013.12.004

[27] Washiyama M, Sasaki Y, Hosokawa T, Nagumo S. Anti-inflammatory constituents of Sappan Lignum. Biol Pharm Bull. 2009;**32**(5):941-944. doi:10.1248/bpb.32.941

[28] Yang BO, Ke CQ, He ZS, Yang YP, Ye Y. Brazilide A, a novel lactone with an unprecedented skeleton from *Caesalpinia sappan*. Tetrahedron Lett. 2002;**43**(9):1731-1733. doi:10.1016/S0040-4039(02)00109-0

[29] Namikoshi M, Hiroyuki T, Mariko T, Nuno S, Faculty O, Sagamiko P. Homoisoflavonoids and related compounds. III. Phenolic constituents of *Caesalpinia japonica* SIEB. et ZUCC. Chem Pharm Bull (Tokyo). 1987;**35**(9):3568-3575.

[30] Chen P, Yang J-S. Flavonol galactoside caffeiate ester and homoisoflavones from *Caesalpinia millettii* HOOK. et ARN. Chem Pharm Bull (Tokyo). 2007;**55**(4):655-657. doi:10.1248/cpb.55.655

[31] Maheswara M, Siddaiah V, Venkata Rao C. Two new homoisoflavonoids from *Caesalpinia pulcherrima*. Chem Pharm Bull (Tokyo). 2006;**54**(8):1193-1195. doi:10.1248/cpb.54.1193

[32] Billah MM, Islam R, Khatun H, et al. Antibacterial, antidiarrhoeal, and cytotoxic activities of methanol extract and its fractions of *Caesalpinia bonducella* (L.) Roxb leaves. BMC Complement Altern Med. 2013;**13**(1):101. doi:10.1186/1472-6882-13-101

[33] Magalhães IL, Chagas F, Pinto L, Filho RB. Chemical Constituents from Caesalpinia férrea: Identification and ^{1}H and ^{13}C Resonance Assignment. Am J Anal Chem. 2014;**5**(July):688-694. doi:10.4236/ajac.2014.510077

[34] Bahia MV, Dos Santos JB, David JP, David JM. Biflavonoids and other phenolics from *Caesalpinia pyramidalis* (Fabaceae). J Braz Chem Soc. 2005;**16**(6 B):1402-1405. doi:10.1590/S0103-50532005000800017

[35] Zanin JLB, Massoni M, Freitas GC De, et al. Caesalpinioflavone, a New Cytotoxic Biflavonoid Isolated from *Caesalpinia pluviosa* var. peltophoroides. J Braz Chem Soc. 2015;**26**(4):804-809.

[36] Dawson J, Walters M. Uric acid and xanthine oxidase: Future therapeutic targets in the prevention of cardiovascular disease? Br J Clin Pharmacol. 2006;**62**(6):633-644. doi:10.1111/j.1365-2125.2006.02785.x

[37] Sticher O. Natural product isolation. Nat Prod Rep. 2008;**25**(3):517-554. doi:10.1039/b700306b

Chapter 6

Biocatalysis of Rutin Hexadecanedioate Derivatives: Effect of Operating Conditions on Acylation Performance and Selectivity

Manel Slimane, Aude Cordin, Mohamed Ghoul and
Chebil Latifa

Additional information is available at the end of the chapter

http://dx.doi.org/10.5772/67621

Abstract

Rutin was enzymatically acylated with hexadecanedioic acid, in *tert*-amyl alcohol, by an immobilized lipase from *Candida antartica* "Novozym 435". The effect of different techniques of water removal, temperature, concentration of rutin and diacid/rutin molar ratio was investigated. The obtained results indicated that drying the media by adding the molecular sieves in the outer loop of the reactor was the most efficient method leading to water content lower than 200 ppm. The highest performances (conversion yield and initial rate) were reached at 90°C, 131 mM of rutin, and 118 mM of acid. Depending on the water content and the diacid/rutin molar ratio, only rutin 4‴-hexadecanedioate or both rutin 4‴-hexadecanedioate and dirutin 4‴, 4‴-hexadecanedioate were synthesized.

Keywords: biocatalysis, lipase, rutin, hexadecanedioic acid, water activity

1. Introduction

Flavonoids are benzo-γ-pyrone derivatives widely distributed in plants kingdom [1–4]. They have been shown to possess a wide range of biological activities including antiviral, anti-allergic, anti-inflammatory, anti-tumor properties [5, 6]. These properties of flavonoids are mainly due to their antioxidant activities based on their ability to act as hydrogen or electron donors [7]. The magnitude of this activity depends on their structure. Flavonoids with high biological activities are used in food preparations, cosmetics or pharmaceuticals [8–10]. However, their integration into several preparations is affected by their poor solubility in very hydrophilic and very hydrophobic solvents [11, 12]. In order to overcome this drawback, different ways of functionalization are described in the literature. One of them is the acylation reaction, with the aim to enhance both the solubility in hydrophobic media and the biological activities of flavonoids [13]. This reaction can be carried out via enzymatic or chemical route. Enzymatic reactions are preferred to chemical ones due to their regio-selectivity toward polyhydroxylated compounds like rutin [14, 15]. Rutin is a flavonol glycoside and one of the most studied flavonoids in the literature. Several papers dealt with improvement of the solubility and the biological activities of rutin by enzymatic acylation under a wide range of operating conditions. These studies showed that the performance (conversion yield, regioselectivity, etc.) of this reaction was affected by several factors (reaction media, solubility and nature of the substrates, operating conditions, enzyme concentration, etc.). Lue et al. [16] reported that the use of ionic liquids as acylation medium enhances the solubility of substrates and gives high conversion yields of rutin. Zheng et al. [17] observed better rutin conversion rates when acylation was conducted under ultrasound radiations. This improvement was attributed by these authors to the enhancement of lipase activity. Water content of the medium has a strong effect on acylation reaction. It can affect both the solubility of the substrates and the activity of the biocatalyst. In the case of rutin, Ardhaoui et al. [18] reported that the highest rutin conversion yield was reached with water content less than 200 ppm (76%).

The most used enzyme for rutin acylation reactions is lipase B from *Candida antarctica*. This enzyme has shown high performances in terms of conversion yield and enantioselectivity [19].

The chain length or the nature of the acyl donor can also affect the conversion yield of rutin and the regioselectivity of the reaction [16, 18, 20, 21]. Conversion yields arise from 48 to 87% with the length of carbon chain [18]. Dicarboxylic acids, another class of fatty acids, can be used for the acylation of flavonoids. These compounds have two advantages compared to monocarboxylic acids. In one hand, they exhibit more flexibility due to their second carboxyl group. In the second hand, according to their structure, they have bacteriostatic and bactericidal properties against a variety of aerobic and anaerobic bacteria [22]. However, few data are available concerning their use as acyl donors with flavonoids. Only Theodosiou et al. [23] and Ardhaoui et al. [18] reported the enzymatic modification of flavonoids with diacids without any optimization of operating conditions of this reaction. The aim of the present work was to study the enzymatic synthesis of rutin hexadecanedioate by Novozym 435 in organic medium. The effect of several factors such as drying techniques, temperature, rutin concentration, molar ratio rutin/hexadecanedioic acid, concentration of biocatalyst, and its reuse were investigated. The behavior of initial rate, productivity, conversion rates, and the regioselectivity of the reaction were quantified and discussed.

2. Materials and methods

2.1. Enzyme and chemicals

Immobilized lipase B from *Candida antarctica* (Novozym 435, 7000 PLU/mg: propyl laurate units synthesized per gram of catalyst) was purchased from Novozymes AS, Bagsvaerd, Denmark. *tert*-amyl alcohol was supplied by Merck (France). Rutin was furnished by Sigma-Aldrich (France). Hexadecanedioic acid (Cathay Biotechnology, China) was used as acyl donor. The molecular sieves 4Å, used to dry solvents and reactions media, were provided by Acros organis (France).

2.2. Rutin ester synthesis

The enzymatic syntheses of rutin hexadecanedioate were carried out in a stirring batch reactor (250 ml) from "Pilotes Systèmes" (France) or Wheaton® reactors (USA). Agitation speed was varied from 300 to 500 rpm. Rutin (65–196 mM) was dissolved in 250 ml of dried *tert*-amyl alcohol at different temperatures (60, 80, and 90°C). The hexadecanedioic acid concentration was adjusted to obtain a di-acid/rutin molar ratio in the range of 0.05–20 in the solution. Esterification reactions were started by the addition of Novozym 435 (10, 30, and 50 g/L). The reaction was stopped after 50–72 h by removing the biocatalyst. Blank samples, containing all components except the enzyme preparation, were carried out in tandem with the enzymatic trials.

Different drying techniques were used: (i) drying with molecular sieves added to the bulk medium or (ii) introduced in an external loop. In this last case, solvent and water were evaporated under vacuum, crossing in a fist part molecular sieves as vapor phase then as liquid phase (**Figure 1a**). Another configuration of external loop consists in the water removal only in liquid phase (**Figure 1b**).

Water content was 800–1000, 400–550, and 200–300 ppm, respectively with molecular sieves added to the bulk medium, the use of drying in vapor and liquid phases and drying only through liquid phase.

2.3. Analytical procedure

2.3.1. *Karl Fisher analysis*

The water content of the reaction medium was determined by a coulometric Karl Fisher apparatus (KF 737II coulometer) Metrohm (France). The reagent was Hydranal-Coulomat AG-H (Riedel-de-Haën, France).

2.3.2. *High-performance liquid chromatography analysis*

The substrate and product concentrations were determined by high-pressure liquid chromatography in external calibration. Analysis were carried out at 55°C in a system (Alliance 2690 Waters) composed of a column (Symmetry® C18, 4.6 × 250 mm, 5 μm, Waters, France), a UV detector (250 and 350 nm, Waters 2487, France) and a ELSD (Evaporative Light Scattering detector, Altech 2000, France). The various compounds were separated using water (0.1% acetic acid)/methanol (0.1% acetic acid) solutions: 0 min (70/30), 5 min (0/100), 10 min (0/100), 12 min (70/30), 15 min (70/30).

2.3.3. *Purification and determination of the chemical structure of rutin esters*

Rutin esters were purified by liquid–liquid extraction. The residual flavonoid was removed at 60°C under agitation during 45 min with a water/heptane solution (2/3, v/v), while, the flavonoid esters were separated from the acyl donor by using acetonitrile (50 mL) at 40°C during 20 min of agitation. The flavonoid esters solution was concentrated by solvent evaporation under reduced pressure for injection in a preparative HPLC (Waters, France). A column RP18

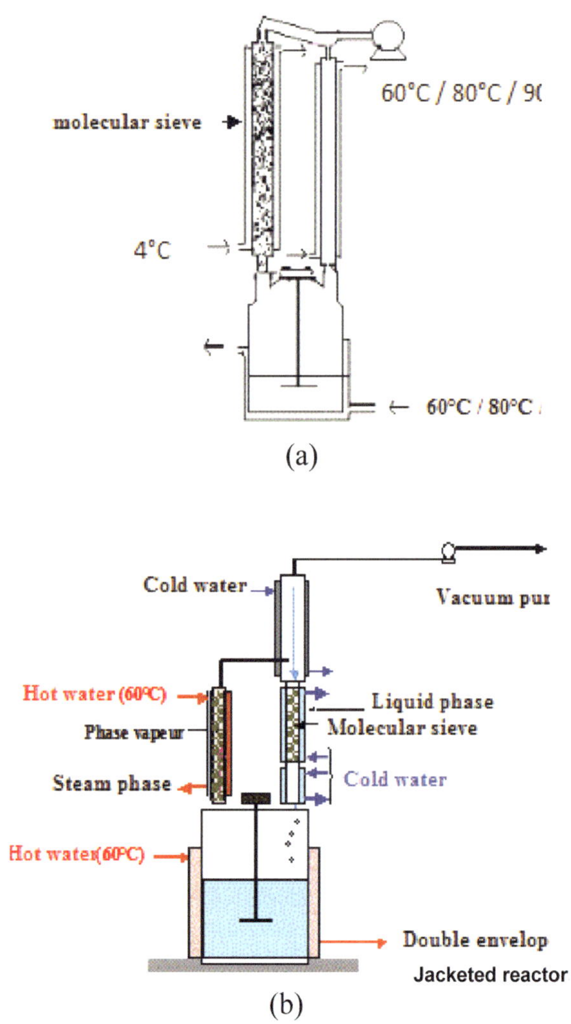

Figure 1. Drying techniques used to remove water from the reaction media. (a) Drying the media by using molecular sieves in the outer loop of the reactor in liquid phase. (b) Drying the media by using molecular sieves in the outer loop of the reactor in vapor and liquid phase.

(30 × 100 mm, 5 μm, Waters XTerra®, France) and a UV detector (350 nm, Waters 2487, France) were used for separate and analyze esters. A gradient of water and acetonitrile with 0.1% acetic acid at a flow rate of 18 mL/min was applied: 0 min (70/30), 10 min (20/80), 12 min (20/80), 13 min (70/30), 15 min (70/30). The medium was diluted 2.5 times in the starting phase (water with 0.1% acetic acid/acetonitrile with 0.1% acetic acid, 70/30) and 850 μl of this solution were injected for each batch. Eighteen batches were produced. Rutin hexadecanedioate and dirutin hexadecanedioate are obtained with a purity ≥95%. This purification method was adapted from Ardhaoui et al. [18].

http://dx.doi.org/10.5772/67621

The chemical structures of the purified dirutin hexadecanedioate and rutin hexadecanedioate were determined by 1H NMR and 13C NMR in DMSO-d6 using a Brücker AM 400 at 400 MHz and at 100 MHz, respectively.

2.4. Determination of conversion rate, initial rate, productivity, and selectivity

2.4.1. Conversion rate

The conversion rate of rutin and acid was calculated from concentrations given by HPLC analysis.

$$\text{Conversion } (\%) = ([S]i - [S]f) * 100/[S]i \tag{1}$$

[S]i: initial substrate concentration (mmol/L)

[S]f: final substrate concentration (mmol/L)

2.4.2. Initial rate of monorutin hexadecanedioate formation

The initial rate was calculated during the first three hours of the synthesis reaction of rutin hexadecanedioate by taking the slope of the kinetic linear portion.

2.4.3. Productivity

The productivity is given by the following expression:

$$\text{Productivity } (g/(L/h)) = \text{Mass of ester formed divided by the working volume of the reactor and by the duration of the reaction} \tag{2}$$

2.4.4. Selectivity

The selectivity is given by the following expression:

$$\text{Selectivity } (\%) = [\text{monorutin}]\,t * 100/([\text{monorutin}]\,t + [\text{dirutin}]\,t) \tag{3}$$

Where, [monorutin] t and [dirutin] t are molar concentrations at t (time) of mono and dirutin esters.

3. Results and discussion

3.1. Influence of water removal techniques

The presence of a minimum amount of water is necessary for maintaining the catalytic activity of lipase. But, high concentration of water can favor the hydrolysis reaction.

Different water removal techniques were investigated. In the presence of molecular sieves, two alternatives have been tested. The molecular sieves were introduced either in the bulk medium or in the outer loop of the reactor. The performance of these reactions carried out with or without water removal were evaluated and compared.

The results are reported in **Figure 3a–c**. The reaction carried out with high water content leads to the lowest conversion yield, initial rate, and productivity while, reactions conducted with water removal are characterized by high conversion yield, initial rate, and productivity. The highest conversion rates of rutin were almost similar for the two configurations of outer loop, 79 and 68%, respectively. In the presence of high water content, only monorutin formation was observed, while at low water content both mono- and di-rutin esters were synthesized. HPLC and NMR analyses showed that the synthesized mono and dirutin esters are respectively rutin 4‴-hexadecanedioate and dirutin 4‴, 4‴-hexadecanedioate (**Figure 2**).

The improved performance of the acylation reaction with a water removal system has already been described by several authors [24–26], but its influence on the selectivity of the reaction has never been mentioned before. The water removal by the two outer loop configurations gave similar results and presents the advantage of avoiding the abrasion of the enzyme.

Due to its implementation facility, the water removal by vapor and liquid phase (**Figure 3b**) was selected as a standard method to investigate the effect of the other factors on this reaction.

3.2. Effect of the temperature on the rutin solubility and esters synthesis

3.2.1. Effect on the solubility

The solubility of substrates is one of the main factors that affect the performance of the acylation reaction. This solubility is drastically influenced by the temperature. For this reason, the effect of the temperature (60, 80, and 90°C) on the solubility of rutin in *tert*-amyl alcohol was evaluated in the first step. In all cases, solubility increases with temperature. At the equilibrium

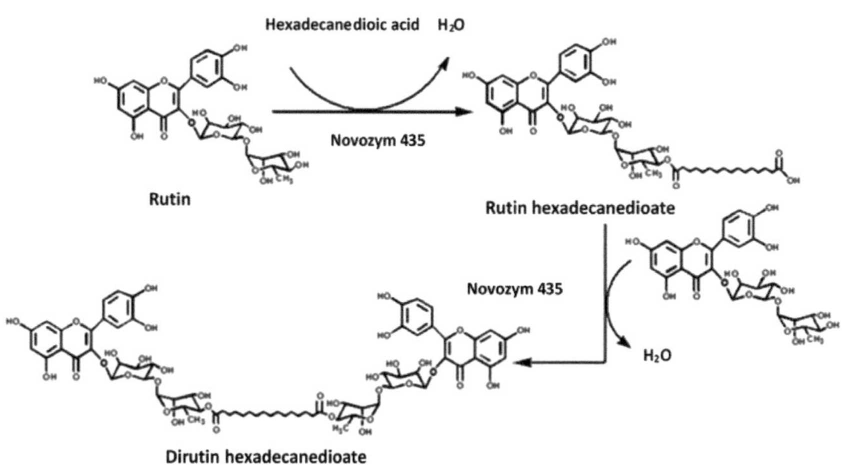

Figure 2. Synthesis of rutin hexadecanedioate and dirutin hexadecanedioate.

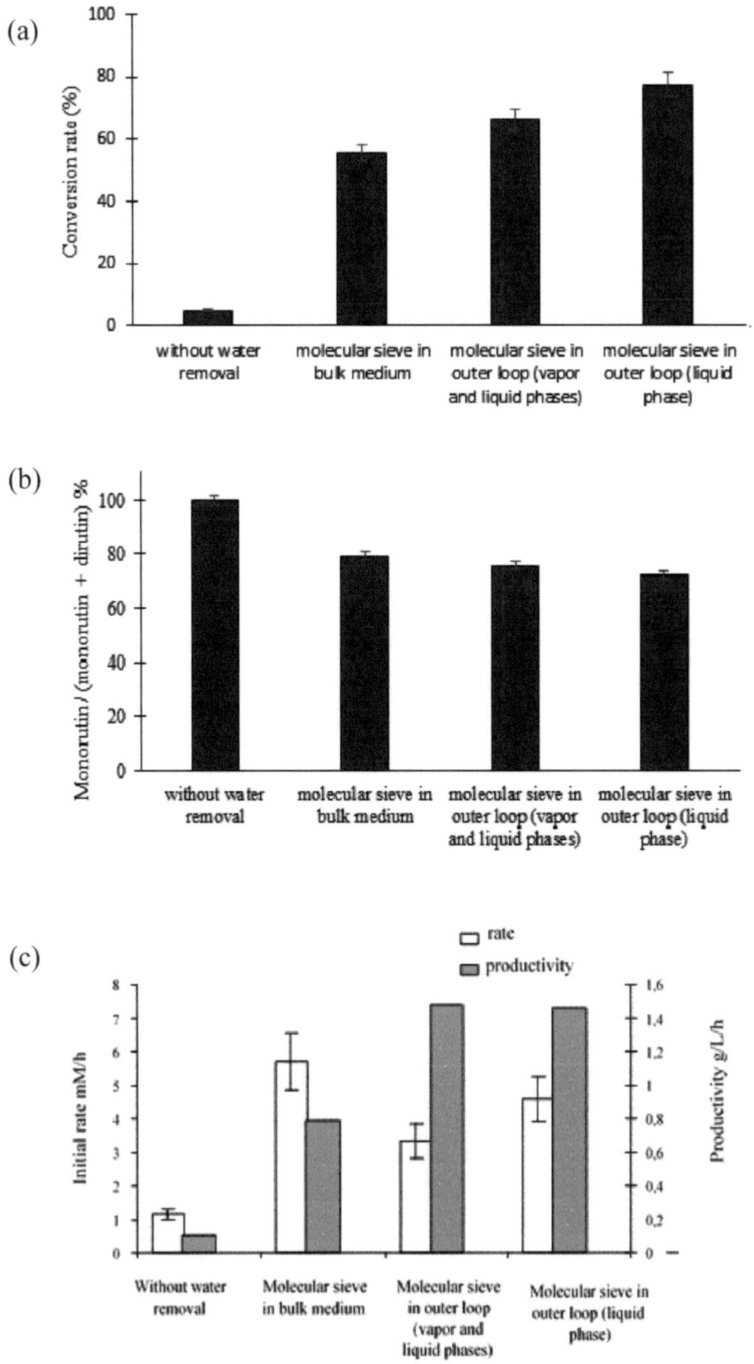

Figure 3. Effect of drying techniques on the esterification of rutin. Reactions carried out with 131 mM of rutin, 117.9 mM of acid, 30 g/L of Novozym 435 at 90°C. (a) Conversion rate of rutin at 50 h. (b) Selectivity of the reaction at 50 h. (c) Initial rate of rutin monoester formation and productivity at 50 h.

(10 h), the concentrations of rutin in the bulk medium are 42.3 ± 2.1 mM, 81.5 ± 4.0 mM, and 102.2 ± 5.1 mM, respectively for 60, 80, and 90°C. Fatty acids are totally soluble in *tert*-amyl alcohol at the studied concentrations and temperatures.

3.2.2. Effect on rutin esters synthesis

To investigate the effect of temperature on conversion rate, initial rate, and productivity, three set points of temperature (60, 80, and 90°C) were studied. Meanwhile, the other factors were kept constant at 65 mM of rutin, 58.5 mM of diacid, and 30 g/L of Novozym 435. NMR analysis showed that the two formed products are rutin 4‴-hexadecanedioate and dirutin 4‴, 4‴-hexadecanedioate (**Figure 2**).

Theodosiou et al. [23] reported similar results during their studies of enzymatic acylation of silybin by different dicarboxilic acids. They observed that at 50°C and after 96 h of incubation, both mono- and di-esters of silybin were synthetized but they identified only the monoester by NMR.

The behavior of the kinetics of rutin esters formation and the productivity of the reaction are summarized in **Figure 4a** and **b**. The highest performances were obtained at 90°C with 38.4 mM of total esters and a productivity of 0.78 g/L/h at 50 h, while the lowest values were found at 60°C with 20.8 mM and 0.38 g/L/h, respectively. Using esculin and palmitic acid as substrates, Lue et al. [16] reported similar effect of temperature. The observed results could be explained by the increase of rutin solubility, the decrease of the medium viscosity, and thus the increase of the mass transfer rate at high temperature. The increase of the temperature favors also the water removal and then the shift of the equilibrium to the product formation. These results suggested that the temperature has to be maintained as high as possible (≥80°C).

3.3. Effect of rutin concentration

During the acylation reaction of rutin by hexadecanedioate acid, three concentrations of rutin were tested (65, 131, and 196 mM) in the presence of hexadecanedioate acid (58.5, 117.9, and 176.4 mM, respectively), and 30 g/L of Novozym 435, in *tert*-amyl-alcohol at 90°C.

Figure 5 reports the obtained results concerning conversion yields (**Figure 5a**), selectivity, productivity (**Figure 5b**), and initial rate of rutin acylation (**Figure 5c**).

It appears that the initial rates of rutin acylation as well as the productivity increase with substrate concentrations (from 2.88 to 5.61 mM/h and from 0.78 to 2.56 g/(L/h), respectively), while conversion yields of both substrates (70% of rutin and 66% of hexadecanedioic acid) and selectivity of the reaction (76%) remain almost unchanged.

At 90°C, whatever the initial concentration, the rutin is totally soluble. Therefore, the behavior of the conversion rate and selectivity is rather due to the effect of molar ratio and not due to the solubility.

3.4. Effect of diacid/rutin molar ratio

The hydrophobicity of the reaction medium varies depending on the diacid/rutin molar ratio. Consequently, the selectivity of the reaction could be affected. This assumption was checked

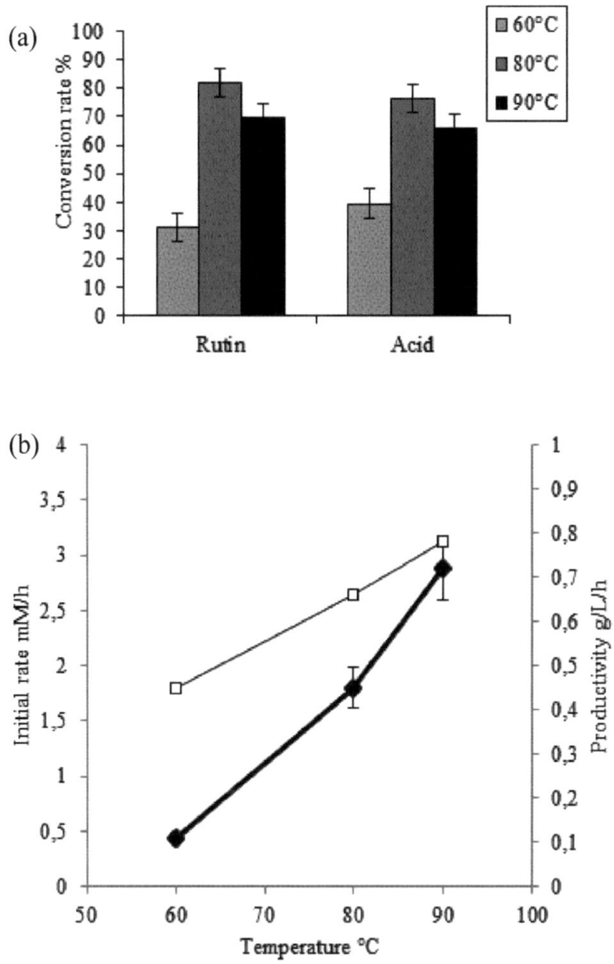

Figure 4. Effect of temperature on the performance of rutin hexadecanedioate synthesis. Reactions carried out with 65 mM of rutin, 58.5 mM of acid, and 30 g/L of Novozym 435 at 60, 80, and 90°C. (a) Effect of temperature on the conversion rate of rutin and hexadecanedioic acid at 50 h. (b) Effect of temperature on the initial rate (■) and productivity (□) at 50 h.

during a preliminary study with a diacid/rutin molar ratio of 0.05, 0.9, and 20. The obtained results showed that only rutin monoester was synthesized in the presence of a diacid/rutin molar ratio of 20 and both mono- and di-esters are produced with the two other lower ratios. In all cases, the rutin diester formation appears only after 5 h of incubation while the acid was completely depleted from the medium after 30–40 hours independently to molar ratio values. After the depletion of the fatty acid concentration, the monorutin ester became a substrate for the dirutin ester synthesis. At the end of the reaction (50 h), a plateau was reached (**Figure 6a–c**). These results suggested that rutin and diacid are better substrates to the enzyme than monorutin ester.

The effect of the molar ratio was investigated in several works. Similar results were observed by Ma et al. [27] during the acylation of isoorientin and isovitexin by Novozym 435.

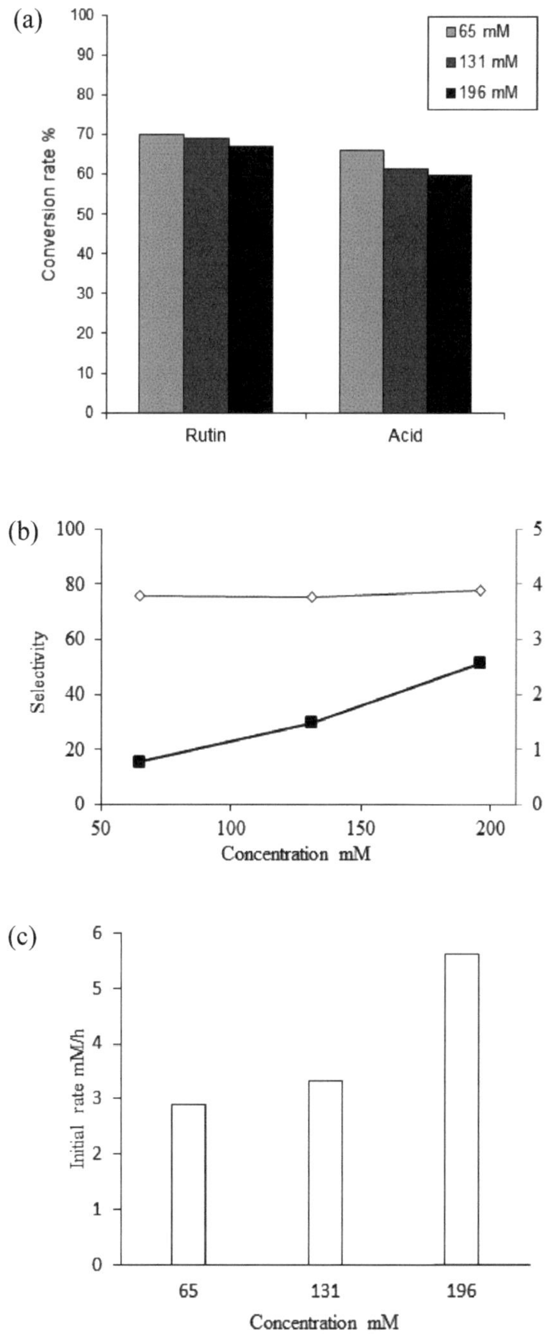

Figure 5. Effect of rutin concentration on the performance of rutin acylation. Reactions performed at 65, 131, 196 mM of rutin with 58.5, 117.9 and 176.4 mM, respectively of acid with 30 g/L of Novozym 435 at 90°C. (a) Rutin and hexadecanedioic acid conversion rate at 50 h. (b) Selectivity (□) and productivity at (■) 50 h. (c) Initial rate of rutin acylation.

http://dx.doi.org/10.5772/67621

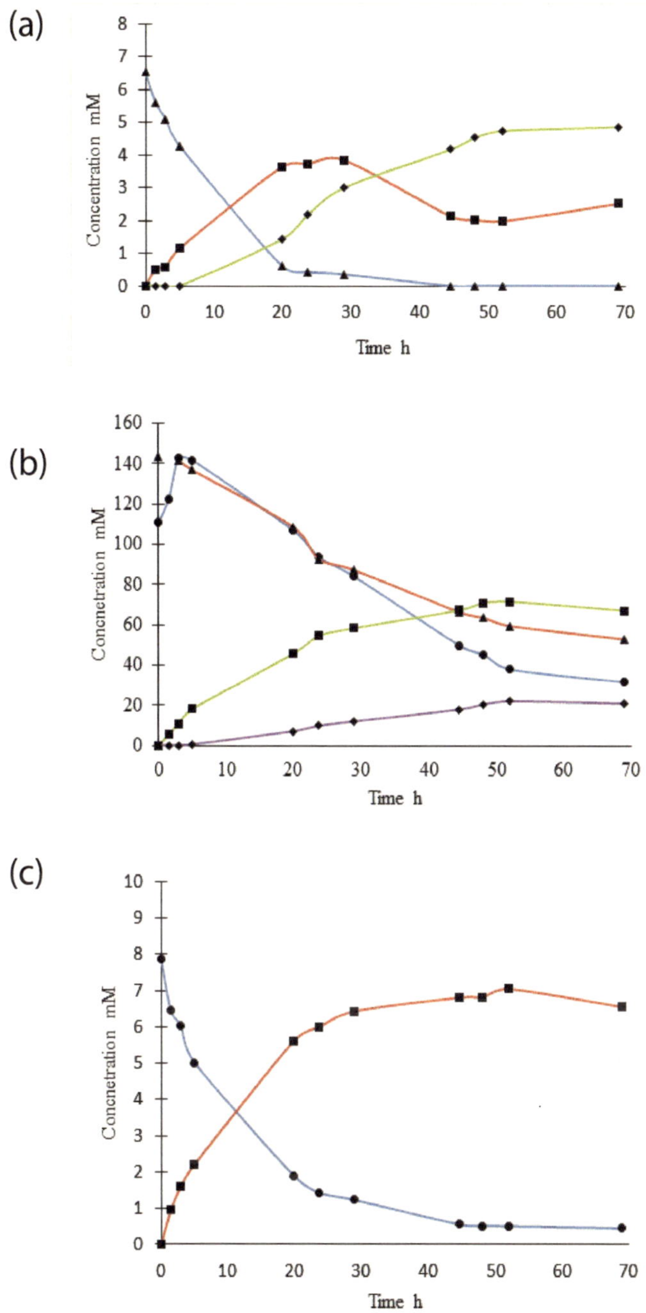

Figure 6. Kinetics of rutin esterification using different initial rutin concentrations. (a) Kinetics of rutin (131 mM) esterification reaction with hexadecanedioic acid (0.05 eq) by Novozym 435 (30 g/L) at 90°C, acid (▲), rutin hexadecanedioate (■) dirutin hexadecanedioate (♦). (b) Kinetics of rutin (131 mM) esterification reaction with hexadecanedioic acid (0.9 eq) by Novozym 435 (30 g/L) at 90°C, rutin (•) acid, (▲) rutin hexadecanedioate (■), dirutin hexadecanedioate (♦). (c) Kinetics of rutin (65 mM) esterification reaction with hexadecanedioic acid (20 eq) by Novozym 435 (30 g/L) at 90°C, rutin (•), rutin hexadecanedioate (■).

4. Conclusion

The effects of several factors on the acylation reaction of rutin by hexadecanedioate acid were investigated. The reactions conducted at 80 - 90°C showed high performances. The increase of rutin concentration from 65 mM to 196 mM led to the formation of mono ester (rutin 4‴-hexadecanedioate) and diester (dirutin 4‴, 4‴-hexadecanedioate). The molar ratio diacid/rutin (0.05 to 20) affects the selectivity of the reaction. In fact, in presence of excess of rutin both monoester and diester were synthesized. However, with diacid excess only the monoester was obtained. The results showed also that the water content of the media is a crucial factor. Drying the media by adding the molecular sieves in the outer loop of the reactor was the most efficient technique leading to water content lower than 200 ppm. In these conditions, the highest performances (conversion yield, initial rate) were reached at 90°C, 131 mM of rutin, 118 mM of acid, and 20 g/L of biocatalyst. At the equilibrium (50 hours), conversion yields of acid and rutin were respectively 73 and 74%, initial rate of monorutin ester formation was 2.20 mM/h and productivity was 1.44 g/(L/h). Depending on the water content and the diacid/rutin molar ratio, only mono- or both mono- and di-rutin esters were synthesized. For water content higher than 800 ppm, only rutin 4‴-hexadecanedioate was produced. At lower water content (<400 ppm), both rutin 4‴-hexadecanedioate and dirutin 4‴, 4‴-hexadecanedioate were observed. Higher values of diacid/rutin molar ratio favor the formation of monorutin ester.

Depending on the target application, the level of water content and molar ratio can be used to modulate the production of only mono- or both mono- and di-ester. The acylation reaction took place only on the glycosidic part, which is the main quality to preserve or enhance the biologic activity of flavonoids. In a further study, physicochemical and biological activity of synthesized esters will be evaluated.

Author details

Manel Slimane[1], Aude Cordin[2], Mohamed Ghoul[1] and Chebil Latifa[1]*

*Address all correspondence to: latifa.chebil@univ-lorraine.fr

1 Reactions and Process Engineering Laboratory, University of Lorraine, Vandœuvre-lès-Nancy, France

2 Compiègne University of Technology, Research Center, Compiegne Cedex, France

References

[1] Arora, A.; Nair, M.G.; Strasburg, G.M. Structure-activity relationships for antioxidant activities of a series of flavonoids in a liposomal system. Free Radic. Biol. Med. 1998; *24*, 1355-1363. doi: 10.1016/S0891-5849(97)00458-9

[2] Pietta, P.G. Flavonoids as antioxidants. J. Nat. Prod. 2000; *63*, 1035-1042. doi:10.1021/np9904509

[3] Rice-Evans, C.A.; Bagchi, D. Nutritional proanthocyanidins, flavonoids, and related phenols. Antioxid. Redox Signal. 2001; *3*, 939-940. doi:10.1089/152308601317203486

[4] Burda, S.; Oleszek, W. Antioxidant and antiradical activities of flavonoids. J. Agric. Food Chem. 2001; *49*, 2774-2779. doi:10.1021/jf001413m

[5] Di Carlo, G.; Mascolo, N.; Izzo, A.A.; Capasso, F. Flavonoids: old and new aspects of a class of natural therapeutic drugs. Life Sci. 1999; *65*, 337-353. doi:10.1016/S0024-3205(99)00120-4

[6] Middleton, E.; Kandaswami, C.; Theoharides, T.C. The effects of plant flavonoids on mammalian cells: implications for inflammation, heart disease, and cancer. Pharmacol. Rev. 2000; *52*, 673-751

[7] Le, K.; Chiu, F.; Ng, K. Identification and quantification of antioxidants in Fructus lycii. Food Chem. 2007; *105*, 353-363. doi:10.1016/j.foodchem.2006.11.063

[8] Foti, M.; Piattelli, M.; Baratta, M.T.; Ruberto, G. Flavonoids, coumarins, and cinnamic acids as antioxidants in a micellar system. Structure-activity relationship. J. Agric. Food Chem. 1996; *44*, 497-501. doi:10.1021/jf950378u

[9] Bohm, H.; Boeing, H.; Hempel, J.; Raab, B.; Kroke, A. Flavonols, flavones and anthocyanins as native antioxidants of food and their possible role in the prevention of chronic diseases. Z. Ernahrung. 1998; *37*, 147-163

[10] Mihara, R.; Barry, K.M.; Mohammed, C.L.; Mitsunaga, T. Comparison of antifungal and antioxidant activities of *Acacia mangium* and *A. auriculiformis* heartwood extracts. J. Chem. Ecol. 2005; *31*, 789-804. doi:10.1007/s10886-005-3544-x

[11] Kuleshova, L.N.; Hofmann, D.W.M.; Boese, R. Lattice energy calculation: a quick tool for screening of cocrystals and estimation of relative solubility. Case of flavonoids. Chem. Phys. Lett. 2013; *564*, 26-32. doi:10.1016/j.cplett.2013.02.008

[12] Chebil, L.; Humeau, C.; Anthoni, J.; Dehez, F.; Engasser, J.-M.; Ghoul, M. Solubility of flavonoids in organic solvents. J. Chem. Eng. Data. 2007; *52*, 1552-1556. doi:10.1021/je7001094

[13] Viskupicova, J; Ondrejovic, M; Maliar, T. Enzyme-mediated preparation of flavonoid esters and their applications. In: Prof. Deniz Ekinci (Ed.), Biochemistry. InTech; 2012. doi:10.5772/34174

[14] Ardhaoui, M.; Falcimaigne, A.; Ognier, S.; Engasser, J.; Moussou, P.; Pauly, G.; Ghoul, M. Effect of acyl donor chain length and substitutions pattern on the enzymatic acylation of flavonoids. J. Biotechnol. 2004a; *110*, 265-272. doi:10.1016/j.jbiotec.2004.03.003

[15] Chebil, L.; Humeau, C.; Falcimaigne, A.; Engasser, J.-M.; Ghoul, M. Enzymatic acylation of flavonoids. Process Biochem. 2006; *41*, 2237-2251. doi:10.1016/j.procbio.2006.05.027

[16] Lue, B.-M.; Guo, Z.; Xu, X. Effect of room temperature ionic liquid structure on the enzymatic acylation of flavonoids. Process Biochem. 2010; *45*, 1375-1382. doi:10.1016/j.procbio.2010.05.024

[17] Zheng, M.-M.; Wang, L.; Huang, F.-H.; Guo, P.-M.; Wei, F.; Deng, Q.-C.; Zheng, C.; Wan, C.-Y. Ultrasound irradiation promoted lipase-catalyzed synthesis of flavonoid esters with unsaturated fatty acids. J. Mol. Catal. B-Enzym. 2013; *95*, 82-88. doi:10.1016/j.molcatb.2013.05.028

[18] Ardhaoui, M.; Falcimaigne, A.; Engasser, J.-M.; Moussou, P.; Pauly, G.; Ghoul, M. Acylation of natural flavonoids using lipase of *Candida antarctica* as biocatalyst. J. Mol. Catal. B-Enzym. 2004; *29*, 63-67. doi:10.1016/j.molcatb.2004.02.013

[19] De los Rios, A.P.; van Rantwijk, F.; Sheldon, R.A. Effective resolution of 1-phenyl ethanol by *Candida antarctica* lipase B catalysed acylation with vinyl acetate in protic ionic liquids (PILs). Green Chem. 2012; *14*, 1584-1588. doi:10.1039/c2gc35196j

[20] Huo, Q.; Gu, X.; Lin, Q.; Sun, Y.; Xu, C.; Xiao, L. Selective esterification and antioxidant activity of rutin. Asian J. Chem. 2011; *23*, 3413-3416

[21] Viskupicova, J.; Danihelova, M.; Ondrejovic, M.; Liptaj, T.; Sturdik, E. Lipophilic rutin derivatives for antioxidant protection of oil-based foods. Food Chem. 2010; *123*, 45-50. doi:10.1016/j.foodchem.2010.03.125

[22] Jung, H.W.; Tschaplinski, T.J.; Wang, L.; Glazebrook, J.; Greenberg, J.T. Priming in systemic plant immunity. Science. 2009; *324*, 89-91. doi:10.1126/science.1170025

[23] Theodosiou, E.; Loutrari, H.; Stamatis, H.; Roussos, C.; Kolisis, F.N. Biocatalytic synthesis and antitumor activities of novel silybin acylated derivatives with dicarboxylic acids. New Biotech. 2011; *28*, 342-348. doi:10.1016/j.nbt.2011.01.006

[24] Gayot, S.; Santarelli, X.; Coulon, D. Modification of flavonoid using lipase in non-conventional media: effect of the water content. J. Biotechnol. 2003; *101*, 29-36. doi:10.1016/S0168-1656(02)00286-9

[25] Kontogianni, A.; Skouridou, V.; Sereti, V.; Stamatis, H.; Kolisis, F.N. Regioselective acylation of flavonoids catalyzed by lipase in low toxicity media. Eur. J. Lipid Sci. Technol. 2001; *103*, 655-660. doi:10.1002/1438-9312(200110)103:10<655::AID-EJLT655>3.3.CO;2-O

[26] Kontogianni, A.; Skouridou, V.; Sereti, V.; Stamatis, H.; Kolisis, F.N. Lipase-catalyzed esterification of rutin and naringin with fatty acids of medium carbon chain. J. Mol. Catal. B-Enzym. 2003; *21*, 59-62. doi:10.1016/S1381-1177(02)00139-X

[27] Ma, X.; Yan, R.; Yu, S.; Lu, Y.; Li, Z.; Lu, H. Enzymatic acylation of isoorientin and isovitexin from bamboo-leaf extracts with fatty acids and antiradical activity of the acylated derivatives. J. Agric. Food Chem. 2012; *60*, 10844-10849. doi:10.1021/jf303595e

Section 2

Flavonoid Synthesis and Their Role in Plants

Chapter 7

The Flavonol-Anthocyanin Pathway in Blackberry and Arabidopsis: State of the Art

Enrique Gutierrez, Ana García-Villaraco Velasco,
Jose Antonio Lucas, F. Javier Gutierrez-Mañero and
Beatriz Ramos-Solano

Additional information is available at the end of the chapter

http://dx.doi.org/10.5772/67902

Abstract

Flavonols and anthocyanins are plant secondary metabolites with an increasing interest due to their beneficial effects on human health. They are present in all plants, participating in plant protection against biotic and abiotic stresses. However, only some plant species accumulate them in relevant amounts, as is the case for berries. Among the health benefits reported is prevention of metabolic syndrome, s, including prevention of insulin resistance associated to type 2 diabetes. Therefore, there is a big interest to improve contents on plant foods to benefit health through the diet, as well as to obtain them for functional ingredients for food supplements. In fulfillment of this objective, a deep study about their biosynthetic pathway has been carried out in model plants, where the genome is available. However, not all species that accumulate them in high amounts have their genome sequenced, as is the case for blackberry. Transcriptomic approaches have been undertaken to gain knowledge of its specific biosynthetic pathway and regulatory elements, aiming to improve bioactive contents in the edible parts. Furthermore, determining the regulatory pathways will help to improve yields and in vitro production. For this purpose, a review on elicitors used to trigger this pathway is presented.

Keywords: flavonols, anthocyanins, blackberry, Rubus, Arabidopsis, elicitors, transcription factors (MYB)

1. Introduction

Blackberry fruits are an important source of bioactive compounds, among which are flavonols, anthocyanins, and catechins. These compounds exhibit beneficial effects on human health, which is the main reason they have become so popular nowadays, especially because they are easy to include in diets, at least in developed countries where incidence of pathologies is high. Among these health benefits are their general antioxidant effects, as all these compounds can scavenge reactive oxygen species (ROS), anti-inflammatory and antimicrobial effect, glucose metabolism, and leukocyte migration (effects in both inflammation and cancer) [1, 2]. Catechins and anthocyanins have also proved their ability to improve endothelial function, hypertension, coronary heart disease, obesity, insulin resistance, as well as glucose and lipid metabolism [3, 4]. Hence, including blackberry on human diet is beneficial for health, since a high content in flavonoids has been reported, and, in addition to its beneficial effects, it lacks adverse or secondary effects. Moreover, these compounds are also responsible for major organoleptic, nutritive, and processing characteristics of feed, food, and beverages, and impact many agronomical crop traits [5, 6].

In addition to the relevance for human health, flavonols, anthocyanins, and catechins play a key role on plant physiology. As secondary metabolites, they play a wide variety of functions, mainly helping the plant to adapt to the environment, among which the following have been reported, namely (i) UV radiation and oxidative stress protection, (ii) pathogen interactions (pathogen resistance), (iii) protection from herbivore, (iv) allelopathy, (v) nodulation (symbiosis), (vi) auxin transport, and (vii) they also attract different organism for pollination because of the color of the flower [1, 7, 8].

Despite the agricultural and biological importance of the genus *Rubus*, knowledge of their genetics and genome is very limited. Hence, in order to gain knowledge about the metabolism of these compounds, different approaches need to be undertaken, including the physiological, metabolic, and transcriptomic levels. One of the most important factors affecting the quality of the fruit is its content in phenolic compounds [6]. The synthesis of these compounds mainly depends on the phenylpropanoid and its derivative pathways, which starts with the phenylalanine, as many others (**Figure 1**). Furthermore, validation of the mechanisms controlling this biosynthetic pathway needs to be referred to model plants; among which, the most commonly used is *Arabidopsis thaliana*. However, the use of Arabidopsis is limited since it lacks edible fruits, so for studies related to fruits, either strawberry or tomato is used as reference.

Rubus sp. is a woody plant that belongs to *Rosaceae* family, in which other plants such as strawberry (*Fragaria vesca* L.), raspberry (*Rubus idaeus* L.), cherry tree (*Prunus avium* L.), or apple (*Malus domestica*) among others are enclosed; all of them belong to a wide group of plants commonly called *berries*. They are known because of their high accumulation of secondary metabolites in the fruit, which behave as functional components, beneficial for human health. Those compounds are mainly polyphenols like flavonols, anthocyanins, and catechins, strong natural antioxidants [9–11].

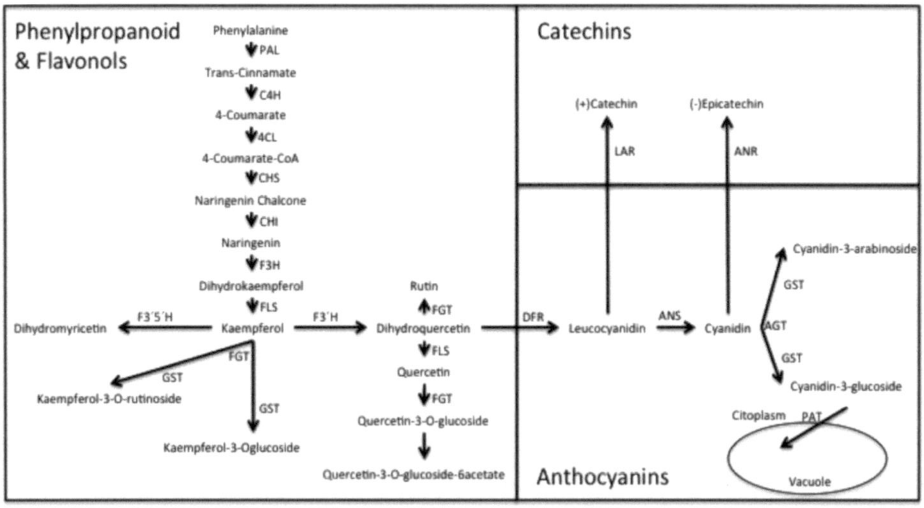

Figure 1. Phenylpropanoid pathway.

Rubus L. genus is composed of 600–800 species approximately, distributed all around the world in very different habitats, and classified into 12 different subgenus that are in turn divided into different groups. The most abundant species in Europe is the *Rubus* spp. subgenus (Eubatus Focke) that is divided in more than 130 species; their natural habitat and distribution is in the woods (**Table 1**).

Subgenus	Number of species
Anoplobatus (Focke)	6
Chamaebatus (Focke)	5
Chamaemorus (Hill)	1
Comaropsis (Rich.)	2
Cylactis (Raf.)	14 (4 series)
Dalibarda (L.)	5
Dalibardastrum (Focke)	4
Idaeobatus (Focke)	117 (9 sections)
Lampobatus (Focke)	10
Malachobatus (Focke)	115 (7 sections)
Orobatus (Focke)	19
Rubus L. (=Eubatus Focke)	132 (6 sections)

Table 1. *Rubus* subgenus.

Plants that belong to the subgenus *Rubus* spp. are typical wild species and are usually handpicked in the season. However, since it became an important plant for agriculture, there has been an increasing interest in improving the size of the fruit, the organoleptic properties, fruit yield, and get rid of the thorns, since they constitute a nuisance for harvest. To achieve these objectives, classic crossbreeding has resulted in development of many commercial varieties to favor a given trait that benefits production in each geographical location. Among these cultivars are "Ashton Cross" that is vigorous and thorny, "Bedford Giant" that in addition to these two traits shows a good yield; "Black satin," also vigorous but thornless; "thornless evergreen" that provides a thornless plant, high yield and high quality fruits; "Fantasia" that produces very large fruits and finally in this shortlisted group is "Loch Ness," that is a thornless cultivar with very large fruits and semierect canes, which is the cultivar used in this study (**Table 2**) [12].

Rubus spp. Var. Loch Ness is a high yielding thornless tetraploid ($4n = 28$) blackberry, and one of the most widely cultivated varieties. However, despite its high-added economic value and as a source of bioactive compounds, its genome has not been sequenced yet. Therefore, other strategies need to be used to gain knowledge of the production and health-related benefits.

The aim of this chapter is to review the literature about blackberry and report the state of the arts about this plant species. As the genome is not reported, data about the core genes in the biosynthetic pathway as well as regulatory genes are referred to as the model plant *A. thaliana*. Also, structure of the bioactives which is responsible for health benefits as well

	Variety	**Anthocyanins (mg eq cyanidin/100g FW)**	**Flavonols (mg eq catechin/100g FW)**
Thorny	Darrow	99.33 ± 48.32	4.23 ± 0.48
	Early Wilson	64.76 ± 23.68	3.21 ± 0.38
	Gazda	27.97 ± 13.71	6.52 ± 0.82
	Lesniczanka	96.63 ± 32.18	5.36 ± 0.45
	Zagroda	143.66 ± 52.59	4.98 ± 0.52
Thornless	Black Beaty	179.46 ± 57.84	3.06 ± 0.40
	Black Satin	175.52 ± 53.97	1.58 ± 0.31
	Chester Thornless	200.34 ± 65.58	3.68 ± 0.46
	Hull Thornless	105.39 ± 31.08	0.82 ± 0.35
	Loch Ness	220.11 ± 81.07	2.39 ± 0.28
	Orkan	142.42 ± 44.01	2.26 ± 0.25
	Smoothstern	186.55 ± 58.94	2.32 ± 0.33
	Tayberry	177.84 ± 56.20	1.7 ± 0.32
	Thornless	147.46 ± 44.02	1.07 ± 0.33

Table 2. Phenolic compound contents of thorny and thornless blackberries (mg/100g FW) [13].

as their qualitative and quantitative composition in berry fruit and in model plant are presented, so a relationship between composition and core and regulatory genes description is envisaged and vice versa. Finally, the physiological role of these secondary metabolites for plants is also presented, and finally highlights the relevance of this pathway of plant secondary metabolism, as well as its biotechnological potential.

2. Chemical structure

Flavonols, anthocyanins, and catechins are molecules belonging to a wider group of secondary metabolites, the flavonoids. Flavonoids represent a large subgroup of a phenolic class of plant specialized metabolites, which are found in almost every plant in the nature. The basic flavan skeleton that forms all flavonoids is a 15-carbon phenylpropanoid core (C6-C3-C6 system), which is arranged into two aromatic rings (A and B) linked by a heterocyclic pyran ring (C). They are characterized by the presence of a double bond between C-2 and C-3, and the attachment of the B ring to C-2. According to the oxidation status and saturation of the heterocyclic ring, flavonoids are categorized into flavonols, flavones, catechins, flavanones, anthocyanins, and isoflavonoids [1]. The most abundant compounds present in blackberry (also in berries) are flavonols, anthocyanins, and catechins (**Figure 2**).

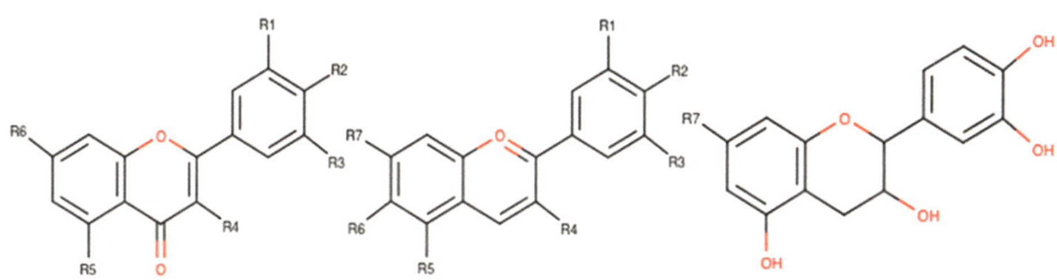

Position	Common substituents
R4 (Flavonol, anthocyanin)	Rutin, arabinose, glucose, galactose, OH, Me, OMe
R1, R2, R3 (Flavonol, anthocyanin)	H, OH, OMe, Me
R5, R6 (Flavonol)	H, OH, OMe, Me
R5, R6, R7 (Anthocyanin)	H, OH, OMe, Me
R7 (Catechin)	OH, Glucose

Figure 2. Flavonol, anthocyanin, and catechin molecular structures and common substituents.

Flavonols have a 3-hydroxyflavone (IUPAC name: 3-hydroxy-2-phenylchromen-4-one) as the main structure. The diversity of these compounds is derived from the different positions of the hydroxyl groups of the phenolic ring that are usually glycosylated and can undergo further modifications like acylations; in this group, the three main families are derived from kaempferol (4′OH), quercetin (3′, 4′, 5′OH) and rutin (3′, 4′OH).

Anthocyanins are mainly glycosylated as well, being the aglycon the anthocyanin molecule. The chemical structure of this aglycone is the flavylium ion (2-phenyl-benzopirilo) that has a benzopyran aromatic ring, and a phenolic ring. There are six different families within this group, namely cyanidin, pelargonidin, delphinidin, malvinidin, peonidin, and petunidin. As in the case of flavonoids, the greatest source of chemical diversity is the number and position of sugars for glycosylation. Acylation is another main biochemical mechanism leading to diverse anthocyanin molecules in *Arabidopsis* [14, 15]. Up to date, several enzymes have been characterized to catalyze these acylation reactions, using either malonyl-CoA or *p*-coumaroyl-CoA as substrates to transfer the malonyl or *p*-coumaroyl groups to cyanin structures [16]. Diversity can be further increased transferring sinapoyl groups to cyanins to form sinapoylated cyanins [17].

Catechins have two benzene rings (A-, B-) and a dihydropyran heterocyclic ring (C) with a hydroxyl group over carbon 3. As a result of this structure, catechins have four diasteroisomers, two with *trans* configuration called catechin ((+)-catechin and (−)-catechin), and two with *cis* configuration called epicatechin ((+)-epicatechin and (−)-epicatechin). These catechins can further polymerize to form proanthocyanins, in which the diversity of structures relies on the number of monomers that polymerize and the type of bonds that stabilize them.

3. Biosynthetic pathway and regulation

Biosynthesis of the flavone backbone is originated from the phenylpropanoid pathway followed by the flavonoid biosynthetic branch (**Figure 1**). The phenylalanine ammonia-lyase (PAL) deaminates the phenylalanine, being converted into trans-cinnamate, after that a hydroxyl group is introduced on the phenyl ring by cinnamic acid 4-hydroxylase (C4H), being the trans-cinnamate converted into 4-coumarate. The carboxyl group of p-4-coumarate is then activated to form 4-coumarate-CoA (by a thioester bond), catalyzed by 4-coumarate-CoA ligase (4CL). This product, 4-coumarate-CoA, is substrate for different enzymes, so it represents a branching point of the pathway to either stilbenes of flavonoids. In this case, 4-coumarate-CoA is then condensed with three units of malonyl-CoA by the chalcone synthase (CHS, first enzyme of the flavonoid pathway), forming the naringenin chalcone (flavonone), which is transformed into naringenin by the chalcone synthase (CHI). Naringenin is hydroxylated by flavonone-3-hydroxylase (F3H) being converted in dihydrokaempferol, which is then hydroxylated by flavonoid-3′-hydroxylase (F3′H) and transformed in dihydroquercetin or by flavonoid-3′-5′-hydroxylase (F3′5′H) to form dihydromyricetin. Flavonols are synthesized at this point by the flavonol synthase (FLS), which introduces a double bond between C2 and C3 in either of the three above-mentioned molecules forming kaempferol, quercetin, or myricetin, respectively. Dihydroquercetin is reduced by dihydroflavonol reductase (DFR) to obtain leucocyanidin; similarly, dihydrokaempferol is transformed in leucopelargonidin

http://dx.doi.org/10.5772/67902

and dihydromyricetin in leucodelphinidin. Anthocyanins are synthetized at this point by the anthocyanidin synthase (ANS) obtaining cyanidin, pelargonidin, or delphinidin, respectively. Catechins include (+)-catechin and (−)-epicatechin; (+)-catechin is obtained when leucocyanidin reductase (LAR) reduces leucocyanidin, and (−)-epicatechin is obtained when anthocyanidin reductase (ANR) reduces cyanidin [1, 7, 8].

All these aglycons are highly apolar, so they are immediately glycosylated to increase polarity, in order to be stored in vacuoles or translocated throughout the plant, hence glycosiltransferases are very important for glycosylation as well as transport mechanisms. In *Arabidopsis*, three genes, *TT12*, *TT19*, and *AHA10*, have been functionally characterized to be associated with the transport of anthocyanins. However, these enzymes show different levels of specificity for flavonoids [18]. Anthocyanins are stored in the central vacuole of cells, so they need to be transported from the cytosol to the vacuole. Two major hypotheses have been proposed to solve this transport: either transporter-mediated or vesicle-mediated transport [19–21].

Plants have depicted a system in which all these enzymes are extremely well organized in the different compartments within cells, in order to improve efficiency of these natural products' synthesis. Successive enzymes are arranged in imaginary units termed metabolons, anchored to the ER membrane, ensuring channeling of the intermediate precursors in the complex without diffusing to the cytosol, avoiding metabolic interferences [18, 22].

A. thaliana is a good model species for the identification of genes controlling flavonoid metabolism (**Table 3**) [23], because all pathway core genes of anthocyanins have been molecularly, genetically, and biochemically characterized in this plant. On the other hand, it is amenable to both molecular and classical genetic analysis [24, 25].

On the other hand, homologues to all core genes in the flavonol-anthocyanin pathway have been identified in *Rubus* sp. Var Loch Ness [26]. Also, most genes corresponding to the MYB transcription factors have also been identified with similar functions. Interestingly, MYB12 [26] that was originally identified as a key flavonol-specific transcriptional activator in *A. thaliana* [27] and in

Enzyme	Regulator
CHS (chalcone synthase)	*MYB12, MYB11, MYB111*
CHI (chalcone isomerase)	*MYB12, MYB11, MYB111*
F3H (flavonol 3 hydroxylase)	*MYB12, MYB11, MYB111*
F3'H (flavonol 3'hydroxylase)	*MYBL2*
FLS (flavonol synthase)	*MYB12, MYB111, MYB11*
DFR (dehydroflavonol reductase)	*TT2, TT8, TTG1, MYBL2*
ANS (anthocyanidin reductase)	*TT1, TT2, TT8, TTG1, TT16*
ANR (anthocyanidin synthase)	*TT2, TTG1*
LAR (leucoanthocyanidin reductase)	*TT2, TT8, TTG1*

Table 3. List of the flavonol-anthocyanin pathway core and regulatory genes in *A. thaliana* [23].

other plant species such as tomato [28] has not been found in *Rubus* [26, 29] suggesting a different control mechanism of the flavonol-anthocyanin pathway in this plant species.

Currently, dihydrokaempferol and dihydroquercetin are the only two dihydroflavonol molecules identified in *Arabidopsis* [18]. Flavonoids have been analyzed using liquid chromatography-mass spectrometry (LC-MS) and/or nuclear magnetic resonance (NMR). Briefly, anthocyanins and glycosylated kaempferol flavonols are mostly found in leaves [17], whereas seeds contain epicatechin, PAs, and larger amounts of glycosylated quercetin flavonols [30, 31]. Interestingly, arabidopsis seeds contain large amounts of PAs similar to those present in other crop seeds or fruits (**Table 4**) [32–34].

Group	Compound
Flavonols	Quercetin-rhamnoside-hexoside
	Quercetin-hexoside-rhamnoside
	Quercetin-3-O-rhamnoside
	Quercetin-rhamnoside dimer 1
	Quercetin-rhamnoside dimer 2
	Quercetin-rhamnoside dimer 3
	Quercetin-rhamnoside dimer 4
	Quercetin-di-rhamnoside
	Quercetin-3-O-glucoside
	Kaempferol
	Kaempferol-rhamnoside
	Kaempferol-rhamnoside-hexoside
	Kaempferol-3, 7-di-O-rhamnoside
	Kaempferol-3-O-glucoside-7-O-rhamnoside
	Isorhamnetin-hexoside-rhamnoside
	Isorhamnetin-di-rhamnoside
	Isorhamnetin-rhamnoside
Anthocyanins	Procyanidin dimer
	Procyanidin trymer
	Procyanidin tetramer
	Procyanidin pentamer
	Procyanidin hexamer
	Procyanidin heptamer
Catechins	Epicatechin

Table 4. Compounds identified by LC-MS-MS in seed extracts of wild type *A. thaliana* [31].

The metabolic profile of flavonols and anthocyanins in blackberry fruits is formed by the flavonols kempferol and quercetin and their respective derivatives, while cyanidin derivatives are the unique anthocyanidins present. Interestingly, catechins and epicatechins are also present, especially upon fruit ripening [26, 35]. The specific composition for blackberries obtained from *Rubus spp*. Var Loch Ness appears in **Table 5**.

Group	Compound
Flavonols	Rutin
	Kaempferol-glucoside
	Quercetin-glucoside
	Kaempferol-rutinoside
	Quercetin-3-O-glucoside-6″-acetate
Anthocyanins	Cyanidin-3-glucoside
	Cyanidin-3-arabinoside
Catechins	(+)-Catechin
	(−)-Epicatechin
	Epicatechin isomer

Table 5. Compounds identified in *Rubus* spp. Var. Loch Ness fruit by LC-MS-IT-ToF [26].

4. Flavonol, anthocyanin, and catechin functions in plants

Plants are sessile organisms, due to this fact they have developed different methods for protection against the stressful conditions of the surrounding, including abiotic and biotic stimuli. The most important mechanism is the production of secondary metabolites, like flavonols, anthocyanins, and catechins [36, 37].

There is an increase in the production of these compounds under adverse or stressing conditions, such as intense UV radiation, heat, drought, and salt stress, presence of heavy metals, herbivores, insects, nematodes, etc., because reactive oxygen species (ROS), the natural products obtained from metabolic reactions, play a relevant role in cell signaling and homeostasis. In certain situations, as previously described, ROS levels can undergo a mild increase, triggering defensive responses as SAR or ISR [38], or suffer a dramatic increase that results in cell damage (lipids, DNA, and protein structures) if not controlled.

Living beings have different methods to get rid of these ROS, enzymatic and nonenzymatic; but if the amount is too high the organism cannot transform all of them, causing the damages described before. Phenylpropanoids and flavonoids, in which flavonols, anthocyanins, and catechins are enclosed, are the nonenzymatic antioxidants known to have high antioxidant activities, because of their capacity to directly quench ROS, thanks to the hydroxyl group present in rings A and B; they also interfere over the enzymatic systems composed of cyclo-oxygenase (in animals only), lipoxygenase, glutathione S-transferase, and xanthine oxidase, which is the other system in charge of ROS removal, together with the SOD-APX and the ascorbate-glutathione cycle enzymes that contribute to ROS control [39]. Hence, these compounds are involved in fine tuning of defensive and adaptive metabolisms, integrating all the external information, to optimize plant energetic resources for survival.

Flavonoids are known to be nonessential regulators auxin transport, modulating different transporters such as PIN proteins, and the transporter superfamily (ABCB) [40–42] proteins involved in their transport along the plant. It has been demonstrated that changes in flavonols accumulation lead to changes in auxin transport, therefore changes in auxin distribution [43], and the corresponding changes in plant physiology.

4.1. Abiotic stress

Abiotic stress is defined as the negative impact caused by the nonliving factors in the plant. Under adverse conditions, like intense UV radiation, heat, drought, and salt stress, presence of heavy metals, etc., there is a high increase of the reactive oxygen species (ROS) that lead to signal transduction to activate plant defense or to oxidative damage, as described above.

The UV radiation causes a stressful situation for plants [44], which is handled in two ways. First of all flavonoids and other pigments present mainly in the outer parts of the plant (epidermis and mesophyll tissues) absorb and considerably reduce the amount of radiation; the second one would consist in decreasing the effect of ROS caused by the radiation by scavenging of ROS [7]. Among flavonols, the main compound related to light absorption is kaempferol 3-O-glucoside because of its monohydroxy B-ring, and the flavonol with the greatest antioxidant properties is quercetin 3-O-glucoside, because of its dihydroxy B-ring. It has been shown that upon different UV exposure, synthesis of phenolic compounds is increased [45]. This may be the primary mechanism of response, which can be followed by others such as accumulation of pigments or lignification processes. Hence, flavonoids and anthocyanins are involved in protection against oxidative stress due to high UV radiation.

The impact of drought and salt stress on flavonoid biosynthesis has been studied in *A. thaliana* [46, 47]. An increase of glycosides of quercetin, cyanidin, and kaempferol during drought stress has been reported, being kaempferol glycosides the most significantly increased [47]. Although the behavior of flavonoids during these types of stresses is still not well documented, this evidence their role against salt and drought stress.

These studies carried out in *A. thaliana* are very convenient to elucidate the mechanism of action of these flavonoids and to see the flavonoid profile. However they cannot be directly extrapolated to evaluate behavior in fruit production, fruit quality, or fruit endurance, as *A. thaliana* does not have edible fruits. For these purposes, other model plants are used, such as tomato or strawberry among the berries.

Concerning fruit quality, there is a great concern in the endurance of the fruit after harvest; it is one of the most important traits for commercial value and economic profit. The relationship between the overripening and the antioxidant properties has been evaluated [48]. A study on tomato overexpressing AtMYB12, the transcription factor activating the flavonol anthocyanin pathway, showed a notable increase in flavonoid biosynthesis, as well as its antioxidant capacity. The high anthocyanin and high flavonol profiles resulted in a longer, more durable shelf life, comparing with control plants, indicating that the endurance is directly correlated with this profile. Based on this data, it seems that the overripening time is determined by the oxidative damage of the fruit under changing conditions [49]. Therefore, an increase in flavonoids and anthocyanins is related to better fruit quality during the postharvest period.

Another stress factor is the levels of heavy metals. As a consequence of industrial development, pollution with heavy metals has dramatically increased. Heavy metals toxicity can result from different mechanisms, the first one is the generation of ROS by Fenton reaction and autoxidation [7], blocking of essential functional groups in biomolecules, and displacement of essential metal ions from biomolecules. Cadmium and other metals provoke a depletion of GSH and inhibit mainly the glutathione reductase (among other enzymes implicated in the ROS cycle) [45]; in consequence, the plant has to increase dramatically other antioxidants such as flavonoids in order to keep a normal the normal homeostasis of the plant cells. Flavonoids are known to form specific union with heavy metals, providing a great adaptation method to heavy metals toxicity autoxidation [7]. Based on these characteristics, some applications derived from these studies have been proposed to improve survival of plants in hostile environments, for example, increasing flavonoid synthesis to allow plant growth in the presence of heavy metals, so soil detoxification can be achieved by phytoremediation [50].

4.2. Biotic stress

Flavonoids are important molecules for plant adaptation under adverse conditions, among which defense to biotic stress is included. These molecules have a nonspecific mechanism of action; their effect is partly derived from their antioxidant properties, because of the ROS generated by plants when they are attacked by some pathogen. Flavonoids are involved in the earliest defense mechanism and the programmed cell death, and they have been found in necrotic and adjacent cells to pathogen invasion in the hypersensitive response [46, 51].

Their role in defense is not limited to the hypersensitive response, since consistent with their ability to chelate metals, they are able to inhibit some pathogen enzymes, mainly those involved in digesting the cell wall by chelating metals, blocking, or retarding pathogen invasion [52]. Different studies have shown that there are different mechanisms of action against pathogen infection, inhibition of cellulases, pectinases, and xylanases, chelation of metal ions that belongs to cell membranes and enzymes, and more general detoxifying cells of ROS [45].

They can also affect bacterial DNA synthesis, by interacting with DNA gyrases, as the B ring of flavonoids can form hydrogen bonds with nucleic acid bases, or by direct interaction with the ATP binding site of the gyrase, leading to an inhibition of the synthesis of new DNA. This may be their method of protection against virus [45].

Antifungal properties have also been proved for flavonoids; these properties depend on their structure, for example dihydroquercetin has proved to be much more active against

Fusarium sp. infections than other types of flavonols, and it is believed that is due to the hydroxyl groups [53]. In addition to the antifungal effects reported in plant, some have shown that certain compounds like phenols, phenolic acids, flavonoids, and isoflavonoids inhibit pathogen proliferation in the rizhosphere, preventing root infections.

4.3. Other functions

Flavonoids play a very important role in symbiotic bacteria relations. Bacteria belonging to the family Rhizobiaceae include several genera, each of them specific to a legume species. Rhizobiaceae are capable of fixing nitrogen for the plant; in exchange they obtain photosyntates. First, they need to establish the symbiotic relationship and form the nodule; in this process, flavonoids are key since these bacteria are attracted by these flavonoids that are specific signals for each rhizobia-legume couple. There are studies of different plants growing in soils with low nitrogen concentration that induces the accumulation of flavonoids [54]. Based on this fact, knowledge of the specific flavonoids that enhance symbiosis establishment could be applied to field production of legumes, in low productivity soils, to enhance nodulation, which in turn, will enhance yield in developing areas. This goal could be achieved at a low cost and easily implemented in local areas therefore contributing to food security, as marked by the FAO.

Connecting with this improvement in production and also with their natural physiological role, flavonoids provide color, taste, and fragrance to the fruit and seeds, and also play an important role in pollination, because these characteristics attract insects [45]. Although these characteristics may attract some organisms can also deter some others, in the cases of herbivores and some nematodes, avoiding to be eaten by these living beings.

5. Flavonols', anthocyanins', and catechins' health properties

Flavonoids are known because of their beneficial effect on human health. It has been known for long time and mainly attributed to its antioxidant potential. Compounds able to scavenge free radicals are in general beneficial for health. As all living beings, humans also produce ROS and there is also a system to get rid of these free radicals to prevent damage of the cells (DNA, lipids, and proteins). These damages are called oxidative damages which have been related to carcinogenesis, neurodegeneration, atherosclerosis, diabetes, and aging; however, the precise underlying mechanisms for these health benefits are starting to be unraveled.

Flavonoids have received increasing attention due to their anti-inflammatory, antimicrobial, and anticancer activities. Structural-functional relationship analyses identified luteolin as one of the most potent inhibitors of xanthine oxidase, a key enzyme in ROS production. Reduction of ROS by apigenin prevents endothelial damage during acute inflammation and restores mitochondrial function. Most of the anti-inflammatory and antimicrobial activities attributed to flavones seem to be centered on their ability to regulate the Toll receptor (TLR)/NFκB axis. This is a central pathway in the host-pathogen interplay in mammals, responsible for the expression of inflammatory mediators, including tumor necrosis factor (TNFα), interleukin-1 (IL-1β) and cyclooxygenase-2 (COX-2), an enzyme mediating the conversion of arachidonic acid to prostaglandins. Notably, great similarities are found between the mammalian TLR/NFκB and plant pathogen defense pathways, suggesting that flavones may regulate evolutionary conserved targets [55]. It has also been reported that in animal models, apigenin reduces the phosphorylation of the NFκB p65 subunit, required for its transcriptional activity. Inhibition of p65 phosphorylation reduces the expression of inflammatory cytokines, limiting the cell damage characteristic of acute inflammation [56]. Other flavones inhibit COX-2 by halting NFkB nuclear localization [57]. Overall, glycosides show less anti-inflammatory activity than aglycones, probably a consequence of their reduced cellular absorption [58]. Recent studies identified additional mechanisms responsible of the anti-inflammatory activity of flavones, including the regulation of noncoding RNAs. Large microRNA screenings showed that apigenin reduces microRNA155 (miR155) expression, a main inflammatory regulator miR155 binds to 3′-UTR regions of several inflammatory cytokines, suggesting an additional mechanism by which flavones can restore homeostasis during acute inflammation, independent of their anti-oxidant activity [1].

Consistent with the ability of flavones to regulate inflammation, interventions with the Mediterranean diet, which is rich in flavonoids, showed improved cardiac function, reduced hypertension and obesity [59, 60]. Flavones also affect leukocyte migration, with very specific targets, deeply affecting cancer and inflammation [61, 62]. Flavones ability to reduce cell migration has great impact on cancer, suggesting alternative therapeutic approaches to reduce metastasis. The anticarcinogenic effect of flavones is given in part by their ability to induce DNA damage, and is accompanied by cell cycle arrest at G1 or G2, depending on the particular cell type. Interestingly, the ability of apigenin to induce cell death in cancer cells is independent of ROS production [63] supporting a beneficial role of flavones independent of their anti-oxidant activity.

Identification of the direct targets will highly contribute to understand the molecular mechanism related to flavones and health. The use of PD-Seq (phage display high-throughput sequencing), a novel approach for small target identification, identified several targets, suggesting that dietary compounds, unlike pharmaceuticals, may target several molecules [64]. This statement encourages the use of healthy plant-based foods or extracts, rich in polyphenols but with a complex mixture of compounds that will contribute to prevent the onset of disease by reaching many small targets simultaneously. Under this rationale, a study of naturally healthy fruits or plant materials is seriously encouraged to prevent the onset of disease.

As flavonoids, anthocyanins' health-promoting effects have been frequently linked to their high antioxidant activities. However, there is increasing evidence reporting that some of their biological effects may be related to their ability to modulate mammalian cell signaling pathways [65, 66]. Anthocyanins also offer protection against certain age-related degenerative diseases cancers, cardiovascular disease [9, 55, 67, 68]; anti-inflammatory activity [69], promotion of visual acuity [70], and hindering obesity and diabetes [71, 72] have also been reported as beneficial effects of these compounds.

In addition to the many target-specific effects of each compound detailed above, effects are more complex to evaluate when any of these phenolics are delivered through the diet in a complex food matrix. The variability of effects relays in two points: on the one hand, natural variability in composition and on the other hand, variability in absorption at the individual level. It has been estimated that only 5–10% of the total polyphenol intake is absorbed in the small intestine. Currently, it is estimated that 500–1000 different microbial species inhabit the gastrointestinal tract. However, they do not seem to be ubiquitous but reflect the interpersonal differences in the gut microbial community [73]. Consequently, apart from the interindividual variation in daily intake of polyphenols, interindividual differences in the composition of the gut microbiota may lead to differences in bioavailability and bioefficacy of polyphenols and their metabolites [74, 75].

The other factor that will condition effects on health is intimately associated with the sessile nature of plants. Plants have to overcome environmental changes by changing their chemical composition, synthetizing metabolites that will contribute to a better adaptation to changes in abiotic factors of to fight back biotic challenges. Since environmental conditions are variable along the year, and flavonols and antocyanins play a role in adaptation to UV stress, it may be anticipated that concentration in plant will be higher in spring and summer when light hours and intensity are higher. Hence, fruits produced in winter or in summer will presumably have different concentrations, as has been demonstrated in blackberries [76]. Moreover, given their role in plant defense, their levels may also fluctuate depending on disease prevalence along a given season, and therefore, health benefits will be different, since the dose is different. Consequently, any attempt to modulate the amplitude of these fluctuations will result in enhanced fruit quality, more reliable in terms of health benefits.

In order to achieve this goal, understanding the metabolic pathway and its regulation is a milestone on the way to develop varieties in which the main regulators are overexpressed to ensure a high and constant, or low variability, fruit bioactive contents. This goal may be achieved through crossbreeding or by the means of metabolic engineering in plants [77] or through elicitation of secondary metabolism with external agents such as beneficial bacteria or derived molecules [76, 78] or even other chemical molecules such as salicylic acid.

6. Elicitation

An alternative to new varieties is using beneficial bacteria or parts of the same as a tool to trigger plant metabolism in field production in order to cause a mild-biotic stress in the plant that smoothly and constantly triggers secondary metabolism to achieve constant concentrations of bioactives in the edible fruits [77, 78]. The ability of many beneficial bacteria to trigger plant metabolism in different species has been reported and there is increasing evidence of bacterial derived elicitors with the same effect [79]. These bacterial derived molecules are termed as MAMPs (microbe-associated molecular patterns) and can be either structural molecules from the bacterial cell wall, or derived metabolites that bacteria release to trigger the plant in their intimate relationship. Each MAMP or strain has to be evaluated for each plant species, since genome-genome specificity has been occasionally reported.

To support the value of this strategy in the study of blackberry metabolism, a transcriptome analysis from field grown blackberry fruits was performed to study this plant species; contigs were obtained and blasted to the genome of *Fragaria*, the model plant within the Rosaceae finding 73.5% similarity with *Fragaria vesca* subspecies *vesca* [80]. Once the gene information was available, field grown blackberry plants were inoculated with a beneficial bacteria strain (*Pseudomonas fluorescens* N21.4) as a biostimulant, aiming to trigger flavonoid biosynthesis as part of an induced systemic response (ISR). Fruits were studied in three states along maturation, carrying on a bioactive characterization and studying core and regulatory gene expression. As a result, the concentration of flavonoids increased in the fruit along with maturation over the noninoculated controls; hence, core and regulatory genes were characterized and their expression was studied demonstrating genes likely involved in controlling the activity of pathway branches, associated with enhanced accumulation of anthocyanins, catechins, and flavonols in developing fruits of blackberry [81].

These results prove that elicitation is a very interesting tool to achieve better fruit quality in terms of higher amounts of bioactives, as flavonoid biosynthesis can be modulated without genetic manipulation. Using this method, we can achieve two different goals at the same time: first we increase plant defense by a natural way without using any chemicals, which are harmful for the environment and humans also, therefore contributing to environmentally friendly agricultural practices. The second one would be obtaining a food with reliable beneficial effects for human health, since delivering the bacteria through the roots will attenuate fluctuations in bioactive compounds. This second statement is nowadays one of the world's great concerns, since achieving food security refers to both ends of society: those that are hungry should have enough amounts of healthy and nutritious foods, as well as those that have a lot of food, but unbalanced and low quality. Therefore, healthy eating is becoming really popular as population is trying to have a good diet to prevent development of disease rather than healing, so to decrease the intake of different drugs in a long time period. Both this two goals can be achieved by the application of these rhizobacterias that will increase these secondary metabolites by a natural method. Beyond health-related issues in food production, these elicitors and elicitation technology [80] can be used to unravel metabolic pathways and their regulation for further application in metabolic engineering and cell cultures.

Acknowledgements

This Project was funded by Ministerio de Economía y Competitividad: AGL-2013-45189-R. Grant reference: BES-2014-069990.

Author details

Enrique Gutierrez, Ana García-Villaraco Velasco, Jose Antonio Lucas, F. Javier Gutierrez-Mañero and Beatriz Ramos-Solano*

*Address all correspondence to: bramsol@ceu.es

Faculty of Pharmacy, Universidad San Pablo-CEU Universities, Boadilla del Monte, Madrid, Spain

References

[1] Jiang N, Doseff AI, Grotewold E. Flavones: From biosynthesis to health benefits. Plants (Basel, Switzerland). 2016;5(2).

[2] Kaume L, Howard LR, Devareddy L. The blackberry fruit: A review on its composition and chemistry, metabolism and bioavailability, and health benefits. Journal of Agricultural and Food Chemistry. 2012;60(23):5716-27.

[3] Savini I, Catani MV, Evangelista D, Gasperi V, Avigliano L. Obesity-associated oxidative stress: Strategies finalized to improve redox state. International Journal of Molecular Sciences. 2013;14(5):10497-538.

[4] Smeriglio A, Barreca D, Bellocco E, Trombetta D. Chemistry, pharmacology and health benefits of anthocyanins. Phytotherapy Research. 2016;30(8):1265-86.

[5] Winkel-Shirley B. Biosynthesis of flavonoids and effects of stress. Current Opinion in Plant Biology. 2002;5(3):218-23.

[6] Winkel-Shirley B. Flavonoid biosynthesis. A colorful model for genetics, biochemistry, cell biology, and biotechnology. Plant Physiology. 2001;126(2):485-93.

[7] Falcone Ferreyra ML, Rius SP, Casati P. Flavonoids: biosynthesis, biological functions, and biotechnological applications. Frontiers in Plant Science. 2012;3:222.

[8] Mouradov A, Spangenberg G. Flavonoids: a metabolic network mediating plants adaptation to their real estate. Frontiers in Plant Science. 2014;5:16.

[9] Hou DX, Fujii M, Terahara N, Yoshimoto M. Molecular mechanisms behind the chemopreventive effects of anthocyanidins. Journal of Biomedicine and Biotechnology. 2004(5):321-5.

[10] Samorodovabiank GB. Flavonoids as natural antioxidants of ascorbic acids of fruits and berries. Biochemistry-Moscow. 1965;30(2):213.

[11] Wu PP, Ma GZ, Li NH, Deng QG, Yin YY, Huang RQ. Investigation of in vitro and in vivo antioxidant activities of flavonoids rich extract from the berries of *Rhodomyrtus tomentosa*(Ait.) Hassk. Food Chemistry. 2015;173:194-202.

[12] Yang L. The new royal-horticultural-society dictionary of gardening. New York Times Book Review. 1992;20:3200.

[13] Kolniak-Ostek J, Kucharska AZ, Sokol-Letowska A, Fecka I. Characterization of Phenolic Compounds of Thorny and Thornless Blackberries. Journal of Agricultural and Food Chemistry. 2015;63(11):3012-21.

[14] Nakayama T, Suzuki H, Nishino T. Anthocyanin acyltransferases: specificities, mechanism, phylogenetics, and applications. Journal of Molecular Catalysis B-Enzymatic. 2003;23(2-6):117-32.

[15] Yonekura-Sakakibara K, Saito K. Function, structure, and evolution of flavonoid glycosyltransferases in plants. Recent Advances in Polyphenol Research. 2014;4:61-82.

[16] Fraser CM, Thompson MG, Shirley AM, Ralph J, Schoenherr JA, Sinlapadech T, et al. Related *Arabidopsis* serine carboxypeptidase-like sinapoylglucose acyltransferases display distinct but overlapping substrate Specificities. Plant Physiology. 2007;144(4):1 986-99.

[17] Luo J, Nishiyama Y, Fuell C, Taguchi G, Elliott K, Hill L, et al. Convergent evolution in the BAHD family of acyl transferases: identification and characterization of anthocyanin acyl transferases from *Arabidopsis thaliana*. Plant Journal. 2007;50(4):678-95.

[18] Shi M-Z, Xie D-Y. Biosynthesis and metabolic engineering of anthocyanins in *Arabidopsis thaliana*. Recent Patents on Biotechnology. 2014;8(1):47-60.

[19] Gomez C, Conejero G, Torregrosa L, Cheynier V, Terrier N, Ageorges A. In vivo grapevine anthocyanin transport involves vesicle-mediated trafficking and the contribution of anthoMATE transporters and GST. Plant Journal. 2011;67(6):960-70.

[20] Pourcel L, Irani NG, Lu YH, Riedl K, Schwartz S, Grotewold E. The formation of anthocyanic vacuolar inclusions in *Arabidopsis thaliana* and implications for the sequestration of anthocyanin pigments. Molecular Plant. 2010;3(1):78-90.

[21] Zhao J, Dixon RA. The 'ins' and 'outs' of flavonoid transport. Trends in Plant Science. 2010;15(2):72-80.

[22] Jorgensen K, Rasmussen AV, Morant M, Nielsen AH, Bjarnholt N, Zagrobelny M, et al. Metabolon formation and metabolic channeling in the biosynthesis of plant natural products. Current Opinion in Plant Biology. 2005;8(3):280-91.

[23] Routaboul JM, Dubos C, Beck G, Marquis C, Bidzinski P, Loudet O, et al. Metabolite profiling and quantitative genetics of natural variation for flavonoids in *Arabidopsis*. Journal of Experimental Botany. 2012;63(10):3749-64.

[24] North G. Current biology at 20. Current Biology. 2010;20(24):R1049-R51.

[25] Somerville C, Koornneef M. Timeline—a fortunate choice: the history of *Arabidopsis* as a model plant. Nature Reviews Genetics. 2002;3(11):883-9.

[26] Garcia-Seco D, Zhang Y, Gutierrez-Manoro FJ, Martin C, Ramos-Solano B. RNA-Seq analysis and transcriptome assembly for blackberry (*Rubus* spp. Var. Lochness) fruit. BMC Genomics. 2015;16:11.

[27] Mehrtens F, Kranz H, Bednarek P, Weisshaar B. The Arabidopsis transcription factor MYB12 is a flavonol-specific regulator of phenylpropanoid biosynthesis. Plant Physiology. 2005;138(2):1083-96.

[28] Cle C, Hill LM, Niggeweg R, Martin CR, Guisez Y, Prinsen E, et al. Modulation of chlorogenic acid biosynthesis in *Solanum lycopersicum*; consequences for phenolic accumulation and UV-tolerance. Phytochemistry. 2008;69(11):2149-56.

[29] Chen Q, Yu HW, Tang HR, Wang XR. Identification and expression analysis of genes involved in anthocyanin and proanthocyanidin biosynthesis in the fruit of blackberry. Scientia Horticulturae. 2012;141:61-8.

[30] Kerhoas L, Aouak D, Cingoz A, Routaboul JM, Lepiniec L, Einhorn J, et al. Structural characterization of the major flavonoid glycosides from *Arabidopsis thaliana* seeds. Journal of Agricultural and Food Chemistry. 2006;54(18):6603-12.

[31] Routaboul JM, Kerhoas L, Debeaujon I, Pourcel L, Caboche M, Einhorn J, et al. Flavonoid diversity and biosynthesis in seed of *Arabidopsis thaliana*. Planta. 2006;224(1):96-107.

[32] Almeida JRM, D'Amico E, Preuss A, Carbone F, de Vos CHR, Deiml B, et al. Characterization of major enzymes and genes involved in flavonoid and proanthocyanidin biosynthesis during fruit development in strawberry (*Fragaria × ananassa*). Archives of Biochemistry and Biophysics. 2007;465(1):61-71.

[33] Auger B, Marnet N, Gautier V, Maia-Grondard A, Leprince F, Renard M, et al. A Detailed survey of seed coat flavonoids in developing seeds of *Brassica napus* L. Journal of Agricultural and Food Chemistry. 2010;58(10):6246-56.

[34] Buendia B, Gil MI, Tudela JA, Gady AL, Medina JJ, Soria C, et al. HPLC-MS Analysis of proanthocyanidin oligomers and other phenolics in 15 STRAWBERRY CULTIVARS. Journal of Agricultural and Food Chemistry. 2010;58(7):3916-26.

[35] Jaakola L, Maatta K, Pirttila AM, Torronen R, Karenlampi S, Hohtola A. Expression of genes involved in anthocyanin biosynthesis in relation to anthocyanin, proanthocyanidin, and flavonol levels during bilberry fruit development. Plant Physiology. 2002;130(2):729-39.

[36] Dixon RA, Paiva NL. Stress-induced phenylpropanoid metabolism. Plant Cell. 1995;7(7): 1085-97.

[37] Zandalinas SI, Sales C, Beltran J, Gomez-Cadenas A, Arbona V. Activation of secondary metabolism in citrus plants is associated to sensitivity to combined drought and high temperatures. Frontiers in Plant Science. 2017;7:17.

[38] Pieterse CMJ, Zamioudis C, Berendsen RL, Weller DM, Van Wees SCM, Bakker P. Induced Systemic Resistance by Beneficial Microbes. In: VanAlfen NK, editor. Annual Review of Phytopathology, Vol 52. Annual Review of Phytopathology. 52. Palo Alto: Annual Reviews; 2014. pp. 347-75.

[39] Pyngrope S, Bhoomika K, Dubey RS. Reactive oxygen species, ascorbate-glutathione pool, and enzymes of their metabolism in drought-sensitive and tolerant indica rice (*Oryza sativa* L.) seedlings subjected to progressing levels of water deficit. Protoplasma. 2013;250(2):585-600.

[40] Bandyopadhyay A, Blakeslee JJ, Lee OR, Mravec J, Sauer M, Titapiwatanakun B, et al. Interactions of PIN and PGP auxin transport mechanisms. Biochemical Society Transactions. 2007;35:137-41.

[41] Geisler M, Blakeslee JJ, Bouchard R, Lee OR, Vincenzetti V, Bandyopadhyay A, et al. Cellular efflux of auxin catalyzed by the Arabidopsis MDR/PGP transporter AtPGP1. Plant Journal. 2005;44(2):179-94.

[42] Santelia D, Henrichs S, Vincenzetti V, Sauer M, Bigler L, Klein M, et al. Flavonoids redirect PIN-mediated polar auxin fluxes during root gravitropic responses. Journal of Biological Chemistry. 2008;283(45):31218-26.

[43] Kuhn BM, Geisler M, Bigler L, Ringli C. Flavonols accumulate asymmetrically and affect auxin transport in *Arabidopsis*. Plant Physiology. 2011;156(2):585-95.

[44] Schmitz-Hoerner R, Weissenbock G. Contribution of phenolic compounds to the UV-B screening capacity of developing barley primary leaves in relation to DNA damage and repair under elevated UV-B levels. Phytochemistry. 2003;64(1):243-55.

[45] Mierziak J, Kostyn K, Kulma A. Flavonoids as important molecules of plant interactions with the environment. Molecules. 2014;19(10):16240-65.

[46] Nabbie F, Shperdheja O, Millot J, Lindberg J, Peethambaran B. Flavonoid profiling by LC-MS IN 14-3-3v knockout mutants of *Arabidopsis thaliana* during drought stress conditions. Planta Medica. 2014;80(10):834.

[47] Nakabayashi R, Mori T, Saito K. Alternation of flavonoid accumulation under drought stress in *Arabidopsis thaliana*. Plant Signaling & Behavior. 2014;9(8):e29518.

[48] Bhagwan A, Reddy YN, Rao PV, Mohankumar KC. Shelf life extensión of tomato fruits by postharvest antioxidant application. Journal of Applied Horticulture. 2000; 2: 88-91.

[49] Martin C, Zhang Y, De Stefano R, Robine M, Butelli E, Bulling K, et al. Different ROS-scavenging properties of flavonoids determine their abilities to extend shelf life of tomato. Free Radical Biology and Medicine. 2015;86:S11-S.

[50] Fernandez S, Poschenrieder C, Marceno C, Gallego JR, Jimenez-Gamez D, Bueno A, et al. Phytoremediation capability of native plant species living on Pb-Zn and Hg-As mining wastes in the Cantabrian range, north of Spain. Journal of Geochemical Exploration. 2017;174:10-20.

[51] Lattanzio V, Lattanzio V, Cardinali A. Phytochemistry: Advances in Research. Filippo Imperato. 2006: 23-67. ISBN: 81-308-0034.

[52] Treutter D. Significance of flavonoids in plant resistance and enhancement of their biosynthesis. Plant Biology. 2005;7(6):581-91.

[53] Skadhauge B, Thomsen KK, von Wettstein D. The role of the barley testa layer and its flavonoid content in resistance to *Fusarium* infections. Hereditas. 1997;126(2):147-60.

[54] Ibrahim MH, Jaafar HZE, Rahmat A, Rahman ZA. Involvement of nitrogen on flavonoids, glutathione, anthocyanin, ascorbic acid and antioxidant activities of Malaysian medicinal plant *Labisia pumila Blume* (Kacip Fatimah). International Journal of Molecular Sciences. 2012;13(1):393-408.

[55] Renaud S, Delorgeril M. Wine, alcohol, platelets, and the French paradox for coronary heart-disease. Lancet. 1992;339(8808):1523-6.

[56] Haghparast A, Farzin H, Toroghi R. Knockdown of p65 subunit of NF-kB transcription factor downregulates the induction of pro-inflammatory cytokines in human lung epithelial cells infected with influenza H9N2 virus. European Journal of Immunology. 2016;46:816-816.

[57] Lai CS, Li SM, Chai CY, Lo CY, Dushenkov S, Ho CT, et al. Anti-inflammatory and antitumor promotional effects of a novel urinary metabolite, 3′,4′-didemethylnobiletin, derived from nobiletin. Carcinogenesis. 2008;29(12):2415-24.

[58] Hostetler G, Riedl K, Cardenas H, Diosa-Toro M, Arango D, Schwartz S, et al. Flavone deglycosylation increases their anti-inflammatory activity and absorption. Molecular Nutrition & Food Research. 2012;56(4):558-69.

[59] Garcia M, Shook J, Kerstetter J, Kenny A, Bihuniak J, Huedo-Medina T. The efficacy of the mediterranean diet on obesity outcomes: a meta-analysis. FASEB Journal. 2015;29:1.

[60] Martinez-Gonzalez MA. Benefits of the Mediterranean diet beyond the Mediterranean Sea and beyond food patterns. BMC Medicine. 2016;14:3.

[61] Hendricks JJA, Alblas J, van der Pol SMA, van Tol EAF, Dijkstra CD, de Vries HE. Flavonoids influence monocytic GTPase activity and are protective in experimental allergic encephalitis. Journal of Experimental Medicine. 2004;200(12):1667-72.

[62] Henkels KM, Frondorf K, Gonzalez-Mejia ME, Doseff AL, Gomez-Cambronero J. IL-8-induced neutrophil chemotaxis is mediated by Janus kinase 3 (JAK3). FEBS Letters 1667-72.

[63] Arango D, Parihar A, Villamena FA, Wang LW, Freitas MA, Grotewold E, et al. Apigenin induces DNA damage through the PKC delta-dependent activation of ATM and H2AX causing down-regulation of genes involved in cell cycle control and DNA repair. Biochemical Pharmacology. 2012;84(12):1571-80.

[64] Arango D, Morohashi K, Yilmaz A, Kuramochi K, Parihar A, Brahimaj B, et al. Molecular basis for the action of a dietary flavonoid revealed by the comprehensive identification of apigenin human targets. Proceedings of the National Academy of Sciences of the United States of America. 2013;110(24):E2153-E62.

[65] Meiers S, Kemeny M, Weyand U, Gastpar R, von Angerer E, Marko D. The anthocyanidins cyanidin and delphinidin are potent inhibitors of the epidermal growth-factor receptor. Journal of Agricultural and Food Chemistry. 2001;49(2):958-62.

[66] Williams RJ, Spencer JPE, Rice-Evans C. Flavonoids: Antioxidants or signalling molecules? Free Radical Biology and Medicine. 2004;36(7):838-49.

[67] Joseph JA, Shukitt-Hale B, Denisova NA, Bielinski D, Martin A, McEwen JJ, et al. Reversals of age-related declines in neuronal signal transduction, cognitive, and motor behavioral deficits with blueberry, spinach, or strawberry dietary supplementation. Journal of Neuroscience. 1999;19(18):8114-21.

[68] Seeram NP, Adams LS, Hardy ML, Heber D. Total cranberry extract versus its phytochemical constituents: Antiproliferative and synergistic effects against human tumor cell lines. Journal of Agricultural and Food Chemistry. 2004;52(9):2512-7.

[69] Shin WH, Park SJ, Kim EJ. Protective effect of anthocyanins in middle cerebral artery occlusion and reperfusion model of cerebral ischemia in rats. Life Sciences. 2006;79(2):130-7.

[70] Matsumoto H, Nakamura Y, Tachibanaki S, Kawamura S, Hirayama M. Stimulatory effect of cyanidin 3-glycosides on the regeneration of rhodopsin. Journal of Agricultural and Food Chemistry. 2003;51(12):3560-3.

[71] Bispo K, Amusquivar E, Garcia-Seco D, Ramos-Solano B, Gutierrez-Manero J, Herrera E. Supplementing diet with blackberry extract causes a catabolic response with increments in insulin sensitivity in rats. Plant Foods for Human Nutrition. 2015;70(2):170-5.

[72] Tsuda T, Horio F, Kitoh J, Osawa T. Protective effects of dietary cyanidin 3-O-beta-D-glucoside on liver ischemia-reperfusion injury in rats. Archives of Biochemistry and Biophysics. 1999;368(2):361-6.

[73] Cardona F, Andres-Lacueva C, Tulipani S, Tinahones FJ, Queipo-Ortuno MI. Benefits of polyphenols on gut microbiota and implications in human health. Journal of Nutritional Biochemistry. 2013;24(8):1415-22.

[74] Cerda B, Periago P, Espin JC, Tomas-Barberan FA. Identification of urolithin A as a metabolite produced by human colon microflora from ellagic acid and related compounds. Journal of Agricultural and Food Chemistry. 2005;53(14):5571-6.

[75] Gross G, Jacobs DM, Peters S, Possemiers S, van Duynhoven J, Vaughan EE, et al. In Vitro Bioconversion of polyphenols from black tea and red wine/grape juice by human intestinal microbiota displays strong interindividual variability. Journal of Agricultural and Food Chemistry. 2010;58(18):10236-46.

[76] Butelli E, Titta L, Giorgio M, Mock HP, Matros A, Peterek S, et al. Enrichment of tomato fruit with health-promoting anthocyanins by expression of select transcription factors. Nature Biotechnology. 2008;26(11):1301-8.

[77] Ramos-Solano B, Garcia-Villaraco A, Gutierrez-Manero FJ, Lucas JA, Bonilla A, Garcia-Seco D. Annual changes in bioactive contents and production in field-grown blackberry after inoculation with *Pseudomonas fluorescens*. Plant Physiology and Biochemistry. 2014;74:1-8.

[78] Gutiérrez-Mañero FJ, García-Villaraco A, Lucas JA, Gutiérrez E, Ramos-Solano B. Inoculant/elicitation technology to improve bioactive/phytoalexin contents in functional foods. International Journal of Current Microbiology and Applied Sciences. 2015;4:224-241.

[79] Algar E, Gutierrez-Mañero FJ, Bonilla A, Luca JA, Radzki, Ramos-Solano B. Pseudomonas Fluorescens N21.4 metaolites enhance secondary metabolism isoflavones in soybean (*Glycine max*) calli cultures. Journal of Agricultural and Food Chemistry. 2012;60:11080-11087.

[80] Garcia-Seco D, Zhang Y, Gutierrez-Manero FJ, Martin C, Ramos-Solano B. Application of *Pseudomonas fluorescens* to blackberry under field conditions improves fruit quality by modifying flavonoid metabolism. PLOS One. 2015;10(11):23.

[81] Luo J, Butelli E, Hill L, Parr A, Niggeweg R, Bailey P, et al. AtMYB12 regulates caffeoyl quinic acid and flavonol synthesis in tomato: expression in fruit results in very high levels of both types of polyphenol. Plant Journal. 2008;56(2):316-26.

Chapter 8

Flavonoid Accumulation Behavior in Response to the Abiotic Stress: Can a Uniform Mechanism Be Illustrated for All Plants?

Hakan Cetinkaya, Muhittin Kulak,
Muhammet Karaman, Halide Sedef Karaman and
Ferudun Kocer

Additional information is available at the end of the chapter

http://dx.doi.org/10.5772/68093

Abstract

This review concentrates on two aspects of how total flavonoid content and individual flavonoid compounds change with the perception of environmental stress and the subsequent changes in those metabolites after post-harvest conditions are of the main points of the study. Hereby, along with this study, the flavonoid synthesis or their accumulation with their importance in plants and then in humans is briefly described. According to the literature cited herein, it seems that a universal mechanism concerned with flavonoid accumulation in response to the abiotic stress factors cannot be illustrated. Flavonoid accumulation exhibits different reactions to the different stressors. Flavonoid accumulation behavior not only varies depending on the developmental stage, species and even cultivars of the same species but also post-harvest processes.

Keywords: total flavonoid, abiotic stress, post-harvest processes

1. Introduction

Phenolic compounds are secondary metabolites derived from pentose phosphate, shikimate and phenylpropanoid pathways in plants [1], and a wide range of functions including participation in the regulation of growth and developmental processes and interactions with biotic and abiotic environmental stimuli have been attributed to the those phytochemicals [2]. Of those compounds, flavonoids comprise the large and common group of plant phenolics with more than 5000 different described flavonoids in six major subclasses, including flavones, flavonols, flavanones, flavanols, anthocyanidins and isoflavones [3].

Carbon skeleton of flavonoids occurs from combining of two phenyl ring and a propane chain. Rings of 2-phenyl benzopyran consisting 15 carbons are referred as A, B and C-rings [4] (**Figure 1**). Flavonoids structure's diversity can be classified according to do both major classification and oxidation level [5]. Additionally type, number and binding positions of substitutions binding to aromatic rings cause flavonoids structure's diversity [6].

Since plants are open systems and do not exist in a vacuum, they are continuously interacted with their biotic and non-biotic surroundings. In order to explore what kind of mechanisms underlining the defense against changing environmental conditions and other physiological and biochemical processes for the plant are still great concern of the researchers. Like all living and non-living things in the universe, with each step upwards in the life span, novel properties concerned with quality and quantities of the flavonoid content depending on the environmental, ontogenetic, annual and diurnal variations, which are not present at the current stage of the plant may emerge but it is worthy to underline that these effects are species dependent. The post-harvest practices such as "from wild to domestication," "from fields to shelves" and "from shelves to pharmacy" are also great interest of the consumers for the sustainable healthier life conditions. Hence, universal and uniform mechanisms with respect to the production, accumulation or secretion of the flavonoid have not been proposed yet (**Figure 2**). Two aspects of flavonoid content and their individual compounds can be discussed. One is the content which is directly dependent on plant species itself and with its responses against abiotic stress conditions. This can be simplified as "plant health." The later one is about the changes, which are related to the human consumption. This second aspect can be also simplified as "human health."

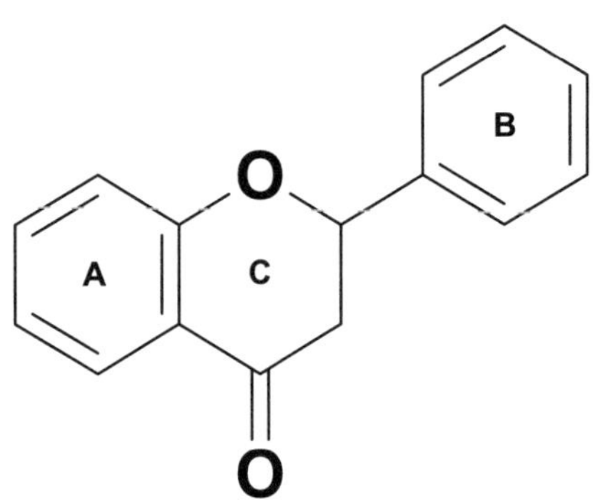

Figure 1. The basic structure of flavonoids.

1.1. Abiotic stress challenges regarding with quantity versus quality: two sides of the coin

As sessile organisms, plants are often exposed to various environmental stress factors. Hence, plants must regulate their growth and development in response to ever-changing

http://dx.doi.org/10.5772/68093

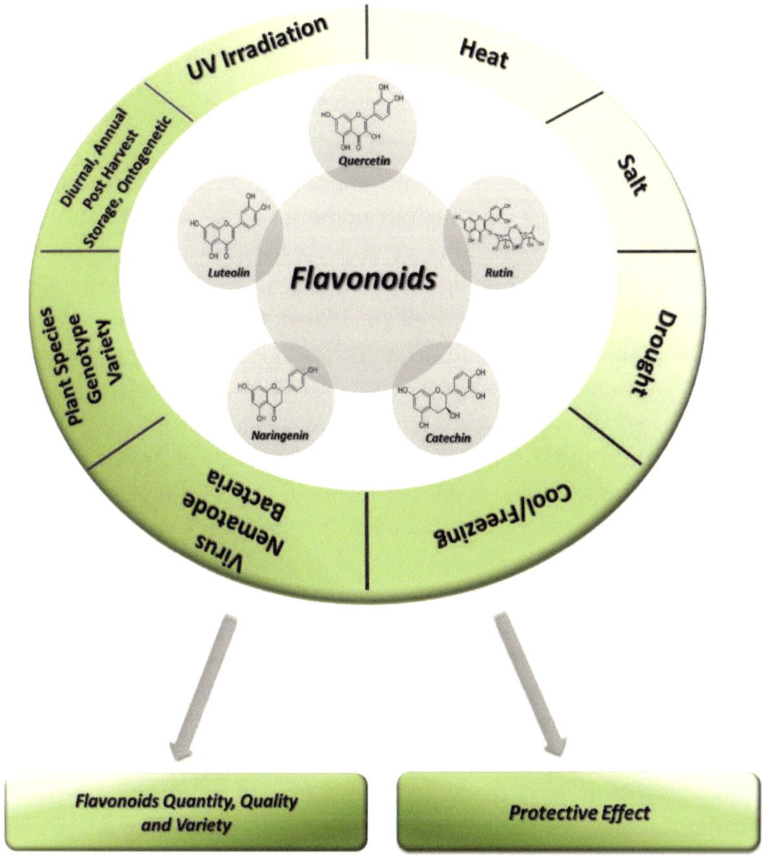

Figure 2. Biotic and abiotic factors affecting flavonoid content and composition in plants.

environmental conditions and their stimuli. Once plants cannot tolerate or overcome the unfavorable environmental conditions, plant growth and development are likely adversely influenced and subsequently significant loss of crop yields [7, 8]. Along with the stress conditions, plant behavior may change with respect to the secondary metabolite synthesis, production, secretion and storage when subjected to the abiotic stress factors [9]. Some secondary metabolite synthesis, enzyme activities and soluble substance accumulation were positively influenced by abiotic stress conditions. These are considered as consequences of plant adaptive strategies concerned with establishment of some changes allowing to the plant to sustain its life under ever-changing conditions.

Many results concerned with total flavonoid and their individual compounds in response to the different stressors. For total flavonoid content, increases were determined [2, 10–14, 16, 17, 19, 20] whereas decreases were found [15, 18, 21] under different stress conditions.

Based on the literature review, we cannot deduce and explain the flavonoid accumulation or their compound profile using one simple sentence. The stress effect is compound specific. A uniform mechanism for compound profile variation cannot be illustrated [22]. Furthermore, the flavonoid accumulation is likely dependent stress factors, frequency, duration and timing.

1.2. Is it adaptive strategy to sacrifice the primary metabolites through increases in secondary metabolite production against stress conditions or high efficiency use of secondary metabolite biosynthesis pathway?

As previously mentioned, pentose phosphate, shikimate and phenylpropanoid pathways are of the three pathways in plants, which are responsible for biosynthesis of phenolic compounds [1]. Shikimic acid is a key intermediate in the synthesis of both aromatic amino acids and phenylpropanoids, and oxidative pentose phosphate pathway is of the precursors for the biosynthesis of aromatic amino acids, lignin and flavonoids [23]. Regulation and expression of the genes on the pathways have been well elucidated, but the pathway compartmentation is not yet known [24]. Some of the synthesis-associated genes and enzymes involved in phenolics biosynthesis were characterized in Arabidopsis (*Arabidopsis thaliana*), maize (*Zea mays*) and petunia (*Petunia hybrid*). Also *Fragaria* spp. has been studied for their genes and enzymes ([25–30]; cited by [24]). In order to determine which pathway is preferred for biosynthesis secondary metabolites under abiotic stress factors, the expression of protein or enzymes associated with synthesis of secondary metabolites in the pathways should be determined and then compared with the control group-not stressed group. Determination of the long or short distance metabolic pathways or high or low energy cost pathways in response to the stress is also great concern to understand plant behavior and signaling.

Stressors bring about quantitative and qualitative changes in plant metabolites. Of those, in general, biosynthesis of proteins in the plant leaves is suppressed, triggering the changes at gene expression levels and subsequently the synthesis of new proteins. For the lipid content and composition, the disturbances concerned with fatty acid composition, especially changes in fatty acid carbon chains. The variations in the lipid composition influence membrane lipids and transport functions of membranes. Furthermore, accumulation of the compatible solutes is of the responses against drought, high temperature or high salinity, maintaining the osmotic adjustment and turgor regulation [31].

Plant secondary metabolites have been considered or often referred to as metabolites which are not fundamentals for sustainability of basic plant life processes. However, the crucial and wide range roles of secondary metabolites have been understood. The accumulation of phenylpropanoids increased in response to the environmental stress including pathogen attack, UV-radiation, high light, nutrient deficiency etc. According to Bryant et al. [32] hypothesis, an exchange occurs between carbon and biomass production or formation of defensive secondary metabolites, proposing that secondary metabolites are involved in protective processes of plants in response to stressors. For example, phenyl amide formation and accumulation of anthocyanin and polyamines have been reported as a response to the environmental stresses [33, 34].

1.3. Over accumulation of flavonoid versus reactive oxygen species? Non-enzymatic antioxidant system but any relations with the enzymatic antioxidant system (SOD, CAT, APX)?

Flavonoids are secondary metabolites synthesized by general phenylpropanoid pathway in plants [35]. They have been considered as a secondary (non-enzymatic) reactive oxygen species scavenging system in plants and humans [36]. Flavonoids exhibit direct scavenging of reactive oxygen species [36] one of the ways scavenging reactive oxygen species, flavonoids can easily donate hydrogen atom. Thus, while reactive oxygen species are inactivated by flavonoids, flavonoids return to phenoxyl radical [37]. Flavonoids phenoxyl radical can react with other free radicals and then acquiring a stable quinone structure [38]. The other way of scavenging reactive oxygen species, flavonoids return to phenoxyl radical by donating hydrogen atom at the first step. At the second step, phenoxyl radical scavenge other high reactive radical (R•) by radical-radical termination. Flavonoid phenoxyl radical is highly stabile radical due to presence of a resonance structure redistributed the unpaired electron on the aromatic core [39] **(Figure 3)**.

1.4. The possible protective defense roles of flavonoids in response to UV light have been documented in many studies but what happens if the flavonoid and other pigments cannot completely block the sunlight transmission?

UV light from sunlight is primarily required to perform photosynthesis as basic function and developmental process such as de-etiolation, phototropism and flowering of the plants [40, 41]. But interestingly, the UV light causes damage to DNA, protein and cell membranes of the plants, because, as sessile organisms, plants are more exposed to the UV-light. Subsequently, normal growth and development of plants are retarded [41]. Short-wavelength UV light is grouped into three categories. Of those, UV-A (315–400 nm) directly reaches the earth's surface, and UV-B (280–315 nm) and UV-C (100–280 nm) are blocked by the ozone layer. However, a small quantity of UV-B reaches the earth's surface because of ozone layer depletion and subsequently causes DNA damages [42].

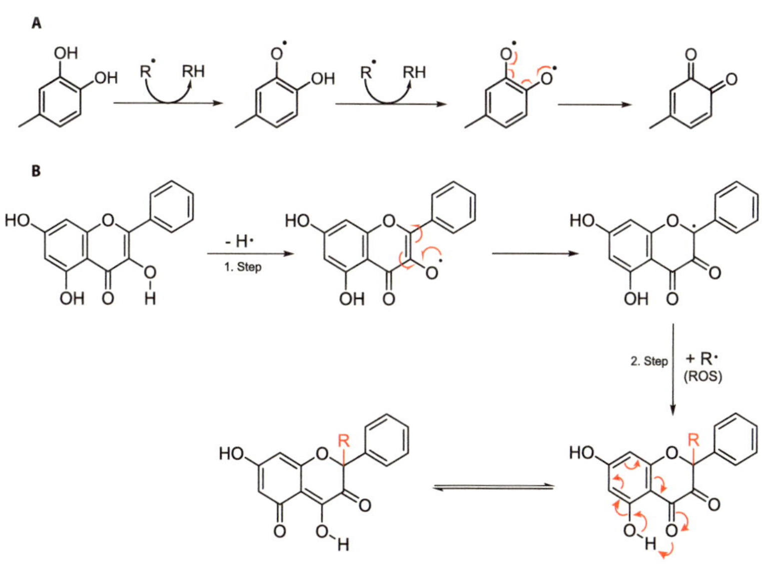

Figure 3. Scavenging of ROS by flavonoids, reproduced from Pietta [38] (A) and reproduced from Amic [39] (B).

UV-B has highest energy of UV light that reaches the earth surface [43]. Although the high level of UV-B causes damage to biomolecules, low level of UV-B regulates morphology, development, phycology and biochemical compositions [44]. While long wavelengths UV light-induced regulation is provided with photoreceptors including phototropins, neochromes, phytochromes, rhodopsins and cryptochromes [45] (**Figure 5**), UV-B-induced regulation is provided with UV RESISTANCE LOCUS 8 (UVR8) receptor protein [46]. UVR8 directly absorbs UV-B radiation and induces the transcription of flavonoids biosynthesis genes by orchestrating UV protective gene expression responses [47].

Flavonoids are synthesized with phenylpropanoid pathway in plant and the pathway includes enzymes such as phenylalanine ammonia lyase (PAL), 4-coumaroyl: CoA ligase (4CL), Chalcone Synthase (CHS), Chalcone Isomerase (CHI), Flavone Synthase (FS) and Dihydroflavonol-4-Reductase (DFRA) [48]. CHS is key enzymes for flavonoid biosynthesis pathway. CHS catalysis condensation reaction of Coumaroyl CoA and Malonyl CoA.

Upregulation of CHS genes transcription in response to the several stressors was reported to induce flavonoid biosynthesis [49]. UVR8 protein interacts with the WD40-repeat domain of COP1 after perception of UV-B light that is one of the stressors [50]. Consequently, UVR8-COP1 complex leads to activation of HY5 gene expression [47]. HY5 proteins as a transcription factor play an enhancing role for UV-B induced-CHS gene expression during seedling development by binding to a conserved G-box sequence [51, 52]. Thus, flavonoids that accumulate in upper epidermis layer specially absorb a large amount of 280–340 nm wavelengths [53]. Thus, the flavonoids accumulated in upper epidermis layer protect the internal tissues of leaves and stems against UV-B. Since the synthesis of kaempferol is deficient in chalcone flavone isomerase mutant tt4 *A. thaliana*, the plant exhibits high sensitivity to UV light [54]. Along with the absorption of UV light with chromophore group of flavonoids, flavonoids may undergo a transformation. While flavonoids with aromatic chromophore absorb light in the 250 nm region of UV spectra, flavonoids contain carbonyl that are conjugated with the aromatic ring chromophore absorb light in the 350 nm region of UV spectra. The transformation of flavonoids after of UV light absorption in vitro conditions is illustrated in **Figure 4** [55].

It is worthy to note that the flavonoids are not unique functional UV-blocker absorbing all UV-B irradiation but the other protective roles of flavonoids cannot be ignored in spite of deficient in absorption of all UV-B irradiation since UV-B induced increase in the quantity of flavonoids has been reported, suggesting that flavonoids may exhibit functions including signal molecules, antioxidant molecules, defensive compounds, allelochemicals [56] after UV-B exposure in plant.

When the UV light cannot be completely blocked via defense system apparatus of the plants, UV light reaches to DNA, resulting in formation of cyclobutene pyrimidine dimers (CPDs) and 6-4 photoproducts (6-4PPs) on DNA [57, 58]. There are two mechanisms repairing the photoproduct that can inhibit transcription and replication and induce mutations [48]: first one is photolyases enzyme, and the other one is nucleotide excision repair mechanism [60]. CPD Photolyase gene expression is regulated by a wide spectrum of light, including far-red, red and blue light [61]. But recent studies reported that photolyase gene expression was regulated by UVR8 receptor protein and the regulation mechanism remains poorly understood [62]. In order to exhibit DNA repair activity, photolyase enzyme needs UV-A light [63] but

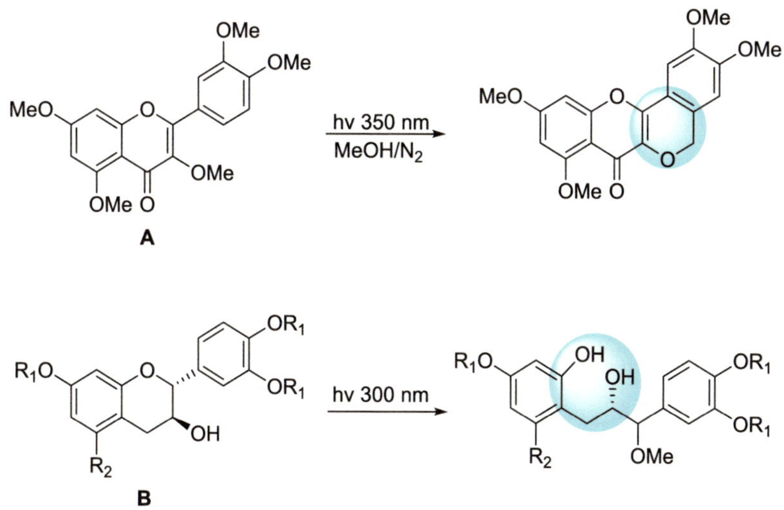

Figure 4. Photochemistry of quercetin pentamethyl ether (A) and photochemistry of flavan-3-ols (B) [55].

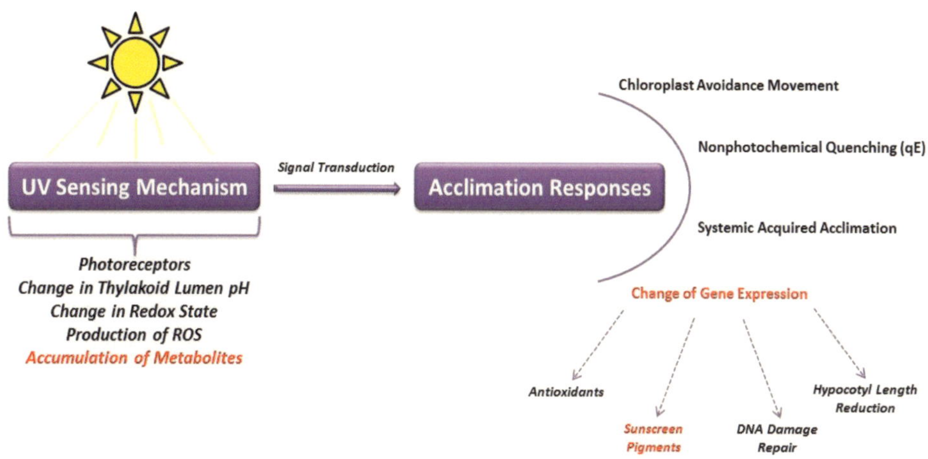

Figure 5. UV light perception, signaling and responses in Arabidopsis (scheme adapted from Li et al. [45]).

the light is not required for the activity of dark repair called nucleotide excision repair mechanism. Recently, photolyase and nucleotide excision repair mechanism in *A. thaliana* were well described [60].

1.5. The fate of the flavonoid-enriched crop plants through the food chain: terminal

Any direct and indirect biotic or abiotic stressors or their combinations at certain time or simultaneously influence the phytochemistry and subsequently the changes orchestrate the plant protection and plants' biological activities. Herewith, phytochemistry of a plant can be regarded as protective roles for plants against stressors and health-promoting properties for humans. The quality of the crop plants is a combination attributed to their composition and contents that shape the commodity value for human consumption. Since humans are, in general, considered to be at the top of the food chain, the terminal of the flavonoid-enriched/poor crop plants would be the human, resulting the health standards.

Nothing stays the same as its former form and the changes are inevitable for all living and non-living things. Therefore, numerous studies—as listed in **Table 1** but not limited in this chapter—have been performed in order to keep the stability or dynamic changes of flavonoid content and its compound. Of those studies, atmosphere conditions are of great interest for long-term storage and subsequently essential for keeping biological value of the crop. In the study reported on *Allium cepa* var. *calonicum* Backer) by [64] (see the detail in **Table 1**), the highest content after storage at conditions with gas composition of 5% CO_2 + 5% O_2 was achieved. Two major compound—quercetin 3,4′-di-O-glucoside and quercetin 4′-O-glucoside (spiraeoside)—exhibited an increase [64]. Effect of carbon dioxide-enriched atmosphere on total

Storage conditions/ cultivars/harvest times	**Total flavonoid content**	**Plant species**	**Researchers**
Different storage temperatures (0, 2, 4, and 6 + 2°C) + 5%, 10%, 20% or 0.03% CO_2	Total flavonoid content	*Phoenix dactylifera* L.	[54]
Storage conditions were at 50, 25, 4, and −20°C	Total flavonoid content	*Anemopsis californica*	[59]
Freeze and thermal drying	Total flavonoid content	*Oxycoccus palustris Pers.*	[60]
Stability testing at different temperatures	Flavonoid glycosides	*Calendula officinalis* and *Betula* sp.	[61]
Storage at 27°C for 9 days	Total flavonoid content	*Paluma cultivar*	[62]
Temperature and storage time	quercetin-3-rutinoside, quercetin-3-glucoside, quercetin-3-D-galactoside	*Sorbus aucuparia*	[63]
Subunit parts of the rhizome during the thermal drying process under treatment temperatures ranging from 40 to 120°C	Mangiferin, iristectorigenin A, irigenin, irilone dichotomitin	*Belamcanda. chinensis* (L.) DC.	[64]
Cultivars and storage conditions	Total flavonoids	*Pistachia vera* L.	[65]
Normal atmosphere and 0% CO_2 + 21% O_2, (2) 5% CO_2 + 5% O_2, 5% CO_2 + 2% O2, 2% CO_2 + 5% O_2, 2% CO_2 + 2% O_2	Quercetin 3,4′-di-O-glucoside, quercetin 3-O-glucoside (isoquercetin), quercetin 4′-O-glucoside (spiraeoside)	*Allium cepa* var. *calonicum* Backer	[64]
At ambient temperature (about 25 ± 2°C) in a refrigerator (4 ± 0.2°C) and sampling days 0, 2, 4, 6, 8, 10, 12, 14	Total flavonoid content	*Juglans sigillata*	[67]

Table 1. Continue

Table 1. Continued

Storage conditions/ cultivars/harvest times	Total flavonoid content	Plant species	Researchers
Different cultivars	Spiraeoside (quercetin-4′-O-β-D-glucoside), rutin and quercetin	*Allium cepa*	[70]
Different plant parts, developmental storage and storage durations (1, 2, 3 and 4 days)	Total flavonoid content	*Clinacanthus nutans* (Burm. f.)	[66]
Storage at 6, 16 and 25°C for 6 days.	Flavonol	*Fragaria ananassa* Duch.	[68]
Different temperatures 25 ± 2°C (room temperature) and 10 ± 1°C (refrigerator) at different time of intervals (1st, 5th and 10th day)	Total flavonoid content	*Brassica rapa* L.	[71]
Storage for 0–7 months at 25 and 37°C	Total flavonoid content	*Oryza sativa* (milled rice)	[72]
Light ((photosynthetically active radiation (PAR) level of 56 ± 0.5 μmol m^{-2} s^{-1} (H); 31 ± 0.2 μmol m^{-2} s(L), or in dark (D). and maturity (0–5% red, 20% red, 50% red, 80% red, 100% red)	Ellagic acid, quercetin, kaempferol and cyanidin 3-glucoside	*Rubus ideaus* L.	[69]
Storage for 7, 15 and 30 days at 4, 22 and 35°C	Catechin, epicatechin, procyanidins B1-B4 and total flavonoids	Cocoa powder	[73]
Industrially squeezed, pasteurized, concentrated and stored under refrigeration (4°C) and at room temperature (20°C)	Flavanone-7-O-glycosides, fully methoxylated flavones	*Citrus clementina* Hort. ex Tan. *C. reticulata* Blanco × *C. sinensis* Osb., *C. sinensis*	[74]
Cultivar and storage conditions	Total flavonoid content	*Malus domestica* Borkh.	[75]

Table 1. Various studies concerned with the post-harvest processes and different cultivars influence on flavonoid content.

flavonoid content changes in *Phoenix dactylifera* L. fruit in response cold storage was tested and the fruits stored under low temperature conditions (0°C) or relatively high CO_2 concentration (20% CO_2) was reported not to exhibit any chilling or CO_2 injury symptoms. Modified conditions have been reported to extend not only the date storability and then fruit quality but also magnify the maintenance of fruit quality in response to the cold temperature storage [65].

Furthermore, the influence of different plant parts, developmental storage and storage durations (1, 2, 3 and 4 days) [66], different temperature and sampling days [67], different storage days [68], Light (photosynthetically active radiation (PAR) level and maturity [69] has been examined to indicate that the there is no constant stability or dynamics of flavonoid content and its compounds in quantity and quality.

2. Conclusion

As a conclusion, since plants are open systems and do not exist in a vacuum, they are continuously interacted with their biotic and non-biotic surroundings. Based on the literatures cited in the present chapter, a universal mechanism with respect to the accumulation behavior of flavonoid cannot be illustrated even the flavonoids commonly exhibit a tendency toward increase in response to the unfavorable conditions. Up to our best research, flavonoid accumulation behavior varies depending on the developmental stage, species and even cultivars of the same species. It also exhibits different reaction to the different stressors.

Beyond physiological aspect for the plants for their survival mechanism, plants are also sources for other living organisms. The quality and then biological efficacy of the flavonoid containing crops are great issue for human beings. According to the literature cited herein, the fate of the flavonoids containing herbal products including bulbs, leaves, fruits etc is influenced by the storage temperatures, storage time, modified storage conditions, cultivars, different parts and subunit parts of the plant, light and maturity.

Author details

Hakan Cetinkaya[1]*, Muhittin Kulak[2], Muhammet Karaman[3], Halide Sedef Karaman[4] and Ferudun Kocer[5]

*Address all correspondence to: hcetinkaya67@gmail.com

1 Department of Horticulture, Faculty of Agriculture, Kilis 7 Aralık University, Kilis, Turkey

2 Department of Biology, Faculty of Arts and Sciences, Kilis 7 Aralık University, Kilis, Turkey

3 Department of Chemistry, Faculty of Arts and Sciences, Kilis 7 Aralık University, Kilis, Turkey

4 Department of Chemistry, Faculty of Sciences, Ataturk University, Erzurum, Turkey

5 University-Industry-Public Collaboration, Development-Research and Application Centre, Kahramanmaraş Sütçü Imam University, K.Maraş, Turkey

References

[1] Randhir, R., Y.-T. Lin, and K. Shetty, Phenolics, their antioxidant and antimicrobial activity in dark germinated fenugreek sprouts in response to peptide and phytochemical elicitors. Asia Pac J Clin Nutr, 2004. **13**: pp. 295-307.

[2] Hassan, M.A., et al., Effects of salt and water stress on plant growth and on accumulation of osmolytes and antioxidant compounds in cherry tomato. Not Bot Horti Agrobo 2015. **43**(1): pp. 1-11.

[3] Ross, J.A. and C.M. Kasum, Dietary flavonoids: bioavailability, metabolic effects, and safety. Annu Rev Nutr, 2002. **22**: pp. 19-34.

[4] Huvaere, K. and L.H. Skibsted, Flavonoids protecting food and beverages against light. J Sci Food Agric, 2015. **95**(1): pp. 20-35.

[5] Formica, J.V. and W. Regelson, Review of the biology of Quercetin and related bioflavonoids. Food Chem Toxicol, 1995. **33**(12): pp. 1061-80.

[6] Cook, N.C. and S. Samman, Flavonoids—chemistry, metabolism, cardioprotective effects, and dietary sources. Nutr Biochem, 1996. **7**: pp. 66-7.

[7] Hamrouni, I., H.B. Salah, and B. Marzouk, Effects of water-deficit on lipids of safflower aerial parts. Phytochemistry, 2001. **58**(2): pp. 277-80.

[8] Ozkan, A. and M. Kulak, Effects of water stress on growth, oil yield, fatty acid composition and mineral content of *Sesamum indicum*. J Anim Plant Sci, 2013. **23**: pp. 1686-90.

[9] Cetinkaya, H., M. Koc, and M. Kulak, Monitoring of mineral and polyphenol content in olive leaves under drought conditions: application chemometric techniques. Ind Crops Prod, 2016. **88**: pp. 78-84.

[10] Yang, Y., et al., Influence of drought on oxidative stress and flavonoid production in cell suspension culture of Glycyrrhiza inflata Batal. Z Naturforsch C, 2007. **62**(5-6): pp. 410-6.

[11] Halimeh, R., et al., Effect of drought interactions with ascorbate on some biochemical parameters and antioxidant enzymes activities in *Dracocephalum moldavica* L. Middle East J Sci Res, 2013. **13**(4): pp. 522-531.

[12] Basahi, J.M., I.M. Ismail, and I.A. Hassan, Effects of enhanced UV-B radiation and drought stress on photosynthetic performance of lettuce (*Lactuca sativa* L. *Romaine*) Plants Annu Res Rev Biol, 2014. **4**(11): pp. 1739-56.

[13] Márquez-García, B., M.Á. Fernández-Recamales, and F. Córdoba, Effects of cadmium on phenolic composition and antioxidant activities of *Erica andevalensis*. J Bot, 2012. **2012**: pp. 1-6.

[14] Ahmed, H.R., et al., Soil contamination with heavy metals and its effect on growth, yield and physiological responses of vegetable Crop plants (turnip and lettuce). J Stress Physiol Biochem 2013. **9**(4): pp. 145-62.

[15] Dudjak, J., et al., Effect of cadmium on flavonoid content in young barley (*Hordeum sativum* L.) plants. Plant Soil Environ, 2005. **50**(11): pp. 471-7.

[16] Chutipaijit, S., S. Cha-Um, and K. Sompornpailin, Differential accumulations of proline and flavonoids in indica rice varieties against salinity. Pak J Bot, 2009. **41**(5): pp. 2497-506.

[17] Abdallah, S.B., et al., Salt stress (NaCl) affects plant growth and branch pathways of carotenoid and flavonoid biosyntheses in *Solanum nigrum*. Acta Physiol Plant, 2016. **38**(72): pp. 1-13.

[18] Rezazadeh, A., et al., Effect of salinity on phenolic composition and antioxidant activity of artichoke (*Cynara scolymus* L.) Res J Med Plants, 2012. **6**(3): pp. 245-52.

[19] Spayd, S.E., et al., Separation of sunlight and temperature effects on the composition of *Vitis vinifera* cv. Merlot Berries. Am J Enol Vitic, 2002. **53**: pp. 171-82.

[20] Tarara, J.M., et al., Berry temperature and solar radiation alter acylation, proportion, and concentration of anthocyanin in Merlot grapes. Am J Enol Vitic., 2008. **59**: pp. 235-47.

[21] Yuan, Y., et al., High temperature effects on flavones accumulation and antioxidant system in Scutellaria baicalensis Georgi cells. Afr J Biotechnol, 2011. **10**(26): pp. 5182-92.

[22] Rajabbeigi, E., et al., Interaction of drought stress and UV-B radiation—impact on biomass production and flavonoid metabolism in lettuce (*Lactuca sativa* L.). J Appl Bot Food Qual, 2013. **86**(1): pp. 190-7.

[23] Hopkins, W. G., and N.P.A. Hüner, Introduction to plant physiology (No. Ed. 4). John Wiley and Sons, The University of Western Ontario, Hoboken, NJ, USA, 2009.

[24] Ring, L., et al., Metabolic interaction between anthocyanin and lignin biosynthesis is associated with peroxidase FaPRX27 in strawberry fruit. Plant Physiol, 2013. **163**(1): pp. 43-60.

[25] Almeida, J.R.M., E. D'Amico, A. Preuss, F. Carbone, C.H. de Vos, B. Deiml, F. Mourgues, G. Perrotta, T.C. Fischer, A.G. Bovy, et al., Characterization of major enzymes and genes involved in flavonoid and proanthocyanidin biosynthesis during fruit development in strawberry (Fragaria × ananassa). Arch Biochem Biophys, 2007. **465**: pp. 61-71.

[26] Griesser, M., T. Hoffmann, M.L. Bellido, C. Rosati, B. Fink, R. Kurtzer, A. Aharoni, J. Muñoz-Blanco, and W. Schwab, Redirection of flavonoid biosynthesis through the down-regulation of an anthocyanidin glucosyltransferase in ripening strawberry fruit. Plant Physiol, 2007. **146**: pp. 1528-39.

[27] Griesser, M., F. Vitzthum, B. Fink, M.L. Bellido, C. Raasch, J. Munoz-Blanco, and W. Schwab, Multi-substrate flavonol O-glucosyltransferases from strawberry (Fragaria × ananassa) achene and receptacle. J Exp Bot, 2008. **59**: pp. 2611-25.

[28] Lunkenbein, S., M.L. Bellido, A. Aharoni, E.M.J. Salentijn, R. Kaldenhoff, H.A. Coiner, J. Muñoz-Blanco, and W. Schwab, Cinnamate metabolism in ripening fruit. Characterization of a UDP-glucose:cinnamate glucosyltransferase from strawberry. Plant Physiol, 2006. **140**: pp. 1047-58.

[29] Lunkenbein, S., H. Coiner, C.H. de Vos, J.G. Schaart, M.J. Boone, F.A. Krens, W. Schwab, and E.M. Salentijn, Molecular characterization of a stable antisense chalcone synthase phenotype in strawberry (Fragaria × ananassa). J Agric Food Chem, 2006. **54**: pp. 2145-53.

[30] Schwab, W., T. Hoffmann, G. Kalinowski, and A. Preuß, Functional genomics in strawberry fruit through RNAi-mediated silencing. Genes Genomes Genomics, 2011. **5**: pp. 91-101.

[31] Seyed, Y., S. Lisar, R. Motafakkerazad, M.M. Hossain, and I.M.M. Rahman. Water Stress in Plants: Causes, Effects and Responses, Water Stress, Prof. Ismail Md. Mofizur Rahman (Ed.), InTech: Rijeka, Croatia, 2012, doi: 10.5772/39363. Available from: http://www.intechopen.com/books/water-stress/water-stress-in-plants-causes-effects-and-responses

[32] Bryant, J.P., F.S.I. Chapin, and D.R. Klein, Carbon/nutrient balance of boreal plants in relation to vertebrate herbivory. Oikos 1983; **40**: pp. 357-68.

[33] Winkel-Shirley, B. Flavonoid biosynthesis, A colorful model for genetics, biochemistry, cell biology and biotechnology. Plant Physiol 2001; **26**: pp. 485-93; PMID: 11402179; doi:10.1104/pp.126.2.485

[34] Akula, R., and G.A. Ravishankar, Influence of abiotic stress signals on secondary metabolites in plants. Plant Signal Behav, 2011. **6**(11), pp. 1720-31.

[35] Winkel-Shirley, B., Flavonoid biosynthesis. A colorful model for genetics, biochemistry, cell biology, and biotechnology. Plant Physiol, 2001. **126**(2): pp. 485-93.

[36] Treml, J. and K. Smejkal, Flavonoids as potent scavengers of hydroxyl radicals. Comp Rev Food Sci Food Saf, 2016. **15**(4): pp. 720-38.

[37] Prochazkova, D., I. Bousova, and N. Wilhelmova, Antioxidant and prooxidant properties of flavonoids. Fitoterapia, 2011. **82**(4): pp. 513-23.

[38] Pietta, P.G., Flavonoids as antioxidants. J Nat Prod, 2000. **63**(7): pp. 1035-42.

[39] Amic, D., et al., Structure-radical scavenging activity relationships of flavonoids. Croat Chem Acta, 2003. **76**(1): pp. 55-61.

[40] Jordan, B.R., Molecular response of plant cells to UV-B stress. Funct Plant Biol, 2002. **29**(8): pp. 909-16.

[41] Jansen, M.A.K., V. Gaba, and B.M. Greenberg, Higher plants and UV-B radiation: Balancing damage, repair and acclimation. Trends Plant Sci, 1998. **3**(4): pp. 131-5.

[42] Hu, Z.B., T. Cools, and L. De Veylder, Mechanisms used by plants to cope with DNA damage. Annu Rev Plant Biol, 2016. **67**: pp. 439-62.

[43] Britt, A.B., Repair of DNA damage induced by solar UV. Photosynth Res, 2004. **81**(2): pp. 105-12.

[44] Frohnmeyer, H. and D. Staiger, Ultraviolet-B radiation-mediated responses in plants. Balancing damage and protection. Plant Physiol, 2003. **133**(4): pp. 1420-8.

[45] Li, Z., et al., Sensing and responding to excess light. Annu Rev Plant Biol, 2009. **60**: pp. 239-60.

[46] Heijde, M. and R. Ulm, UV-B photoreceptor-mediated signalling in plants. Trends Plant Sci, 2012. **17**(4): pp. 230-7.

[47] Brown, B.A., et al., A UV-B-specific signaling component orchestrates plant UV protection. Proc Natl Acad Sci U S A, 2005. **102**(50): pp. 18225-30.

[48] Winkel-Shirley, B., Flavonoid biosynthesis. A colorful model for genetics, biochemistry, cell biology, and biotechnology. Plant Physiol, 2001. **126**(2): pp. 485-93.

[49] Dao, T.T.H., H.J.M. Linthorst, and R. Verpoorte, Chalcone synthase and its functions in plant resistance. Phytochem Rev, 2011. **10**(3): pp. 397-412.

[50] Favory, J.J., et al., Interaction of COP1 and UVR8 regulates UV-B-induced photomorphogenesis and stress acclimation in Arabidopsis. EMBO J, 2009. **28**(5): pp. 591-601.

[51] Endt, D.V., J.W. Kijne, and J. Memelink, Transcription factors controlling plant secondary metabolism: what regulates the regulators? Phytochemistry, 2002. **61**(2): pp. 107-14.

[52] Ang, L.H., et al., Molecular interaction between COP1 and HY5 defines a regulatory switch for light control of Arabidopsis development. Mol Cell, 1998. **1**(2): pp. 213-22.

[53] Greenberg, B.M., et al., Morphological and physiological responses of *Brassica napus* to ultraviolet-B radiation: photomodification of ribulose-1,5-bisphosphate carboxylase/oxygenase and potential acclimation processes. J Plant Physiol, 1996. **148**(1-2): pp. 78-85.

[54] Li, J.Y., et al., Arabidopsis flavonoid mutants are hypersensitive to Uv-B irradiation. Plant Cell, 1993. **5**(2): pp. 171-9.

[55] Sisa, M., et al., Photochemistry of flavonoids. Molecules, 2010. **15**(8): pp. 5196-245.

[56] Treutter, D., Significance of flavonoids in plant resistance: a review. Environ Chem Lett, 2006. **4**(3): pp. 147-57.

[57] Law, Y.K., et al., Sequence-dependent thymine dimer formation and photoreversal rates in double-stranded DNA. Photochem Photobiol Sci, 2013. **12**(8): pp. 1431-9.

[58] Song, J., M.G. Kemp, and J.H. Choi, Detection of the excised, damage-containing oligonucleotide products of nucleotide excision repair in human cells. Photochem Photobiol, 2016 **93**: pp. 192-198.

[59] Sancar, A., Photolyase and cryptochrome blue-light photoreceptors. Adv Protein Chem, 2004. **69**: pp. 73-100.

[60] Canturk, F., et al., Nucleotide excision repair by dual incisions in plants. FEBS J, 2016. **283**: p. 18.

[61] Li, N., et al., UV-B-induced CPD photolyase gene expression is regulated by UVR8-dependent and -independent pathways in Arabidopsis. Plant Cell Physiol, 2015. **56**(10): pp. 2014-23.

[62] Brown, B.A. and G.I. Jenkins, UV-B signaling pathways with different fluence-rate response profiles are distinguished in mature Arabidopsis leaf tissue by requirement for UVR8, HY5, and HYH. Plant Physiol, 2008. **146**(2): pp. 576-88.

[63] Sancar, A., Mechanisms of DNA repair by photolyase and excision nuclease (Nobel Lecture). Angew Chem Int Ed, 2016. **55**(30): pp. 8502-27.

[64] Pudzianowska, M., M. Gajewski, J.L. Przybył, A. Buraczyńska, O. Gaczkowska, M. Matuszczak, and M. Dziechciarska. Influence of storage conditions on flavonoids content and antioxidant activity of selected shallot (*Allium cepa* var. *ascalonicum* Backer) hybrid cultivars. Veg Crops Res Bull, 2012. **77**: pp. 101-11.

[65] El-Rayes, D.A., Effect of carbon dioxide-enriched atmosphere during cold storage on limiting antioxidant losses and maintaining quality of 'Barhy' date fruits. J Meteorol Environ Arid Land Agric Sci, 2009. **20**(1): pp. 3-22.

[66] Raya, K.B., S.H. Ahmad, S.F. Farhana, M. Mohammad, N.E. Tajidin, and A. Parvez, Changes in phytochemical contents in different parts of *Clinacanthus nutans* (Burm. f.) lindau due to storage duration. Bragantia, 2005. **74**(4), pp. 445-52.

[67] Zhang, W.E., C.L. Wang, B.B. Shi, and X.J. Pan, Effect of storage temperature and time on the nutritional quality of walnut male inflorescences. J Food Drug Anal, 2016. http://dx.doi.org/10.1016/j.jfda.2016.05.010

[68] Cordenunsi, B.R., M.I. Genovese, J.R.O. do Nascimento, N.M.A. Hassimotto, R.J. dos Santos, and F.M. Lajolo, Effects of temperature on the chemical composition and antioxidant activity of three strawberry cultivars. Food Chem, 2005. **91**(1), pp. 113-21.

[69] Wang, S.Y., C.T. Chen, and C.Y. Wang, The influence of light and maturity on fruit quality and flavonoid content of red raspberries. Food Chem, 2009. **112**(3), pp. 676-84.

[70] Lachman, J., D. Pronek, A. Hejtmánková, J. Dudjak, V. Pivec, and K. Faitová, Total polyphenol and main flavonoid antioxidants in different onion (*Allium cepa* L.) varieties. Hortic Sci, 2003. **30**(4), pp. 142-7.

[71] Azizuddin, A. Qadeer, and S. Ghafoor, Effects of postharvest storage conditions on physico-chemical analysis and antioxidant capacity of *Brassica rapa* L. Bangladesh J Bot, 2016. **45**(1), pp. 85-92.

[72] Thanajiruschaya, P., W. Doksaku, P. Rattanachaisit, and J. Kongkiattikajorn, Effect of storage time and temperature on antioxidant components and properties of milled rice. KKU Res J, 2010. **15**(9), pp. 843-51.

[73] Mrmošanin, J.M., A.N. Pavlović, J.N. Veljković, S.S. Mitić, S.B. Tošić, and M.N. Mitić, Effect of storage temperature and thermal processing on catechins, procyanidins and total flavonoids stability in commercially available cocoa powders. Facta Universitatis, Series Physics, Chemistry and Technology, 2014. **13**(1), pp. 39-49.

[74] Sentandreu, E., J.L. Navarro, and J.M. Sendra, Effect of technological processes and storage on flavonoids content and total, cumulative fast–kinetics and cumulative slow-kinetics antiradical activities of citrus juices. Eur Food Res Technol, 2007. **225**(5-6), p. 905.

[75] Matthes, A., and M. Schmitz-Eiberger Polyphenol content and antioxidant capacity of apple fruit: effect of cultivar and storage conditions. J Appl Bot Food Qual, 2012. **82**(2), pp. 152-57.

Chapter 9

Biosynthesis and Biomimetic Synthesis of Flavonoid Diels-Alder Natural Products

Shah Bakhtiar Nasir, Jia Ti Tee,
Noorsaadah Abd Rahman and Chin Fei Chee

Additional information is available at the end of the chapter

http://dx.doi.org/10.5772/intechopen.68781

Abstract

This chapter describes the biosynthesis and biomimetic synthesis of naturally occurring flavonoid Diels-Alder adducts found either from the family Moraceae or Zingiberaceae. The main topics addressed are biosynthetic studies by employing *Morus alba L.* cell cultures through feeding experiments of various exogenous substrates and putative precursors, as well as a various biomimetic approach for the chemical syntheses of flavonoid Diels-Alder natural products.

Keywords: biomimetic, flavonoid, Diels-Alder, Cycloaddition, biosynthesis

1. Introduction

The flavonoid Diels-Alder natural products are mainly found from the families of Moraceae and Zingiberaceae. Since the majority of these compounds are discovered from the Moraceae, they are often referred as mulberry Diels-Alder flavonoids or mulberry Diels-Alder type adducts. These secondary metabolites exhibit promising biological activities against hypertension, HIV, tuberculosis, anti-inflammation and cancers [1–7]. Thus far, more than 140 of these Diels-Alder type flavonoids have been discovered from nature (**Figure 1**). The structural complexity and promising bioactivities of these flavonoid Diels-Alder natural products have stimulated research interest into their biosynthesis and chemical synthesis.

The Diels-Alder type flavonoids are considered to be formed through an enzymatic Diels-Alder reaction between a dehydroprenyl diene and a chalcone dienophile (**Scheme 1**) [8]. The diene is usually derived from a flavonoid, such as flavone, flavanone, flavonol, flavanonol, or from a

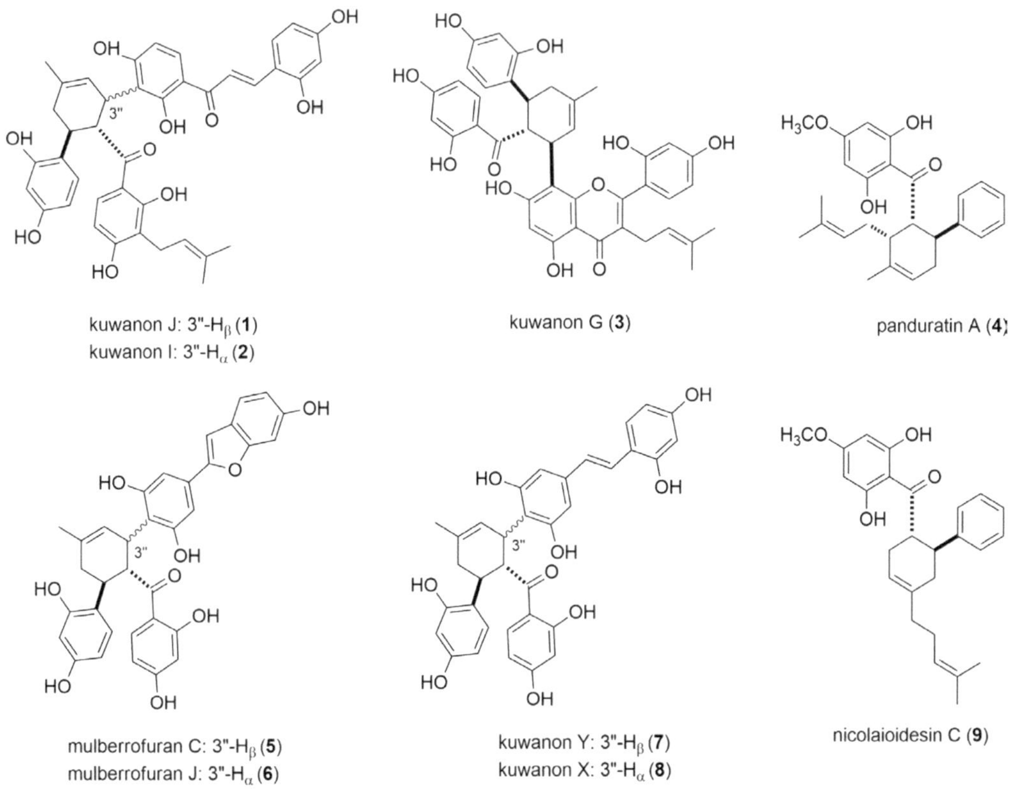

Figure 1. Examples of flavonoid Diels-Alder natural products.

monoterpene, such as myrcene and β-*trans*-ocimene. The dienophile of this class of Diels-Alder compounds is exclusively derived from a chalcone derivative. Subsequent oxidation and cyclization steps of these flavonoid Diels-Alder adducts can result in more complex structures. The Diels-Alder adducts bearing the *cis-trans* stereochemistry on the cyclohexenyl ring would be derived through an *endo* transition state (**12**), whereas the *trans-trans* stereochemistry arises

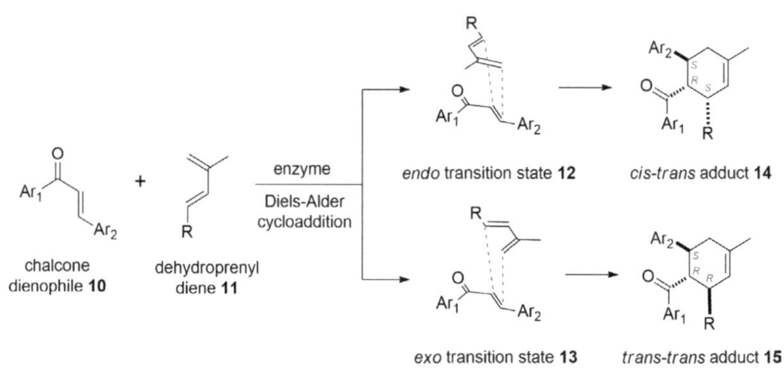

Scheme 1. Stereochemistry on the cyclohexene ring of flavonoid Diels-Alder natural products.

http://dx.doi.org/10.5772/intechopen.68781

from the *exo* transition state (**13**) (**Scheme 1**) [8]. The stereochemistry of these adducts, including the absolute configuration on the cyclohexene ring, has been explicitly confirmed by circular dichroism (CD) spectroscopic evidence [9] and X-ray crystallographic analysis [10, 11]. The unique structural features and diverse activities of these adducts have recently aroused much interest of synthetic and medicinal chemistry. The main topics addressed in this chapter are biosynthesis and biomimetic synthesis of flavonoid Diels-Alder natural products and about 40 references are cited. As the flavonoid Diels-Alder natural products are composed of a diverse family of secondary metabolites, other subclasses where the dienophile is not a chalcone (e.g. mongolicin B, -E, sanggenon B, -R, -S, dimoracin, mulberrofuran H, meroterpene, pauferrol A derivatives, etc) are not covered in this chapter.

2. Biosynthesis of the flavonoid Diels-Alder natural products

Although the biosynthesis of the flavonoid Diels-Alder natural products that derived from a monoterpene is not well-studied [12, 35], it is hypothesized that a Diels-Alder reaction between a chalcone dienophile and a monoterpene (β-*trans*-ocimene or myrcene) would lead to the direct formation of these adducts (**Figure 2**).

The biosynthesis of the mulberry Diels-Alder flavonoids has been intensively studied by Professors Taro Nomura and Shinichi Ueda. The biosynthetic studies of these adducts were carried out in the callus tissues of *Morus alba L* [13]. In their pioneering studies, the callus tissues induced from the leaves or seedlings were cultivated and subjected to selection over a period of 9 years for cell strains with high-pigment productivity [14]. Extraction of these high pigmented cell cultures resulted in isolation of six Diels-Alder adducts, kuwanons J (**1**), Q (**23**), R (**24**), V (**25**), mulberrofuran E (**26**), and chalcomoracin (**27**) along with morachalcone A (**28**), isobavachalcone (**29**), and moracin C (**30**) (**Figure 3**) [15–18].

The structures of metabolites **1**, **23**–**27** suggested that they are either the Diels-Alder adducts from a prenylchalcone and a dehydroprenylchalcone or the Diels-Alder adducts from a prenylchalcone and a dehydroprenyl-2-arylbenzofuran. Nomura and co-workers hypothesized that kuwanon J (**1**) was an adduct of morachalcone A (**28**) and dehydroprenylmorachalcone A. Kuwanon Q (**23**) was an adduct of isobavachalcone (**29**) and dehydroprenylmorachalcone

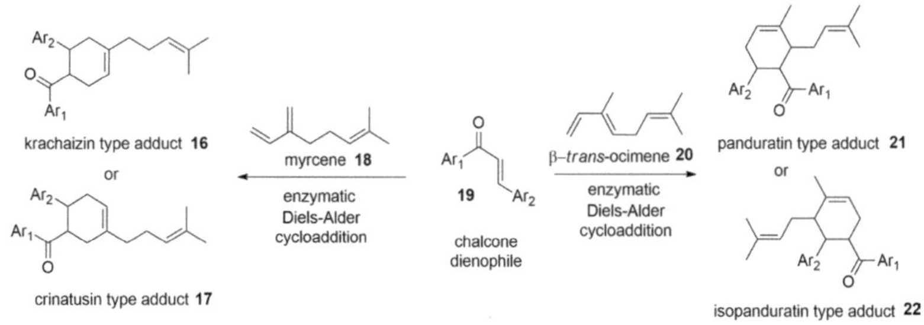

Figure 2. Plausible biosynthesis of flavonoid Diels-Alder natural products that derived from a monoterpene.

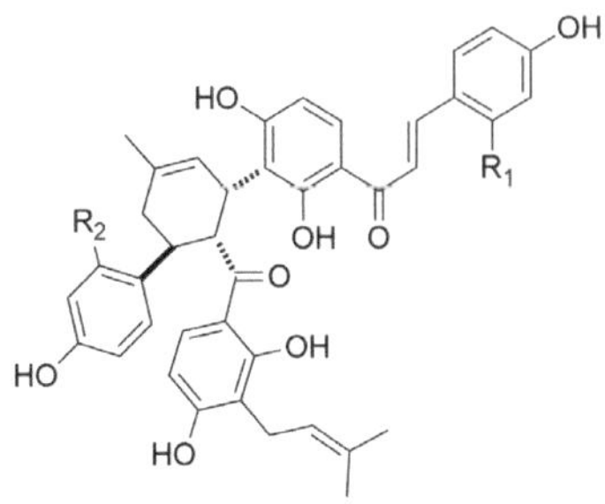

kuwanon J: $R_1 = R_2 = OH$ (**1**)
kuwanon Q: $R_1 = OH$; $R_2 = H$ (**23**)
kuwanon R: $R_1 = H$; $R_2 = OH$ (**24**)
kuwanon V: $R_1 = R_2 = H$ (**25**)

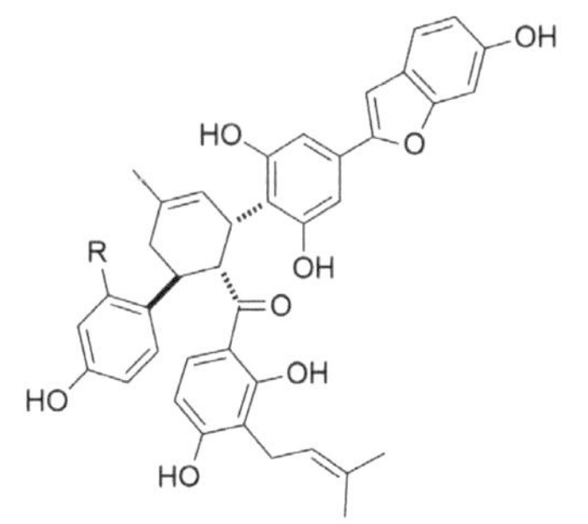

mulberrofuran E: R = H (**26**)
chalcomoracin: R = OH (**27**)

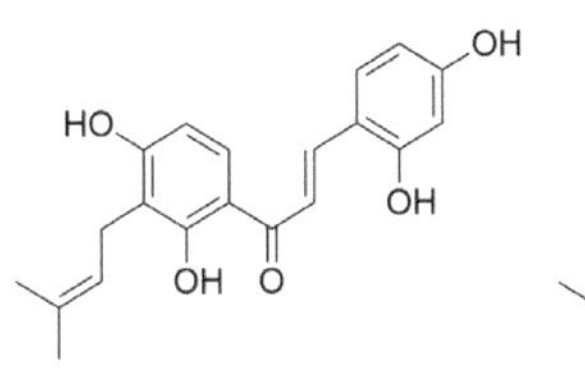

morachalcone A (**28**)

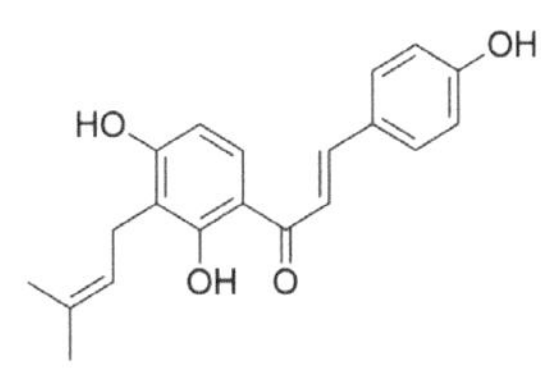

isobavachalcone (**29**)

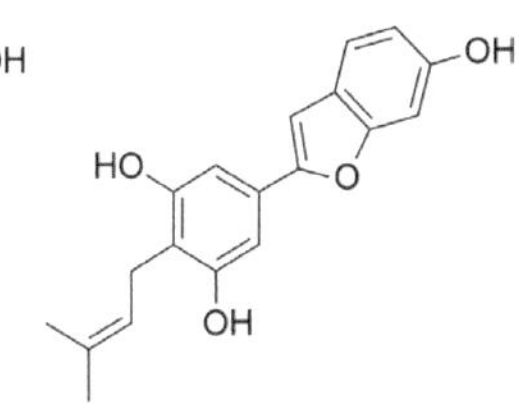

moracin C (**30**)

Figure 3. Metabolites isolated from the *Morus alba* cell cultures [15–18].

A. Kuwanon R (**24**) was an adduct of morachalcone A (**28**) and dehydroprenylisobavachalcone. Kuwanon V (**25**) was an adduct of isobavachalcone (**29**) and dehydroprenylisobavachalcone. Chalcomoracin (**27**) was an adduct of morachalcone A (**28**) and dehydroprenylmoracin C. Mulberrofuran E (**26**) was an adduct of isobavachalcone (**29**) and dehydroprenylmoracin C. It is interesting that these Diels-Alder metabolites and their monomeric precursors (morachalcone A, isobavachalcone and moracin C) were isolated from *M. alba* cell cultures. In addition, the callus tissue can produce 100 times more mulberrofuran E and chalcomoracin than the intact plant [15–17]. The biosynthetic studies of these Diels-Alder adducts were further examined through feeding experiments of various exogenous substrates and putative precursors to the *M. alba* cell cultures.

2.1. Feeding experiments with ^{13}C-labeled acetate to the *Morus alba* cell cultures

Acetate is an important carbon source for biosynthesis studies in *M. alba* cell cultures. Feeding experiments of [1–^{13}C]-, [2–^{13}C]-, or [1, 2–$^{13}C_2$]-acetates to the *M. alba* cell cultures resulted in the highly ^{13}C-enriched aromatic carbons of chalcomoracin (**27**) and kuwanon J (**1**), indicating that both **27** and **1** are derived from two molecules of cinnamoylpolyketide precursors (**Figure 4**) [19]. From the labeling patterns, the chalcone moiety (**34**) of both chalcomoracin (**27**) and kuwanon J (**1**) is hypothesized to be derived via deoxygenation at C-5 of the cinnamoylpolyketide precursor **31**, followed by Claisen condensation and aromatization (**Figure 5**) [20]. The 2-arylbenzofuran moiety (**36**) of **27** and **1** is hypothesized to be derived by the Aldol condensation at C-3 and C-8 of the cinnamoylpolyketide precursor **32**, followed by decarboxylation and aromatization (**Figure 5**) [19].

http://dx.doi.org/10.5772/intechopen.68781

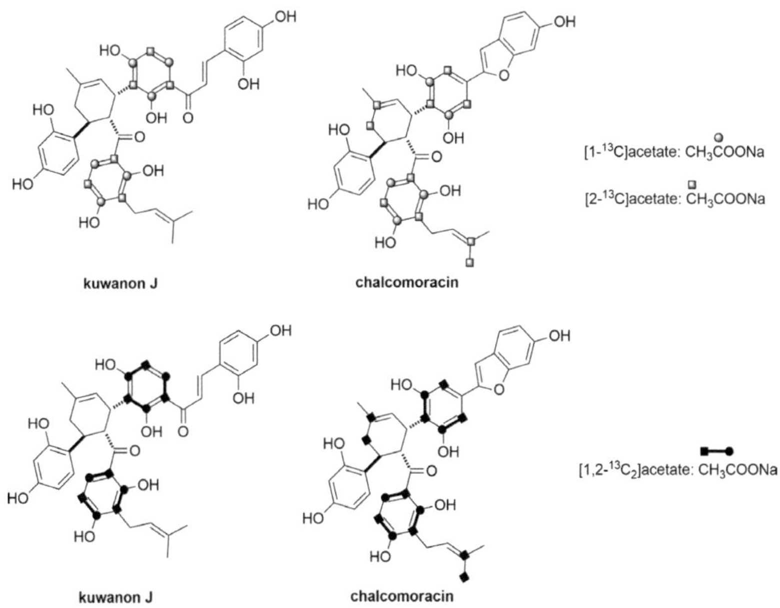

Figure 4. ^{13}C-labeling patterns of Kuwanon J and chalcomoracin from [1-^{13}C]-, [2-^{13}C]-, or [1, 2-$^{13}C_2$]acetate [19].

However, unlike the aromatic carbons, the isoprene units of chalcomoracin were marginally labeled (~0.4% enrichment) [19]. On the basis of ^{13}C–^{13}C spin coupling in the ^{13}C-NMR spectrum, the labeling of [2–^{13}C] acetate was incorporated into the starter acetate carbons in the biosynthesis of the isoprene unit of chalcomoracin (**27**). On the contrary, the [1–^{13}C] acetate was not incorporated in the isoprene unit of chalcomoracin (**27**) [19]. These findings suggested that a tricarboxylic acid (TCA) cycle was involved in the biosynthesis of the isoprenyl unit of chalcomoracin [8]. The rational of this hypothesis was derived from the ^{13}C-labeling experiments. In the experiment with [2–^{13}C] acetate, the contiguous ^{13}C labels can be derived from the methyl groups of the intact acetate administered by way of at least two passages through the TCA cycle

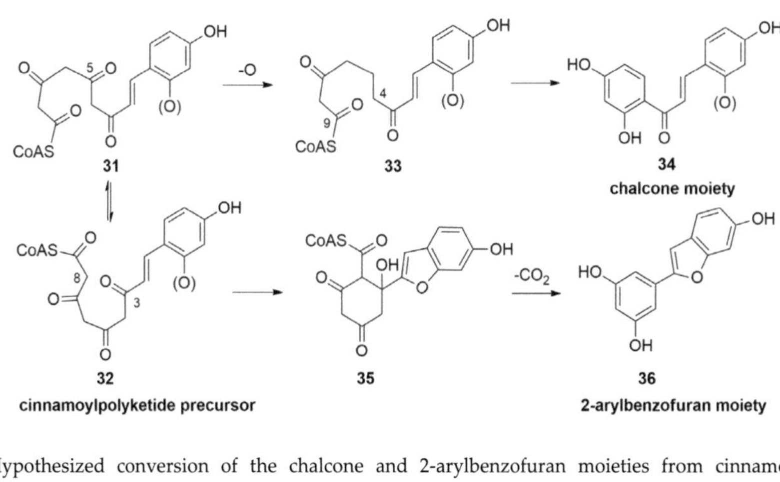

Figure 5. Hypothesized conversion of the chalcone and 2-arylbenzofuran moieties from cinnamoylpolyketide precursor [19].

[19]. In the experiment with [1–^{13}C] acetate, the ^{13}C label was not found in the isoprenyl unit, presumably due to the removal of carbon dioxide during passage through the TCA cycle (**Figure 6**).

This hypothesis was reinforced by the feeding experiment with [2–^{13}C] acetate in a pulsed manner (three times, every 12 h) to the *M. alba* cell cultures [21]. The result from this experiment enabled the identification of the satellite peaks based on the ^{13}C–^{13}C spin coupling between carbons at C-25′′ and C23′′, C-7′′ and C-1′′, C-23′′ and C-24′′ as well as C-6′′ and C-1′′ of chalcomoracin. The ^{13}C-enrichment at C-7′′ and C-25′′ occurred after the first and third [2–^{13}C] acetate administrations but not at the second administration suggested the isomerization between the 3,3-dimethylallyl and 3-methylbutadienyl groups (**Figure 7**) [8]. The coupling patterns of the central carbons (C-1′′ and C-23′′) appeared as doublet signal instead of the doublet of doublet signal indicated that these central carbons are independently coupled with the adjacent methyl carbons. Nomura *et al.* hypothesized that the independent ^{13}C-labeling pattern at the isoprenyl unit might due to the transfer of ^{13}C-labeling from *cis*-methyl to *trans*-methyl through the diene formation (**Figure 7**) [8, 21]. Taken together, these findings gave conclusive evidence on the diene formation from the isoprenyl moiety for the Diels-Alder cycloaddition reaction. Thus, the feeding experiment with ^{13}C-labeled acetate revealed that the Diels-Alder adducts chalcomoracin and kuwanon J are biosynthesized through the [4 + 2] cycloaddition reaction between two cinnamoylpolyketide-derived molecules [8].

2.2. Feeding experiments with methoxychalcone and prenylated flavone precursors

Based on the fact that methoxychalcone or methoxy-substituted Diels-Alder adducts have not been found in the *M. alba* cell cultures, therefore involvement of these precursors in the construction of the Diels-Alder adducts would be an important evidence for the enzymatic intermolecular Diels-Alder reaction in *M. alba* cell cultures.

Indeed, feeding methoxychalcone **37** to the cell cultures yielded prenylchalcone **38** and Diels-Alder adducts **40–43** (**Figure 8**) [22]. The formation of the prenylchalcone **38** from methoxychalcone **37** in the cell cultures indicated that isoprenylation occurs after the formation of chalcone skeleton from cinnamoylpolyketide precursor.

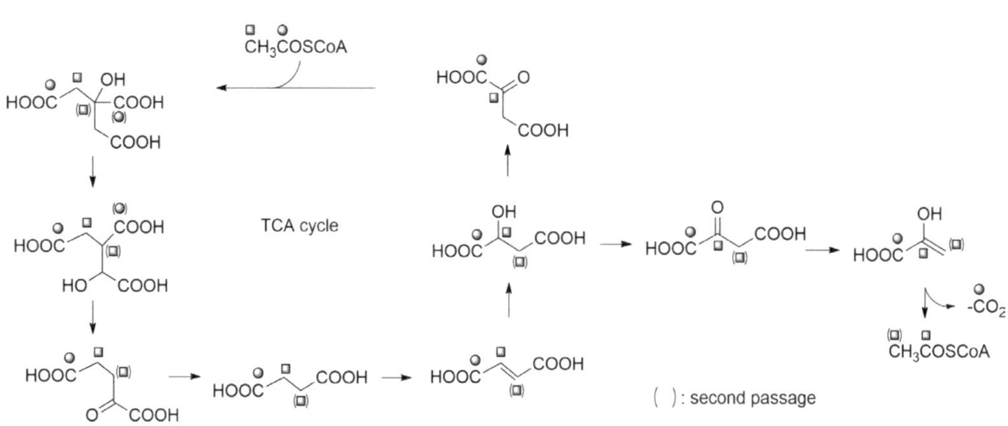

Figure 6. Formation of reorganized [1, 2-$^{13}C_2$]acetate through the TCA cycle [19].

http://dx.doi.org/10.5772/intechopen.68781

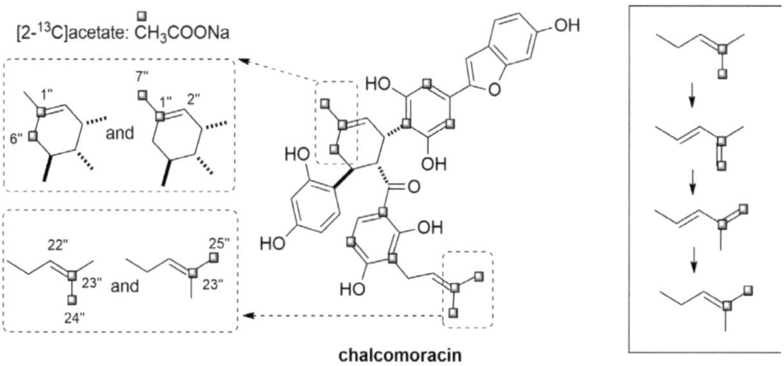

Figure 7. Two independent ^{13}C-labeling patterns at the isoprenyl units of chalcomoracin and the transfer of the ^{13}C-labeling from *cis*-methyl carbon to *trans*-methyl carbon through the diene formation [8, 19].

The metabolites **40–43** revealed that the methoxychalcone **37** was incorporated into the Diels-Alder adducts. Interestingly, when the synthetic prenylchalcone **38** was fed to the cell cultures, the same Diels-Alder metabolites **40–43** were isolated. Similarly, the feeding experiment of trimethoxychalcone **39** afforded the Diels-Alder metabolite **44** [22]. Taken together, these results suggested that both the requisite diene and dienophile can be derived from the same chalcone precursor. For example, dehydrogenation of the prenyl unit of chalcone **38**, followed by

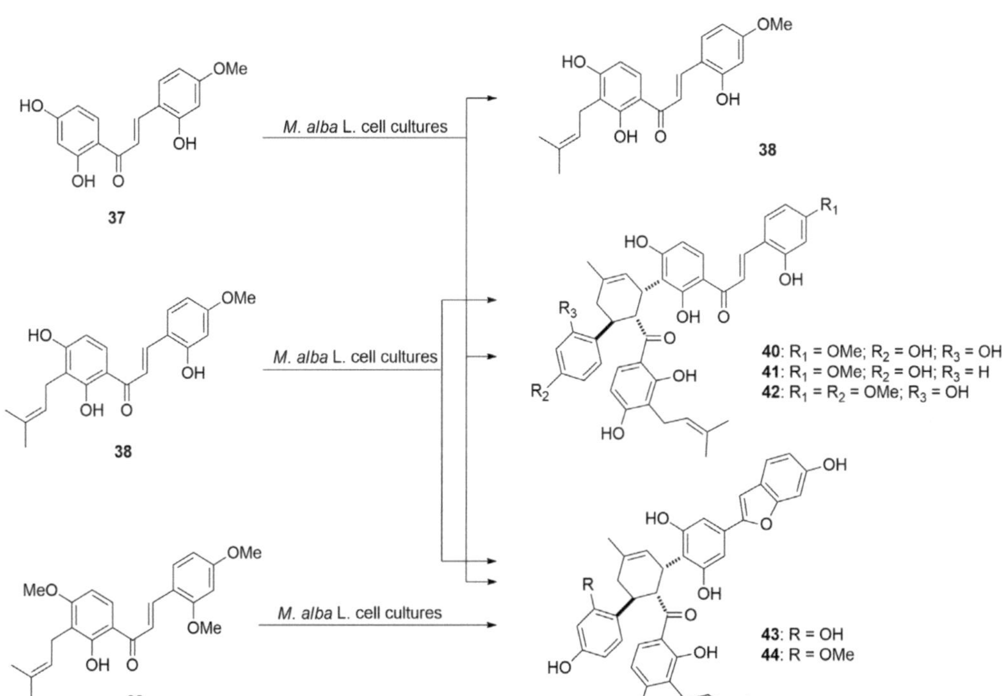

Figure 8. Feeding experiments of methoxychalcone derivatives to the *M. alba* cell cultures [22].

intermolecular [4 + 2] cycloaddition reaction with the *α*, *β*-double bond of another chalcone **38** leads to the formation of the Diels-Alder adduct **42** (**Figure 8**).

In addition, all these Diels-Alder metabolites derived from the methoxychalcone precursors were optically active and have the same stereochemistry as that of chalcomoracin (**27**) and kuwanon J (**1**). The results based on the feeding experiments of methoxychalcone derivatives revealed that the [4 + 2] cycloaddition reaction in the *M. alba* cell cultures is an enzymatic process.

Nomura *et al.* further attempted the synthesis of Diels-Alder natural product, artonin I (**46**) by using *M. alba* cell cultures (**Figure 9**) [23]. Although it is theoretically possible that artonin I could be derived from a chalcone dienophile (morachalcone A **28**) and a prenylflavone diene (**45**), but precursor of **45** (artocarpesin **47**) has not been found in *M. alba* cell cultures. Indeed, feeding **47** to the *M. alba* cell cultures resulted in the isolation of artonin I (**46**) through dehydrogenation of the prenyl group of **47** followed by the enzymatic [4 + 2] cycloaddition reaction with an endogenously generated morachalcone A **28**. This is the first example of a natural product's structure elucidation through enzymatic synthesis by using *M. alba* cell cultures [8].

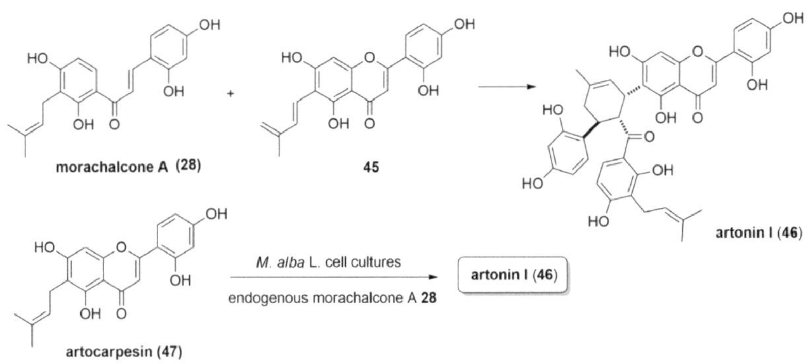

Figure 9. Biosynthesis of artonin I by administration of artocarpesin to the *M. alba* cell cultures [23].

3. Biomimetic synthesis of the flavonoid Diels-Alder natural products

The Diels-Alder cycloaddition reaction which named after Otto Paul Hermann (1876–1954) and Kurt Alder (1902–1958) was discovered during their studies on the reaction of benzoquinone and cyclopentadiene in 1928. Today, this cycloaddition reaction is a well-known method that is widely used to synthesize a six-membered cyclic compound in a regio- and stereocontrolled way. The following section discusses the use of this powerful synthetic methodology to prepare flavonoid Diels-Alder natural products based on the biosynthesis models.

3.1. Thermal conditions

During the early studies of the Diels-Alder cycloaddition reaction, the reaction was essentially carried out under thermal conditions owing to the simplicity of the experimental setup

http://dx.doi.org/10.5772/intechopen.68781

and the efficiency of the thermal process. Today, thermal promoted Diels-Alder cycloaddition reaction remains the first line approach for the construction of a six-membered cyclic compound, including that of flavonoid Diels-Alder natural products [24–30].

In 2010, Rizzacasa and co-workers reported the synthesis of racemic methyl ether derivatives of chalcomoracin, mongolicin F, mulberrofurans C and J *via* thermal Diels-Alder reaction (180°C in toluene) between chalcone dienophiles (**39** and **49**) and a dehydroprenyl-2-arylbenzofuran diene (**48**) (**Scheme 2**). The thermal Diels-Alder reaction resulted in a mixture of *endo*- and *exo*-diastereomers in almost equal quantity [24].

Rizzacasa and co-workers also reported a similar strategy for the synthesis of (±)-kuwanon I and J hexamethyl ethers. They hypothesized that the presence of an *ortho*-phenol group in the chalcone dienophile was essential for the Diels-Alder cycloaddition reaction. However, attempts to deprotect the methyl ethers of these Diels-Alder adducts using various demethylating agents were unsuccessful [25].

Rahman and co-workers utilized the thermal-promoted Diels-Alder reaction to synthesize (±)-dorsterone, (±)-kuwanon V and (±)-morusalbanol A pentamethyl ethers based on the biosynthesis models [27, 29, 30].

3.2. High pressure conditions

Although the thermal-promoted Diels-Alder reaction provides a rapid entry to flavonoid Diels-Alder adducts, this method may not be successful due to the instability of the diene or dienophile under a high-temperature condition. This limitation can be overcame using a high-pressure system for the Diels-Alder reaction.

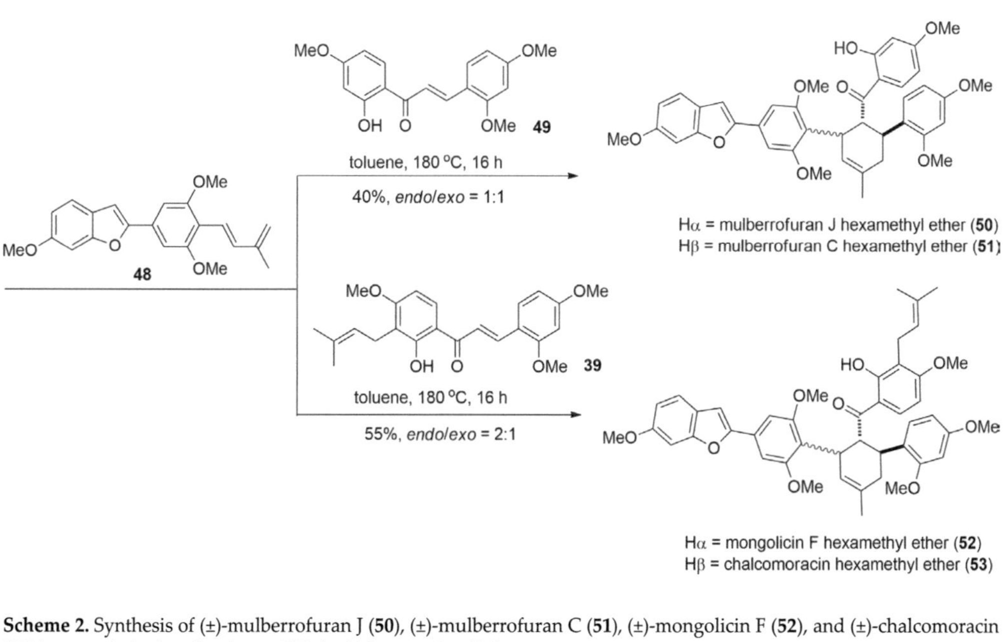

Scheme 2. Synthesis of (±)-mulberrofuran J (**50**), (±)-mulberrofuran C (**51**), (±)-mongolicin F (**52**), and (±)-chalcomoracin (**53**) hexamethyl ethers by thermal Diels-Alder reaction [24].

In 2013, Mcleod and co-workers utilized this strategy to synthesize (±)-panduratin A (**4**) and (±)-4-hydroxypanduratin A (**58**) [31]. Instead of late-stage Diels-Alder cycloaddition to synthesize the cyclohexenyl core of **4** and **58**, they initiated the biomimetic Diels-Alder reaction in an early stage by using methyl cinnamate (**54**) and β-*trans*-ocimene (**20**) (**Scheme 3**). High-pressure Diels-Alder reaction between **54** and **20** in dichloromethane at 19 kbar at room temperature gave a mixture of (±)-panduratin I (**56**) and (±)-panduratin H (**57**) in 1:2.9 ratio in 93% yield after 3 days. Subsequent transformations of panduratin H **57** afforded the natural products (±)-panduratin A and (±)-4-hydroxypanduratin A in three more further steps.

3.3. Single electron transfer initiated Diels-Alder reaction

In 1960 when Yates and Eaton first reported the acceleration of the Diels-Alder reaction by Lewis acid catalysts, a variety of Lewis acid catalysts have been developed to accelerate the reaction [32].

Porco and co-workers developed a Lewis acid catalyst system that composed of multiple components (CoI_2/*o*-phenanthroline/ZnI_2/Bu_4NBH_4) for the [4 + 2]-cycloaddition reaction between 2′-hydroxychalcone dienophiles and various simple dienes [33]. They hypothesized that the mechanism of this catalytic system was a single electron transfer initiated process (**Scheme 4**).

According to their report, the role of $CoI_2$and $Bu_4NBH_4$was hypothesized to be an electron donor [33]. As outlined in **Scheme 4**, coordination of ZnI_2 activated the carbonyl of 2′-hydroxychalcone **59** to form complex **62**. In the presence of electron donors, complex **62** may undergo metal-ion-promoted single electron transfer to generate a chalcone radical anion **63**. The regioselective addition of **63** to the diene should generate a stabilized, allylic radical **64** which may undergo ring-closing cyclization to produce ketyl intermediate **65**. Loss of ZnI_2 from **65** and subsequent single electron transfer to another complex **62** may

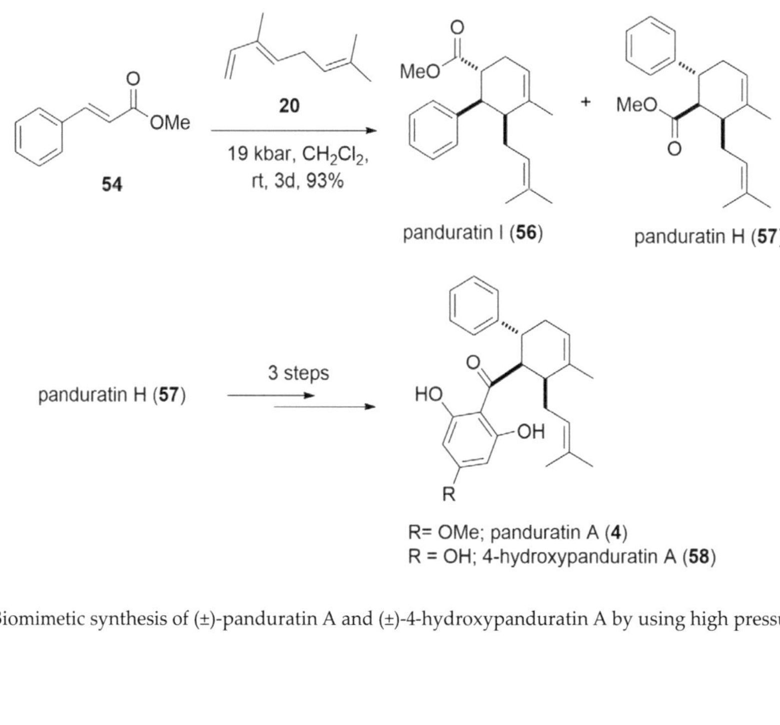

Scheme 3. Biomimetic synthesis of (±)-panduratin A and (±)-4-hydroxypanduratin A by using high pressure conditions [31].

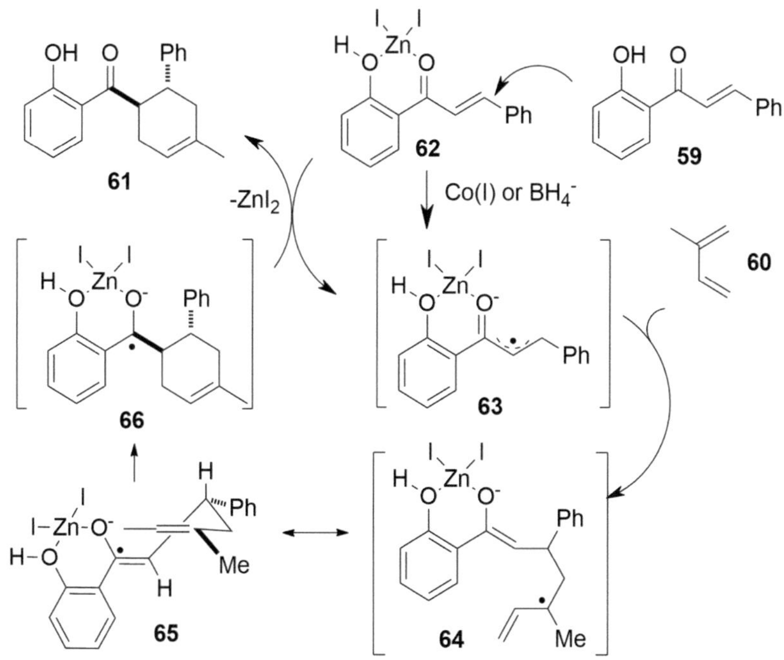

Scheme 4. Proposed mechanism for an electron transfer-initiated Diels-Alder cycloaddition reaction [33].

afford cycloadduct **61**, thereby restarting the catalytic cycle [33]. Following this mechanistic studies, Porco *et al.* further established the total synthesis of (±)-nicolaioidesin C (**9**) by using myrcene as a diene (**Scheme 5**) [33].

Rahman and co-workers used the thermal-promoted as well as single-electron-transfer-initiated Diels-Alder reaction to compare the efficiency of the biomimetic synthesis of (±)-kuwanon V (**71**) and (±)-doresterone (**70**) methyl ethers [27]. Thermal Diels-Alder cycloaddition between dienophile **69** and diene **68** in a pressure tube at 160°C for 18 h afforded **70** (*exo*-adduct) and **71** (*endo*-adduct) in 55% yield in a 1.5:1 ratio (**Scheme 6**). A comparable result (48% yield, 1.7:1 ratio) was obtained by using the single electron transfer initiated Diels-Alder reaction (ZnI_2, Bu_4NBH_4, CoI_2, 1, 10-phenanthroline in 60:10:10:10 mol%).

Recently, Valentina *et al.* reported the synthesis of (±)-kuwanol E and the heptamethyl ether derivative of (±)-kuwanol Y by using a combination of thermal conditions and Lewis acid

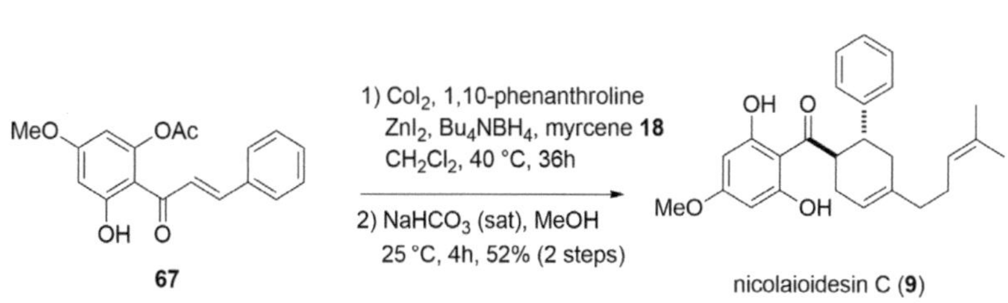

Scheme 5. Biomimetic synthesis of (±)-nicolaioidesin C (**9**) [33].

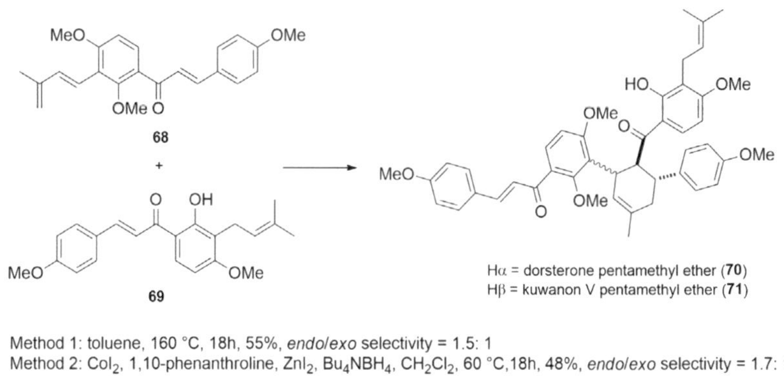

Scheme 6. Biomimetic synthesis of (±)-dorsterone and (±)-kuwanon V pentamethyl ethers [27].

catalyst [34]. The key synthetic step involved a borane tetrahydrofuran mediated biomimetic intermolecular Diels-Alder cycloaddition reaction. It is noteworthy that the *endo*/*exo* diastereoselectivity of the reaction was proven to be temperature-controlled.

3.4. Chiral ligand-Brønsted acid catalysis

The first asymmetric synthesis of flavonoid Diels-Alder natural products was reported by Palomo and co-workers in 2010 (**Scheme 7**). They employed a recoverable chiral auxiliary ((1 *R*)-(+)-camphor) in the asymmetric synthesis of nicolaioidesin C (**9**) [35]. First, the biomimetic Diels-Alder reaction between myrcene **18** and α'-hydroxy enone dienophile **72** was

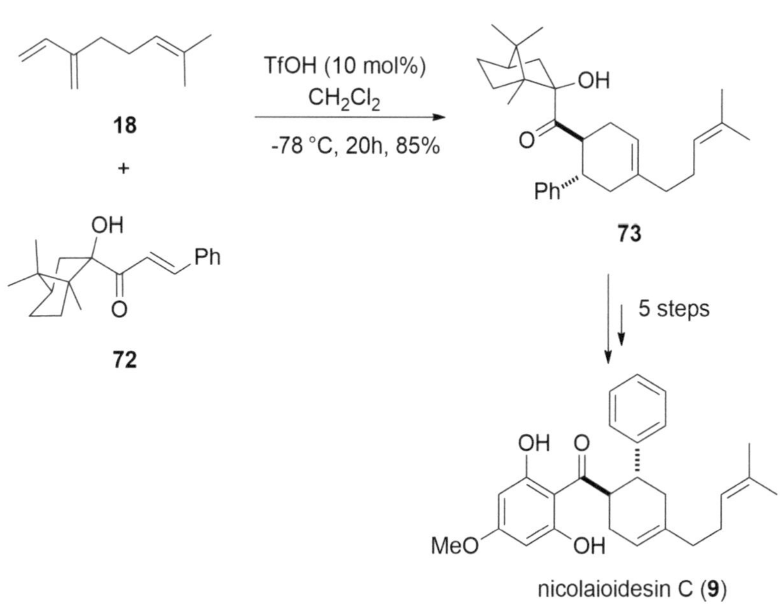

Scheme 7. Asymmetric biomimetic synthesis of (-)-nicolaioidesin C (**9**) [35].

http://dx.doi.org/10.5772/intechopen.68781

catalyzed by triflic acid at -78°C in dichloromethane to afford an enantiomeric enriched intermediate **73** in 85% yield. Subsequent transformation of the intermediate **73** in five further steps afforded (-)-nicolaioidesin C (**9**).

3.5. Silver nanoparticles catalyzed dehydrogenative Diels-Alder reaction

In 2010, Porco and co-workers discovered that silver (0) nanoparticles (AgNp) could effectively catalyze the Diels-Alder cycloaddition reaction [36]. The AgNP was prepared from a 3:1 molar ratio of $AgBF_4/Bu_4NBH_4$ in CH_2Cl_2 and then coated with silica gel. The solid product was filtered and then calcinated at 220°C to give AgNP. A proposed catalytic cycle was showed in **Scheme 8** [36]. It was hypothesized that proton removal and single electron transfer from the absorbed chalcone **59** to the silver nanoparticles may generate the AgNP-stabilized phenoxyl radical intermediate **74** which is in resonance with the radical **75**. A proposed concerted Diels-Alder reaction between the radical intermediate **74/75** and diene **60** provides **76** which generates **55** via back electron transfer (BET) and protonation [36]. A final desorption step gave the Diels-Alder adduct **61**. Porco and co-workers hypothesized that this silver

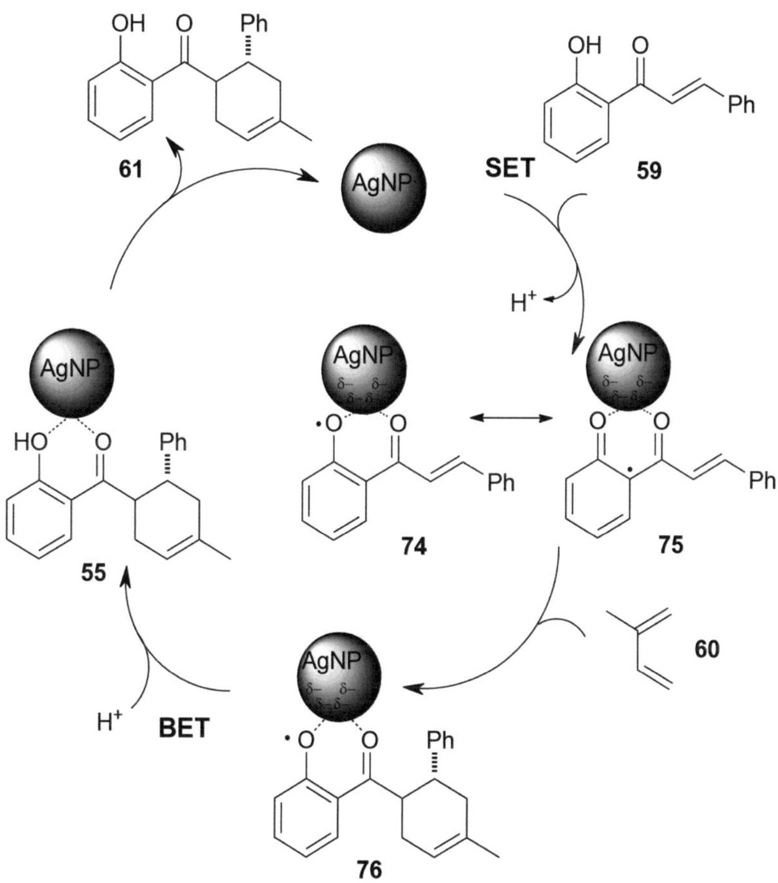

Scheme 8. Proposed mechanism for the silver nanoparticles-catalyzed Diels-Alder reaction [36].

nanoparticle (AgNp) may serve as 'electron shuttle' catalysts by accepting and returning a single electron from and to the substrate [36].

Following the mechanistic studies, Porco *et al.* utilized AgNP for the biomimetic syntheses of (±)-panduratin A (**Scheme 9**) [36] and (±)-sorocenol B (**Scheme 10**) [37]. Inspired by the aforementioned biosynthesis studies, Porco and co-workers found that the AgNP can also be used to promote dehydrogenation of the prenyl group of a flavonoid to form the requisite diene for the Diels-Alder reaction with a 2′-hydroxychalcone dienophile. Such tandem reactions were successfully employed for the synthesis of (±)-brosimone A and (±)-brosimone B (**Scheme 11**) [38].

3.6. Chiral ligand-Lewis acid complex mediated Diels-Alder reaction

In 2014, Lei and Wulff *et al.* reported the first enantioselective total synthesis of (-)-kuwanon I (**2**), (+)-kuwanon J (**1**), (-)-brosimone A (**86**) and (-)-brosimone B (**84**) by using chiral ligand-Lewis

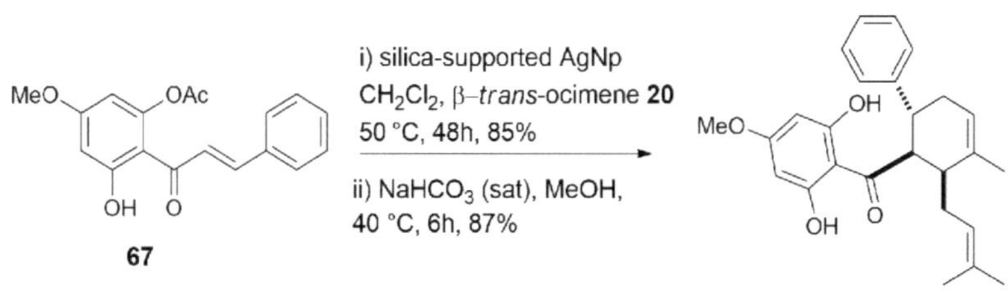

Scheme 9. Synthesis of (±)-panduratin A (**4**) [36].

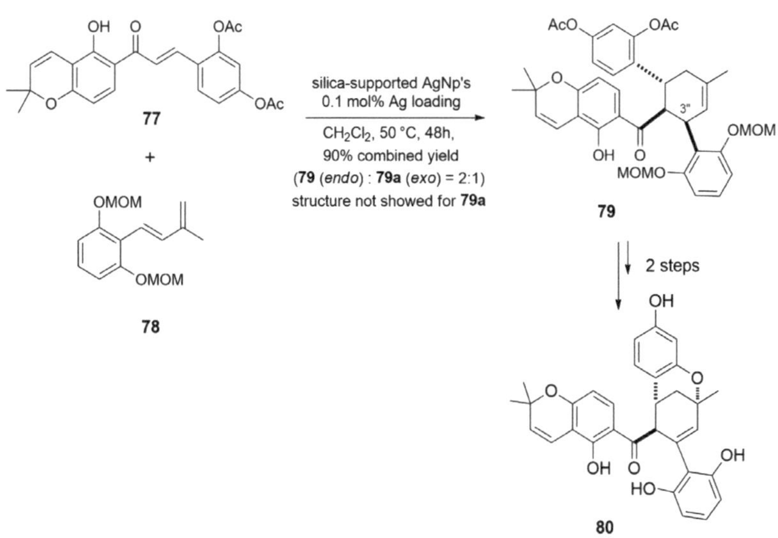

Scheme 10. Biomimetic synthesis of (±)-sorocenol B [37].

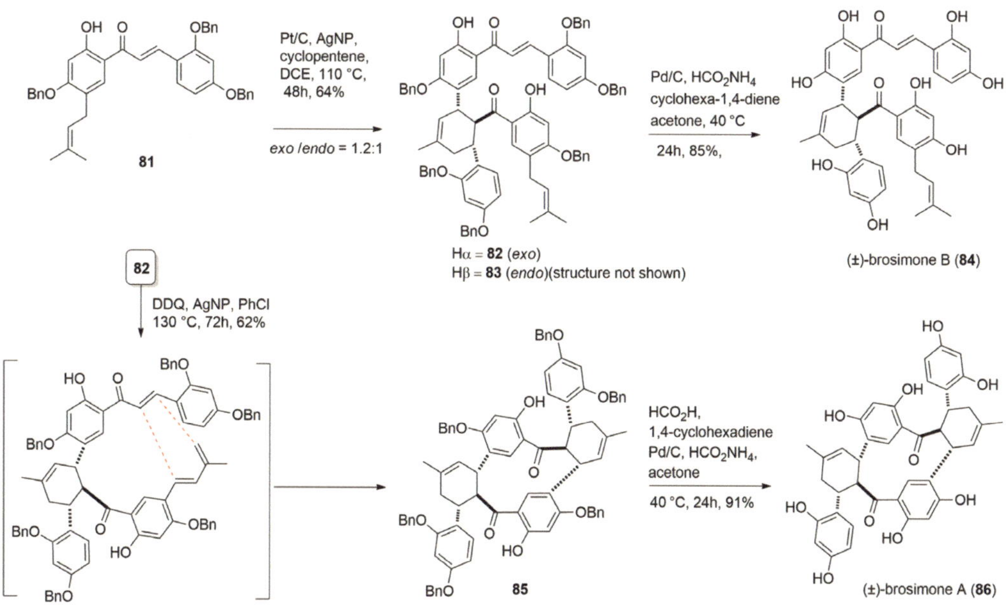

Scheme 11. Biomimetic synthesis of (±)-brosimone A and (±)-brosimone B [38].

acid complex. This complex was prepared by coordination of an axially chiral ligand such as VANOL or VAPOL to borane [39].

Scheme 12 shows the mechanism proposed by Lei and co-workers for the enantioselective Diels-Alder reaction [39]. The mechanism was proposed to proceed through the formation of a chiral boron complex **88**, followed by formation of a tetracoordinate boron complex **89** with 2′-hydroxychalcone dienophile. Subsequently, Diels-Alder reaction between the chiral complex **89** and a diene afforded a mixture of *endo*/*exo* diastereomers in high enantiomeric excess. Lei and co-workers proposed that the enantioselective Diels-Alder reaction may be induced by the following factors [39, 40].

(a) The coordination bond between boron and dienophile which may lower the energy of LUMO.

(b) The mobility of dienophile may be reduced upon complexation.

(c) The π-π stacking between the chiral ligand and dienophile shielding one face of the chalcone dienophile from attack by the diene.

Following the mechanistic studies, the (*S*)-VANOL-borane complex was efficiently used to mediate the synthesis of (-)-kuwanon I (**2**), (+)-kuwanon J (**1**), (-)-brosimone A (**84**) and (-)-brosimone B (**85**) [39]. Asymmetric Diels Alder reaction for these molecules was summarized in **Schemes 13–15**.

Based on the reported results, the chiral ligand strongly influences the enantioselectivity of the cycloaddition reaction. A 2.5 equivalent of (*R*)-VANOL is required for the optimal formation of

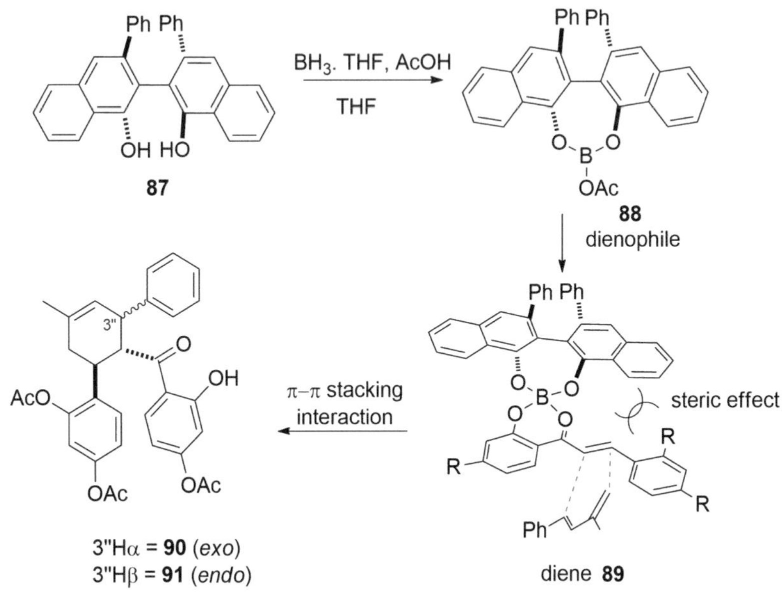

Scheme 12. Proposed mechanism for the chiral ligand-Lewis acid complex mediated enantioselective Diels-Alder reaction [39, 40].

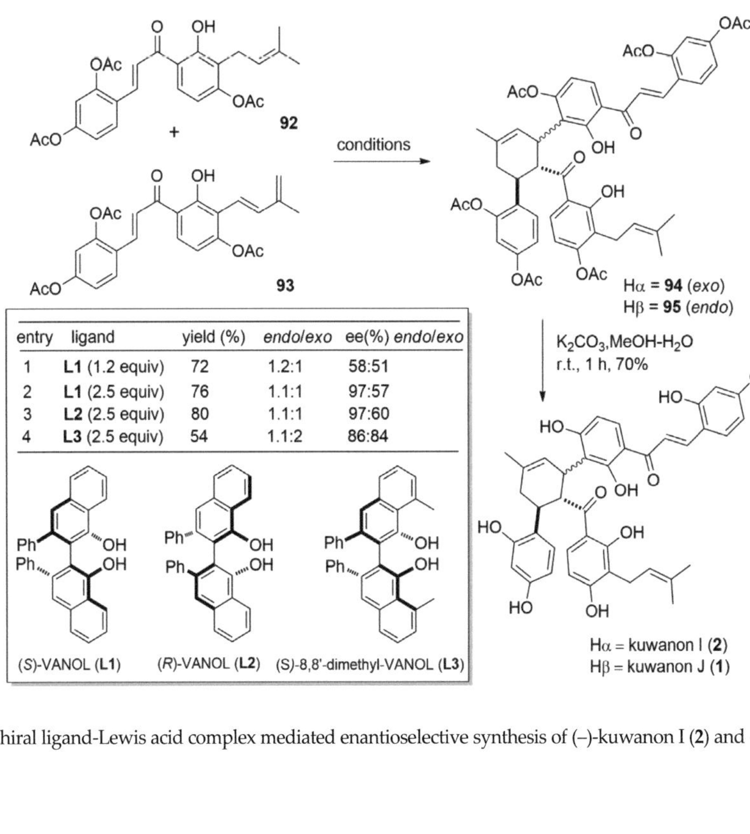

entry	ligand	yield (%)	*endo/exo*	ee(%) *endo/exo*
1	**L1** (1.2 equiv)	72	1.2:1	58:51
2	**L1** (2.5 equiv)	76	1.1:1	97:57
3	**L2** (2.5 equiv)	80	1.1:1	97:60
4	**L3** (2.5 equiv)	54	1.1:2	86:84

Scheme 13. Chiral ligand-Lewis acid complex mediated enantioselective synthesis of (–)-kuwanon I (**2**) and (+)-kuwanon J (**1**) [39].

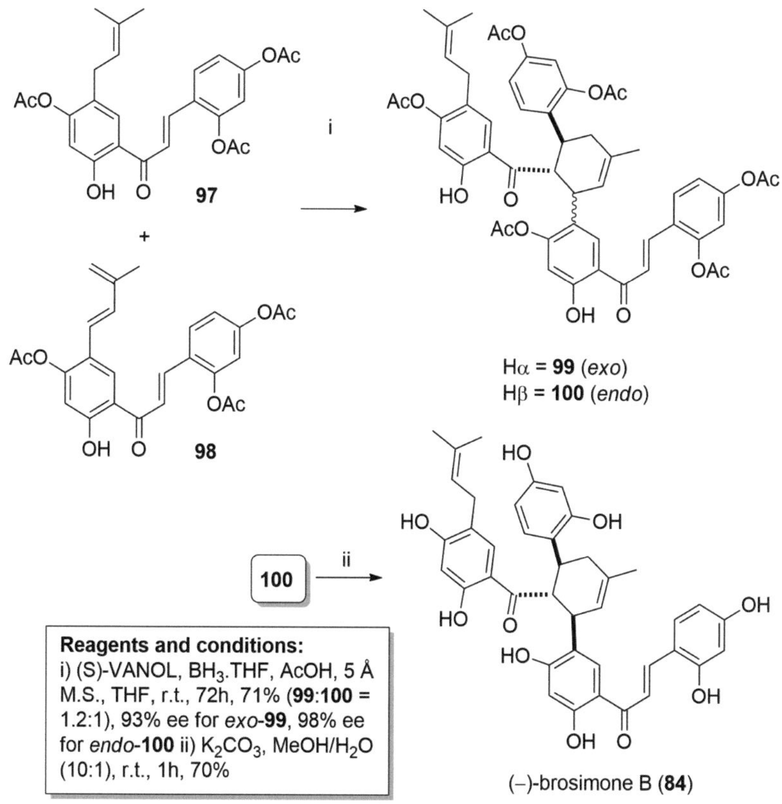

Scheme 14. Enantioselective synthesis of (–)-brosimone B [39].

kuwanon J precursor *endo*-**95** (97% ee, 1.1:1 *endo*/*exo*), whereas similar amount of (*S*)-8, 8′-dimethyl-VANOL is required for the optimal formation of kuwanon I precursor *exo*-**94** (84% ee, 1.2:1 *exo*/*endo*). Finall, deprotection of the acetate group of *endo*-**95** and *exo*-**94** furnished the desired natural products (-)-kuwanon J (**1**) and (+)-kuwanon I (**2**), respectively (**Scheme 13**) [39].

The synthetic routes for (–)-brosimone B (**84**) and (–)-brosimone A (**86**) were showed in **Schemes 14** and **15**, respectively. For (–)-brosimone B (**84**), cycloaddition reaction between dienophile **97** and diene **98** using (*S*)-VANOL gave a mixture of diastereomers **99** and **100** in 71% yield in a 1.2:1 ratio. Remarkably, excellent enantiomeric excess (*ee*) values for both compounds were obtained (98% *ee* for **100**, 93% *ee* for **99**). Deprotection of the acetyl groups of **100** gave (-)-brosimone B in 70% yield (**Scheme 14**) [39, 40].

The diene **98** was also used in the synthesis of brosimone A (**86**) in a one-pot inter-/intramolecular Diels-Alder cycloaddition cascade strategy (**Scheme 15**). The (*S*)-VANOL-borane complex efficiently mediated the cycloaddition reaction to give a mixture of three diastereomers **101–103** (**Scheme 15**). Deprotection of the adduct **103** gave (-)-brosimone A (**86**) in 70% yield [39, 40].

In 2016, Porco and co-workers reported the syntheses of the flavonoid Diels-Alder natural products sanggenon C (**108**) and sanggenon O (**109**) by using a combination of silver nanoparticles (AgNP) and a BINOL-borate catalyst (**Scheme 16**) [41].

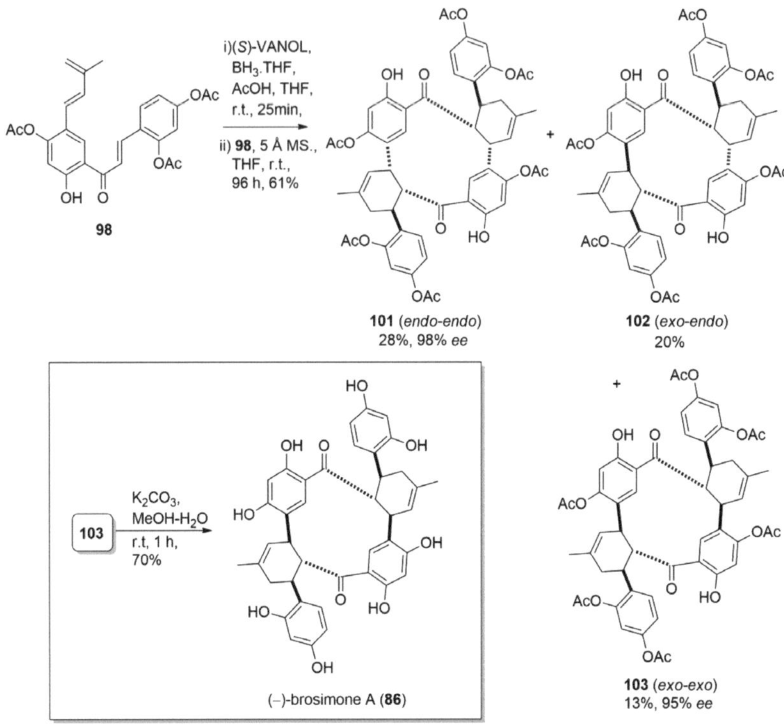

Scheme 15. Enantioselective synthesis of (–)-brosimone A [39].

A catalytic amount of triphenylborate ($B(OPh)_3$) and (*R*)-3,3′-dibromoBINOL was used to mediate the asymmetric Diels-Alder reaction between diene precursor **104** and dienophile **105** (**Scheme 16**). In the first step, the diene precursor **104** underwent a retro 6π-electrocyclisation followed by a formal 1,7 hydrogen shift process to afford the requisite diene functionality. Reaction of this diene with dienophile **105** in the present of a catalytic amount of chiral

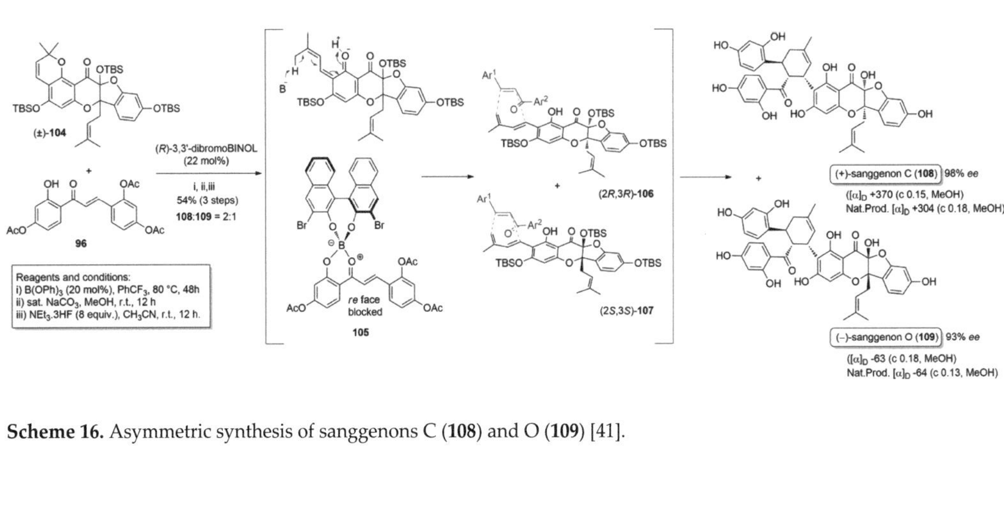

Scheme 16. Asymmetric synthesis of sanggenons C (**108**) and O (**109**) [41].

http://dx.doi.org/10.5772/intechopen.68781

BINOL-borate complex ((*S*)-3,3′-dibromoBINOL/triphenylborate) afforded a mixture of cycloadducts, which after deprotection gave sanggenon C (**108**) and sanggenon O (**109**) in 2:1 ratio of 98 and 93% ee, respectively. The use of AgNP gave a racemic mixture of **108** and **109**.

In conclusion, this chapter has provided an overview of biosynthesis and biomimetic synthesis of flavonoid Diels-Alder natural products. Intensive biosynthesis studies led by Nomura *et al.* have provided important information for the enzymatic formation of these natural products. In particular, information from the diene formation and the feeding experiments have paved the way for an exploration of chemical synthesis of these natural products. Finally, with the innovative chemical strategies, enantiomerically pure flavonoid Diels-Alder natural products were made possible for further biological activities evaluation.

Author details

Shah Bakhtiar Nasir[1], Jia Ti Tee[1], Noorsaadah Abd Rahman[1] and Chin Fei Chee[2]*

*Address all correspondence to: cheechinfei@um.edu.my

1 Department of Chemistry, Faculty of Science, University of Malaya, Kuala Lumpur, Malaysia

2 Nanotechnology and Catalysis Research Centre, University of Malaya, Kuala Lumpur, Malaysia

References

[1] Nomura T, Hano Y, Ueda S. Chemistry and biosynthesis of natural diels-alder type adducts from moraceous plants. In: Attaur R, editor. Studies in Natural Products Chemistry. Vol. **17**. Part D. Elsevier, Amsterdam; 1995. pp. 451-478

[2] Yang Y, et al. The latest review on the polyphenols and their bioactivities of Chinese Morus plants. Journal of Asian Natural Products Research. 2014;**16**(6):690-702

[3] Dai SJ, et al. New Diels-alder type adducts from Morus Macroura and their anti-oxidant activities. Chemical and Pharmaceutical Bulletin. (Tokyo). 2004;**52**(10):1190-1193

[4] Fukai T, Kaitou K, Terada S. Antimicrobial activity of 2-arylbenzofurans from Morus species against methicillin-resistant *Staphylococcus aureus*. Fitoterapia. 2005;**76**(7-8):708-711

[5] Zheng ZP, et al. Tyrosinase inhibitory constituents from the roots of Morus nigra: A structure-activity relationship study. Journal of Agricultural and Food Chemistry. 2010;**58**(9):5368-5373

[6] Zhang QJ, et al. Three new cytotoxic Diels-Alder-type adducts from *Morus australis*. Chemistry & Biodiversity. 2007;**4**(7):533-1540

[7] Phung TX, et al. Chalcone-derived Diels-Alder adducts as NF-kappaB inhibitors from *Morus alba*. Journal of Asian Natural Products Research. 2012;**14**(6):596-600

[8] Nomura T, Hano Y, Fukai T. Chemistry and biosynthesis of isoprenylated flavonoids from Japanese mulberry tree. Proceedings of the Japan Academy. Series B. 2009;**85**(9):391-408

[9] Nomura T. Phenolic compounds of the mulberry tree and related plants. Fortschritte der Chemie organischer Naturstoffe. 1988;**53**:87-201

[10] Rama Rao AV, et al. Structures of albanols a and b, two novel phenols from Morus alba bark. Tetrahedron Letters. 1983;**24**(29):3013-3016

[11] Hano Y, et al. Absolute configuration of natural Diels-Alder Type adducts from the Morus Root Bark. Heterocycles. 1988;**27**(10):2315-2325

[12] Tan EC, et al. Proteomic analysis of cell suspension cultures of Boesenbergia rotunda induced by phenylalanine: Identification of proteins involved in flavonoid and phenylpropanoid biosynthesis pathways. Plant Cell, Tissue and Organ Culture (PCTOC). 2012;**111**(2):219-229

[13] Nomura T, Hano Y. Isoprenoid-substituted phenolic compounds of moraceous plants. Natural Product Reports. 1994;**11**(2):205-218

[14] Ueda S, et al., Kuwanon J. a new Diels-Alder adduct and chalcomoracin from callus culture of morus alba L. Chemical and Pharmaceutical Bulletin. 1982;**30**(8):3042-3045

[15] Ueda S, Matsumoto J, Nomura T. Four new natural Diels-Alder type adducts, Mulberrofuran E, Kuwanon Q, R, and V from callus culture of Morus Alba L. Chemical and Pharmaceutical Bulletin. 1984;**32**(1):350-353

[16] Ikuta J, et al. Constituents of Morus alba L. cell cultures. (1). Structures of four new natural Diels-Alder type adducts, Kuwanons J, Q, R, and V. Chemical and Pharmaceutical Bulletin. 1986;**34**:2471-2478

[17] Takasugi M, et al. Chalcomoracin, a natural Diels–Alder Adduct from diseased mulberry. Chemistry Letters. 1980;**9**(12):1573-1576

[18] Takasugi M, et al. Moracin C and D, new phytoalexins from diseased mulberry. Chemistry Letters. 1978;**7**(11):1239-1240

[19] Hano Y, Nomura T, Ueda S. Biosynthesis of chalcomoracin and Kuwanon J, the Diels-Alder type adducts, in Morus Alba L. cell cultures. Chemical and Pharmaceutical Bulletin. 1989;**37**(2):554-556

[20] Ayabe S-I, Udagawa A, Furuya T. NAD(P)H-dependent 6′-deoxychalcone synthase activity in Glycyrrhiza echinata cells induced by yeast extract. Archives of Biochemistry and Biophysics. 1988;**261**(2):458-462

[21] Hano Y, et al. Dynamic participation of primary metabolites in the biosynthesis of chalcomoracin and β-sitosterol in Morus alba cell cultures. Naturwissenschaften. 1992;**79**(4):180-182

[22] Hano Y, Nomura T, Ueda S. Biosynthesis of optically active Diels-Alder type adducts revealed by an aberrant metabolism of O-methylated precursors in Morus alba cell cultures. Journal of the Chemical Society, Chemical Communications. 1990;**(8)**:610-613

[23] Hano Y, et al. A novel way of determining the structure of artonin I, an optically active Diels-Alder type adduct, with the aid of an enzyme system of *Morus alba* cell cultures. Journal of the Chemical Society, Chemical Communications. 1992;**(17)**:1177-1178

[24] Gunawan C, Rizzacasa MA. Mulberry Diels-Alder adducts: Synthesis of chalcomoracin and mulberrofuran C methyl ethers. Organic Letters. 2010;**12**(7):1388-1391

[25] Boonsri S, et al. Synthetic studies towards the mulberry Diels-Alder adducts: H-bond accelerated cycloadditions of chalcones. Organic & Biomolecular Chemistry. 2012;**10**(30):6010-6021

[26] Chee CF, et al. An efficient synthesis of (±)-panduratin A and (±)-isopanduratin A, inhibitors of dengue-2 viral activity. Tetrahedron Letters. 2010;**51**(3):495-498

[27] Chee CF, et al. Synthesis of (±)-kuwanon V and (±)-dorsterone methyl ethers via Diels–Alder reaction. Tetrahedron Letters. 2011;**52**(15):1797-1799

[28] Jung EM, Lee YR. ChemInform abstract: First concise total syntheses of biologically interesting Nicolaioidesin C (V), Crinatusin C1 (IX) and Crinatusin C2 (X). ChemInform. 2008;**39**(43):1199-1204

[29] Tee JT, et al. Model studies on construction of the oxabicyclic [3.3.1] core of the mulberry Diels–Alder adducts morusalbanol A and 441772-64-1. Tetrahedron Letters. 2015;**56**(36):5082-5085

[30] Tee JT, et al. A strategy toward the biomimetic synthesis of (±)-Morusalbanol A Pentamethyl ether. Synthesis. 2016;**48**(14):2263-2270

[31] Pasfield LA, et al. Synthesis of (±)-panduratin A and related natural products using the high pressure Diels–Alder reaction. Asian Journal of Organic Chemistry. 2013;**2**(1):60-63

[32] Yates P, Eaton P. Acceleration of the Diels-Alder reaction by aluminium chloride. Journal of the American Chemical Society. 1960;**82**(16):4436-4437

[33] Cong H, et al. Electron transfer-initiated Diels-Alder cycloadditions of 2′-hydroxychalcones. Journal of the American Chemical Society. 2008;**130**(29):9214-9215

[34] Iovine V, et al. Total synthesis of (±)-Kuwanol E. Journal of Natural Products. 2016; **79**(10):2495-2503

[35] Bañuelos P, et al. (1R)-(+)-Camphor and acetone derived α'-hydroxy enones in asymmetric Diels–Alder reaction: Catalytic activation by Lewis and Brønsted acids, substrate scope, applications in syntheses, and mechanistic studies. The Journal of Organic Chemistry. 2010;**75**(5):1458-1473

[36] Cong H, et al. Silver nanoparticle-catalyzed Diels-Alder cycloadditions of 2′-hydroxychalcones. Journal of the American Chemical Society. 2010;**132**(21):7514-7518

[37] Cong H, Porco JA, Jr. Total synthesis of (+/-)-sorocenol B employing nanoparticle catalysis. Organic Letters. 2012;**14**(10):2516-2519

[38] Qi C, et al. Biomimetic dehydrogenative Diels-Alder cycloadditions: Total syntheses of brosimones A and B. Angewandte Chemie International Edition. 2013;**52**(32):8345-8348

[39] Han J, et al. Enantio selective biomimetic total syntheses of kuwanons I and J and brosimones A and B. Angewandte Chemie International Edition. 2014;**53**(35):9257-9261

[40] Han J, et al. Recent advances in the total synthesis of prenylflavonoid and related Diels–Alder natural products. Synthesis. 2015;**47**(11):519-1533

[41] Qi C, et al. Asymmetric syntheses of the flavonoid Diels–Alder natural products sanggenons C and O. Journal of the American Chemical Society. 2016;**138**(3):798-801

Chapter 10

Flavonoids in Agriculture

Mariana Palma-Tenango,

Marcos Soto-Hernández and

Eva Aguirre-Hernández

Additional information is available at the end of the chapter

http://dx.doi.org/10.5772/intechopen.68626

Abstract

Flavonoids are compounds that are secondary metabolites, but which play an important role in the biological activities of plants. They can be responsible for the color of flowers and fruits and for the attraction of pollinators. They also participate in plant-microorganism symbiosis. These relationships can be used to naturally control weeds and insect pests and reduce stress and diseases in order to increase crop yield. To improve the understanding of the different biological systems where flavonoids are involved in their symbiotic relationships and in plant physiology, tools such as metabolomics are used, which give a broader picture and allow to search for strategies to solve problems specific to the agricultural sector.

Keywords: flavonoids, agriculture, plants, metabolomics

1. Introduction

Flavonoids are secondary metabolites of plants; this group of phenolic compounds includes approximately 4500 compounds [1]. They are classified into different subgroups (**Figure 1**). Their nuclear structure includes carbons C6-C3-C6, and the diversity of flavonoids depends on the position of the aromatic ring [2].

Flavonoids and flavones are the most common in plants, whereas flavanones, flavanols, dihydroflavones, and dihydrochalcones have a limited distribution. Flavonoids are present in plant tissues in relatively high concentrations in free forms (aglycones) or conjugated with sugar molecules (glycosides) [5].

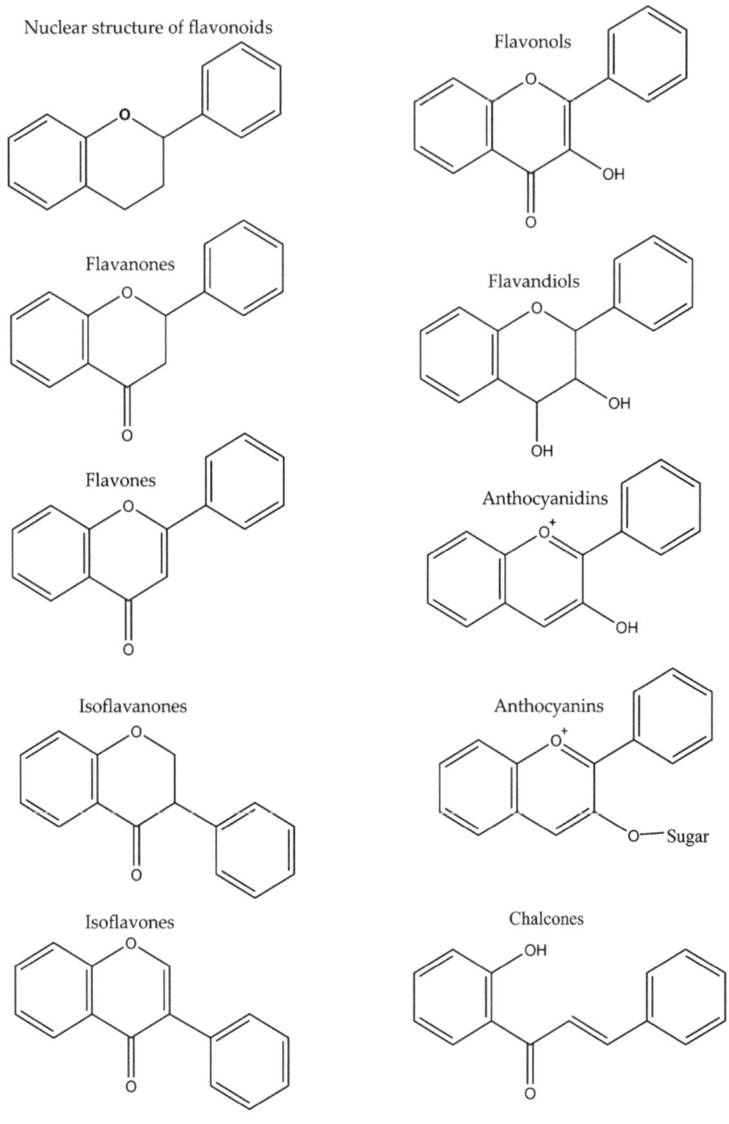

Figure 1. Structure of the different types of flavonoids (modified from Refs. [3, 4]).

Flavonoids are polyphenolic compounds found in all vascular and non-vascular plants [6]. They are important in the diet of humans because they possess a high nutritional value; besides, they are used as effective medicines in the treatments of certain diseases, and therefore, they are called nutraceuticals. Nowadays, a diet rich in fruits and vegetables is recommended to contribute to the prevention and treatment of cardiovascular diseases, diabetes, cancer, chronic inflammatory disorders, and degenerative diseases [7].

Some examples of nutraceutical flavonoids are the following: isoflavones (genistein, daidzein) obtained from celery, soybeans, and other legumes. Isoflavones have antitumor, anticancer, and antioxidant action, and they improve immune response, lower cardiovascular risk, and menopausal symptoms. Quercetin is contained in onion, citrus fruits, broccoli, red grapes, apple, and cherries. Kaempferol is present in broccoli and radishes [8]. Anthocyanins are present in red wine and fruits. These flavonoids are important due to their antioxidant, anticancer, and antithrombotic properties, as well as their ability to lower blood cholesterol [9].

Their concentration can vary among species: in fruits, vegetables, and medicinal plants, ranges from 0 to 6125.6 mg kg^{-1} have been reported. For example, 1720.5 mg kg^{-1} has been observed in spinach, 3575.4 mg kg^{-1} in strawberry, and 459.9 mg kg^{-1} in apple [10].

2. Function of flavonoids in plants

In plants, flavonoids play important roles in many biological processes (**Figure 2**). They participate in seed development and growth [11], fruit growth and ripening [12], pollen tube germination [13], and hormone transport [6]. Flavonoids respond to biotic and abiotic factors by providing antioxidant properties; they prevent damage caused by fungi, viruses, bacteria, and herbivores; function as chemical messengers in association with mycorrhizae and bacteria; act as chemical attractants to pollinating animals; and have allelopathic functions [14, 15].

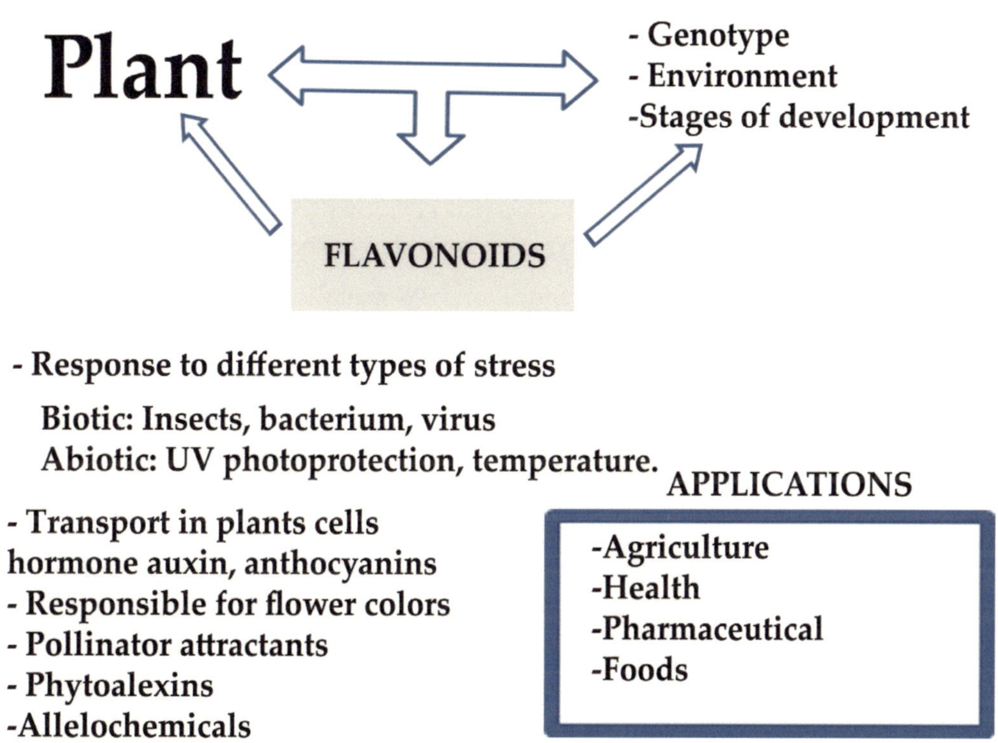

Figure 2. Main functions of flavonoids in plants (modified from Refs. [17–19]).

They also participate in pigment and color differences in flowers, fruits, and seeds [3]. For example, flavonols are related with yellow, flavanols with ochre to brown, and anthocyanins with red to purple [4]. In the case of corn kernels and petunia flowers, anthocyanins and proanthocyanidins are mainly responsible for the pigments. These accumulate in the vacuole or cell wall. In corn, accumulation occurs by vacuolar sequestration of anthocyanins [16].

As part of their defense strategy, plants induce systems of antioxidant activity, reactive oxygen species (ROS), enzymatic and non-enzymatic, soluble in water and in lipids, located in different cell compartments. The enzymatic ROS system consists of several enzymes such as superoxide dismutase (SOD), catalase (CAT), ascorbate peroxidase (APX), monodehydroascorbate reductase (MDAR), dehydroascorbate reductase (DHAR), glutathione peroxidase (GPX), and glutathione reductase (GR). Non-enzymatic antioxidants include pigments (carotenoids, anthocyanidins), vitamins (A, C, E), and flavonoids, among others [19]. The synthesis of the latter compounds by the plants is related to environmental biotic and abiotic stress factors. Plants subjected to conditions of severe stress accumulate dihydroxy B-ring substituted flavonoids, which are effective eliminators of ROS.

Flavonoids activate a network of events, including stress-induced morphogenesis, which protects plants from unexpected lesions of different origins [15]. Therefore, flavonoids play an important role in the protection of biological systems against the harmful effects of oxidative processes on macromolecules, they are important to catalyze electron transport and to eliminate reactive oxygen, especially in the form of superoxide anions, hydroxyl radicals, lipid peroxides, or hydroperoxides. In this way, they block the deleterious activity of these substances on the cells [15, 20].

3. Flavonoids: allelopathy and its applications in agriculture

Allelopathy is defined as the direct or indirect effect of secondary compounds produced by a donor plant on a recipient plant. This type of relationship can be beneficial or harmful [21].

Currently, to solve multiple problems in agriculture, allelopathy is being considered as a natural control of weeds and insect pests and to reduce stress and diseases, in order to increase crop yield [22, 23]. Weeds are the species that compete the most with crops, and for their management, aqueous extracts are used, such as natural herbicides from sorghum, sunflower, eucalyptus, and rice, among others; a greater efficacy is obtained when they are mixed together than alone [24]. On the other hand, it has been demonstrated that exudates of roots of rice plants reduce the attack of fungi of the genus *Fusarium* in melon. In addition, plants of *Brassica napus* L., incorporated into the soil, decrease the population of certain nematodes in orchards [23]. These examples show the potential of some allelochemicals to generate herbicides, fungicides, insecticides, and nematicides.

Knowledge of allelopathy can be valuable to improve crop rotation proposals [22]. Recent publications highlight the role of flavonoids in allelopathy, involved in soil interactions, since they have been identified in significant concentrations in many bioactive root exudates. Simple phenols and flavonoids are released by the decomposition of plant tissues as leachates and by the process of microbial degradation and transformation in the soil [22, 24].

In many legumes, it has been shown that the flavonoids quercetin and kaempferol, free and glycosylated, commonly released by germinating seeds and roots, persist for days in the soil and possess an important phytoinhibitory activity, stimulating seed germination at low concentrations but inhibiting seedling growth at high concentrations. These compounds are also present in leaf extracts of walnut trees [25].

The activity of several flavonoids is highly concentration-dependent; some of these compounds may be inhibitory or stimulatory, depending on the availability of the concentration in the soil/water solution in the rhizosphere [21]. Other allelopathic flavonoids are lutonarin, saponarin, and isovitexin, as well as catechin and cyanidin, luteolin 7-Ob-glucuronide, neochamaejasmin, mesoneochamaejasmin, chamaejasmenin, genkwanol, daphnodorin, and dihydrodaphnodorin, among others [21, 26, 27]. Allelochemicals directly and indirectly affect plants; indirect effects include soil alteration, physicochemical properties, changes in microbial populations, and differential nutrient availability for plants. The direct action is the physiological and biochemical changes generated during plant growth and development [22, 28]. Scientific studies have demonstrated several mechanisms of action of flavonoids when applied exogenously on plants. Such as changes in membrane permeability and inhibition of plant nutrient absorption, inhibition of cell division, elongation and submicroscopic structure, effects on photosynthesis and respiration of the plant, consequences on different enzymatic functions and activities, effects on the synthesis of endogenous hormones and proteins, and the disruption of adenosine triphosphate (ATP) formation [28].

4. Metabolomics: flavonoids as a tool in agriculture

It has been described that over 7000 natural products belong to the flavonoid family and to analyze them, physicochemical methods as nuclear magnetic resonance (NMR) and mass spectrometry have been used [29], but nowadays the most important issue is concerned with the identification of free and conjugated forms of these compounds. Metabolite profiling is an essential tool to analyze the effects of pathway engineering approaches; in this sense, the metabolomics approach used to solve this problem as the liquid-chromatography-mass spectrometry has supported the quantitative and qualitative analysis of flavonoids. In addition, straightforward and efficient methods approach toward rapid flavonoid identification by combining simple high-performance liquid chromatography (HPLC) and NMR methods, facilitating the analysis of flavones and flavanones [30].

The literature describes a lot of information on metabolomics in areas of medicinal plants, chemosystematics, adulteration of plants, and so on, but it is scarce in agriculture [31]. Flavonoids are a diverse group in agricultural crops, and there are studies using metabolomic tools to analyze the content of flavonoids using different analytical techniques (**Table 1**).

Red tomato contains flavonoids mainly in the peel; through omic studies, it may be possible to modify the pattern of flavonoids in the pulp [42]. Metabolic profiling represents a useful tool to characterize varieties with functional markers, such as flavonoids. For example, metabolite profiling allowed the characterization of Italian tomato landraces and different fruit types [43].

Crop	Plant organs	Analytical technique	Compound	Reference
Soybean	Leaves	(RP)-HPLC and ^{1}H NMR	Naringenin, rutin, quercetin, kaempferol and its glucosides, and total flavonoids	[32]
Red tomato	Fruit	HPLC/DAD and LC/NMR, LC/MS, and LC/MS/MS	Naringenin chalcone and rutin	[33]
Red and yellow raspberry cultivars	Fruit	UPLC/QqQ-MS/MS	Flavan-3-ol, quercetin-3,4-diglucoside, quercetin, kaempferol-3-glucuronide, rutin, and the conjugates of isorhamnetin	[34]
Avocado	Fruit	MS/MS LC-DAD-ESI-TOF MS	Catechin and epicatechin Rutin, Naringin, kaempferol, apigenin, luteolin, isorhamnetin	[35]
Maize	Kernels	LC-MS/MS	Apigenin, luteolin, methyl chrysoeriol, malvidin, pentose, rhamnose, selgin, tricin, chalcone synthase, chalcone isomerase	[36]
Rice	Leaf and bran	LC-QTOF-MS	Tricin, tricin 7-O-rutinoside, and tricin 7-O-β-ᴅ-glucopyranoside	[37]
Wheat	Flag leaf	UPLC-QTOF MS/MS	Isovitexin, isoorientin, Isoschaftoside, quercetin 3-rutinoside-7-glucoside and methylisoorientin-2″-O-rhamnoside	[38]
Grape	Berries	LC-MS	Quercetin and derivatives, kaempferol and isorhamnetin	[39]
Broad bean	Pods	UHPLC-ESI-qTOF-MS^{2}	Glycosylated flavonoids	[40]
Tea	Leaves	UPLC-Q-TOF MS	Flavan-3-ols, flavonols and their glucosides	[41]

Table 1. Metabolomic studies of flavonoids in agricultural crops.

Another study related with metabolomic profiling is the analysis of flavonoid distribution in three *Momordica* species, where 13 flavonoids were found in a special pattern [44].

The novel approach of genetic metabolomics referred to as metabolite profiling combined with quantitative trait locus (QTL) analysis was applied to detect flux control points in flavonoids biosynthesis of *Populus*. It was found that flavonoid profile can be used for QTL analysis to reveal loci that control the flux of their biosynthesis [45].

Plant metabolism is disrupted by several types of stress. Flavonoids are involved in the response toward abiotic stress [46]; a good model to understand how flavonoids contribute to the mitigation of oxidative and drought stress is *Arabidopsis thaliana*. Another example is the metabolomic analysis of tea, where the effects of light intensity and temperature on the metabolites in tea grown in the shade were evaluated; they found that most flavonoids (flavan-3-ols, flavonols, and their glucosides) decreased significantly in the shading treatments. Their study also showed a greater effect of temperature on galloylation of catechins than light intensity [41].

The growing stage of plants has been studied in connection with flavonoid synthesis, an interesting metabolomic study is the analysis of soybean leaves [47], it showed significant changes in the content of flavonoids and isoflavonoids, and kaempferol derivatives were used as markers. The results demonstrated that metabolite production changed depending on the growing stage, and they mention that the information can be useful to understand physiological characterization and suggest an optimal harvesting time of this crop.

Metabolic analysis has the potential to generate a complete vision of metabolic networks and has revealed multiple detection, quantification, and analysis strategies to evaluate numerous metabolites such as flavonoids. This tool attempts to integrate compound and metabolite analyses along with other biological data overview of plants, for example, phenotypic, morphological, and genetic data [48].

5. Symbiosis flavonoids-microorganisms

Flavonoids are considered signaling compounds in plant-microorganisms symbiosis; their function is signaling in response to pathogens, bacteria, fungi. Their participation is important in the nodulation of roots, where they are secreted to the rhizosphere [21], in the case of legumes under low nitrogen conditions, and where their interaction is specific with gram-negative bacteria called rhizobia that are nitrogen fixers [49]. They are also involved in the transcription of genes for the biosynthesis of rhizobial signaling molecules called Nod factors, which are perceived by the plant to allow symbiotic root infection [49]. Therefore, they can stimulate or inhibit rhizobial Nod gene expression. Root exudates can be flavonoids that participate as signaling compounds in the arbuscular mycorrhizal symbiosis [50]. Root exudation of flavonoids increases or decreases depending on the response to the symbiotic and pathogenic interaction of the plant and microbes.

It has been shown that certain flavonoids are stimulants for the germination of ectomycorrhizal fungi spores [50]. Depending on the arbuscular mycorrhizal fungi colonizers of the roots and the stage of symbiosis development [51].

The flavonoid role in the rhizosphere (**Table 2**) is important in agriculture because flavonoid interaction in the symbiosis with microorganisms can be a tool for nitrogen fixation in soils

Participation	Flavonoid	Species	Symbiosis type	Cite
Stimulant for germination of spores	Coumestrol, medicarpin, ononin, formononetin, daidzein, genistein, biochanin A and 4′,7-dihydroxyflavanone, 4,4′-dihydroxy-2′-methoxychalcone	*Medicago sativa* L. cv. Sitel	Fungus	[51]
	hesperidin, morin, rutin, quercetin, naringenin, genistein, and chrysin	*Suillus bovinus*	Fungus	[52]
Chemo attractant response from the rhizobia	Luteolin and apigenin	–	Bacterium	[53]
Auxin transport inhibitors	Kaempferol	*Medicago truncatula*	–	[54]

Table 2. Role of flavonoids in the rhizosphere.

poor in macronutrients. It can be a strategy to stimulate beneficial bacteria or inhibit harmful bacteria and fungi. It can also be considered in the genetic improvement of plant species.

6. Importance of flavonoids in pest control

Every day, the demand for natural products obtained from plants increases to be used as pest control agents. New pesticides are being developed using flavonoids, as they are an alternative to synthetic pesticides. They can inhibit enzymatic activity and prevent the growth of larvae of different insect species [55]. Some flavonoids interfere in the process of moulting and reproduction of several insects, that is, they inhibit the formation of juvenile hormone (ecdysone). Flavonoids inhibit transcription of ecdysone receptor-dependent genes (EcR) [56].

It has been reported that some types of flavonoids have had an effect on agricultural pests with ovicidal effect, oviposition, fecundity, mortality, weight reduction, and emergence of adults [57, 58]. Quercetin, rutin, and naringin showed positive effects for the control of nymphs and adults of the aphid *Eriosoma lanigerum* Hausmann. These can be used as an insecticide in the integrated management programs of this aphid [59]. Some flavonoids can influence agricultural pests depending on the concentration applied; if they are low, they do not affect them [60]; therefore, it is necessary to test the minimum concentrations for the flavonoid to have an insecticidal effect.

7. Concluding remarks

Throughout this chapter, the importance and interest in the knowledge of flavonoids as a big group of natural products, present in different organisms, have been emphasized. They have been widely investigated, from their biosynthesis, analytical techniques for their analysis, to their biological activity. They are compounds with different responses to biotic and abiotic factors, which can influence the agricultural sector as attractants for pollinators, as pigment to flowers and fruits, allelochemical functions, symbiosis with beneficial organisms, and as pest control.

The importance of flavonoids lies not only in the functions in plants but also in different therapeutic and nutraceutical applications. It is a group that is present in a wide variety of plant species, in different concentrations. In addition, its beneficial effects on health have been demonstrated and one of the most notable examples is its antioxidant activity.

However, the information linked to agriculture is scattered. Therefore, the purpose of this chapter is to provide an overview of their applications and show that metabolomics is an effective tool for research linked to metabolic processes of flavonoids with various agricultural crops.

Author details

Mariana Palma-Tenango[1], Marcos Soto-Hernández[2]* and Eva Aguirre-Hernández[1]

*Address all correspondence to: msoto@colpos.mx

1 Department of Ecology and Natural Resources, Laboratory of Phytochemistry, Faculty of Science, Universidad Nacional Autónoma de México, Ciudad Universitaria, Coyoacán, México

2 Department of Botany, Laboratory of Phytochemistry, Colegio de Postgraduados, Campus Montecillo, Texcoco, México

References

[1] Croteau R, Kutchan T, Lewis N. Secondary metabolites. In: Buchanan B, Gruissem W, Jone R, editors. Biochemistry Molecular Biology of Plants. Sommerset New Jersey, American Society of Plant Physiologists ed. 2000. pp. 1250-1318. DOI: 10.1016/j.phytochem.2011.10.011

[2] Heim KE, Tagliaferro AR, Bobilya DJ. Flavonoid antioxidants: Chemistry, metabolism and structure-activity relationships. Journal of Nutritional Biochemistry. 2002;**13**:572-584. DOI: http://doi.org/10.1016/S0955-2863(02)00208-5

[3] Falcone FM, Rius SP, Casati P. Flavonoids: Biosynthesis, biological functions, and biotechnological applications. Frontiers in Plant Science. 2012;**3**:1-16. DOI: http://doi.org/10.3389/fpls.2012.00222

[4] Petrussa E, Braidot E, Zancani M, Peresson C, Bertolini A, Patui S, et al. Plant flavonoids-biosynthesis, transport and involvement in stress responses. International Journal of Molecular Sciences. 2013;**14**(7):14950-14973. DOI: http://doi.org/10.3390/ijms140714950

[5] Di Carlo G, Mascolo N, Izzo A, Cappasso F. Flavonoids old and new aspect of a class of natural therapeutic drugs. Life Sciences. 1999;**65**(4):337-353

[6] Peer WA, Murphy AS. Flavonoids and auxin transport: Modulators or regulators? Trends in Plant Science. 2007;**12**:556-563. DOI: http://doi.org/10.1016/j.tplants.2007.10.003

[7] Ghasemzadeh A, Ghasemzadeh N. Review flavonoids and phenolic acids: Role and biochemical activity in plants and human. Journal of Medicinal Plants Research. 2011;**5**: 6697-6703

[8] Zeng S, Wu M, Zou C, Liu X, Shen X, Hayward A, et al. Comparative analysis of anthocyanin biosynthesis during fruit development in two Lycium species. Physiologia Plantarum. 2014;**150**:505-516

[9] Rivera G, Bocanegra G, Mongec A. Traditional plants as source of functional foods a review. Journal of Food. 2010;**8**:159-167

[10] Sultana B, Anwar F. Flavonols (kaempferol, quercetin, myricetin) contents of selected fruits, vegetables and medicinal plants. Food Chemistry. 2008;**108**:879-884. DOI: http://doi.org/10.1016/j.foodchem.2007.11.053

[11] Song C, Xiang DB, Yan L, Song Y, Zhao G, Wang YH, et al. Changes in seed growth, levels and distribution of flavonoids during tartary buckwheat seed development. Plant Production Science. 2016;**1008**:1-10. DOI: http://doi.org/10.1080/1343943X.2016.1207485

[12] Andreotti C, Ravaglia D, Ragaini A, Costa G. Phenolic compounds in peach (*Prunus persica*) cultivars at harvest and during fruit maturation. Annals of Applied Biology. 2008;**153**:11-23. DOI: http://doi.org/10.1111/j.1744-7348.2008.00234.x

[13] Ylstra B, Touraev A, Moreno RM, Stöger E, Van T, Vicente O, et al. Flavonols stimulate development, germination, and tube growth of tobacco pollen. Plant Physiology. 1992;**100**: 902-907. DOI: http://doi.org/10.1104/pp.100.2.902

[14] Wink M. Introduction: Biochemistry, physiology and ecological functions of secondary metabolites. Annual Plant Reviews. 2010;**40**:1-19

[15] Agati G, Azzarello E, Pollastri S, Tattin M. Flavonoids as antioxidants in plants: Location and functional significance. Plant Science. 2012;**196**:67-76

[16] Koes R, Verweij W, Quattrocchio F. Flavonoids: A colorful model for the regulation and evolution of biochemical pathways. Trends in Plant Science. 2005;**10**(5):236-242. DOI: http://doi.org/10.1016/j.tplants.2005.03.002

[17] Winkel-Shirley B. Biosynthesis of flavonoids and effects of stress biosynthesis of flavonoids and effects of stress. Current Opinion in Plant Biology. 2002;**5**:218-223. DOI: http://doi.org/10.1016/S1369-5266(02)00256-X

[18] Cesco S, Mimmo T, Tonon G, Tomasi N, Pinton R, Terzano R, et al. Plant-borne flavonoids released into the rhizosphere: Impact on soil bio-activities related to plant nutrition. A review. Biology and Fertility of Soils. 2012;**48**(2):123-149. DOI: http://doi.org/10.1007/s00374-011-0653-2

[19] Dinakar C, Djilianov D, Bartels D. Photosynthesis in desiccation tolerant plants: Energy metabolism and antioxidative stress defense. Plant Science. 2012;**182**:29-41

[20] Xu C, Sullivan JH, Garrett WM, Caperna TJ, Natarajan S. Impact of solar ultraviolet-B on the proteome in soybean lines differing in flavonoid contents. Phytochemistry. 2008;**69**(1):38-48

[21] Weston LA, Mathesius U. Flavonoids: Their structure, biosynthesis and role in the rhizosphere, including allelopathy. Journal of Chemical Ecology. 2013;**39**(2):283-297. DOI: http://doi.org/10.1007/s10886-013-0248-5

[22] Farooq M, Bajwa AA, Cheema SA, Cheema ZA. Application of allelopathy in crop production. International Journal of Agriculture and Biology. 2013;**15**:1367-1378

[23] Muzell TM, Vidal RA, Balbinot JA, Von HB, Da Silva SF. Allelopathy: Driving mechanisms governing its activity in agriculture. Journal of Plant Interactions. 2016;**11**(1):53-60

[24] Cho MH, Lee SW. Phenolic phytoalexins in rice: Biological functions and biosynthesis. International Journal of Molecular Sciences. 2015;**16**(12):29120-29133

[25] Hai Z, Jin-Ming G, Wei-Tao L, Jing-Cheng T, Xing-Chang Z, Zhen-Guo J, et al. Allelopathic substances from walnut (*Juglans regia* L.). Allelopathy Journal. 2008;**21**:425-432

[26] Beninger CW, Hall JC. Allelopathic activity of luteolin 7-O-β-glucuronide isolated from *Chrysanthemum morifolium* L. Biochemical Systematics and Ecology. 2005;**33**(2):103-111

[27] Yan Z, Guo H, Yang J, Liu Q, Jin H, Xu R, et al. Phytotoxic flavonoids from roots of *Stellera chamaejasme* L. (Thymelaeaceae). Phytochemistry. 2014;**106**:61-68

[28] Li ZH, Wang Q, Ruan X, Pan CD, Jiang DA. Phenolics and plant allelopathy. Molecules. 2010;**15**(12):8933-8952. DOI: doi:10.3390/molecules15128933

[29] Maciej S, Piotrr K. Liquid chromatographic-mass spectrometric analysis of flavonoids. In: Weckwerth W, Kahl G, editors. The Handbook of Plant Metabolomics. Germany: John Wiley Sons; 2013. pp. 197-212

[30] Blunder M, Orthaber A, Bauer R, Bucar F, Kunert O. Efficient identification of flavones, flavanones and their glycosides in routine analysis via off-line combination of sensitive NMR and HPLC experiments. Food Chemistry. 2017;**218**:600-609. DOI: http://doi.org/10.1016/j.foodchem.2016.09.077

[31] Ibarra-Estrada E, Soto-Hernández RM, Palma-Tenango M. Metabolomics as a Tool in Agriculture. In: Prasain J, editor. Croatia, Metabolomics-Fundamentals and Applications. INTECH; 2016. DOI: 10.5772/66485

[32] Yun DY, Kang YG, Yun B, Kim EH, Kim M, Park JS, et al. Distinctive metabolism of flavonoids between cultivated and semiwild soybean unveiled through metabolomics approach. Journal of Agricultural and Food Chemistry. 2016;**64**:5773-5783

[33] Le Gall G, Dupont MS, Mellon FA, Davis AL, Collins GJ, Verhoeyen ME, et al. Characterization and content of flavonoid glycosides in genetically modified tomato (*Lycopersicon esculetum*) fruits. Food and Chemistry. 2003;**51**:2438-2446

[34] Carvalho E, Franceschi P, Feller A, Palmieri L, Wehrens R, Martens S. A targeted metabolomics approach to understand differences in flavonoid biosynthesis in red and yellow raspberries. Plant Physiology and Biochemistry. 2013;**72**:79-86

[35] Hurtado-Fernandez E. Avocado (*Persea americana*): complementary of different omics technologies for its metabolic characterization [thesis]. Granada Spain: Universidad de granada; 2014. p. 544. Available from: http://hdl.handle.net/10481/34097

[36] Wen W, Li D, Li X, Gao Y, Li W, Li H, et al. Metabolome-based genome-wide association study of maize kernel leads to novel biochemical insights. Nature Communications. 2014;**5**:3438. DOI: http://doi.org/10.1038/ncomms4438

[37] Yang Z, Nakabahashi R, Mori T, Takamatsu S, Kitanaka S, Saito K. Metabolome analysis of *Oryza sativa* (rice) using liquid chromatography-mass spectrometry for characterization

organ specific of flavonoids with anti-inflammatory and anti-oxidant activity. Chemical and Pharmaceutical Bulletin. 2016;**64**:952-956

[38] Zhang Y, Ming X, Wang X, Liu J, Huang B, Guo X, et al. UPLC-QTOF analysis reveals metabolic changes in the flag leaf of wheat (*Triticum aestivum* L.) under low-nitrogen stress. Plant Physiology and Biochemistry. 2017;**111**:30-38. DOI: http://doi.org/10.1016/j.plaphy.2016.11.009

[39] Ruocco S, Stefanini M, Stanstrup J, Perenzoni D, Mattivi F, Vrhovsek U. The metabolomic profile of red non-V. vinifera genotypes. Food Research International. 2017:1-10. DOI: http://doi.org/10.1016/j.foodres.2017.01.024

[40] Abu-Reidah IM, Ar′ráes-Román D, Waras I, Fernández-Gutiérrez A, Segura-Carretero A. UHPLC/MS2-based approach for the comprehensive metabolite profiling of bean (*Vicia faba* L.) by-products: A promising source of bioactive constituents. Food Research International. 2017;**93**:87-96. DOI: http://doi.org/10.1016/j.foodres.2017.01.014

[41] Zhang Q, Shi Y, Ma L, Yi X, Ruan J. Metabolomic analysis using ultra-performance liquid chromatography-quadrupole-time of flight mass spectrometry (UPLC-Q-TOF MS) uncovers the effects of light intensity and temperature under shading treatments on the metabolites in tea. Plos ONE. 2014;**9**:11. DOI: http://doi.org/10.1371/journal.pone.0112572

[42] Bovy A, Schijlen E, Hall RD. Metabolic engineering of flavonoids in tomato (*Solanum lycopersicum*): The potential for metabolomics. Metabolomics. 2007;**3**:399-412

[43] Baldina S, Picarella M, Troise AD, Pucci A, Ruggieri V, Ferracane R, et al. Metabolite profiling of Italian tomato landraces with different fruit type. Frontiers in Plant Science. 2016;**7**:2-13

[44] Madala NE, Piater L, Dubery I, Steenkamp P. Distribution patterns of flavonoids from three Momordica species by ultra-high performance liquid chromatography quadrupole time of flight mass spectrometry: A metabolomics profiling approach. Revista Brasileira de Farmacognosia. 2016;**26**:507-513

[45] Morrel K, Goeminne G, Storme V, Sterck L, Ralph J, Coppieters W, et al. Genetic metabolomics of flavonoids biosynthesis in Populus: A case study. The Plant Journal. 2006;**47**:224-237

[46] Nakabayashi R, Mori T, Saito K. Alteration of flavonoids accumulation under drought stress in *Arabidopsis thaliana*. Plant Signaling and Behavior. 2014;**9**(8):e29518

[47] Song H, Ryu HW, Lee KJ, Jeong Y, Kim DS, Oh SR. Metabolomics investigation of flavonoid synthesis in soybean leaves depending of the growth stage. Metabolomics. 2014;**10**:833-841

[48] Dixon RA, Gang DR, Charlton AJ, Fiehn O, Kuiper HA, Reynolds TL, et al. Applications of metabolomics in agriculture. Journal of Agricultural and Food Chemistry. 2006;**54**: 8984-8994. DOI: http://doi.org/10.1021/jf061218t

[49] Liu CW, Murray J. The role of flavonoids in nodulation host-range specificity: An update. Plants. 2016;**5**:1-13. DOI: http://doi.org/10.3390/plants5030033

[50] Steinkellner S, Lendzemo V, Langer I, Schweiger P, Khaosaad T, Toussaint JP, et al. Flavonoids and strigolactones in root exudates as signals in symbiotic and pathogenic plant-fungus interactions. Molecules. 2007;**12**(7):1290-1306. DOI: http://doi.org/10.3390/12071290

[51] Larose G, Chênevert R, Moutoglis P, Gagné S, Piché Y, Vierheilig H. Flavonoid levels in roots of *Medicago sativa* are modulated by the developmental stage of the symbiosis and the root colonizing arbuscular mycorrhizal fungus. Journal of Plant Physiology. 2002;**159**(12):1329-1339. DOI: http://doi.org/10.1078/0176-1617-00896

[52] Kikuchi K, Matsushita N, Suzuki K, Hogetsu T. Flavonoids induce germination of basidiospores of the ectomycorrhizal fungus *Suillus bovinus*. Mycorrhiza. 2007;**17**(7):563-570. DOI: http://doi.org/10.1007/s00572-007-0131-8

[53] Phillips DA. Flavonoids: Plant signals to soil microbes. In: Ibrahim RK, Stafford HA, editors. Phenolic Metabolism in Plants. 1st ed. Colorado, North America: Springer US; 1992. pp. 201-216

[54] Zhang J, Subramanian S, Stacey G, Yu O. Flavones and flavonols play distinct critical roles during nodulation of *Medicago truncatula* by *Sinorhizobium meliloti*. The Plant Journal. 2009;**57**:171-183

[55] Kim JS, Kwon CS, Son KH. Inhibition of ά-glucosidase and ά-amylase by luteolin, a flavonoid. Bioscience, Biotechnology, and Biochemistry. 2000;**64**:2458-2461

[56] Oberdorster E, Clay MA, Cottam DM, Wilmot FA, McLachlan JA, Milner MJ. Common phytochemicals are ecdysteroid agonists and antagonists: A possible evolutionary link between vertebrate and invertebrate steroid hormones. Journal of Steroid Biochemistry and Molecular Biology. 2001;**77**:229-238

[57] Salunke BK, Kotkar HM, Mendki PS, Upasani SM, Maheshwari VL. Efficacy of flavonoids in controlling *Callosobruchus chinensis* (L.) (Coleoptera: Bruchidae) a post-harvest pest of grain legumes. Crop Protection. 2005;**24**:888-893

[58] Goławska S, Sprawka I, Łukasik I, Goławski A. Are naringenin and quercetin useful chemicals in pest-management strategies? Journal of Pest Science. 2014;**87**(1):173-180

[59] Ateyyat M. Impact of flavonoids against woolly apple aphid, eriosoma lanigerum (Hausmann) and its sole parasitoid *Aphelinus mali* (Hald.). Journal of Agricultural Science. 2012;**4**:227-236

[60] Monique S, Simmonds J. Importance of flavonoids in insect-plant interactions: Feeding and oviposition. Phytochemistry. 2001;**56**:245-252

Section 3

Flavonoid Roles in Metabolism and Health

Chapter 11

Antioxidant Capacity of Anthocyanin Pigments

Julia Martín, Eugenia Marta Kuskoski,
María José Navas and Agustín G. Asuero

Additional information is available at the end of the chapter

http://dx.doi.org/10.5772/67718

Abstract

Anthocyanins are a family of natural pigments classified into the group of flavonoids, considered to be responsible for the color and taste of many fruits and vegetables, i.e. berries. Anthocyanins are common components of the human diet. Besides their interest as colorant because of their coloring properties, the study of anthocyanin compounds stems from their wide applicability in the prevention and even in the treatment of various human diseases. However, various aspects of the pharmacological roles of anthocyanins remain in the dark, having still several obstacles to the development of robust diets or prescribing lines on consumption of anthocyanins. The chemical structure of anthocyanins determines in large measure its capacity and efficacy as an antioxidant agent. In this study, the following aspects are reviewed: the antioxidant effect of anthocyanin pigments; the oxidative stress, the bioavailability after intake and biological aspects of anthocyanins, the method for measuring the antioxidant activity of anthocyanins, the relationship between structure and activity; and the influence of the anthocyanins in the antioxidant activity of wines. Finally an overview of some potential uses in food industry is attempted mainly focusing in the anthocyanin encapsulation topic. Attention has been paid to the more recent publications in the field.

Keywords: anthocyanins, antioxidant, biological properties, wines, encapsulation

1. Introduction

Fruits and vegetables supply a number of micronutrients, such as minerals, fibres and vitamins, as well as a whole series of compounds called phytochemicals, among which are the secondary metabolites of a phenolic nature, called polyphenols [1–3]. Phenolic compounds have attracted the attention of researchers for decades [4–7]. This was initially due to their physiological importance to plants, mainly relating to pigmentation and flavour [8, 9] and, more recently, because of their free radical scavenging capacity, which, among other biological effects, increases antioxidant activity and prevents cellular oxidation [10, 11].

The flavonoids (**Figure 1**) constitute the largest group of phenols and are considered to be responsible for the colour and taste of many fruits and vegetables. More than 9000 flavonoid structures have been described, with formula, references and biological information [12, 13]. These include more than 600 different anthocyanins that are widely distributed among at least 27 families, 73 genera and innumerable species. It has been shown that, of the flavonoids studied, around 5000 have antioxidant activity [4, 5, 14].

Anthocyanins, the largest group of phenolic pigments, are found in red wine, some cereals, root vegetables and red fruits. The red, blue and purple colours (**Figure 2**) of most fruits, flowers and leaves are due to anthocyanins. They are glycosides (water-soluble molecules) of aglycons called anthocyanidins and effective donors of hydrogen. A wide variety of anthocyanins are produced by the higher plants via modification of the six common anthocyanin aglycons (cyanidin, delphinidin, pelargonidin, malvidin, peonidin and petudinin) present in nature. A summary of previous history with references to the pioneers in this field of work has been given [5, 14]. Apart from their physiological role in plants, anthocyanins are regarded as important components in human nutrition [5, 14–16]. It has been stated that the consumption of the anthocyanins is of the order of 200 mg/day, a high amount if compares with the intake of other dietary flavonoids [5]. A possible association between consumption of anthocyanins and quality of the diet is admitted [17], although there are currently no recommendations regarding their dietary intake. A glass of red wine provides around 115 mg of polyphenols, contributing towards a total intake of phenolic compounds of 1171 mg/person/day [18, 19]. The antioxidant activity of anthocyanins is depending to a large extent with their chemical structure: number and position of the hydroxyl groups and the conjugated double bonds, as well as on the presence of electron donors in the structural ring [5, 20].

Figure 1. Some selected samples containing anthocyanins.

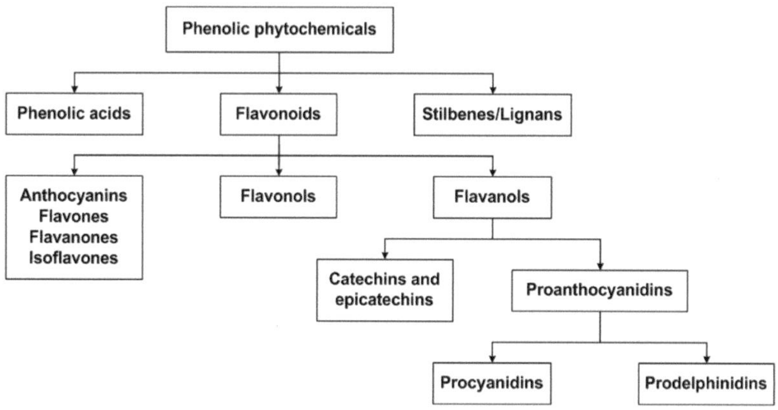

Figure 2. Type of phytochemicals [5].

Numerous epidemiological studies have confirmed the influence of the consumption of antioxidants contained in fruits, vegetables and grains [21–24]. Some beverages [25, 26], such as wine, tea and coffee, have received considerable attention due to their protective effects against the oxidative damage related to various chronic diseases, including cancer, reducing the risk of contracting these diseases by 30–50% [27]. The principal cause of death in the Western world is related to chronic diseases such as coronary heart disease or heart attacks. Low plasmatic levels of vitamin E and vitamin C have been shown to increase the risk of angina pectoris among the population of Scotland [28, 29]. This is attributed, to a great extent, to the low consumption of foodstuffs rich in micronutrients, vitamins and antioxidants, combined with the general lifestyle.

In agreement with the *French Paradox* [30, 31] and other studies [32, 33] undertaken about the European population (WHO Project MONICA, MOLI-SANS, FLORA and ATHEAN EU Projects), the components of the Mediterranean diet [34–38], especially vitamins and polyphenols, are the factors responsible for the low incidence of coronary heart disease in these populations [39–41]. The moderate consumption of red wine [42–46] is another factor closely linked to this low incidence, as the phenolic compounds have a cumulative effect. A diet rich in fruits and vegetables increases by itself the antioxidant capacity of the plasma and the level of plasmatic polyphenols [45]. These factors are increased when supplemented by the intake of red wine. Consumption of wine in moderate amounts has also proved to be beneficial [47] to the skeletal system lowering the risk of loss of mass and fractures. What is clear is that a high consumption of fruits or vegetables rich in antioxidants is related to a decrease in cardiovascular diseases and cancer [41].

Anthocyanins have an antioxidant potential twice that other known antioxidants, such as (+)–catechin and other compounds like vitamin E, synthetic antioxidants such as BHA (butylated hydroxyanisole) and BHT (butylated hydroxytoluene), compounds widely used in food technology [13, 48–51] that have undesirable effects on the enzymes of the human body. The apparent capacity of the strongly polarized anthocyanins to regenerate lipophilic antioxidants like vitamin E could be because they have similar properties to vitamin C, such as protecting the biomembranes from peroxidation, by effectively trapping the peroxyl radicals.

Using the oxygen radical absorbance capacity (ORAC) method, Wang and Goodman [52] evaluated the antioxidant capacity of 14 anthocyanins and obtained values more than 3.5 times greater than those for trolox (a synthetic antioxidant similar to vitamin E). Kuskoski et al. [51], using the ABTS method in purified and isolated patterns of anthocyanins, found an activity twice that of trolox and also confirmed the influence of the structure or the combination of anthocyanins on the antioxidant capacity.

Sources rich in anthocyanins are very interesting options as functional foods [53–58]. Here, the oxidative process, the antioxidant effect and the biological properties of the anthocyanin pigments, described in last years, are reviewed. Furthermore, the most commonly used chemical methods to determine the antioxidant capacity of the anthocyanins are outlined. An overview of the bioavailability of anthocyanins, the metabolism after their intake and their presence and influence in red wine is also given. Finally, an overview of some potential uses in food industry is attempted mainly focusing in the anthocyanin encapsulation topic.

The fertility field of flavonoids antioxidants (e.g. anthocyanins) has grown exponentially in recent decades in such a way that a number of areas are involved such as nutrition, food processing, physiology, biochemistry, pharmacology and analytical chemistry affecting foods and health. Emphasis in this contribution is given in most recent reviews and references. Some 150 journals are cited from the fields of food science and technology, nutrition, chemistry (analytical) and biochemistry, engineering, agriculture, medicine, pharmacy, biology, physiology and clinic. Taking into account that thousands of references are available, the authors apologize for those they may have overlooked or inadvertently omitted. For older references please consult, for example, some reviews [4–7] published on 2012 and the excellent monograph of Andersen and Markhan [59].

2. Oxidative process

The process of oxidation has been studied for many years [60], because of the importance it has both for the organisms and the foodstuffs. In live organisms, the oxidative metabolism is essential for the survival of the cells. Oxidation is related to the production of energy associated with the degradation of glucans, lipids and proteins, to the detoxification of many xenobiotics and to the immune response through some of the free radicals (FR) generated [61].

Oxygen is associated with the conditions for aerobic life and is the motive force for the maintenance of the metabolism and cellular viability, but, at the same time, it is responsible for the formation of partially reduced mediators with high reactivity, known as reactive oxygen species (ROS). The majority of ROS are FR, that is, active molecular species with a separated electron at a higher energy level, which, therefore have paramagnetic properties, providing them with high reactivity [20, 62].

The systems of antioxidant protection have to act on the substrates susceptible to oxidation in a controlled way to maintain the physiological equilibrium of the organism. The protective effect of some enzymes, such as superoxide dismutase (SOD), catalase and glutation peroxidase, may start when an excess of FR is produced. If this excess cannot be neutralized, oxidation of the lipidic membrane, the low-density lipoproteins (LDLs), the protein cellular components, DNA and enzymes can occur, thereby destroying them [63, 64].

It is worth emphasizing that arteriosclerosis is currently defined as chronic inflammation of the vascular system, triggered by a specific inflammatory agent, the oxidized LDL. The LDLs are very small particles made up by lipids, cholesterol and proteins, with the function of transporting cholesterol and lipids from the blood to the adipose and muscular tissue and, in general, to all cells of the body [65, 66]. However, the LDLs can be oxidized by the FR, affecting, consequently, the molecules of cholesterol and fatty acids that constitute each LDL. The oxidized LDLs are involved in the pathogenesis of coronary heart diseases [67, 68].

Environmental, dietary or physiological factors can provoke an imbalance in favour of oxidation, causing what is known as oxidative stress [69–71]. Whether the oxidation or the oxidative stress, in particular, is either a primary cause or a side effect of many chronic diseases and of the phenomenon of ageing itself has been a scientific debate prompted over the last few decades. Therefore, many efforts and resources have been devoted to finding out the role oxidants play in hindering oxidation, thus resulting in either the prevention or the retardation of the oxidative stress [72].

An excessive production of ROS, particularly hydroxyl radicals, can easily initiate the process of oxidation of the LDLs. In turn, they contribute to a greater or lesser degree to the onset of coronary heart diseases, rheumatoid arthritis, inflammatory diseases, cancer, renal diseases, pancreatitis, multiple sclerosis, Parkinson's disease, cataracts, diabetes, pulmonary disorders and all diseases related to cellular ageing [73]. The intake of dietary antioxidants, that is exogenous antioxidants (**Figure 3**), is very important [74], and some compounds of this family, that is vitamin E, β-carotene and phenolic compounds, are only synthesized by plants [27, 31, 34, 35]. Therefore, it is important to maintain a balance between oxidants and antioxidants. It is worth bearing in mind that over a lifetime, as the individual ages, this balance tilts in favour of the oxidants [75].

In foods, oxidation can be one of the main causes of alterations leading to rancidity, deterioration and loss of nutritional, commercial and organoleptic quality (colour, taste, smell and texture), besides being a possible health risk to the consumer. For this reason, the food industry, by improving the preparation of the products and by using antioxidants, is trying to prevent and slow down the process of deterioration, in order to offer the consumer a safer deadline for use, which guarantees the quality of the food product [76–80].

However, according to studies carried out in vivo during the last two decades, FR and ROS are no longer seen only as [71] destructive factors but also (and perhaps first of all) as messengers involved in intracellular signalling. So, there has been a substantial change [10] in the conception of these processes in both normal and pathological conditions. Ideas about the role of FR in the functioning of cells and organisms have been revised, resulting in a new concept of redox equilibrium. Oxidative stress is then viewed as [72] a modulation of thiol redox reactions, involved mainly in signalling pathways. On this way, nonradical oxidants (enzymatically generated hydrogen peroxide, other peroxides, quinones, etc.) play a basic role [10] in the oxidation of thiols for the sake of signalling, the formation of free radical intermediates being not necessary. The common conviction of the beneficial effect that the phenolic plants exert on the improvement of health is being revised [64].

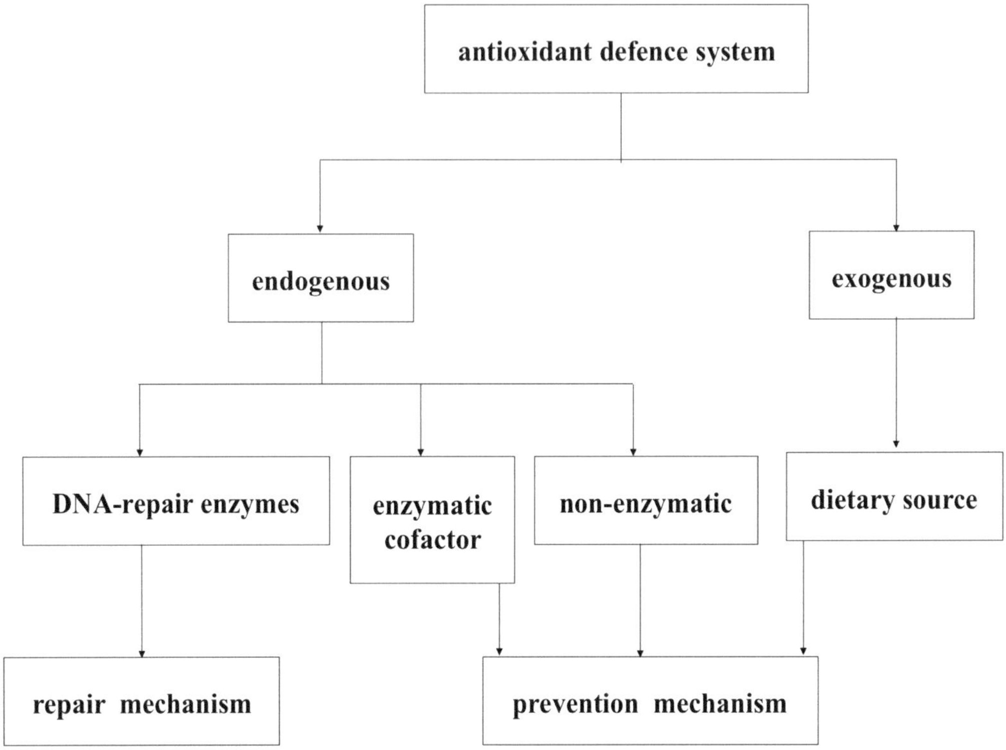

Figure 3. Summary of antioxidant defence system [74].

The potential health benefits of natural antioxidants, while interesting, seem to escape our basic understanding of biological oxidation processes. Oxidation balances both very beneficial, even crucial, outcomes with decidedly negative impacts. This suggests that moderation in the use of some antioxidants may be advisable. Note that a large measure of the biological oxidation occurring in the body is essential for extracting energy from food and is highly adaptive, depending on health status. Apart from energetics, oxidation supports immunological integrity. While the bulk of epidemiological evidence supports the nutritional/health value of fruits and vegetables [4, 11, 12, 19, 22, 33, 39, 40, 50], the doses of individual components they contain, such as specific antioxidants that may contribute to improved health and reduced risk of certain diseases, remain uncertain.

3. Bioavailability and metabolism of the anthocyanins

The existing knowledge concerning with the absorption, distribution, metabolism and excretion (ADME) of anthocyanin compounds (including their decomposition within the gastrointestinal lumen) has been the subjects of several recent reviews [81–95]. In general, few comparative studies have been undertaken about their metabolism, physiological availability or biotransformation after intake in comparison with the number of studied devoted to absorption and distribution. Little information is also available on the effects of food matrix on anthocyanin bioavailability, particularly food matrices of the usual diet [92].

In general, anthocyanins are considered to have a remarkably low bioavailability (relatively low as well in comparison with that of other flavonoids), on the basis of the levels detected in human blood after ingestion [81]. This fact contrasts with the health-promoting properties [81, 83, 84, 90] of anthocyanins, suggesting bioavailability and their interaction with other components present. Anthocyanins appear to be rapidly absorbed in the stomach and small intestine [89] and removed, being in the plasma and urine where reach low maximal concentrations [90]. After oral administration, anthocyanins follow a particular pattern different from other flavonoids [84]. A 20–25% of intact anthocyanins were detected in plasma few minutes after intake [86]. Kinetic studies have shown that anthocyanins have a rapid distribution and appearance in blood that is compatible with a tricompartmental model. Elimination takes place mainly through bile. Anthocyanins could be absorbed from the stomach as well as intestines where they undergo decomposition catalysed by microbiota. Bacterial action is capable of hydrolysing anthocyanins or aglucons into simpler phenolic compounds, which can be absorbed and still maintain free phenolic groups, retaining part of the reducing capacity of the original molecule. Active transport may play a role in the absorption of anthocyanins from the stomach as well as in their transfer within the kidney or liver [84]. The metabolic destination of the anthocyanins can differ depending on their aglucon structure, as well as on the tissue where they are metabolized (intestine or liver).

However, the persistence of anthocyanin metabolites, phenolic acid breakdown products (which could be responsible for the health benefits associated with anthocyanins) suggests enterohepatic recycling, leading to prolonged residence time, and supports the notion that anthocyanins are far more bioavailability than previously suggested [81, 88, 92]. However, the compounds as well as the molecular mechanisms involving all those biological events [83] still remain under exploited. The ability to cross membranes, pH effect, digestive enzymes, microbiota, biliary acids and food matrix are critical factors, which may contribute to this apparent paradox [86]. There are many doubts if the effect is due to the native compounds or other forms, their mechanism or which factors have crucial impact on bioavailability [86]. To clear the access both native and metabolized forms in vivo and to distinguish their different biological roles have been a very challenging task. Accumulative evidence, which is emerging, suggests multiple roles [92] explaining the apparent incongruity (poor absorption). Compared with other flavonoids, much remains to be discovered [94, 95] about details and mechanisms of anthocyanin absorption and transport. The activity of anthocyanins could be associated with the ability to elicit cell adaptive responses involving the transcription factor Brf2 by affecting the "nucleophilic one" of the organism [89]. Recent studies on the bioavailability topic are summarized in **Table 1** [82, 96–108].

Comments	References
Pharmacokinetic trial to evaluate the bioavailability of anthocyanins and colonic polyphenol metabolites after consumption of aronia berry extract in plasma and urine	[96]
Pharmacokinetic characterization and bioavailability of strawberry anthocyanins relative to meal intake	[97]
Bioavailability studies and anticancer properties of malvidin-based anthocyanins, pyranoanthocyanins and nonoxonium derivatives	[98]
Effect of red cabbage fermentation on anthocyanin bioavailability and plasma antioxidant capacity in humans	[99]
Bioavailability of red raspberry anthocyanins and ellagitannins: new insights	[82]
Bioavailability and uptake of anthocyanins and their metabolites from grape/blueberry juice and smoothie in vivo and in vitro	[100]
Tissue bioavailability and intake of tart cherry anthocyanins	[101]
Confirmation and identification of tart cherry anthocyanins in several target tissues of healthy rats	[102]
Bioactive anthocyanins in 'Queen Garnet' plum: maturity and bioavailability	[103]
Use of anthocyanins as bioactive colourants in lipstick formulations	[104]
Application of the developed flavonoid-poor menu meals to the study of the bioavailability of bilberry anthocyanins as model flavonoids	[105]
Anthocyanin stability, mucus binding, and uptake into epithelial cells in healthy individuals that retained red grape or chokeberry juice in the mouth	[106]
Absorption and bioavailability of anthocyanins across the gastrointestinal mucosa	[107]
Effects of processing sour cherry fresh fruit to the final juice product on the content of anthocyanins and other related polyphenols	[108]

Table 1. Bioavailability of anthocyanins.

4. Biological activity of the anthocyanins

Establishing the biological activities of phytochemicals, flavonoids and polyphenol is dependent on the complete understanding of their intake, absorption, metabolism and excretion; however, to date, this had only realized for a limited few structures [109]. The increasing evidence of potential therapeutic effects that present anthocyanin compounds has boosted the interest in the knowledge of their biochemistry and biological effects during the last two decades [95, 110–112]. Biological properties of anthocyanins depend on their bioavailability. The chemical structure of anthocyanins [113] determines their rate and extent of intestinal absorption and nature of the metabolites in the plasma. The growing and current interest in the study of anthocyanin compounds [114, 115] stems from their wide applicability in the prevention and even in the treatment of various human diseases. They could also be used in the control of the viruses that cause immunodeficiency, such as the causal agent of AIDS, and they have a strong activity against the influence A and B viruses as well as against the herpes virus [116]. Though many articles have been devoted to varying biological effects of anthoxyanins, only a limited number of studies deal with their antimicrobial activity [117].

The favourable effects of anthocyanins on improvement of vision in humans (increase in visual acuity), one of the first reported, were described in 1966, which prompted their introduction into ophthalmology [56, 57]. It continues to be an interesting field of study due to the prevalence of myopia in today's society [118]. Although these effects are not completely understood [119], it has been confirmed that cyanidin helps regeneration of rhodopsin. Anthocyanins have been associated with substances that strengthen the capillaries, reinforce the action of vitamin C and favour the accumulation of this vitamin in the liver and in the suprarenal glands. Blackcurrant anthocyanins inhibit transient myopia, reduce eye fatigue, improve dark adaptation and enhance retinal blood flow with glaucoma [56]. Anthocyanin-rich bilberry extract has a protective effect on visual function during retinal inflammation [116].

Anthocyanins have been shown to be effective in the prevention of arteriosclerosis and cardiovascular diseases [25, 40, 41, 72]. Commercial extracts of *Vaccinium myrtillus* (bilberry) [120, 121] contain glucosides of delphinidin and cyanidin and, since 1977, have been used to inhibit platelet aggregation [122] because of their preventive effect in the initial stage of the formation of thrombi, in the treatment of some diseases related to poor microcirculation resulting from capillary fragility, and also to prevent the oxidation of the LDLs [123–125].

Moreover, it has been demonstrated that these preparations accelerate the spontaneous process of cicatrization and that they have a preventive and curative activity against gastroduodenal ulcers induced in rats. These effects are probably due to their influence on the biosynthesis of mucopolysaccharides [126], which improves the efficacy of the gastric mucous layer and increases the base substance of the connective tissue and of the capillaries.

Another described effect is the inhibition in vitro that certain anthocyanins have on the porcine pancreatic elastase [127]. This enzyme attacks fibres and collagen, playing an important role in some pathologies, such as arteriosclerosis, emphysema and rheumatoid arthritis. Beneficial effects have also been described in experiments with diabetes, with a substantial reduction observed in the sugar concentration in urine and plasma of rats treated with the anthocyanin pigments of grapes [128]. It is suggested that anthocyanins act by reducing the biosynthesis of collagen, lipoproteins and glycoproteins, as well as reducing the activity of elastase and adenosine deaminase, which are both known to be high in diabetic patients.

Anthocyanins are recognized for their various [67, 129] pharmacological and medicinal properties. They are antimutagenic, anti-inflammatory and vasotonic [111, 112, 130, 131]. They protect against radiation, are chemoprotective against the toxicity of platinum, are used in therapy against cancer and are hepatoprotective against carbon tetrachloride. They also have other effects due to several actions of a variety of enzymes and metabolic processes.

There are various patented pharmaceutical preparations [132] containing flavylium salts and anthocyanins for the treatment of wounds, gastroduodenal ulcers, inflammation of the mouth and throat, vascular diseases and other diseases linked with the lipidic and the glyceric acid metabolisms. More recently, they have been used in the treatment of circulatory diseases.

Some studies specify the anticarcinogenic effect of anthocyanins [12, 13, 19, 111, 112, 133–135]. They inhibit the growth of carcinogenic cells that provoke colon cancer, induce the apoptosis effect, have the capacity to inhibit in vitro the growth of cells that cause tumours in humans and are even able to act as modulators of the macrophages in the immune response [89, 112].

Anthocyanins are effective against cytotoxicity, lipidic peroxidation, and as protectors of DNA, by forming co-pigments of DNA-anthocyanins. Moreover, anthocyanins have cellular antioxidant mechanisms comparable to or greater than other micronutrients, such as vitamin E. The capacity of the anthocyanins for stabilizing triple-helical complexes of DNA [136] by forming complexes of anthocyanins-DNA [137] is well established.

Pharmacokinetics of anthocyanins has recently reviewed [85, 113, 138, 139]. The most recent papers published on the subject are summarized in **Table 2** [96, 101, 140–151]. Anthocyanins are metabolized to a structurally diverse range of metabolites that exhibit dynamic kinetic profiles. A multicompartmental (theoretical physiologically based) pharmacokinetic (PBMK) model has been proposed [138] in order to describe the anthocyanins fate in vivo. Understanding the elimination kinetics of these metabolites is key to the design of future studies [152] concerning with their utility in dietary intervention or as therapeutics for disease risk reduction.

Comments	References
Pharmacokinetic trial to evaluate the bioavailability of anthocyanins and colonic polyphenol metabolites after consumption of aronia berry extract in plasma and urine	[96]
Evaluation of the protective effects of protocatechuic acid	[140]
Effects of black raspberry extract and protocatechuic acid on DNA adduct formation and mutagenesis in rat oral fibroblasts	[141]
Influence of ethanol on the bioavailability and pharmacokinetics of blackberry anthocyanins	[142]
Pharmacokinetic trial to evaluate the of nanoencapsulation of a phenol extract from grape pomace on human plasma	[143]
Pharmacokinetic characterization of anthocyanins in overweight adults on the basis of meal timing	[97]
Determination of cyanidin 3-glucoside in rat brain, liver and kidneys: a short-term pharmacokinetic study	[144]
Pharmacokinetics, bioavailability and regional brain distribution of polyphenols from apple-grape seed extract mixture and bilberry extract	[145]
Evaluation of changes in metabolic parameters, and in cardiovascular and liver structure and function in rat due to administration of either cyanidin 3-glucoside or Queen Garnet plum juice	[146]
Bioavailability and uptake of anthocyanins and their metabolites from an anthocyanins-rich grape/blueberry juice and smoothie in vivo and in vitro	[101]
Effects of anthocyanins and their corresponding anthocyanidins on the expression levels of organic anion transporting polypeptides in primary human hepatocytes	[147]
Effect of flavan-3-ols and anthocyanins against inflammatory-related diseases	[148]
Anthocyanin pharmacokinetics and dose-dependent plasma antioxidant pharmacodynamics by intake of Montmorency tart cherries in healthy humans	[149]
Pharmacokinetics of the metabolites of cyanidin-3-glucoside	[150]
Abundance and persistence of metabolites of anthocyanins in human urine	[151]

Table 2. Selected papers on pharmacokinetics of anthocyanins in the 2014–2016 period.

In words of Kay [152], 'These studies on (flavonoid) metabolism and biological activity of metabolites mark a new beginning in phytochemical research and, in this respect, this work is in its infancy'. Phenol-Explorer web database gathers polyphenol metabolites [153] identified in human and animal biofluids, from 221 publications.

5. Methods for measuring the antioxidant activity of anthocyanins

Although a plethora of biological actions has been ascribed to flavonoids, their antioxidant activity, in particular, has recently attracted much attention. Anthocyanins behave as antioxidants by a variety of ways, including direct trapping of ROS, inhibition of enzymes responsible for superoxide anion production, chelation of transition metals involved in processes forming radicals and prevention of the peroxidation process by reducing alkoxy and peroxy radicals.

There are a variety of methods for measuring antioxidant activity, either in vitro or in vivo (greater complexity involved) or a combination of both. The number of reviews published on the matter reflects the transcendence of this hot topic and its richness. Selected reviews found in the literature from 2000 up to the present time are summarized in **Table 3** [154–204]. The most common chemical methods used for measurement in vitro of antioxidant activity of polyphenolic compounds (e.g. anthocyanins) are shown in **Table 4** [197–241]. Methodological contributions are preferably cited in **Table 4** instead of specific practical applications. Both conceptual and technical problems limiting the use and validity of three commonly used [119] assays TEAC/ABTS*+, DPPH and ORAC have been subject of recent revision. Some reviews dealing with the DPPH [208, 212, 214], ORAC [223] and CUPRAC [239, 240] assays have also been the subject of recent treatments. However, the aspects concerning with the assay chemistry, standardization and report of the antioxidants determination have not been solved after 25 years of intense study [199].

Antioxidant activity is always measured in an indirect way as a response (of the antioxidants present in the sample) to induced oxidation [192, 160, 173]. For foodstuffs, there is a range of methods for determining antioxidant activity. These can vary from those that evaluate the inhibition of lipidic peroxidation by the antioxidants and quantify the products as peroxides, hydroperoxides and products resulting from decomposition measured by the thiobarbituric acid reactive substances (TBARS) assay [171], to methods that determine the content of free fatty acids, polymer content, viscosity, absorptivity at 232 and 268 nm, colour and physiological measurements in vivo, such as measuring the products from oxidation of the LDLs, or indirect indicators of lipidic oxidation. Alternatively, antioxidant activity can be evaluated by measuring the immunological response to antigens (the products of lipidic oxidation). Though solvent effect is a vital parameter [203] exerting an influence on the chemical behaviour of antioxidant compounds, the information concerning about its role on the antioxidant capacity is relatively scarce.

There are some drawbacks to assays in vivo. The interpretation of changes in the antioxidant activity of the plasma can be complicated because of the possibility of producing adaptability in response to an increase in oxidative stress. However, assays in vitro can also have their drawbacks, such as the interactions between samples and reagents.

Content	References
Antioxidant activity/capacity measurement: classification, physicochemical principles, mechanisms and electron transfer-based assays	[154]
Antioxidant activity/capacity measurement: hydrogen atom transfer-based, mixed-mode and lipid peroxidation assays	[155]
Antioxidant activity/capacity measurement: reactive oxygen and nitrogen species scavenging assays, oxidative stress biomarkers and chromatographic/chemometric assays	[156]
Recent applications for in vitro antioxidant activity assay	[157]
Evaluation of procedures for assessing anti- and pro-oxidants in plant samples	[158]
Capacity of antioxidants to scavenge multiple reactive oxidants and to inhibit plasma lipid oxidation induced by different biological oxidants	[159]
Analytical methods applied to antioxidant and antioxidant capacity assessment in plant-derived products	[160]
Advantages and limitations of commons testing methods for antioxidants	[161]
A comprehensive overview on the biology behind some reactive molecules and the means for their detection	[162]
Potentiometric study of antioxidant activity: development and prospects	[163]
Methods for determining the efficacy of radical-trapping antioxidants	[164]
Electrochemical methods for total antioxidant capacity	[165]
The role of consumption of dietary bioactives on the prevention of adverse health	[166]
Synthetic and natural phenolic antioxidants: mode of action, health effects, degradation products and toxicology	[167]
Up-to-date overview of methods available for measuring antioxidant activity	[168]
Use of metallic nanoparticles and quantum dots as novel tools for reliable assessment of antioxidant activity in food and biological samples	[169]
Review on in vivo and in vitro methods evaluation of antioxidant activity	[170]
IUPAC technical report: methods of measurement and evaluation of natural antioxidant capacity/activity	[171]
Evaluating the antioxidant capacity of natural products: a review on chemical and cellular-based assays	[172]
Application of free radical diphenylpicrylhydrazyl to estimate the antioxidant capacity of food samples	[173]
Application of both stationary and flow electrochemical methods for analysis of antioxidant properties of plant and clinical samples	[174]
Phenol-based antioxidants and the in vitro methods used for their assessment	[175]
Main components in the foodstuffs and beverages: antioxidant methods, chemical and kinetic basis	[176]
Estimation of antiradical properties of antioxidants using DPPH assay	[177]
Evaluation of antioxidants: scope, limitations and relevance of assays	[178]
A comprehensive review of cupric reducing antioxidant capacity methodology	[179]
Overview of the importance and mechanism of action of antioxidants, as well as of the methods of assessment of the antioxidant capacity	[180]
Methods for evaluating the potency and efficacy of antioxidants	[181]
A comprehensive review of chemical methods to evaluate antioxidant ability	[182]
Assessment of antioxidant capacity in vitro and in vivo	[183]

http://dx.doi.org/10.5772/67718

Content	References
Direct measurement of the total antioxidant capacity of foods: the 'QUENCHER' approach	[184]
Flow injection-based methods for fast screening of antioxidant capacity	[185]
Flow injection-based systems for determination of scavenging capacity against biologically relevant reactive species of oxygen and nitrogen	[186]
Antioxidant assays for plant and food components	[187]
Oxygen radical antioxidant capacity and trolox equivalent antioxidant capacity assays comparison to estimate the total antioxidant capacity of food products	[188]
How to standardize the multiplicity of methods to evaluate natural antioxidants	[189]
The use of total antioxidant capacity, total antioxidant capacity test, as a biomarker of disease in biochemistry, medicine, food and nutritional sciences	[190]
Critical review of the most commonly used methods for in vitro determination of antioxidant capacity	[191]
Updated methodology to determine antioxidant capacity in plant foods, oils and beverages	[192]
Methods to measure the antioxidant defence system	[193]
Popular methods commonly used for testing antioxidant activity in vitro: reliability, efficiency, accessibility and biological relevance	[194]
Methods for measuring antioxidant activity and assays for measuring overall reducing capacity	[195]
Model systems of the evaluation of antioxidants in three types of foods: bulk oil, oil-in-water emulsions, and muscle foods	[196]
The multifaceted aspects of antioxidants and the basic kinetic models of inhibited autoxidation	[197]
Application of various chemical methods to determine antioxidant activity in fruit pulp	[198]
Overview of cell culture models for antioxidant research	[199]
Standardized methods for the determination of antioxidant capacity and phenolics in foods and dietary supplements	[200]
Review of methods to determine chain-breaking antioxidant activity in food	[201]
Methods of measuring antioxidant activity particularly as they relate to lipid oxidation	[202]
Methods used to evaluate the free radical scavenging activity in foods and biological systems	[203]
Methodologies for evaluation of total antioxidant activities in complex mixtures	[204]

Table 3. Published reviews on used methods for measurement of antioxidant activity.

The in vivo antioxidant potential of anthocyanins can be measured by reducing the serum concentration of the reactive substance to thiobarbituric acid (TBARS assay) or by increasing the resistance to oxidation in the plasma of the lipidic peroxidation caused by 2,2′-azobis (2-amidinopropane) hydrochloride (AAPH) or by Cu^{2+}.

Most in vitro measurements of the antioxidant activity of anthocyanins involve the following factors: calculating the rate and range of the decrease of the substance in assay or the oxygen consumption, the formation of products from oxidation and the formation or decline of the number of FR. Detection can be carried out by inhibition of fluorescence, chemoluminiscence, oxygen consumption or absorbance, the evolution of which is related to the end product.

Method	Detection	Measurement/oxidant	References
Radical $ABTS^{•+}$	Reduction of absorbance of the radical cation in an aqueous medium at 414 nm (or 645, 734 or 815 nm)	TEAC value, antioxidant activity equivalent to trolox (μmol/g)	[197–205]
Radical $DMPD^{•+}$	Reduction of absorbance to 505 nm	Expressed in μmol equivalent to trolox (TEAC) by g of sample	[206, 207]
Radical $DPPH^{•+}$	Reduction of absorbance to 517 nm	Expressed in EC_{50} (quantity of antioxidant required to reduce to 50% of the initial concentration of DPPH) or in TEAC	[208–219]
FRAP	Increase of the absorbance to 593 nm	Expressed in μmol of equivalents of reduced ferric ion (Fe^{2+}) by g of sample or a value equivalent to a pattern	[203, 204, 220]
ORAC-PE	Reduction of fluorescence (β-phycoerythrin)	μmol equivalent to trolox (TEAC) by g of sample	
ORAC-FL	Reduction of fluorescence (fluorescein)	μmol equivalent to trolox (TEAC) by g of sample	[221–236]
ORAC-PGR	Reducon of fluorescence (pyrogallol red)	μmol equivalent to trolox (TEAC) by g of sample	
CUPRAC	Absorbance measurement of the Cu(I)-neocuproine chelate	μmol equivalent to trolox (TEAC) by g of sample	[199, 237–241]

Abbreviations: ABTS (2,2′-azino-bis (3-ethylbenzothiazolinine-6-sulfonic acid); DMPD (N,N-dimethyl-p-phenylenediamine dihydrochloride); DPPH (2,2-diphenyl-1-picrylhydrazyl); FRAP (ferric reducing ability of plasma); ORAC-PE (oxygen radical absorbance capacity) with β-phycoerythrin; ORAC-FL (oxygen radical absorbance capacity) using fluorescein (3′6′-dihydroxyspiro[isobenzofuran-1[3H], 9′[9H]-xanthen]-3-one), ORAC-PGR (oxygen radical absorbance capacity) with pyrogallol red (pyrogallol sulphone phthalein); TROLOX (6-hydroxy-2,5,7,8-tetramethilcroman-2-carboxylic acid).

Table 4. Commonly used methods for measurement in vitro of antioxidant activity.

FR can be generated by various chromogenic compounds, such as azo ABTS (2,2′-azino-bis(3-ethyl-benzthiazoline)-6-sulphonic acid), DMPD (N,N-dimethyl-p-phenylenediamine dihydro-chloride), DPPH (2,2-diphenyl-1-picrylhydrazyl), FRAP (ferric reducing ability of plasma) and DMPO (5,5-dimethyl-1-pyrroline N-oxide). Inhibition of oxidation can be measured by the reduction in fluorescence by the ORAC method or by the TRAP (total radical-trapping antioxidant parameter) assay.

Currently, ABTS is one of the methods most frequently used for assays of coloured compounds, like anthocyanins [197], as the radical generated has a maximum absorption at a wavelength of 734 nm, reducing the possibilities of interference of antioxidants that absorb in the red colour zone. The radical $ABTS^{•+}$ can be generated by enzymes (peroxidase, myoglobin) or chemically (manganese dioxide, potassium persulphate or ABAP (2,2′-azobis-2-amidino-propane hydrochloride). The radical, once generated, displays new characteristics with maximums of absorption at 414, 645, 734 and 815 nm.

Kuskoski et al. [51] found a maximum absorption of around 754 nm in an alcoholic medium, and this wavelength was used to determine the antioxidant activity of fruit extracts of baguaçu (*Eugenia umbelliflora* Berg) that are rich in anthocyanin pigments [242].

If compared with other methods of formation of FR, such as DPPH, DMPD and others, the capture reaction time of the radical $ABTS^{\bullet+}$ is fairly rapid, it can range from 1 to 7 min, although according to Re et al. 4 min is sufficient to complete the reaction. Antioxidant data based on ABTS assay are dependent on reaction time because the applied standard compounds (trolox) present a scavenging kinetic profile [200] different from that of polyphenol-rich foods. Studies have been carried out on the effects of molecular structure (molecular weight, number of —OH groups, redox potential) on kinetics and dynamics of [201] the trolox equivalent antioxidant capacity assay with ABTS. Attempt has been made to standardize the method [202] by extrapolating to zero sample concentration.

The chromatic properties of the stable radical cation DPPH were first described [171] by Blois in 1958, who used the radical to measure the antioxidant activity of several natural compounds. Only much later did Brand-Williams et al. develop a technique based on the reduction of the absorbance of the radical $DPPH^{\bullet}$ at 517 nm. This technique has also been applied by other authors with modifications and measurement of absorbance at 515 nm. Results are expressed as IC50 [213], that is, the quantity of antioxidant required to reduce the initial concentration of DPPH to 50%, or as the percentage of interacted DPPH % DDPH = [(Absreferencia − Abse xtracto)/(Absreferencia)] × 100. DPPH assay on food additives and foods and beverages has been subject to interlaboratory study [211, 215]. The DMPH reaction has been revisited and re-evaluated [216–218] and simplified in order to characterize samples of wine origin [210].

The influences of reaction time, DPPH concentration inference and kinetics parameters of bioactive molecules and plant extracts [209] in the reaction with the DPPH radical have been evaluated. A collaborative study on the DDPH assay [215] has been promoted as well as a kinetic-matching approach to express antioxidant capacity in a more standardized way.

The spectrophotometric DMPD method, described by [171, 206, 207] Fogliano et al. in 1999, is similar to the ABTS method. In the presence of an adequate oxidant solution, the radical cation $DMPD^{\bullet+}$ generated has the ability to link hydrogen atoms, causing the discolouration of the solution, producing a reduction of absorbance measured at 515 nm. DMPD cannot be used with hydrophobic antioxidants, as it is only water soluble [171, 206]. DMPD method is not considered suitable for assays of coloured compounds, as interference can occur in the measurements, because they absorb in the same region of the spectrum.

The ferric reducing ability of plasma (FRAP) assay measure the ability of antioxidant to reduce the ferric $[Fe^{3+}\text{–}(TPTZ)^2]^{3+}$ complex to the ferrous $[Fe^{2+}\text{–}(TPTZ)^2]^{2+}$ complex (blue coloured) in acidic medium. It is a simple, reproducible method that can not only be applied to the study of the antioxidant activity of plasma, or in foods and beverages, but also to the study of the antioxidant efficacy of pure compounds with results that are comparable to those of more complex methodologies. It is widely used to determine the antioxidant activity of anthocyanins in different samples. However, the FRAP assay is carried out at a very low pH (3.6), far from the pH found in biological fluids. Nevertheless, this method has the advantage of determining the activity of the antioxidant directly in plasma; it does not depend on an enzymatic or a nonenzymatic method for generating FR and evaluates the antiradical efficacy of plasma. It also does not need the isolation of plasma fragments as is required in LDL.

The assay by fluorescence spectrophotometry known as ORAC was first set up [25, 29, 171] by Cao et al. in 1993 and later modified by Cao et al. in 1995. The ORAC method is based on measuring the decrease of the fluorescence of the proteins β-phycoerythrin and R-phycoerythrin (PE). These proteins have a high fluorescence in the presence of peroxyl radicals generated by the thermic decomposition of the 2′2′azobis (2-amidinopropan) dihydrochloride (AAPH); the decrease is recorded in the presence of antioxidants. It is considered to be a very sensitive method that evaluates the oxidation process from its beginning, although it has the drawbacks of being expensive and time-consuming [194].

However, β-phycoerythrin [219] is photo-unstable and it forms complexes with polyphenols giving, therefore low values of ORAC-PE. For this reason, it is substituted [234–236] by fluorescein (ORAC-FL), which, in contrast, is much more stable, and does not react with polyphenols, making it a much more precise and more economic method. Two alternative solutions have been proposed to decrease a systematic error related to AAPH addition in the fluorescence-based ORAC assay [221].

A simple mathematical model for conversion of ORAC values to mass units [229] has been proposed. ORAC standardization [227] and validation [230] have been attempted. The use of pyrogallol red as a probe [233, 225] for competitive antioxidant assay is a significant improvement. Pyrogallol reacts faster than fluorescein with RCOO* radicals, and its consumption does not present induction times, even in the presence of very reactive oxidants, with the exception of ascorbic acid. First action ORAC assay has been reported both with fluorescein [226] (dextracts from tea, blueberry and grape skins) and pyrogallol red [228] (red wine, fruit juices and iced teas).

A proportional measurement of antioxidant activity is obtained using the ORAC assay, which is currently one of the most commonly used methods for measuring the antioxidant activity of the anthocyanins [218].

Cupric reducing antioxidant capacity (CUPRAC) test is conceptually similar to the FRAP test, but is based on the reduction of Cu^{2+} ions in the presence of neocuproine (2,9-dimethyl-1,10-phenatroline) at pII 7, which involves faster kinetics. The ammonium acetate buffer solution account for the liberated protons in reaction with polyphenols.

The total radical-trapping antioxidant parameter (TRAP) method [171] proposed by Wayner et al. (1985) is based on the measurement of oxygen consumption during a peroxidation reaction of lipids controlled and induced by the thermic decomposition of some substances, such as ABAP or AAPH, which produces a flow of peroxyl radicals at a constant rate that is temperature dependent. These peroxyl radicals initiate a chain of lipoperoxidations. The method has some problems, including being sensitive to temperature and to changes in pH. The storage conditions of the samples are also important due to the liability of some antioxidants; therefore, their immediate analysis is recommended. When this is not possible, it is advisable to rapidly collect plasma for blood samples, to store them at −80°C and to measure them within 3 days. The concentration of proteins or uric acid, because of their high antioxidant power, should be taken into account when describing the results.

It is interesting to mention the fact that electrochemical [243–249] and ESR [250] methods are increasingly being applied to the determination of antioxidant capacity. The kind of technology and free radical generator or oxidant influences the antioxidant capacity measurement. A key factor that helps researchers to choose a given method and to understand the results obtained is the comparison of different analytical methods. In order to gather comprehensible information about the total antioxidant capacity of a food [184], at least two of these tests, and preferably all, should be combined, if possible, taking into account both the arguments for and against, and its applicability. **Table 5** shows selected articles [20, 190, 198, 204, 231, 251–260] in which more than one criterion has been applied to real samples with practical purposes. Advantages and limitations of the most common chemical methods of determination of the antioxidant capacity are compiled in **Table 6** [160, 161, 167, 171–173, 176, 184, 231, 261–263].

Samples	Methods	References
Eight anthocyanidins, seven anthocyanins and two synthesized 4′-hydroxy flaviliums	DDPH, ABTS, hydroxyl radical scavenging activity, FRAP	[20]
Dried fruits and juices from chokeberry	FRAP, ABTS	[251]
Protucatechuates	DPPH, ORAC, CAT	[252]
Six deoxyanthocyanidins and cyaniding-3-glucoside	DPPH, FRAP	[253]
Plant foods	CUPRAC, ABTS	[254]
Anthocyanins from different varieties blueberries	Inhibiting activity on lipid peroxidation, hydroxyl radical scavenging, superoxide anion radical, DPPH	[255]
Plan extracts	DPPH, ABTS, AAPH	[198]
Commercial beverages (wines, beers, soft drinks and waters)	TRAP, TEAC, FRAP	[256]
Popular antioxidants-rich US foods	ABTS, DPPH	[257]
Commercially available vegetable juices (23)	FRAP, DPPH, ABTS	[258]
Anthocyanidins, anthocyanidin-3-glucosides and portisins	DPPH, FRAP	[259]
Plants extracts of industrial interest (30)	DDPH, ABTS, FRAP, FRAP, SOD, ORAC	[260]
Food products	ORAC, TEAC	[231]
Selected small fruits	ABTS, FRAP, DPPH	[204]
Pulps of frozen fruits	ABTS, DPPH, DMPD	[190]

CAT (conjugate autoxidizable assay); SOD (superoxide dismutase assay).

Table 5. Antioxidant capacity of selected samples evaluated using more than one criterion.

Mixed hydrogen atom transfer (HAT) and single electron transfer (SET)		
ABTS	• Inexpensive and easy to use • Applicability in lipid and aqueous phase • Stable to pH • Fast reaction • Can be automated and adapted for use with microplates	• Complex mechanism of reaction • Extra step to generate free radical • Free radical not stable for long periods of time • Not standardised
DPPH	• Simple and highly sensitive • Can be automated • It just needs a UV-vis spectrophotometer to perform • No sample separation is needed	• High price of ABTS reagent • Complex mechanism of reaction • DPPH colour can be lost • Steric accessibility influences the reaction • Sensitive to acidic pH
Single electron transfer (SET)		
DMPD	• Simpler, more productive and less expensive and compared ABTS test	• No data of its stoichiometry with anti-oxidant standard and radical stability are available • DMPD is only soluble in water
FRAP	• Simplicity, speed and robustness • It does not require specialized equipment • It can be performed using automated, semi-automated, or manual methods	• It is nonspecific • Not all antioxidants reduce Fe^{3+} at a rate fast enough to allows its measurement • Compounds that absorbs at the wave-length of the determination may interfere • Requiring an acidic pH
CUPRAC	• Rapid way to study plant extract profiles • Fast enough to oxidize thiol-type antioxidants • Selective • Stable and accessible reagents • Applicable to both hydrophilic and lipophilic antioxidants It is carried out at nearly physi-ological pH values	• FRAP and CUPRAC depend on the reac-tion time • The antioxidant which reduce metal ions may exert pro-oxidant effect under certain conditions • Low correlation between the capacity measured by FRAP or CUPRAC method with that for radical scavenging measured by competition method such as ORAC
Hydrogen atom transfer (HAT)		
ORAC	• Uses biologically relevant free radicals • Simple and standardised • Integrates both degree and time of antioxidant reaction • Determine the capacity of hydrophilic and hydrophobic samples simply • May be performed in thermostated microplates	• Expensive equipment • Data variability can be large across equipment • pH sensitive • Requires long times to quantifies results

Table 6. Advantages and disadvantages of the most commonly chemical methods used for testing the antioxidant activity [160, 161, 167, 171–173, 176, 184, 231, 261–263].

Mechanisms involved in the corresponding chemical reactions are also shown in the table: hydrogen atom transfer, HAT, ability of an antioxidant to reduce radicals by hydrogen donation for ORAC and TRAP assays; single electron transfer, SET, ability of an antioxidant to transfer one electron to reduce any compounds, including metals, carbonyl and radicals for DMPD and FRAP assays. HAT and SET mechanisms may occur together as in ABTS and DPPH assays. The DPPH method is one of the oldest and most frequently used for determining the antioxidant activity of food extracts and single compounds. In comparison with DPPH assay, the ABTS assay estimates more accurately [183, 254] the antioxidant capacity of foods, especially for those contain lycophilic, lipophilic and highly pigmented compounds. However, it has been stated that methods using HAT reactions will be preferred to those with SET reactions because the peroxyl radicals used in the first are the main FR found in lipid oxidation and biological systems [259]. ORAC is the most commonly used total radical-trapping antioxidant assay and the most widely used essay for evaluating antioxidant [172] both in the industry and in the academic institutions. The evaluation of total antioxidant capacity is preferable than the individual antioxidant measurements [74] due to the complexity of food composition and the possibility of synergic interactions among the antioxidant compounds.

6. Antioxidant activity of the anthocyanins

The capacity of phenolic compounds to trap FR depends upon their structure, in particular, of the hydrogen atoms of the aromatic group that can be transferred to the FR [5, 10, 20, 24, 63, 113] and of the capacity of the aromatic compound to cope with the uncoupling of electrons as a result of the surrounding displacement of the electrons-π system. As compared to other antioxidants, research on their health effects started more recently. This late interest in polyphenols is largely explained by the complexity [264] of their chemical structures. The anthocyanin and anthocyanidin health properties are due to their peculiar chemical structure, as they are very reactive towards ROS because of their electron deficiency [265–269]. The antioxidative properties of anthocyanidins have been recently explored; most of the widely distributed anthocyanidins and anthocyanins show more scavenging activity than that of the well-known strong antioxidants trolox and catechol [20]. The physicochemical characteristics of anthocyanins [83, 90, 91], that is structure and size of the molecules (number and position of hydroxyl and methoxyl groups), water solubility and acidity constants, can control their ability to cross biological barriers. Results of antioxidant activity of foods are commonly expressed in TEAC (mmol or μmol/g sample), a capacity equivalent to trolox (a hydrosoluble synthetic antioxidant similar to vitamin E). However, some authors suggest [270] that the results should be expressed in vitamin C equivalent antioxidant capacity (VCEAC in mg/100 g), given that vitamin C is found naturally in some foods, whereas trolox is a synthetic compound.

Quantum chemical computations have recently been performed to study [265] the antioxidative properties of anthocyanidins, quantitative structure activity relationships (QSAR) and mechanisms of action involved such as HAT, SET and SPLET (sequential proton loss electron transfer). Construction and evaluation of QSAR for predicting anthocyanin activity radical scavenging using quantum chemical descriptor have been developed [271] with good prediction efficiency.

3D-QSAR models from 21 anthocyanins based on their ORAC values have been used [272] with prediction (eggplant and radish) purposes. 3D-QASR models have also been developed in a series [273] of anthocyanin derivatives of CYP3AH inhibitors (cytochrome P450).

7. Structure of the anthocyanins

The chemical structure of anthocyanins is appropriate for acting as antioxidants, as they can donate hydrogen or electrons to the FR or trap them and delocalize them in their aromatic structure [5, 10, 20, 265–269]. The structural differences among anthocyanins are related [5, 14, 274, 275] to the number of hydroxyl or methoxyl groups in the anthocyanidin skeleton, the position and the number of bonded sugar residues as well as by the aliphatic or aromatic carboxylates bonded to them. The hydroxylation pattern influences [276, 277] physiological properties such as light absorption and antioxidative activity, which is the base for many beneficial health effects of flavonoids. The hydroxyl groups in positions 3′ and 4′ provide a high stability to the radical formed by trapping FR and displacing the electrons in ring B, as well as the free hydroxyl groups in position 3 of ring C, and in position 5 of ring A, together with the carbonyl group in position 4 (**Figure 4**).

There are three important structural criteria for evaluating the antiradical effectiveness of a compound: (1) the presence of neighbouring hydroxyl groups, that is, in the position of ring B; (2) double bonds at conjugation 4-oxo of ring C; and (3) hydroxyl groups in positions 3 and 5 of ring A.

The aglucons with identical hydroxylation in rings A and C, and a single OH group in ring B (4′-OH), including pelargonidin, malvidin and peonidin (**Figure 4**), have lower antioxidant activity when compared to compounds with groups 3′, 4′ di-OH substituted (e.g. cyanidin) (**Figure 5**).

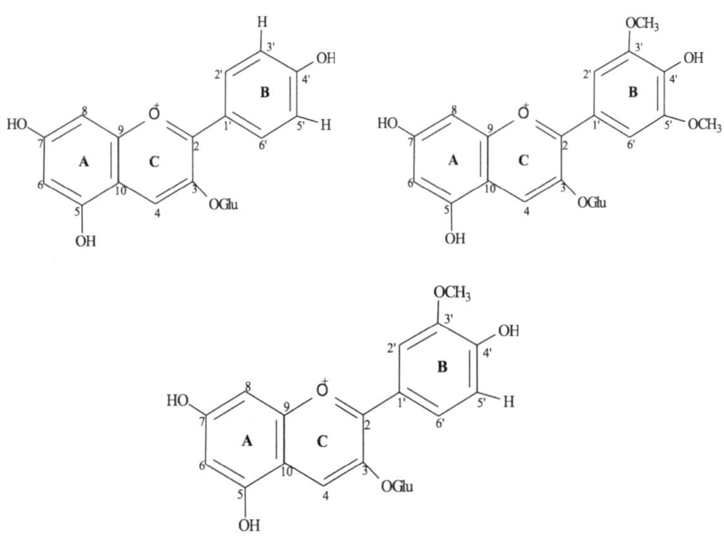

Figure 4. Chemical structure of pelargonidin (top left), malvidin (top right) and peonidin (bottom).

http://dx.doi.org/10.5772/67718

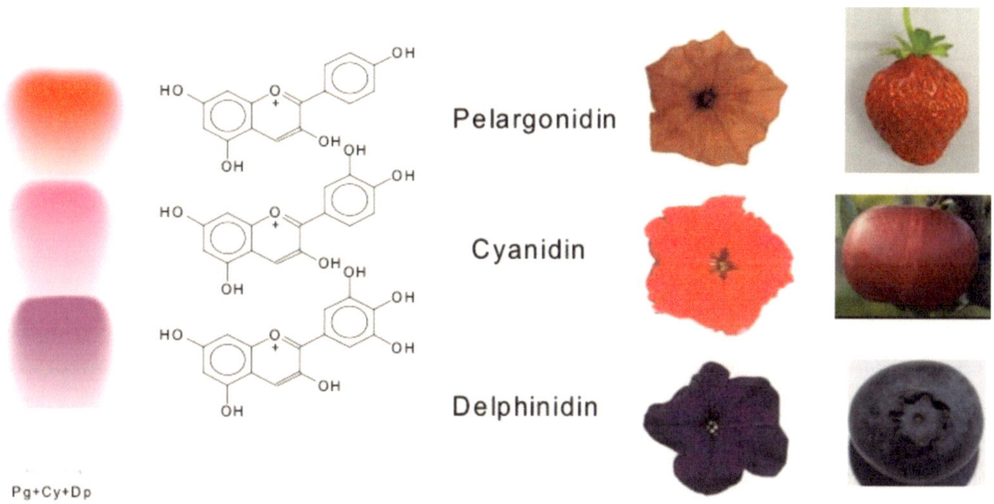

Figure 5. Chemical structures of pelargonidin, cyaniding and delphinidin, their spots on TLC and the colour of plant tissues [276].

Apparently, the OH groups in position 3′ and 4′ of ring B (catechol) are determinants of the antioxidant activity of saturated flavonoids. However, delphinidin is an exception to this principle as it has groups 3′ and 4′ di-OH substituted (**Figure 5**) and still has a low antioxidant activity. The importance of the hydroxyl groups in position 3′ and 4′ of ring B contributes to the high antioxidant capacity also found for flavones [276, 277].

Most flavonoids are found naturally in a glycosylate form, and glycosylation changes the antioxidant activity [5, 278]: for cyanidin, there is an increase; for malvidin, a decrease; and for pelargonidin, no significant effect was shown [137]. Different sugars can have distinct effects on antioxidant activity. For example, in ORAC assays for cyanidin, glycosylation in position 3 of ring C with glucose or rhamnose increases the antioxidant activity, but with galactose, it declines.

The glycosylation (site, type and number of the glycosyl, glycosidic bond type) generally enhances [269] the stability, results in the hypsochromic effect and blueing, decreases the bioavailability and anticancer activity, and decreases, increases, or does not change the antioxidant activity of the anthocyanidins or anthocyanins. Note the diverse and complex chemistry of acyl groups and that their stabilizing effect exerted may be either independent or synergic. However, the acylation decreases the polarity of anthocyanins and creates steric hindrance effects (changing molecular size and spatial structure) to decrease the sensitivity of the anthocyanins to nucleophilic attack [274] and increasing the *in vitro* and *in vivo* chemical stability (though it lowers their apparent absorption) [113]. Nonacylated monoglycosylated anthocyanins have a greater inhibitory effect on human colorectal adenocarcinoma (HT29) cell proliferation [279]; anthocyanins with pelargonidin, triglycoside and/or acylation with cinnamic acid have a lesser effect.

Anthocyanins are more than flavylium cations [280]. In aqueous solutions, equilibrium of at least four other species determined by pH (and temperature) exists [281–285]. Above about pH 2.5, the coloured flavylium cation (only stable at pH ≤ 1, rare in natural environments) form typically hydrates (pH 4–5) to form the colourless hemiacetal (carbinol pseudo-base), followed by ring-opening tautomerization to the light yellow (E)-chalcone, which can isomerize to the (Z)-chalcone. At pH values of 7–8, blue-purple quinoidal anions (which fades in several minutes) are formed. **Figure 6** shows a sample of wine (moderately acid pH 3.5–4.0) at different pH values and corresponds to the graphical abstracts of reference [280]. The state of ionization of the anthocyanins can be an important factor in relation to their antiradical activity. This is corroborated by the fact that the pseudo-base and the quinoidal base of malvidin 3-glucoside, generated at pH 4.0 and pH 7.0, respectively, have differences in antioxidant activity. It is possible play with the colour of anthocyanins [286, 287], for example, complexation with metal ions or with colourless organic molecules (co-pigments) such as hydroxylated benzoic or cinnamic acids. Experiments undertaken with synthetic colourants (Ponceau 4R) have shown that they do not have antioxidant activity, whereas anthocyanin pigments confer an antioxidant activity far greater than that of the synthetic colourants available on the market. This shows that natural pigments besides providing a good source of colour have considerable antioxidant potential. Public concern about synthetic food dyes (suspected to cause adverse effect on health) has increased recently. For this reason, consumers and food manufacturers (i.e. beverage industry) increasingly demand "cleaner" colourants from natural sources [13, 48, 49, 54, 57, 79]. **Table 7** compares [288] the characteristics of both synthetic and natural colourants. Interesting alternatives in food systems to synthetic colourants are acylated anthocyanins [289–293]. A huge variety of hues can be achieved as a function of anthocyanin structure and pH of food matrix. The increasing interest in foods that help to prevent diseases has boosted the market for nutraceutical and/or medicinal food [294]. The term functional food appeared in Japan in the 1980s associated with processed food containing ingredients that affect physiological functions. Identification of health effects provoked by anthocyanins will increase their demanding what would open new perspectives [295] for their use in the food market.

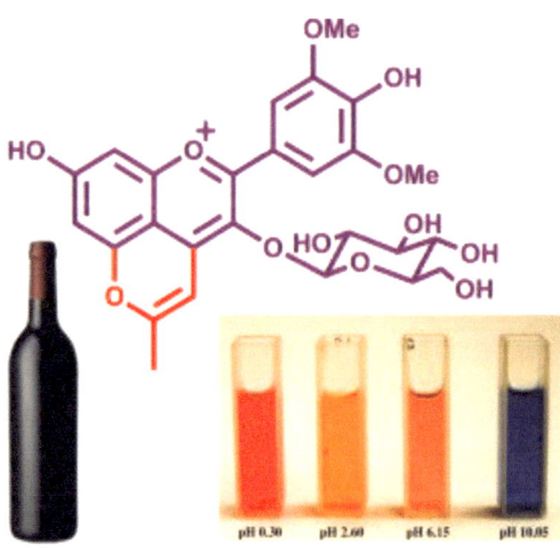

Figure 6. Red wine at various pH values: graphical abstracts of Ref. [285].

Synthetic antioxidants	Natural antioxidants
Inexpensive	Expensive
Widely applied	Usage of some products restricted
Medium to high antioxidant activity	Wide ranging antioxidant activity
Increasing safety concern	Perceived as innocuous substances
Usage of some of them banned	Increasing usage and expanding applications
Low water solubility	Broad range of solubilities
Decreasing interest	Increasing interest
Some of them stored in adipose tissue	Completely metabolized

Table 7. Advantages and disadvantages of natural and synthetic antioxidants commonly used for food protections [288].

8. Influence of the anthocyanins in the antioxidant activity of wines

Anthocyanins are the most abundant polyphenolic compounds in red wines. Red wine is probably the foodstuff that presents the highest diversity of these polyphenolic pigments in their original form and in other derivative structures. Various studies in vitro and in vivo have confirmed that wine has antioxidant properties, mainly attributable to its composition rich in phenolic compounds [296–298], which vary from 1200 to 2400 mg/L [299–303].

Wines, particularly red wines, inhibit platelet aggregation, increase antioxidant capacity in humans and reduce the susceptibility to lipidic peroxidation in plasma [45, 304–306]. Anthocyanins are the pigments responsible for the attractive colour of red wine and are one of the main flavonolic compounds with antioxidant activity, which is why red wine has a greater antioxidant activity than white wine [307–310]. Its antioxidant capacity can be up to 10 times stronger than that of white wine [205].

Nevertheless, alcohol itself has a protective effect as, to some extent, it is a mediator of the increase (close to 50%) of the level of high-density lipoproteins (HDLs) and of the decrease (of around 18%) of low-density lipoproteins (LDLs), such as cholesterol [311, 312]. However, various studies have correlated the effect of the consumption of red wine with a reduction in coronary heart disease, which is more significant than that for beer or other alcoholic drinks [18, 45]. Therefore, this reduction can be attributed to nonalcoholic components present in red wine [304, 307].

The nonalcoholic components of wine, mainly phenolic compounds, are considered to be the primary factor responsible for this protective effect. There is a significant concentration of flavonoids in red wines (>500 mg/L) and a very low one in white wines (<60 mg/L) [313–315]. In a study by Frankel et al. [300], the relative antioxidant activity of 20 Californian wines was mainly correlated with the presence of cyanidin and malvidin 3-glucoside. Similar results were obtained by Aguirre et al. [316] and Rivero-Pérez et al. [317, 318] in Chilean and Spanish red wines, respectively. According to Fernández-Pachón et al. [319], in the ranking of activity, the most active is the anthocyanins and flavan-3-ol, followed by the phenolic and flavonol acids.

Ghiselli et al. [320] studied three polyphenolic subfractions of red wine, evaluating the capacity to trap hydroxyl and peroxyl radicals, the inhibition in vitro of the oxidation of LDLs and platelet aggregation. The fraction containing the anthocyanins proved to be the most effective in its capacity to trap ROS and to inhibit the oxidation of LDLs and platelet aggregation. Anthocyanins are quantitatively the most abundant phenolic subclass in red wine [321, 322]. The other two fractions, containing the phenolic acids and quercitin 3-glucuronide, and procyanidins, catechins and quercitin 3-glucoside, are less active.

Some authors still attribute the antioxidant activity of red wines to all polyphenolic compounds [323–325], not discarding the hypothesis that the different classes of polyphenolic compounds can be more effective and act in a synergistic way. However, according to Fernández-Pachón et al. [319], no synergistic effects were observed among the isolated fractions of red wines (anthocyanins, flavonols and phenolic acids). Galanakis et al. [326] characterized the phenolic content and antioxidant capacity of Cypriot wines. The higher concentrations of phenols did not always reflect higher antioxidant capacity of wines, probably due to the observed antagonistic effect between hydroxycinnamic acid derivatives, flavonols and anthocyanins.

9. Encapsulation of anthocyanins

As it has been mentioned throughout the chapter, anthocyanins are potentially used in food and pharmaceutical industries since their practical applications as natural colourants [13, 49, 58, 76, 79, 327] as well as their potential health benefits to humans [13, 56, 90, 110, 114]. Nevertheless, the incorporation of anthocyanins into food and medical products is a challenging task due to their low stability and susceptibility to degradation [292] towards environmental conditions during processing and storage. In order to prevent these limitations, delivery systems have been developed, and among them, encapsulation [328–334] would appear to be an interesting option.

Encapsulation, developed approximately 60 years ago [335], is a rapidly expanding technology to entrap one substance (active agent, solid, liquid or gas) within another substance (a matrix or a polymeric wall) in the form of micro- and nanoparticles to protect the 'actives' from environmental conditions and their interactions with other components or to control their release [331, 332]. In addition, once encapsulated bioactive compounds are easier to transportation and handling, masking of undesirable flavour and compartmentalization of two or more reactive species [328, 335].

In general, the three-stage process during encapsulation is [335] as follows: (i) formation of the wall around the material; (ii) ensuring that undesired leakage does not occur; and (iii) ensuring that undesired materials are kept out. For that end, different techniques have been studied and applied to encapsulate active agents, some of them successfully applied for anthocyanins including spray-drying, emulsification, ionotropic gelation or coacervation, and thermal gelation [328, 330].

Among the most important factors to take into account when choosing the microencapsulation technique are particle size, physicochemical properties of the core, the process cost and the selection of wall materials. According to the literature, encapsulation by spray-drying is the most economical, simplest and the most applied method (80–90% of encapsulates are spray-dried) for preservation of anthocyanin pigments [332], being maltodextrins as the most used coating material. The use of other techniques than spray-drying [331] still remains an unexplored area. This fact could be explained by the hydrophilic nature of anthocyanins [332, 336], so it is, therefore, a promising area of research.

In order to increase the efficiency and stability by spray-drying, different biopolymers are used [328, 333], most common are natural gums (gum arabic, alginates, etc.), proteins (dairy proteins, soy proteins, gelatine, etc.), carbohydrates (maltodextrins and cellulose derivatives) and/or lipids (waxes, emulsifiers). Some authors [330, 332] have revealed that a combination of other wall materials and other modifiers (as oxygen scavengers, antioxidants, chelating agents, and surfactants) increases the encapsulation efficiencies.

Table 8 [131, 328–342] summarizes some selected reviews on anthocyanin, polyphenol and bioactive compounds encapsulation. In last years, the use of biodegradable polymeric nanoparticles has attracted the interest of researches [337] due to their good biocompatibility, easy design and preparation, structure variations and interesting biomimetic characters.

Content	References
Anthocyanins	
Overview of the most recent studies and patents aimed at enhancing anthocyanin stability in food systems	[328]
Anthocyanin extraction, microencapsulation and release properties during in vitro digestion	[329]
Study on colour stability and microencapsulation from Jamun of anthrocyanin pigment using spray-drying	[330]
Health benefits of anthocyanins and their extraction, characterization, encapsulation and delivery	[331]
Encapsulation of anthocyanins from berry-type fruit species as a technology for improving the stability and/or bioavailability of anthocyanins	[332]
Microencapsulation of anthocyanins with different biopolymers through spray-drying	[333]
Nonthermal stabilization mechanisms of anthocyanins in model and food systems	[334]
Stabilization of cranberry anthocyanins in nutraceutical capsules	[335]
Polyphenols	
Relevant recent studies on biopolymer nanoparticles and natural nanocarriers for nanoencapsulation of phenolic compounds	[338]
The encapsulation methods in plants using protein matrices	[339]
Phenolic-enriched foods: sources and processing for enhanced health benefits	[131]
Using nanoparticles to enhance absorption and bioavailability of phenolic phytochemicals	[340]
Overview on encapsulation of natural polyphenolic compounds	[341]
Overview of encapsulation of widely used polyphenols: effectiveness, variations, developments and trends	[336]
Bioactive substances	
Development of food applications containing micro-encapsulated coffee antioxidants	[342]
Encapsulation of active compounds used in food products by drying technology	[337]

Table 8. Selected reviews on encapsulations of anthocyanins, polyphenols and bioactive substances.

10. Final comments

Anthocyanins [13, 14, 17, 293, 343–346] are members of the flavonoid group of phytochemicals, a group predominant in fruits and vegetables, especially in berries. Recent research raised awareness of the importance [347, 348] of anthocyanins in the diet. Anthocyanin identification is critical in adulteration and profiling [349, 350] studies and in evaluating the quality of crude and processed food. The design of plant products with a high added value allows increasing the synthesis [351] of plant-derived food antioxidants and in particular anthocyanins. In an effort to expand the palette of natural organic colourants (colour additives of food and beverage products), the food industry has launched a search for new products, for example blue colourants [352, 353]. Food, pharmaceutical and nutraceutical industries are interested in [354] clean recovery of valuable compounds. Thus, exploration of more efficient, cost-effective and eco-friendly techniques of polyphenol extraction, that is anthocyanins, from food matrices and waste plant food processing residues (grape fruit, fruits by-products, winery waste materials, by-products) is a challenge [355–360]. In any case, in order to ascertain the nutraceutical potential of bioactive compounds, quantification [359, 361] is required, thus obtaining vital information for future food industrial applications.

Apart from their well-known potential for their practical applications as natural colourants [13, 48, 49, 58, 76, 79, 281, 328], anthocyanins show antioxidant activity and a wide variety of health-promoting properties for human health [12, 56, 81, 85, 90, 111, 112, 120, 130, 264, 343], ranging from cytoprotective, antimicrobial and antitumor activities to neuroprotective, anti-obesity and lipidomic potential. Moreover, epidemiological evidence suggests [12, 111, 112, 362] a direct correlation between anthocyanin intake and a lower incidence of chronic and degenerative diseases.

However, the issue of food antioxidants although important is a controversial topic [11, 64, 72, 363–365]. The plethora of published studies on mechanisms [132] that may mediate therapeutic or chemical chemopreventive effects of dietary constituents contrasts sharply with a scarcity of information on their pharmaceutical and clinical-pharmacological properties. Most of the evidence supporting a therapeutic effect of anthocyanins is in vitro or mechanical in nature, although the number of studies on bioavailability in humans has increased significantly over the past two decades. Anthocyanins show a complex biochemical (more than other compounds of flavonoids type), and there is still much to discover [94, 95, 366] about the biochemical activity and clinical pharmacology of these compounds (stability, bioavailability and formulation of dietary constituents), which constitutes an obstacle [367] to understand their health benefits. As evidence of their therapeutic effects accumulates, it is important to understand the nature [81, 85, 87, 89, 139] of the absorption and metabolism "in vivo" and that such knowledge will enable the development of new food products, both fresh and manufactured with greater therapeutic efficacy [95, 366]. Progress in this field requires a multidisciplinary research carried out by a wide range of professionals: food science and technology scientists, chemists (analytical chemists), nutritionist, physiologists, pharmacist, pharmacologists, engineers, physicians, biologists, genetics, clinics, etc., being a field in which promising progress will be undoubtedly made in the future.

More complete details of the basics of polyphenols and anthocyanins can be seen in previous reviews [4–7, 346, 368] by the authors.

Author details

Julia Martín, Eugenia Marta Kuskoski, María José Navas and Agustín G. Asuero*

*Address all correspondence to: asuero@us.es

Department of Analytical Chemistry, Faculty of Pharmacy, The University of Seville, Seville, Spain

References

[1] Harborne JB, Williams CA. Anthocyanins and other flavonoids. Nat. Prod. Rep. 2001;**18**(3):310-333.

[2] Veitch NC, Grayer RJ. Flavonoids and their glycosides including anthocyanins. Nat. Prod. Rep. 2008;**25**(3):555-611.

[3] Ziyatdinova GK, Budnikov HC. Natural phenolic antioxidants in bioanalytical chemistry: State of the art and prospects of development. Russ. Chem. Rev. 2015;**84**(2):194-224.

[4] Martín Bueno J, Ramos-Escudero F, Sáez-Plaza P, Muñoz AM, Navas MJ, Asuero AG. Analysis and antioxidant capacity of anthocyanin pigments. Part I: General considerations concerning polyphenols and flavonoids. Crit. Rev. Anal. Chem. 2012;**42**(2):102-125.

[5] Martín Bueno J, Sáez-Plaza P, Ramos-Escudero F, Jímenez AM, Fett R, Asuero AG. Analysis and antioxidant capacity of anthocyanin pigments. Part II: Chemical structure, color, and intake of anthocyanins. Crit. Rev. Anal. Chem. 2012;**42**(2):126-151.

[6] Navas MJ, Jiménez-Moreno AM, Martín Bueno J, Sáez-Plaza P, Asuero AG. Analysis and antioxidant capacity of anthocyanin pigments. Part III: An introduction to sample preparation and extraction. Crit. Rev. Anal. Chem. 2012;**42**(4):284-312.

[7] Navas MJ, Jiménez-Moreno AM, Martín Bueno J, Sáez-Plaza P, Asuero AG. Analysis and antioxidant capacity of anthocyanin pigments. Part IV: Extraction of anthocyanins. Crit. Rev. Anal. Chem. 2012;**42**(4):313-342.

[8] Manetas Y. Why some leaves are anthocyanic and why most anthocyanic leaves are red? Flora 2006;**201**(3):163-177.

[9] Grotewold E. The genetics and biochemistry of floral pigments. Annu. Rev. Plant Biol. 2006;**57**:761-780.

[10] Tirzitis G, Bartosz G. Determination of antiradical and antioxidant activity: Basic principles and new insights. Acta Biochim. Pol. 2010;**57**(1), 139-142.

[11] Martins N, Barros L, Ferreira ICFR. In vivo antioxidant activity of phenolic compounds: facts and gaps. Trends Food Sci. Technol. 2016;**28**:1-12.

[12] Fraga CG. (ed.) Plant Phenolics and Human Health: Biochemistry, Nutrition and Pharmacology. New York: Wiley, 2010.

[13] He J, Giusti MM. Anthocyanins: natural colorants with health-promoting properties. Annu. Rev. Food Sci. Technol. 2010;**1**(1):163-187.

[14] Andersen OM, Jordheim M. The anthocyanins. In: Andersen OM, Markham KR (eds) Flavonoids: Chemistry, Biochemistry, and Applications. Boca Raton, FL: Taylor and Francis, 2006. pp. 471-551.

[15] Santos Buelga C, Mateus N, de Freitas V. Anthocyanins. Plant pigments and beyond. J. Agric. Food Chem. 2014;**62**(29):6879-6884.

[16] Fang J. Classification of fruits based on anthocyanin types and relevance to their health effects. Nutrition 2015;**31**(11):1301-1306.

[17] Wallace TC, Giusti MM. Anthocyanins. Adv. Nutr. 2015;**6**:620-622.

[18] Saura-Calixto F, Goñi I. Antioxidant capacity of the Spanish Mediterranean diet. Food Chem. 2006;**94**(3):442-447.

[19] Scalbert A, Manach C, Morand C, Rémésy C, Jiménez L. Dietary polyphenols and the prevention of diseases. Crit. Rev. Food Sci. Nutr. 2005;**45**(4):287-306.

[20] Ali HM, Almagribi W, Al-Rashidi MN. Antiradical and reductant activities of anthocyanidins and anthocyanins, structure-activity relationship and synthesis. Food Chem. 2016;**194**:1275-1282.

[21] Guo X, Yang B, Tan J, Jiang J, Li D. Association of dietary intakes of anthocyanins and berry fruits with risk of type 2 diabetes mellitus: a systematic review and meta-analysis of prospective cohort studies. Eur. J. Clin. Nutr. 2016;**70**(12):1360-1367.

[22] Joseph SV, Edirisinghe I, Burton-Freeman BM. Fruit polyphenols: a review of antiinflammatory effects in humans. Crit. Rev. Food Sci. Nutr. 2016;**56**(3):419-444.

[23] Jacob JK, Tiwari K, Correa-Betanzo J, Misran A, Chandrasekaran R, Paliyath G. Biochemical basis for functional ingredient design from fruits. Annu. Rev. Food Sci. Technol. 2012;**3**:79-104.

[24] Masisi K, Beta T, Moghadasian MH. Antioxidant properties of diverse grain cereals: a review on in vitro and in vivo studies. Food Chem. 2016;**196**:90-97.

[25] Shahidi F, Alasalvar C. Handbook of Functional Beverages and Human Health. Boca Raton, FL: CRC Press, 2016.

[26] Aguilera Y, Martin-Cabrejas MA, de Mejia EG. Phenolic compounds in fruits and beverages consumed as part of the Mediterranean diet: their role in prevention of chronic diseases. Phytochem. Rev. 2016;**15**(3):405-423.

[27] Zhang H, Tsao R. Dietary polyphenols, oxidative stress and antioxidant and anti-inflammatory effects. Curr. Opin. Food Sci. 2016;**8**:33-42.

[28] Cao G, Booth SL, Sadowski JA, Prior RL. Increases in human plasma antioxidant capacity following consumption of controlled diets high in fruits and vegetables. Am. J. Clin. Nutr. 1998;**68**(5):1081-1087.

[29] Wang H, Cao G, Prior RL. Total antioxidant capacity of fruits. J. Agric. Food Chem. 1996;**44**(3):701-705.

[30] Renaud S, Lorgeril M. Wine, alcohol, platelets, and the French paradox for coronary heart disease. Lancet 1992;**339**(8808):1523-1526.

[31] Lippi G, Franchini M, Guidi GC. Red wine and cardiovascular health: the "French Paradox" revisited. Int. J. Wine Res. 2010;**2**:1-7.

[32] Cerletti C, de Curtis A, Bracone F, Digesu C, Morganti AG, Iacoviello L, de Gaetano G, Donati MB. Diethary anthocyanins and health: data from FLORA and ATHENA EU Projects. Br. J. Clin. Nutr. 2017;**83**(1):103-106.

[33] Bonaccio M, Pounish G, Cerletti C, Donati MB, Iacoviello L, de Gaetano G. Mediterranean diet, dietary polyphenols and low grade inflammation: results from the MOLI-SANI study. Br. J. Clin. Pharmacol. 2017;**83**(1):107-113.

[34] Giacosa A, Barale R, Bavaresco L, Faliva MA, Gerbi V, LaVecchia C, Negri E, Opizzi A, Perna S, Pezzotti M, Rondanelli M. Mediterranean way of drinking and longevity. Crit. Rev. Food Sci. Nutr. 2016;**56**(4):635-640.

[35] Delgado-Lista J, Perez-Martinez P, Garcia-Rios A, Perez-Caballero AI, Perez-Jimenez F, Lopez-Miranda J. Mediterranean diet and cardiovascular risk: beyond traditional risk factors. Crit. Rev. Food Sci. Nutr. 2016;**56**(5):788-801.

[36] Guidi L, Penella C, Landi M. Anthocyanins in Mediterranean diet: common and innovative sources. In: Warner LM (ed) Handbook of Anthocyanins, Chapter 1. New York: Nova Science Publishers, 2015. pp. 1-49.

[37] Chiou A, Panagopoulos EA, Karathanos VT. Anthocyanins and other flavonoids in dried fruits of the Mediterranean area. In: Warner LM (ed) Handbook of Anthocyanins, Chapter 15. New York: Nova Science Publishers, 2015. pp. 395-419.

[38] Grosso G, Mistretta A, Frigiola A, Giuttadauria S, Biondi A, Basile F, Vitaglione P, D'Orazio N, Galvano F. Mediterranean diet and cardiovascular risk factors: a systematic review. Crit. Rev. Food Sci. Nutr. 2014;**54**(5):593-610.

[39] Tomé-Carneiro J, Visioli F. Polyphenol-based nutraceuticals for the prevention and treatment of cardiovascular disease: review of human evidence. Phytomed. 2016;**23**(11):1145-1174.

[40] Cassidy A, Bertoia M, Chiuve S, Flint A, Forman J, Rimm EB. Habitual intake of anthocyanins and flavanones and risk of cardiovascular disease in men. Am. J. Clin. Nutr. 2016;**104**(3):587-594.

[41] Liobikas J, Skemiene K, Trumbeckaite S, Borutaite V. Anthocyanins in cardioprotection: a path through mitochondria. Pharmacol. Res. 2016;**113**B:808-815.

[42] Guilford JM, Pezzuto JM. Wine and health: a review. Am. J. Enol. Vit. 2011;**62**(4):471-486.

[43] Rifler JO, Lorcerie F, Durand P, Delmas D, Ragot K, Limagne E, Mazue F, Riedinger JM, d'Athis P, Hudelot B, Prost M, Lizard G, Latrufle N. A moderate red wine intake improves blood lipid parameters and erythrocytes membrane fluidity in post myocardial infarct patients. Mol. Nutr. Food Res. 2012;**56**(2):345-351.

[44] Yoo YJ, Saliba AJ, Prenzler PD. Should red wine be considered as a functional food? Compr. Rev. Food Sci. Food Saf. 2010;**9**(5):530-551.

[45] Garcia-Alonso M, Minihane A-M, Rimbach G, Rivas-Gonzalo JC, de Pascual-Teresa S. Red wine anthocyanins are rapidly absorbed in humans and affect monocyte chemoattractant protein 1 levels and antioxidant capacity of plasma. J. Nutr. Biochem. 2009;**20**(7):521-529.

[46] Higgins LM, Llanos E. A healthy indulgence? Wine consumers and the health benefits of wine. Wine Econ. Policy 2015;**4**(1):3-11.

[47] Kutlesa Z, Budimir MD. Wine and bone health: a review. J. Bone Mineral Metab. 2016;**34**(1):11-22.

[48] Wrolstad RE, Culver CA. Alternatives to those artificial FD&C food colorants. Annu. Rev. Food Sci. Technol. 2012;**3**:59-77.

[49] Martin N, Roriz CL, Morales P, Barros L, Ferreira, IC. Food colorants: challenges, opportunities and current desires of agroindustries to ensure consumer expectation and regulatory practices. Trends Food Sci. Technol. 2016;**52**:1-15.

[50] Eghbaliferiz S, Iranshahi M. Prooxidant activity of polyphenols, flavonoids, anthocyanins and carotenoids: updated review of mechanisms and catalysing metals. Phytother. Res. 2016;**30**(9):1379-1391.

[51] Kuskoski EM, Asuero AG, Troncoso AM, Garcia-Parrilla MC, Fett R. Actividad antioxidante de pigmentos antociánicos. Food Sci. Technol. (Campinas) 2004;**24**:691-693.

[52] Wang W, Goodman MT. Antioxidant property of dietary agents in a human LDL-oxidation in vivo model: interaction of protein binding activity. Nutr. Res. 1999;**19**(2):191-202.

[53] Castro-Acosta ML, Lenihan-Geels GN, Corpe CP, Hall WL. Berries and anthocyanins: promising functional food ingredients with postprandial glycaemia-lowering effects. Proc. Nutr. Soc. 2016;**75**(3):342-355.

[54] Lee J. Rosaceae products: anthocyanins quality and comparisons between dietary supplements and foods. NFS J. 2016;**4**:1-8.

[55] Cheynier V, Gomez C, Ageorges A. Flavonoids: anthocyanins. In: Nollet LML, Toldrá F (eds) Handbook of Analysis of Active Compounds in Functional Foods. Boca Raton, FL: CRC Press, 2012. pp. 379-404.

[56] Tsuda T. Anthocyanins as functional food factors: chemistry, nutrition and health promotion. Food Sci. Technol. Res. 2012;**18**(3):315-324.

[57] Shipp J, Abdel-Aal ESM. Food applications and physiological effects of anthocyanins as functional food ingredients. Open Food Sci. J. 2010;**4**(2):7-22.

[58] Motohashi N, Sakagami H. Anthocyanins as functional food colors. Top. Heterocycl. Chem. 2009;**16**:1-40.

[59] Andersen OM, Markham KR (eds) Flavonoids: Chemistry, Biochemistry and Application. Boca Raton, FL: CRC Press, 2005.

[60] Salvayre R, Negre-Salvayre A, Camaré C. Oxidative theory of artherosclerosis and antioxidants. Biochimie 2016;**125**:281-286.

[61] Choe E, Min DB. Mechanisms of antioxidants in the oxidation of foods. Compr. Rev. Food Sci. Food Saf. 2009;**8**(4):345-358.

[62] Johnson DR, Decker EA. The role of oxygen in lipid oxidation reactions: a review. Annu. Rev. Food Sci. Technol. 2015;**6**:171-190.

[63] Urquiza-Martinez MV, Navarro BF. Antioxidant capacity of food. Free Rad. Antiox. 2016;**6**(1):1-12.

[64] Liu Z-Q. Antioxidants may not always be beneficial to health. Nutrition 2014;**30**(1):131-133.

[65] Zanotti I, Dall'Asta M, Mena P, Mele L, Bruni R, Ray S, del Rio D. Artheroprotective effect of (poly)phenols: a focus on cell cholesterol metabolism. Food Funct. 2015;**6**(1):13-31.

[66] Kapourchali FR, Surendiran G, Goulet A, Moghadasian MH. The role of dietary cholesterol in lipoprotein metabolism and related metabolic abnormalities: a mini-review. Crit. Rev. Food Sci. Nutr. 2016;**56**(14):2408-2415.

[67] Yin H, Xu L, Porter NA. Free radical lipid peroxidation: mechanisms and analysis. Chem. Rev. 2011;**111**(10):5944-5972.

[68] Koyama T, Watanabe H, Ito H. The association of circulating inflammatory and oxidative stress biomarker levels with diagonal earlobe crease in patients with atherosclerotic diseases. J. Cardiol. 2016;**67**(4):347-351.

[69] Pisoschi AM, Pop A. The role of antioxidants in the chemistry of oxidative stress: a review. Eur. J. Med. Chem. 2015;**97**:55-74.

[70] Kaur R, Kaur J, Mahajan J, Kumar R, Aora S. Oxidative stress-implications, source and its prevention. Environ. Sci. Pollut. Res. 2014;**21**(3):1599-1613.

[71] Halliwell B. Biochemistry of oxidative stress. Biochem. Soc. Trans. 2007;**35**(5):1147-1150.

[72] Halliwell B. The antioxidant paradox: Less paradoxical now? Br. J. Clin. Pharmacol. 2013;**75**(3)637-644.

[73] Altiparmak IH, Erkus ME, Gunebakmaz O. Oxidative stress is associated with not only coronary artery disease on statin therapy but also diabetes mellitus and hypertension. Indian Heart J. 2016;**68**(2):194-195.

[74] Barroso MF, de-los-Santos-Álvarez N, Delerue-Matos C, Oliveira MBPP. Towards a reliable technology for antioxidant capacity and oxidative damage evaluation: electrochemical (bio)sensors. Biosens. Bioelectron. 2011;**30**(1):1-12.

[75] Vitale G, Salvioli S, Franceschi C. Oxidative stress and the ageing endocrine system. Nat. Rev. Endocrinol. 2013;**9**(4):228-240.

[76] Mateus N, de Freitas V. Anthocyanins as food colorants, in anthocyanins, biosynthesis, functions and applications. In: Gould K, Davies K, Winifield C (eds), Chapter 9. New York: Springer, 2008. pp. 284-304.

[77] Jiang J, Xiong YL. Natural antioxidants as food and feed additives to promote health benefits and quality of meat products: a review. Meat Sci. 2016;**120**:107-117.

[78] Botelho G, Canas S, Lameiras J. Development of phenolic compounds encapsulation techniques as a major challenge for food industry and for health and nutrition fields. In: Grumezescu A (ed) Nutrient Delivery. Cambridge: Academic Press, 2017. pp. 535-586.

[79] Chung C, Rojanasasithara T, Mutilangi W, McClements DJ. Stability improvement of natural food colours: impact of amino acid and peptide addition of anthocyanin stability in model beverages. Food Chem. 2017;**218**:277-184.

[80] Xiaonan S. Impact of Food Processing on Anthocyanins. PhD Thesis, National University of Singapore. Singapore: Springer, 2015.

[81] Lila M, Burton-Freeman B, Grace M, Kalt W. Unraveling anthocyanin bioavailability for human health. Annu. Rev. Food Sci. Technol. 2016;**7**:375-393.

[82] Ludwig IA, Mena P, Calani L, Borges G, Pereira-Caro G, Bresciani L, del Rio D, Lean MEJ, Crozier A. New insights into the bioavailability of red raspberry anthocyanins and ellagitannins. Free Rad. Biol. Med. 2015;**89**:758-769.

[83] Fernandes I, Faria A, de Freitas V, Calhau C, Mateus, N. Multiple-approach studies to assess anthocyanin bioavailability. Phytochem. Rev. 2015;**14**(6):899-919.

[84] Fang J. Bioavailability of anthocyanins. Drug Metab. Rev. 2014;**46**(4):508-520.

[85] Pojer E, Mattivi F, Johnson D, Stockely S. The case for anthocyanin consumption to promote human health: a review. Compr. Rev. Food Sci. Food Saf. 2013;**12**(5):483-508.

[86] Fernandes I, Faria A, Calhau C, de Freitas V, Mateus N. Bioavailability of anthocyanins and derivatives. J. Funct. Foods 2014;**7**:54-66.

[87] Hribar U, Ulrih P. The metabolism of anthocyanins. Curr. Drug Metab. 2014;**15**(1):3-13.

[88] Carbonell-Capella JM, Buniowska M, Barba FJ, Esteve MJ, Frígola A. Analytical methods for determining bioavailability and bioaccessibility of bioactive compounds from fruits and vegetables: a review. Compr. Rev. Food Sci. Food Saf. 2014;**13**(2):155-171.

[89] Speciale A, Cimino F, Saija A, Canali R, Virgili F, Bioavailability and molecular activities of anthocyanins as modulators of endothelial function. Genes Nutr. 2014;**9**(4):404.

[90] Fernandes I, de Freitas V, Mateus, N. Anthocyanins and human health: how gastric absorption may influence acute human physiology. Nutr. Aging 2014;**2**(1):1-14.

[91] Faria A, Fernandes I, Mateus N, Calhau C. Bioavailability of anthocyanins. In: Raawat KG, Mérillon JM (eds) Natural Products. Berlin: Springer-Verlag, 2013. pp. 2465-2487.

[92] Novotny JA. Anthocyanin bioavailability: past progress and current challenges. In: Emerging Trends in Dietary Components for Preventing and Combating Disease, Chapter 32, ACS Symposium Series, 2012. p. 559-568.

[93] Del Rio D, Borges G, Crozier A. Berry flanonoids and phenolics: bioavailability and evidence or protective effects. Br. J. Nutr. 2010;**104**(S3):S67–S90.

[94] Yang M, Koo S, Song W, Chun O. Food matrix affecting anthocyanin bioavailability: review. Curr. Med. Chem. 2011;**18**(2):291-300.

[95] McGhie TK, Walton MC. The bioavailability and absorption of anthocyanins: towards a better understanding. Mol. Nutr. Food Res. 2007;**51**(6):702-713.

[96] Xie LY, Lee SG, Vance TM, Wang Y, Kim B, Lee JY, Chun OK, Bolling BW. Bioavailability of anthocyanins and colonic polyphenol metabolites following consumption of aronia berry extract. Food Chem. 2016;**211**:860-868.

[97] Sandhu AK, Huang YC, Xiao D, Park E, Edirisinghe I, Burton-Freeman B. Pharmacokinetic characterization and bioavailability of strawberry anthocyanins relative to meal intake. J. Agric. Food Chem. 2016;**64**(24):4891-4899.

[98] Oliveira H, Wu N, Zhang Q, Wang JY, Oliveira J, de Freitas V, Mateus N, He JR, Fernandes I. Bioavailability studies and anticancer properties of malvidin based anthocyanins, pyranoanthocyanins and non-oxonium derivatives. Food Funct. 2016;**7**(5):2462-2468.

[99] Wiczkowski W, Szawara-Nowak D, Romaszko J. The impact of red cabbage fermentation on bioavailability of anthocyanins and antioxidant capacity of human plasma. Food Chem. 2016:**190**:730-740.

[100] Kuntz S, Rudloff S, Asseburg H, Borsch C, Frohling B, Unger F, Dold S, Spengler B, Rompp A, Kunz C. Uptake and bioavailability of anthocyanins and phenolic acids from grape/ blueberry juice and smoothie in vitro and in vivo. Br. J. Nutr. 2015;**113**(7):1044-1055.

[101] Kirakosyan A, Seymour EM, Kaufman P, Bolling S. Tissue bioavailability of Tart Cherry anthocyanins. FASEB J. 2015;**29(**1 Supplement):606-610.

[102] Kirakosyan A, Seymour EM, Wolforth J, McNish R, Kaufman P, Bolling S. Tissue bioavailability of anthocyanins from whole tart cherry in healthy rats. Food Chem. 2015;**171**:26-31.

[103] Netzel ME, Fanning K, Russell D, Stanley R, Topp B. Bioactive anthocyanins in 'Queen Garnet' plum—maturity and bioavailability. In: OHare TJ and Netzel ME (eds) XXIX International Horticultural Congress on Horticulture: Sustaining Lives, Livelihoods and Landscapes (Ihc2014): VI International Symposium on Human Health Effects of Fruits and Vegetables (Favhealth 2014), vol 106. Book Series: Acta Hortic, 2015. pp. 219-222.

[104] Westfall A. In Vitro Evaluation of Biological Activity of Anthocyanin Based Lipstick Formulations. 29th Hayes Graduate Research Forum, 2015.

[105] Sakakibara H, Ichikawa Y, Tajima S, Makino Y, Wakasugi Y, Shimoi K, Kobayashi S, Kumazawa S, Goda T. Practical application of flavonoid-poor menu meals to the study of the bioavailability of bilberry anthocyanins in human subjects. Biosci. Biotechnol. Biochem. 2014;**78**(10):1748-1752.

[106] Kamonpatana K, Failla ML, Kumar P, Giusti MM. Anthocyanin structure determines susceptibility to microbial degradation and bioavailability to the buccal mucosa. J. Agric. Food Chem. 2014;**62**(29):6903-6910.

[107] Fang J. Some anthocyanins could be efficiently absorbed across the gastrointestinal mucosa: extensive presystemic metabolism reduces apparent bioavailability. J. Agric. Food Chem. 2014;**62**(18):3904-3911.

[108] Toydemir G, Capanoglu E, Boyacioglu D, Beekwilder J, de Vos RCH, Hall RD. Sour cherry (*Prunus cerasus* L.) anthocyanins: effects of juice processing on phenolic compounds and bioavailability. In: X International Symposium on Vaccinium and Other Superfruits, vol 1017. Book Series: Acta Hortic, 2014. pp. 387-398.

[109] Kay CD. The future of flavonoid research. Br. J. Nutr. 2010;**104**(S3):S91–S95.

[110] Li D, Wang P, Luo Y, Zhao M, Chen F. Health benefits of anthocyanins and molecular mechanisms: update from recent decade. Crit. Rev. Food Sci. Nutr. 2017;**57**(8):1729-1741.

[111] Riaz M, Zia-Ul-Haq M, Saad B. Anthocyanins and Human Health: Biomolecular and Therapeutical Aspects. New York: Springer, 2016.

[112] Wallace TC, Giusti MM. Anthocyanins in Health and Disease. Boca Raton, FL: CRC Press, 2014.

[113] Prior RL, Wu X. Anthocyanins: structural characteristics that result in unique metabolic patterns and biological activities. Free Rad. Res. 2006;**40**(10):1014-1028.

[114] Smeriglio A, Bareca D, Bellocco E, Trombetta D. Chemistry, pharmacology and health benefits of anthocyanins. Phytother. Res. 2016;**30**(8):1256-1286.

[115] Legua P, Forner-Giner MA, Nuncio-Jáuregui N, Hernández F. Polyphenolic compouns, anthocyanins and antioxidant activity of nineteen pomegranate fruits: a rich source of bioactive compounds. J. Funct. Foods 2016;**23**:628-636.

[116] Ahmad A, Kaleem M, Ahmed Z, Shafiq H. Therapeutic potential of flavonoids and their mechanism of action against microbial and viral infections—a review. Food Res. Int. 2015;**77**:221-235.

[117] Cisowska A, Wojnicz D, Hendrich AB. Anthocyanins as antimicrobial agents of natural plant origin. Nat. Prod. Commun. 2011;**6**(1):149-156.

[118] Ghosh D, Konishi T. Anthocyanins and anthocyanin-rich extracts: role in diabetes and eye function. Asian Pac. J. Clin. Nutr. 2007;**16**(2):200-208.

[119] Miyake S, Takahashi N, Sasaki M, Kobayashi S, Tsubota K, Ozawa Y. Vision preservation during retinal inflammation by anthocyanin-rich bilberry extract: cellular and molecular mechanism. Lab. Invest. 2012;**92**(1):102-109.

[120] Smeriglio A, Monteleone D, Trombetta D. Health effects of *Vaccinium myrtillus* L.: evaluation of efficacy and technological strategies for preservation of active ingredients. Mini Rev. Med. Chem. 2014;**14**(7):567-584.

[121] EUROPEAN MEDICINES AGENCY. Assessment report on *Vaccinium myrthillus* L., fructus recens and *Vaccinium myrtillus* L., fructus siccus. Final. EMA/HMPC/55161/2013, Committee on Herbal Medicinal Products (HMPC), 2015.

[122] Yang Y, Shi Z, Reheman A, Jin W, Li C, Zhu G, Wang Y, Freedman JJ, Ling W, Ni H. Anthocyanins inhibit platelet activation and attenuate thrombus growth in both human ad murine thrombosis models. Blood 2010;**116**(1):3197-3197.

[123] Liu LK, Lee HJ, Shih YW, Chyau CC, Wang CJ. Mulberry anthocyanins extracts inhibit LDL oxidation and macrophage-derived foam cell formation induced by oxidative LDL. J. Food Sci. 2008;**73**(6):4113-4121.

[124] Kao ES, Tseng TH, Lee HJ, Chan KC, Wang CJ. Anthocyanins extracted from Hibiscus attenuate oxidized LDL-mediated foam fell formation involving regulation of CD36 gene. Chem. Biol. Interact. 2009;**179**(2-3):212-218.

[125] Astadi IR, Astuti M, Santoso U, Nugraheni PS. In vitro antioxidant activity of anthocyanins of black soybean seed coat in human low density lipoprotein (LDL). Food Chem. 2009;**112**(3):659-663.

[126] Bone K, Mills S. Principles and Practice of Phytotherapy: Modern Herbal Medicine. Edinburg: Churchill Livingstone, 2013.

[127] Ksonzékova P, Mariychuk R, Eliasova A, Mudronova D, Csank T, Kiraly J Marcincakova D, Pistl J, Tkacikova L. In vitro studies of biological activities of anthocyanin-rich berry extracts on porcine intestinal epithelial cells. J. Sci. Food Agric. 2016;**96**(4):1093-1100.

[128] Jankowski A, Jankowska B, Niedworok J. The effects of anthocyanin dye from grapes on experimental diabetes. Folia Med. Cracov. 2000;**41**(3-4):5-15.

[129] Semaming Y, Pannengpetch P, Chattipakorn SC, Chattipakorn N. Pharmacological properties of protocatechuic acid and its potential roles as complementary medicine. J. Evid. Based Complement. Alternat. Med. 2015;**2015**:11 (ArticleID593902).

[130] McDougall GJ. Phenolic-enriched foods: sources and processing for enhanced health benefits. In: Proceedings of the Nutrition Society, 2016. p. 9.

[131] Vendrame S, Klimis-Zacas D. Anti-inflammatory effect of anthocyanins via modulation of nuclear factor-kB and mitogen-activated protein kinase signaling cascades. Nutr. Rev. 2015;**76**(3):348-358.

[132] Choi TH, Chang KC, Shin SC, Chung JII, Kim HJ, Kim JS. Pharmaceutical composition for wound heating containing anthocyanin extracted from the black soybean seed coat, WO2008126980A1, PCT/KR2008/000513. Industry-Academic Cooperation Foundation Gyeongsang National University, 2008.

[133] Lin B-W, Gong C-C, Song H-F, Cui YY. Effects of anthocyanins on the prevention and treatment of cancer. Br. J. Pharmacol. 2016. doi:10.1111/6ph.13627

[134] Thomasset S, Teller N, Cai H, Marko D, Berry DP, Steward WP, Gescher AJ. Do anthocyanins and anthocyanidins, cancer chemopreventive pigments in the diet, merit development as potential drugs. Cancer Chemother. Pharmacol. 2009;**64**(1):201-211.

[135] Galvano F, Salamone F, Nicolosi A, Vitaglione P. Anthocyanins-based drugs for colon cancer treatment: the nutriotionist's point of view. Cancer Chemother. Pharmacol. 2009;**64**(2):431-432.

[136] Webb MR, Min K, Ebeler SE. Anthocyanins interactions with DNA: intercalation, topoisomerase I inhibition and oxidative reactions. J. Food Biochem. 2008;**32**(5):176-196.

[137] Kong J-M, Chia L-S, Goh N-K, Chia T-F, Brouillard R. Analysis and biological activities of anthocyanins. Phytochem. 2003;**64**(5):923-933.

[138] Celli GB, Ghanem A, Brooks MSL. A theoretical physiologically-based pharmacokinetic approach for modeling the fate of anthocyanins in vivo. Crit. Rev. Food Sci. Nutr. 2016. doi:10.1080/10408398.2015.1104290

[139] Kay CD. Aspects of anthocyanin absorption, metabolism and pharmacokinetics in humans. Nutr. Res. Rev. 2006;**19**(1):137-146.

[140] Hornedo-Ortega R, Alvarez-Fernandez MA, Cerezo AB, Richard T, Troncoso AM, Garcia-Parrilla MC. Protocatechuic acid: inhibition of fibril formation, destabilization of preformed fibrils of amyloid-beta and alpha-synuclein, and neuroprotection. J. Agric. Food Chem. 2016;**64**(41):7722-7732.

[141] Guttenplan JB, Chen KM, Sun YW, Kosinska W, Zhou Y, Kim S, Sung Y, Gowda K, Amin S, Stoner GD, El-Bayoumy K. Effects of black raspberry extract and protocatechuic acid on carcinogen-DNA adducts and mutagenesis, and oxidative stress in rat and human oral cells. Cancer Prev. Res. 2016;**9**(8):704-712.

[142] Marques C, Fernandes I, Norberto S, Sá C, Teixeira D, Freitas V, Mateus N, Calhau Faria A. Pharmacokinetics of blackberry anthocyanins consumed with or without ethanol: a randomized and crossover trial. Mol. Nutr. Food Res. 2016;**60**(11):2319-2330.

[143] Motilva MJ, Macia A, Romero MP, Rubio L, Mercader M, Gonzalez-Ferrero C. Human bioavailability and metabolism of phenolic compounds from red wine enriched with free or nano-encapsulated phenolic extract. J. Funct. Foods 2016;**25**:80-93.

[144] Fornasaro S, Ziberna L, Gasperotti M, Tramer F, Vrhovsek U, Mattivi F, Passamonti S. Determination of cyanidin 3-glucoside in rat brain, liver and kidneys by UPLC/MS-MS and its application to a short-term pharmacokinetic study. Sci. Rep. 2016;**6**:22815.

[145] Chen TY, Kritchevsky J, Hargett K, Feller K, Klobusnik R, Song BJ, Cooper B, Jouni Z, Ferruzzi MG, Janle EM. Plasma bioavailability and regional brain distribution of polyphenols from apple/grape seed and bilberry extracts in a young swine model. Mol. Nutr. Food Res. 2015;**59**(12):2432-2447.

[146] Bhaswant M, Fanning K, Netzel M, Mathai M, Panchal SK, Brown L. Cyanidin 3-glucoside improves diet-induced metabolic syndrome in rats. Pharmacol. Res. 2015;**102**:208-217.

[147] Riha J, Brenner S, Srovnalova A, Klameth L, Dvorak Z, Jager W, Thalhammer T. Effects of anthocyans on the expression of organic anion transporting polypeptides (SLCOs/OATPs) in primary human Hepatocytes. Food Funct. 2015;**6**(3):772-779.

[148] Mena P, Dominguez-Perles R, Girones-Vilaplana A, Baenas N, Garcia-Viguera C, Villano D. Flavan-3-ols, anthocyanins, and inflammation. IUBMB Life 2014;**66**(11):745-758.

[149] Seymour EM, Warber SM, Kirakosyan A, Noon KR, Gillespie B, Uhley VE, Wunder J, Urcuyo DE, Kaufman PB, Bolling SF. Anthocyanin pharmacokinetics and dose-dependent plasma antioxidant pharmacodynamics following whole tart cherry intake in healthy humans. J. Funct. Foods 2014;**11**:509-516.

[150] de Ferrars RM, Czank C, Zhang Q, Botting NP, Kroon PA, Cassidy A, Kay CD. The pharmacokinetics of anthocyanins and their metabolites in humans. Br. J. Pharmacol. 2014;**171**(13):3268-3282.

[151] Kalt W, Liu Y, McDonald JE, Vinqvist-Tymchuk MR, Fillmore SAE. Anthocyanin metabolites are abundant and persistent in human urine. J. Agric. Food Chem. 2014;**62**(18): 3926-3934.

[152] Kay CD. Rethinking paradigms for studying mechanisms of action of plant bioactives. Nutr. Bull. 2015;**40**(4):335-339.

[153] Rothwell JA, Urpi-Sarda M, Boto-Ordonez M, Llorach R, Farran-Codina A, Barupal DK, Neveu V, Manach C, Andres-Lacueva C. Systematic analysis of the polyphenol metabolome using the Phenol-Explorer database. Mol. Nutr. Food Res. 2001;**60**(1):203-211.

[154] Apak R, Özyürek M, Güçlü K, Çapanoglu E. Antioxidant activity/capacity measurement. 1. Classification, physicochemical principles, mechanisms, and electron transfer (ET)-based assays. J. Agric. Food Chem. 2016;**64**(5):997-1027.

[155] Apak R, Özyürek M, Güçlü K, Çapanoglu E. Antioxidant activity/capacity measurement. 2. Hydrogen atom transfer (HAT)-based, mixed-mode (electron transfer (ET)/HAT), and lipid peroxidation assays. J. Agric. Food Chem. 2016;**64**(5):1046-1070.

[156] Apak R, Özyürek M, Güçlü K, Çapanoglu E. Antioxidant activity/capacity measurement. 3. Reactive oxygen and nitrogen species (ROS/RNS) scavenging assays, oxidative stress biomarkers, and chromatographic/chemometric assays. J. Agric. Food Chem. 2016;**64**(5):1028-1045.

[157] Bunaciu AA, Danet AF, Fleschin S, Aboul-Enein HY. Recent applications for in vitro antioxidant activity assay. Crit. Rev. Anal. Chem. 2016;**46**(5):389-399.

[158] Majer P, Vidovic M, Czegeny G, Jovanovic SV, Strid A, Hideg E. Evaluation of procedures for assessing anti- and pro-oxidants in plant samples. Anal. Methods 2016;**8**(28):5569-5580.

[159] Niki E. Antioxidant capacity of foods for scavenging reactive oxidants and inhibition of plasma lipid oxidation induced by multiple oxidants. Food Funct. 2016;**7**(5):2156-2168.

[160] Pisoschi AM, Pop A, Cimpeanu C, Predoi G. Antioxidant capacity determination in plants and plant-derived products: a review. Oxid. Med. Cell. Longe. 2016;**2016**:1-36.

[161] Amorati R, Valgimigli L. Advantages and limitations of common testing methods for antioxidants. Free Rad. Res. 2015;**49**(5):633-649.

[162] Held P. An Introduction to Reactive Oxygen Species Measurement of ROS in Cells. Winooski, VT, USA: BioTek Instruments Inc., Application Guide. 2010.

[163] Li B, Pratt DA. Methods for determining the efficacy of radical-trapping antioxidants. Free Rad. Biol. Med. 2015;**82**:187-202.

[164] Shahidi F, Ambigaipalan P. Phenolics and polyphenolics in foods, beverages and spices: antioxidant activity and health effects—a review. J. Funct. Foods 2015;**18**:820-897.

[165] Shahidi F, Zhong Y. Measurement of antioxidant activity. J. Funct. Foods 2015;**18**:757-781.

[166] Vilela D, González MC, Escarpa A. Nanoparticles as analytical tools for in-vitro antioxidant-capacity assessment and beyond. Trends Anal. Chem. 2015;**64**:1-16.

[167] Alam MN, Bristi NJ, Rafiquzzaman M. Review on in vivo and in vitro methods evaluation of antioxidant activity. Saudi Pharm. J. 2013;**21**(2):143-152.

[168] Apak R, Gorinstein S, Böhm V, Schaich KM, Özyürek M, Güçlü K. Methods of measurement and evaluation of natural antioxidant capacity/activity (IUPAC Technical Report). Pure Appl. Chem. 2013;**85**(5):957-998.

[169] López-Alarcón C, Denicola A. Evaluating the antioxidant capacity of natural products: a review on chemical and cellular-based assays. Anal. Chim. Acta 2013;**763**:1-10.

[170] Craft BD, Kerrihard AL, Amarowicz R, Pegg RB. Phenol-based antioxidants and the in vitro methods used for their assessment. Compr. Rev. Food Sci. Food Saf. 2012;**11**:148-173.

[171] Gülcin I. Antioxidant activity of food constituents: an overview. Arch. Toxicol. 2012;**86**(3):345-391.

[172] Pinchuk I, Shoval H, Dotan Y, Lichtenberg D. Evaluation of antioxidants: scope, limitations and relevance of assays. Chem. Phys. Lipids 2012;**165**(6):638-647.

[173] Pisoschi AM, Negulescu GP. Methods for total antioxidant activity determination: a review. Biochem. Anal. Biochem. 2011;**1**(1):1-11.

[174] Laguerre M, Decker EA, Lecomte J, Villeneuve P. Methods for evaluating the potency and efficacy of antioxidants. Curr. Opin. Clin. Nutr. Metab. Care 2010;**13**(5):518-525.

[175] Liu Z-Q. Chemical methods to evaluate antioxidant ability. Chem. Rev. 2010;**110**(10): 5675-5691.

[176] Niki E. Assessment of antioxidant capacity in vitro and in vivo. Free Rad. Biol. Med. 2010;**49(4)**:503-515.

[177] Gökmen V, Serpen A, Fogliano V. Direct measurement of the total antioxidant capacity of foods: the 'QUENCHER' approach. Trends Food Sci. Technol. 2009;**20**(6):278-288.

[178] Magalhães LM, Santos M, Segundo MA, Reis S, Lima JLFC. Flow injection based methods for fast screening of antioxidant capacity. Talanta 2009;**77**(5):1559-1566.

[179] Magalhães LM, Santos M, Segundo MA, Reis S, Lima JLFC. Automatic flow injection based methodologies for determination of scavenging capacity against biologically relevant reactive species of oxygen and nitrogen. Talanta 2009;**78**(4):1219-1226.

[180] Moon J-K, Shibamoto T. Antioxidant assays for plant and food components. J. Agric. Food Chem. 2009;**57**:1655-1666.

[181] Frankel EN, Finley JW. How to standardize the multiplicity of methods to evaluate natural antioxidants. J. Agric. Food Chem. 2008;**56**(13):4901-4908.

[182] Kusano C, Ferrari B. Total antioxidant capacity: a biomarker in biomedical and nutritional studies. J. Cell. Mol. Biol. 2008;**7**(1):1-15.

[183] Magalhaes LM, Segundo MA, Reis S, Lima JLFC. Methodological aspects about in vitro evaluation of antioxidant properties. Anal. Chim. Acta. 2008;**613**(1):1-19.

[184] Pérez-Jiménez J, Arranz S, Tabernero M, Díaz-Rubio ME, Serrano J, Goñi I, Saura-Calixto F. Updated methodology to determine antioxidant capacity in plant foods, oils and beverages: extraction, measurement and expression of results. Food Res. Int. 2008;**41**(3):274-285.

[185] Somogyi A, Rosta K, Pusztai P, Tulassay Z, Nagy G. Antioxidant measurements. Physiol. Meas. 2007;**28**(4):R41–R55.

[186] MacDonald-Wicks LK, Wood LG, Gar ML. Methodology for the determination of biological antioxidant capacity in vitro: A review. J. Sci. Food Agric. 2006;**86**(13):2046-2056.

[187] Wood LG, Gibson PG, Garg ML. A review of the methodology for assessing in vivo antioxidant capacity. J. Sci. Food Agric. 2006;**86**(13):2057-2066.

[188] Decker EA, Warner K, Richards MP, Shahidi F. Measuring antioxidant effectiveness in food. J. Agric. Food Chem. 2005;**53**:4303-4310.

[189] Huang D, Ou B, Prior RL. The chemistry behind antioxidant capacity assays. J. Agric. Food Chem. 2005;**53**(10):1841-1856.

[190] Kuskoski EM, Asuero AG, Troncoso AM, Mancini-Filho J, Fett R. Aplicación de diversos métodos químicos para determinar actividad antioxidante en pulpa de frutos. Food Sci. Technol. (Campinas). 2005;**25**(4):726-732.

[191] Liu RH, Finley J. Potential cell culture models for antioxidant research. J. Agric. Food Chem. 2005;**53**(10):4311-4314.

[192] Prior RL, Wu X, Schaich K. Standardized methods for the determination of antioxidant capacity and phenolics in foods and dietary supplements. J. Agric. Food Chem. 2005;**53**(10):4290-4302.

[193] Roginsky V, Lissi EA. Review of methods to determine chain-breaking antioxidant activity in food. Food Chem. 2005;**92**(2):235-254.

[194] Antolovich M, Prenzler PD, Patsalides E, McDonald S, Robards K. Methods for testing antioxidant activity. Analyst. 2002;**127**(1):183-198.

[195] Sánchez Moreno C. Review: methods used to evaluate the free radical scavenging activity in foods and biological systems. Food Sci. Technol. Int. 2002;**8**(3):121-137.

[196] Llesuy S, Evelson P, Campos AM, Lissi E. Methodologies for evaluation of total antioxidant activities in complex mixtures. A critical review. Biol. Res. 2001;**34**(2):51-73.

[197] Kuskoski EM, Asuero AG, Troncoso AM, Fett R. Actividad antioxidante de pulpas de frutos tropicales. Aplicación del método ABTS. Alimentaria 2006;**376**:67-70.

[198] Tan JBL, Lim YY. Critical analysis of current methods for assessing the in vitro antioxidant and antibacterial activity of plant extracts. Food Chem. 2015;**172**:814-822.

[199] Schaich KM, Tian X, Xie J. Reprint of "Hurdles and pitfalls in measuring antioxidant efficacy: a critical evaluation of ABTS, DPPH, and ORAC assays. J. Funct. Foods 2015;**14**:111-125.

[200] Magalhaes LM, Barreiros L, Reis S, Segundo MA. Kinetic matching approach applied to ABTS assay for high-throughput determination of total antioxidant capacity of food products. J. Food Compos. Anal. 2014;**33**(2):187-194.

[201] Tian X, Schaich KM. Effects of molecular structure on kinetics and dynamics of the Trolox Equivalent Antioxidant Capacity assay with ABTS^{+*}. J. Agric. Food Chem. 2013;**61**(23):5511-5519.

[202] Barton HJ. A "zero sample concentration approach": standardization of method for the estimation of total antioxidant activity by the use of extrapolation to zero sample concentration. A novel standard. 1. ABTS cation radical scavenging. J. Agric. Food Chem. 2010;**58**(16):8918-8926.

[203] Çelik SE, Özyürek M, Güçlü K, Apak R. Solvent effects on the antioxidant capacity of lipophilic and hydrophilic antioxidants measured by CUPRAC, ABTS/persulphate and FRAP methods. Talanta 2010;**81**(4-5):1300-1309.

[204] Ozgen M, Reese RN, Tulio AZJR, Scheerens J, Miller AR. Modified 2,2-azino-bis-3-ethylbenzothiazoline-6-sulfonic acid (abts) method to measure antioxidant capacity of selected small fruits and comparison to ferric reducing antioxidant power (FRAP) and 2,2′-diphenyl-1-picrylhydrazyl (DPPH) methods. J. Agric. Food Chem. 2006;**54**(4):1151-1157.

[205] Villaño D, Fernandez-Pachon AM, Troncoso AM, Garcia Parrilla MC. The antioxidant activity of wines determined by the ABTS method: influence of sample dilution and time. Talanta 2004;**64**(2):501-509.

[206] Mehdi MM, Rizvi SI. N,N-Dimethyl-p-phenylenediamine dihydrochloride-based method for the measurement of plasma oxidative capacity during human aging. Anal. Biochem. 2013;**436**(2):165-167.

[207] Verde V, Fogliano V, Ritieni A, Maiani G, Morisco F, Caporaso N. Use of N,N-dimethyl-p-phenylenediamine to evaluate the oxidative status of human plasma. Free Rad. Res. 2002;**36**(8):69-873.

[208] Foti MC. Use and abuse of the DPPH(*) radical. J. Agric. Food Chem. 2015;**63**(40):8765-8776.

[209] Fadda A, Serra M, Molinu AMG, Azara E, Barberis A, Sanna D. Reaction time and DPPH concentration influence antioxidant activity and kinetic parameters of bioactive molecules and plant extracts in the reaction with the DPPH radical. J. Food Compos. Anal. 2014;**35**(2):112-119.

[210] Carmona-Jiménez Y, García-Moteno MV, Igartuburu JM, García-Barroso C. Simplification of the DPPH assay for estimating the antioxidant activity of wine by-products. Food Chem. 2014;**165**:198-204.

[211] Shimamura T, Sumikura Y, Yamazaki T, Tada A, Kashiwagi T, Ishikawa H, Matsui T, Sugimoto N, Akyama H, Ukeda H. Applications of the DPPH assay for evaluating the antioxidant capacity of food additives. Interlaboratory evaluation study. Anal. Sci. 2014;**30**(7):717-721.

[212] Pyrzynska K, Pekal A. Application of free radical diphenylpicrylhydrazyl (DPPH) to estimate the antioxidant capacity of food samples. Anal. Methods 2013;**5**(17):4288-4295.

[213] Chen Z, Bertin R, Froldi G. EC_{50} estimation of antioxidant activity in DPPH assay using several statistical programs. Food Chem. 2013;**138**(1):414-420.

[214] Mishra K, Ojha H, Chaudhury NK. Estimation of antiradical properties of antioxidants using DPPH assay. A critical review and results. Food Chem. 2012;**130**(4):1036-1043.

[215] Plank DW, Szpylka J, Sapirstein H, Woollard D, Zapf CM, Lee V, Chen CY, Liu RH, Tsao R, Düsterloh A, Baugh S. Determiation of antioxidant activity in foods and beverages by reaction with 2,2′-diphenyl-1-picrylhydrazyl (DPPH): Collaborative Study First Action 2012.04. J. AOAC Int. 2012;**95**(6):1562-1569.

[216] Dawidowicz AL, Wianowska D, Olszowy H. On practical problems in estimation of antioxidant activity of compounds by DPPH method (problems in estimation of antioxidant activity. Food Chem. 2012;**131**(3):1037-1043.

[217] Xie J, Schaich KM. Re-evaluation of the 2,2-diphenyl-1-picrylhydrazyl free radical (DPPH) assay for antioxidant activity. J. Agric. Food Chem. 2012;**62**(19):4251-4260.

[218] Sharma OP, Bhat TK. DPPH antioxidant assay revisited. Food Chem. 2009;**113**(4):1202-1205.

[219] Kuskoski EM, Asuero AG, Morales MT, Fett R. Frutos tropicais silvestres e polpas de frutas congeladas: atividade antioxidante, polifenóis e antocianinas. Ciência Rural 2006;**36**(4);1283-1287.

[220] Benzie IF, Choi SW. Antioxidants in food content measurement, significance, action, cautions, caveat and research methods. Adv. Food Nutr. 2014;**71**:1-53.

[221] Mellado-Ortega E, Zabalgogeazcoa I, de Aldana BRV, Arellano JB. Solutions to decrease a systematic error related to AAPH addition in the fluorescence-based ORAC assay. Anal. Biochem. 2017;**519**:27-29.

[222] Garzón GA, Manns DC, Riedl K, Schwartz SJ, Padilla-Zakour O. Identification of phenolic compounds in petals of Nasturtium Flowers (*Tropaeolum majus*) by high performance liquid chromatography coupled to mass spectrometry and determination of oxygen radical absorbance capacity (ORAC). J. Agric. Food Chem. 2015;**63**(6):1803-1811.

[223] Prior RL. Oxygen radical absorbance capacity (ORAC): new horizonts in relating dietary antioxidant/bioactives and health benefits. J. Funct. Foods 2015;**18**:797-810.

[224] Dorta E, Fuentes-Lemus E, Aspée A, Atala E, Speisky H, Bridi R, Lissi E, López-Alarcón C. The ORAC (oxygen radical absorbance capacity) index does not reflect the capacity of antioxidants to trap peroxyl radicals. RSC Adv. 2015;**5**(50):39899-39901.

[225] Atala E, Aspée A, Speisky H, Lissi E, López-Alarcón C. Antioxidant capacity of phenolic compounds in acidic medium: a pyrogallol red-based ORAC (oxygen radical absorbance capacity). J. Food Compos. Anal. 2013;**32**(2):116-125.

[226] Ou B, Chang T, Huang D, Prior RL. Determination of total antioxidant capacity by oxygen radical absorbance capacity (ORAC) using fluorescein as the fluorescence probe: first action 2012.23. J. AOAC Int. 2013;**96**(6):1372-1376.

[227] Osakwe ON, Siegel A. A novel standardized oxygen radical absorbance assay for evaluating antioxidant natural products. J. AOAC Int. 2013;**96**(6):1365-1371.

[228] Ortiz R, Antilen M, Speisky H, Aliaga M E, López-Alarcón C, Baugh S. Application of a microplate-based ORAC-Pyrogallol red assay for the estimation of antioxidant capacity: First action 2012.03. J. AOAC Int. 2012;**95**(6):1558-1561.

[229] Stockham K, Paimin R, Orbell JD, Adorno P, Buddhadasa S. Modes of handling oxygen radical absorbance capacity (ORAC) data and reporting values in product labelling. J. Food Compos. Anal. 2011;**24**(4):686-691.

[230] Kevers C, Sipel A, Pincemail J, Dommes J. Antioxidant capacity of hydrophilic food matrices: optimization and validation of ORAC assays. Food Anal. Methods 2014;**7**(2):409-416.

[231] Zulueta A, Esteve MJ, Frígola A. ORAC and TEAC assays comparison to measure the antioxidant capacity of food products. Food Chem. 2009;**114**(1):310-316.

[232] Kohri S, Fujii H, Oowada S, Endoh N, Sueishi Y, Kusakabe M, Shimmei M, Kotake Y. An oxygen radical absorbance capacity-live assay that directly quantifies the antioxidant's scavenging capacity against AAPH-derived free radicals. Anal. Biochem. 2009;**386**(2):167-171.

[233] López-Alarcón C, Lissi E. A novel and simple ORAC methodology based on the interaction of pyrogallol red with peroxyl radicals. Free Rad. Res. 2006;**40**(9):979-985.

[234] Dávalos A, Gómez-Cordovés C, Bartolomé B. Extending applicability of the oxygen radical absorbance capacity (ORAC-fluorescein) assay. J. Agric. Food Chem. 2004;**52**(1):48-54.

[235] Prior RL, Hoang H, Gu L, Wu X, Bacchiocca M, Howard L, Hampsch-Woodill M, Huang D, Ou B. Jacob R. Assays for hydrophilic and lipophilic antioxidant capacity (oxygen radical absorbance capacity ($ORAC_{FL}$)) of plasma and other biological and food samples. J. Agric. Food Chem. 2003;**51**(11):3273-3279.

[236] Ou B, Hampsch-Woodill M, Prior RL. Development and validation of an improved oxygen radical absorbance capacity assay using fluorescein as the fluorescent probe. J. Agric. Food Chem. 2001;**49**(10):4619-4626.

[237] Cardenas A, Gomez M, Frontana C. Electrochemical method to quantify antioxidants employing cupric reducing antioxidant capacity, CUPRAC. Procedia Chem. 2014;**12**:62-65.

[238] Gosmaro F, Bagnati M, Berto S, Bellomo G, Prenesti E. Measurement of total antioxidant capacity of human plasma: Setting and validation of the CUPRAC-BCS method on routine apparatus ADVIA 2400. Talanta 2013;**115**:526-532.

[239] Özyürek M, Guçlü K, Tütem E, Baskan KS, Erçağ E, Celik E, Baki S, Yıldız L, Karaman S, Apak R. A comprehensive review of CUPRAC methodology. Anal. Methods 2011;**3**(11):2439-2453.

[240] Özyürek M, Guçlu K, Apak R. The main and modified CUPRAC methods of antioxidant measurement. Trends Anal. Chem. 2011;**30**(4):652-664.

[241] Campos C, Guzman R, López-Fernández E, Casado A. Evaluation of the copper(II) reduction assay using bathocuproinedisulfonic acid disodium salt for the total antioxidant capacity assessment: the CUPRAC-BSC assay. Anal. Biochem. 2009;**392**(1):37-44.

[242] Kuskoski EM, Vega JM, Rios JJ, Fett R, Troncoso AM, Asuero AG. Characterization of anthocyanins from the fruits of Baguaçu (*Eugenia umbelliflora* Berg). J. Agric. Food Chem. 2003;**51**(18):5450-5454.

[243] Hoyos-Arbeláez J, Vazquez M, Contreras-Calderon J. Electrochemical methods as a tool for determining the antioxidant capacity of foods and beverages: a review. Food Chem. 2017;**221**:1371-1381.

[244] Hilgemann M, Costa Bassetto V, Tatsuo Kubota L. Electrochemical approaches employed for sensing the antioxidant capacity exhibited by vegetal extracts: a review. Comb. Chem. High T. SCR 2016;**16**(2):98-108.

[245] Brainina KhZ, Zaharov AS, Vidrevich MB. Potentiometry for the determination of oxidant activity. Anal. Methods 2016;**8**:5667-5675.

[246] Pisoschi AM, Cimpeanu C, Predoi G. Electrochemical methods for total antioxidant capacity and its main contributors determination: a review. Open Chem. 2015;**13**(1):824-856.

[247] Ivanova AV, Gerasimova EL, Brainina KhZ. Potentiometric study of antioxidant activity: Development and prospects. Crit. Rev. Anal. Chem. 2015;**45**:311-322.

[248] Sochor J, Dobes J, Krystofova O, Ruttkay-Nedechy B, Babula P, Pohanka M, Jurilova T, Zitka O, Adam V, Klejdus B, Kizek R. Electrochemistry as a tool for studying antioxidant properties. Int. J. Electrochem. Soc. 2013;**8**(6):8469-8489.

[249] Jaldappagari S, Hegde AH, Narayan PS, Terada NL, Motohasi N. Electrochemical, bioactive mechanism of interaction and analytical studies of anthocyanins and related compounds. In: Anthocyanins, Structure, Biosynthesis and Health Benefits. Motohashi N. (Ed.), Chapter 3, Tokyo, Japan: Nova Science Publishers, 2012. pp. 35-92.

[250] Barriga-Gonzalez G, Aguilera-Venegas B, Folch-Cano C, Pérez-Cruz F, Olea-Azar, C. Electron spin resonance as a powerful tool for studying antioxidant and radicals. Curr. Med. Chem. 2013;**20**(37):4731-4743.

[251] Oszmianski J, Lachowicz S. Effect of the production of dried fruits and juice from chokeberry (*Aronia melanocarpa* L.) on the content and antioxidative activity of bioactive compounds. Molecules 2016;**21**(8):E1098.

[252] Grajeda-Iglesias C, Salas E, Barouh N, Baréa B, Panya A, Figueroa-Espinoza MC. Antioxidant activity of protocatechuates evaluated by DPPH, ORAC, and CAT methods. Food Chem. 2016;**194**:749-757.

[253] Sousa A, Araujo P, Azevedo J, Cruz L, Fernandes I, Mateus N, de Freitas V. Antioxidant and antiproliferative properties of 3-deoxyanthocyanidins. Food Chem. 2016;**192**:142-148.

[254] Baskan KS, Tutem E, Akyüz E, Apak R. Assessment of the contribution of anthocyanins to the total antioxidant capacities of plant foods. Eur. Food Res. Technol. 2015;**241**(4):529-541.

[255] Li YC, He YT. Anthocyanin content and antioxidant activity of different varieties blueberries. Adv. Mater. Res. 2013;**610-613**:3421-3427.

[256] Queirós RB, Tafulo PA, Sales MG. Assessing and comparing the total antioxidant capacity of commercial beverages: application to beers, wines, waters and soft drinks using TRAP, TEAC and FRAP methods. Comb. Chem. High Throughput Screen. 2013;**16**(1):22-31.

[257] Floegel A, Kim D-O, Chung S-J, Koo SI, Chun OK. Comparison of ABTS/DPPH assays to measure antioxidant capacity in popular antioxidant-rich US foods. J. Food Compos. Anal. 2011;**24**(7):1043-1048.

[258] Wootton-Beard PC, Moran A, Ryan L. Stability of the total antioxidant capacity and total polyphenol content of 23 commercially available vegetable juices after in vitro digestion measured by FRAP, DPPH, ABTS and Folin-Ciocalteu methods. Food Res. Int. 2011;**44**(1):217-224.

[259] Azevedo J, Fernandes I, Faria A, Oliveira J, Fernandes A, de Freitas V, Mateus N. Antioxidant properties of anthocyanidins, anthocyanidin-3-glucosides and respective portisins. Food Chem. 2010;**119**(2):518-523.

[260] Dudonné S, Vitrac X, Coutière P, Woillez M, Mérillon J-M. Comparative study of antioxidant properties and total phenolic content of 30 plant extracts of industrial interest using DPPH, ABTS, FRAP, SOD, and ORAC assays. J. Agric. Food Chem. 2009;**57**(5):1768-1774.

[261] Shalaby EA, Shanab SMM. Antioxidant compounds, assay of determination and mode of action. Afr. J. Pharm. Pharmacol. 2013;**7**(10):528-539.

[262] Badarinath AV, Rao KM, Chetty MS, Ramkanth S., Rajan TVS, Gnanaprakash K. A review on in-vitro antioxidant methods: comparisons, correlations and considerations. Int. J. PharmTech Res. 2010;**2**(2):1276-1285.

[263] Ioannou I, Chaaban H, Slimane M, Ghoul M. Origin of the variability of the antioxidant activity determination of food material. In: Ekinci D (ed) Biotechnology, Biochemistry, Genetics and Molecular Biology, Chap 4. Rijeka: InTech, 2015. pp. 77-92.

[264] Quideau S, Deffieux D, Douat-Casassus C, Pouysegu L. Plant polyphenols: chemical properties, biological activity and synthesis. Angew. Chem. Int. Ed. 2011;**50**(3):586-521.

[265] Lu L, Qiang M, Li F, Zhang H, Zhang S. Theoretical investigation on the antioxidative activity of anthocyanidins: a DFT/B3LYP study. Dyes Pigm. 2014;**103**:175-182.

[266] Estévez L, Otero N, Mosquera RA. A computational study on the acidity dependence of radical-scavenging mechanisms of anthocyanidins. J. Phys. Chem. B 2010;**114**(29):9706-9712.

[267] Guzman R, Santiago C, Sánchez M. A density functional study of antioxidant properties on anthocyanidins. J. Mol. Struct. 2009;**935**(1-3):110-114.

[268] Martinez A. Donator acceptor map of psittacofulvins and anthocyanins: are they good antioxidant substances? J. Phys. Chem. B. 2009;**113**(14):4915-4921.

[269] Estévez L, Mosquera RA. Molecular structure and antioxidant properties of delphidin. J. Phys. Chem. A 2008;**112**(42):10614-10623.

[270] Kim D-O, Lee KW, Lee HJ, Lee CY. Vitamin C equivalent antioxidant capacity (VCEAC) of phenolics phytochemicals. J. Agric. Food Chem. 2002;**50**(13):3713-3717.

[271] Jhin C, Hwang KT. Prediction and radical scavenging activities of anthocyanins applying adaptive neuro-fuzzy inference system (ANFIS) with quantum chemical descriptors. Int. J. Mol. Sci. 2014;**15**(8):14715-14727.

[272] Jing P, Zhao S, Ruan S, Sui Z, Chen L, Jiang L, Qian B. Quantitative studies on structure-ORAC relationships of anthocyanins from eggplant and radish using 3D-QASR. Food Chem. 2014;**145**:365-371.

[273] Shityakov S, Puskas I, Roewer N, Förster C, Broscheit J. Three-dimensional quantitative structure-activity relationship and docking studies in a series of anthocyanin derivatives as cytochrome P450 3A4 inhibitors. Adv. Appl. Bioinf. Chem. 2014;7(1):11-21.

[274] Zhao C-L, Yu Y-Q, Chen Z-J, Wen G-S, Wei F-G, Zheng Q, Wang C-D, Xiao X-L. Stability-increasing effects of anthocyanin glycosyl acylation. Food Chem. 2017;**214**:119-128.

[275] Goulas V, Vicente AR, Manganaris GA. Structural diversity of anthocyanins in fruits. In: Motohashi N (ed) Anthocyanins: Structure, Biosynthesis and Health Benefits. New York: Nova Science Publishers, 2012. pp. 225-250.

[276] Halbwirth H. The creation and physiological relevance of divergent hydroxylation patterns in the flavonoid pathway. Int. J. Mol. Sci. 2010;**11**(2),595-621.

[277] Bors W, Michel C, Stettmaier K. Structure-activity relationships governing antioxidant capacities of plant polyphenols. Methods Enzymol. 2001;**335**:166-180.

[278] Kähkonen MP, Heinonen M. Antioxidant activity of anthocyanins and their aglycons. J. Agric. Food Chem. 2003;**51**(3):628-633.

[279] Jing P, Bomser JA, Schwartz SJ, He J, Magnuson BA, Giusti MM. Structure-function relationships of anthocyanins from various anthocyanin-rich extracts on the inhibition of colon cancer cell growth. J. Agric. Food Chem. 2008;**56**(20):9391-9398.

[280] Pina F, Oliveira J, de Freitas V. Anthocyanins and derivatives are more than flavylium cations. Tetrahedron 2015;**71**(20):3107-3114.

[281] Silva VO, Freitas AA, Maçanita AL, Quina FH. Chemistry and photochemistry of natural plant pigments: the anthocyanins. J. Phys. Org. Chem. 2016;**29**(11):594-599.

[282] Basilio N, Pina F. Chemistry and photochemistry of anthocyanins and related compounds: a thermodynamic and kinetic approach. Molecules 2016;**21**(11):1502.

[283] Pina F. Chemical applications of anthocyanins and related compounds. A source of bioinspiration. J. Agric. Food Chem. 2014;**62**(29):6885-6897.

[284] Pina F, Melo MJ, Laia CAT, Parola AJ, Lima JC. Chemistry and applications of flavylium compounds: a handful of colours. Chem. Soc. Rev. 2012;**41**(2):896-908.

[285] Pina F, Petrov V, Laia CAT. Photochemistry of flavylium systems. An overview of a versatile multistate system. Dyes Pigm. 2012;**92**(2):877-889.

[286] Trouillas P, Sancho-García JC, de Freytas V, Gierschner J, Otyepka M, Dangles O. Stabilizing and modulating color by copigmentation: insights from theory and experiment. Chem. Rev. 2016;**116**(9):4937-4982.

[287] Bimpilas A, Panagopoulou M, Tsimogiannis D, Oreopoulou V. Anthocyanin copigmentation and color of wine: the effect of naturally obtained hydroxycinnamic acids as cofactors. Food Chem. 2016;**197**:39-46.

[288] Valenzuela A, Nieto KS. Synthetic and natural antioxidants: food quality protectors. Grasas y Aceites 1996;**47**(3):186-196.

[289] Ahmadiani N, Robbins RJ, Collins TM, Giusti MM. Molar absorptivity (ε) and spectral characteristics of cyanidin-based anthocyanins from red cabbage. Food Chem. 2016;**197**:900-906.

[290] Wang BC, He R, Li ZM. The stability and antioxidant activity of anthocyanins from blueberry. Food Technol. Biotechnol. 2010;**48**(1):42-49.

[291] Bakowska-Barczak A. Acylated anthocyanins as stable, natural food colorants—a review. Polish J. Food Nutr. Sci. 2005;**14/55**(2):107-116.

[292] Giusti MM, Wrolstad RE. Acylated anthocyanins from edible sources and their applications in food systems. Biochem. Eng. J. 2003;**14**(3):217-225.

[293] Andersen OM, Jordheim M. Anthocyanins. In: Encyclopedia of Life Sciences. Chichester: Wiley, 2010. http://wwwels.net. doi:10.1002/9780470015902.a0001909.pub2

[294] El Sohaimy SA. Functional foods and nutraceutical-modern approach to Food Science. World Appl. Sci. J. 2012;**20**(5):691-708.

[295] EFSA (European Food Safety Autority). Scientific opinion on the re-evaluation of anthocyanins (E163) as a food additive. EFSA J. 2013;**11**(4):3145-3196.

[296] Hosu A, Cristea V-M, Cimpoiu C. Analysis of total phenolic, flavonoids, anthocyanins and tannins content in Romanian red wines: prediction of antioxidant activities and classification of wines using artificial neural networks. Food Chem. 2014;**150**:113-118.

[297] Tenore GC. Anthocyanin profile and their biological properties in popular mediterranean red wines. In: Anthocyanins: Structure, Biosynthesis and Health Benefits. New York: Nova Science Publisher, 2012. pp. 283-298.

[298] Menković N, Živković J, Šavikin K, Godevac D, Zdunić G. Phenolic composition and free radical scavenging activity of wine produced from the Serbian autochthonous grape variety Prokupac—a model approach. J. Serb. Chem. Soc. 2014;**79**(1):11-24.

[299] Rice-Evans CA, Miller NJ, Papaganda G. Antioxidant properties of phenolic compounds. Trends Plant Sci. Rev. 1997;**2**(4):152-159.

[300] Frankel EN, Waterhouse AL, Teissedre PL. Principal phenolic phytochemicals in selected California wines and their antioxidant activity in inhibiting oxidation of human low density lipoproteins. J. Agric. Food Chem. 1995;**43**(4):890-894.

[301] Monagas M, Bartolomé B, Gómez-Cordovés C. Updated knowledge about the presence of phenolic compounds in wine. Crit. Rev. Food Sci. Nutr. 2005;**45**(2):85-118.

[302] Ivanova-Petropulos V, Hermosín-Gutiérrez I, Boros B, Stefova M, Stafilov T, Vojnoski B, Dörnyei Á, Kilár F. Phenolic compounds and antioxidant activity of Macedonian red wines. J. Food Compos. Anal. 2015;**41**:1-14.

[303] Pisano PL, Silva MF, Olivieri AC. Anthocyanins as markers for the classification of Argentinean wines according to botanical and geographical origin. Chemometric modeling of liquid chromatography–mass spectrometry data. Food Chem. 2015;**175**:174-180.

[304] Satué-Gracia MT, Heinonen M, Frankel EN. Anthocyanin as antioxidants on human low-density lipoprotein and lecithin-liposome systems. J. Agric. Food Chem. 1997;**45**(9):3362-3367.

[305] Hayek B, Fuhrman J, Vaya M, Rosenblat P, Belinky R, Coleman Elis A, Aviram M. Reduced progression of atherosclerosis in apolipoprotein E-deficient mice following consumption of red wine, or its polyphenols quercetin or catechin, is associated with reduced susceptibility of LDL to oxidation and aggregation. Arterioscler. Thromb. Vasc. Biol. 1997;**17**(11):2744-52.

[306] Netzel M, Strass G, Bitsch I, Konitz R, Christmann M, Bitsch R. Effect of grape processing on selected antioxidant phenolics in red wine. J. Food Eng. 2003;**56**(2):223-228.

[307] Zheng W, Wang SY. Oxygen radical absorbing capacity of phenolics in blueberries, cranberries, chokeberries, and lingonberries. J. Agric. Food Chem. 2003;**51**(2):502-509.

[308] Gris EF, Mattivi F, Ferreira EA, Vrhovsek U, Filho DW, Pedrosa RC, Bordignon-Luiz MT. Phenolic profile and effect of regular consumption of Brazilian red wines on in vivo antioxidant activity. J. Food Compos. Anal. 2013;**31**(1):31-40.

[309] Jordão AM, Simões S, Correia AC, Gonçalves FJ. Antioxidant activity evolution during portuguese red wine vinification and their relation with the proanthocyanidin and anthocyanin composition. J. Food Process. Preserv. 2012;**36**(4):298-309.

[310] Jordão AM, Correia AC. Relationship between antioxidant capacity, proanthocyanidin and anthocyanin content during grape maturation of touriga nacional and tinta roriz grape varieties. S. Afr. J. Enol. Vitic. 2012;**33**(2):214-224.

[311] Langer RD, Criqui MH, Reed DM. Lipoproteins and blood pressure as biological pathways for effect of moderate alcohol consumption on coronary heart disease. Circulation 1992;**85(3)**:910-915.

[312] Savolainen MJ, Kesaniemi YA. Effects of alcohol on lipoproteins in relation to coronary artery disease. Curr. Opin. Lipidol. 1995;**6**(4):243-250.

[313] Simonetti P, Pietta P, Testolin G. Polyphenol content and total antioxidant potential of selected Italian wines. J. Agric. Food Chem. 1997;**45**(4):1152-1155.

[314] Radovanović B, Radovanović A. Free radical scavenging activity and anthocyanin profile of cabernet sauvignon wines from the balkan region. Molecules 2010;**15**(6):4213-4226.

[315] Bajčan D, Vollmannová A, Šimanský V, Bystrická J, Trebichalský P, Árvay J, Czako P. Antioxidant activity, phenolic content and colour of the slovak cabernet sauvignon wines. Potravinarstvo 2016;**10**(1):89-94.

[316] Aguirre MJ, Chen YY, Isaacs M, Matsuhiro B, Mendoza L, Torres S. Electrochemical behaviour and antioxidant capacity of anthocyanins from Chilean red wine, grape and raspberry. Food Chem. 2010;**121**(1):44-48.

[317] Rivero-Pérez MD, Muñiz P, González-Sanjosé ML. Contribution of anthocyanin fraction to the antioxidant properties of wine. Food Chem. Toxicol. 2008;**46**(8):2815-2822.

[318] Rivero-Pérez MD, Muñiz P, González-Sanjosé ML. Antioxidant profile of red wines evaluated by total antioxidant capacity, scavenger activity, and biomarkers of oxidative stress methodologies. J. Agric. Food Chem. 2007;**55**(14):5476-5483.

[319] Fernández-Pachón MS, Villaño D, García-Parilla MC, Troncoso AM. Antioxidant activity of wines and relation with their polyphenolic composition. Anal. Chim. Acta 2004;**513**(1):113-118.

[320] Ghiselli A, Nardini M, Baldi A, Scaccini C. Antioxidant activity of different penol fractions separated from an Italian red wine. J. Agric. Food Chem. 1998;**46**(2):3641-3647.

[321] Trikas ED, Melidou M, Papi RM, Zachariadis GA, Kyriakidis DA. Extraction, separation and identification of anthocyanins from red wine by-product and their biological activities. J. Funct Foods 2016;**25**:548-558.

[322] Cristino R, Costa E, Cosme F, Jordão AM. General phenolic characterisation, individual anthocyanin and antioxidant capacity of matured red wines from two Portuguese appellations of origins. J. Sci. Food Agric. 2013;**93**(10):2486-2493.

[323] López-Vélez M, Martínez-Martínez F, Del Valle-Ribes C. The study of phenolic compounds as natural antioxidants in wine. Crit. Rev. Food Sci. Nutr. 2003;**43**(3):233-244.

[324] Lee J-H, Talcott ST. Fruit maturity and juice extraction influences ellagic acid derivatives and other antioxidant polyphenolics in muscadine grapes. J. Agric. Food Chem. 2004;**52**(2):361-366.

[325] Zhang L, Li N, Gao X. Phenolic compounds and antioxidant activity of wines fermented using ten blueberry varieties. Am. J. Food Technol. 2016;**11**(6):291-197.

[326] Galanakis CM, Kotanidis A, Dianellou M, Gekas V. Phenolic content and antioxidant capacity of cypriot wines. Czech J. Food Sci. 2015;**33**(2):126-136.

[327] Iwashina T. Contribution to flower colors of flavonoids including anthocyanins: a review. Nat. Prod. Commun. 2015;10(3):129-144.

[328] Cortez R, Luna-Vital DA, Margulis D, Gonzalez de Mejia E. Natural pigments: stabilization methods of anthocyanins for food applications. Compr. Rev. Food Sci. Food Saf. 2017;**16**(1):180-198.

[329] Flores FP, Singh RK, Kong F. Anthocyanin extraction, microencapsulation, and release properties during in vitro digestion. Food Rev. Int. 2016;**32**(1):46-67.

[330] Shwetha BP, Preetha R. Study on color stability and microencapsulation of anthocyanin pigment using spray drying. Biosci. Biotechnol. Res. Asia. 2016;**13**(2):1207-1214.

[331] Yousuf B, Gul K, Wani AA, Singh P. Health benefits of anthocyanins and their encapsulation for potential use in food systems: a review. Crit. Rev. Food Sci. Nutr. 2016;**56**(13):2223-2230.

[332] Robert P, Fredes C. The encapsulation of anthocyanins from berry-type fruits. Trends in foods. Molecules 2015;**20**(4):5875-5888.

[333] Mahdavi SA, Jafari SM, Ghorbani M, Assadpoor E. Spray-drying microencapsulation of anthocyanins by natural biopolymers: a review. Dry Technol. 2014;**32**(5):509-518.

[334] Cavalcanti RN, Santos DT, Meireles MAA. Non-thermal stabilization mechanisms of anthocyanins in model and food systems—an overview. Food Res. Int. 2011;**44**(2):499-509.

[335] Bononi M, Tateo F. Stabilization of cranberry anthocyanins in nutraceutical capsules. Int. J. Food Sci. Nutr. 2007;**58**(2):142-149.

[336] Fang Z, Bhandari B. Encapsulation of polyphenols—a review. Trends Food Sci. Technol. 2010;**21**(10):510-523.

[337] Ray S, Raychaudhuri U, Chakraborty R. An overview of encapsulation of active compounds used in food products by drying technology. Food Biosci. 2016;**13**:76-83.

[338] Esfanjani AF, Jafari SM. Biopolymer nano-particles and natural nano-carriers for nano-encapsulation of phenolic compounds. Colloids Surf. B Biointerfaces 2016;**146**:532-543.

[339] Jia Z, Dumont M-J, Orsat V. Encapsulation of phenolic compounds present in plants using protein matrices. Food Biosci. 2016;**15**:87-104.

[340] Li Z, Jiang H, Xu C, Gu L. A review: using nanoparticles to enhance absorption and bioavailability of phenolic phytochemicals. Food Hydrocoll. 2015;**43**:153-164.

[341] Munin A, Edwards-Lévy F. Encapsulation of natural polyphenolic compounds: a review. Pharmaceutics 2011;**3**(4):793-829.

[342] Aguiar J, Estevinho BN, Santos L. Microencapsulation of natural antioxidants for food application—the specific case of coffee antioxidants—a review. Trends Food Sci. Technol. 2016;**58**:21-39.

[343] Kammerer DR. Anthocyanins. In: Carle R, Schweggert R (eds) Handbook of Natural Pigments in Food and Beverages. Industrial Applications for Improving Food Color, Chapter 3. Amsterdam: Elsevier, 2016. pp. 61-80.

[344] Glover BJ, Martin C. Anthocyanins. Curr. Biol. 2012;**22**(5):R147–R150.

[345] Olivas-Aguirre FJ, Rodrigo-Garcia J, Martinez-Ruiz N del R, Cárdenas-Robles AI, Mendoza-Diaz SO, Álvarez-Parrilla E, González-Aguilar GA, de la Rosa LA, Ramos-Jimenez A, Wall-Medrano A. Cyanidin-3-O-glucoside: physical chemistry, foodomics and health effects. Molecules 2016;**21**(9):E1264.

[346] Kuskoski EM, Fett P, Asuero AG. Anthocyanins: a group of natural pigments, isolation, identification and properties. Alimentaria 2002;**339**:61-74.

[347] Andersen OM, Jordheim M. Basic anthocyanin chemistry and dietary sources. In: Wallace TC, Giusti MM (eds) Anthocyanins in Health and Disease, Chapter 2. Boca Raton, FL: CRC Press, 2014. pp. 13-89.

[348] Martin C, Butelli E, Petroni K, Tonelli C. How can research on plants contribute to promoting human health. Plant Cell 2011;**23**:1685-1689.

[349] Gardana C, Ciappellano S, Marinoni L, Fachechi C, Simonetti P. Bilberry adulteration: identification and chemical profiling of anthocyanins by different analytical methods. J. Agric. Food Chem. 2014;**62**(45):10998-11004.

[350] Picariello G, Ferranti P, Garro G, Manganiello G, Chianese L, Coppola R, Addeo F. Profiling of anthocyanins for the taxonomic assessment of ancient purebred *V. vinifera* red grape varieties. Food Chem. 2014;**146**:15-22.

[351] Passeri V, Koes R, Quattrocchio FM. New challenges for the design of high value plant products: stabilization of anthocyanins in plant vacuoles. Front. Plant Sci. 2016;**7**:153 (pp. 9).

[352] Newsome AG, Culver CA, Breemen RB. Nature's palette: the search for natural blue colorants. J. Agric. Food Chem. 2014;**62**(28):6498-6511.

[353] Sasaki N, Takayama T. Achievements and perspectives in biochemistry concerning anthocyanin modification for blue flower coloration. Plant Cell. Physiol. 2015;**56**(1):28-40.

[354] Roselló-Soto E, Galanakis CM, Brncic M, Orlien V, Trujillo FJ, Mawson R, Knoerzer K, Tiwari BK, Barba FJ. Clean recovery of antioxidant compounds from plant foods,

by-products and algae assisted by ultrasound processing. Modeling approaches to optimize processing conditions. Trends Food Sci. Technol. 2015;**42**(2):134-139.

[355] Ameer K, Shahbaz M, Kwon J-H. Green extraction methods for polyphenols from plant matrices and their byproducts: a review. Compr. Rev. Food Sci. Food Saf. 2017. doi:10.1111/1541-4337.12253

[356] Barba FJ, Zhu Z, Koubaa M, Sant'Ana AS. Green alternative methods for the extraction of antioxidant bioactive compounds from winery wastes and by-products: a review. Trends Food Sci. Technol. 2016;**49**:96-109.

[357] Castro-López C, Rojas R, Sánchez-Alejo EJ, Niño-Medina G, Martínez-Avila GCG. Phenolic compounds recovery from grape fruit and by-products: an overview of extraction methods. In: Morata A, Loira I (eds) Grape and Wine Biotechnology, Chapter 5. Rijeka: InTech, 2016. pp. 103-123.

[358] Miraje SY, Amlepatil NM, Sahoo AK, Mote GV. Anthocyanin extraction from winery waste material: a review. JIPBS 2015;**2**(2):218-221.

[359] da Silva LM, de Figuereido EAT, Ricardo NMPS, Viera IGP, de Figuereido RW, Brasil IM, Gomez CL. Quantification of bioactive compounds in pulps and by-products of tropical fruits from Brazil. Food Chem. 2014;**143**:398-404.

[360] Kammerer DR, Kammerer J, Valet R, Carle R. Recovery of polyphenols from the by-products of plant food processing and application as valuable food ingredients. Food Res. Int. 2014;**65**A:2-12.

[361] Jing P, Giusti MM. Analysis of anthocyanins in biological samples. In: Wallace TC, Giusti MM (eds) Anthocyanins in Health and Disease, Chapter 4. Boca Raton, FL: CRC Press, 2014. pp. 115-139.

[362] Zafra-Stone S, Yasmin T, Bagchi M, Chatterjee A, Vinson JA, Bagchi D. Berry anthocyanins in human health and disease prevention. Mol. Nutr. Food Res. 2007;**51**(6):675-683.

[363] Halliwell B, Guteridge J. Free Radicals in Biology and Medicine, 5th ed. Oxford: University Press, 2015.

[364] Croft KD. Dietary polyphenols: antioxidant or not? Arch. Biochem. Biophys. 2016;**595**:120-124.

[365] Poljsak B, and Milisav I. Oxidized forms of dietary antioxidants: friends or foes? Trends Food Sci. Technol. 2014;**39**(2):155-166.

[366] McGhie TK, Stevenson DE. Bioavailability and bioabsorption of anthocyanins. In: Wallace TC, Giusti MM (eds) Anthocyanins in Health and Disease, Chapter 3. Boca Raton, Fl: CRC Press, 2014. pp. 91-113.

[367] Lila MA. Anthocyanins and human health: an in vitro investigative approach. J. Biomed. Biotechnol. 2004;**2004**(5):306-313.

[368] Martin J, Navas MJ, Jimenez-Moreno AM, Asuero AG. Anthocyanin pigments, importance, sample preparation and extraction. In: Phenolic Compounds—Natural Sources, Importance and Applications. Rijeka: InTech, 2016.

Chapter 12

New Insights Regarding the Potential Health Benefits of Isoflavones

Corina Danciu, Diana Simona Antal,
Florina Ardelean, Aimée Rodica Chiş, Codruţa Şoica,
Florina Andrica and Cristina Dehelean

Additional information is available at the end of the chapter

http://dx.doi.org/10.5772/67896

Abstract

Isoflavones are a class of plant secondary metabolites, with an estrogen-like structure presenting a plethora of biological activities. The chapter discusses important facts about this class of phytoestrogens, from biosynthesis to the latest research about their health benefits. The following major points discussed are: biosynthesis, regulation, isolation, metabolism and bioavailability, isoflavones in diet and intake, and new insights regarding the therapeutic effect including cancer chemoprevention. The chapter ends with a mini review of own research of the anti-inflammatory and chemopreventive activity of isoflavonoid genistein alone and incorporated in modern pharmaceutical formulations. The chapter updates the interested researchers in the field with the latest progress regarding potential health benefits of isoflavones.

Keywords: isoflavones, biosynthesis, regulation, isolation, metabolism, bioavailability, therapeutic effect

1. Introduction

Isoflavonoids, a class of secondary metabolites including over 1000 structures [1], are polyphenolic derivatives of 1,2-diphenylpropane, as opposed to the larger group of flavonoids having a 1,3-diphenylpropane skeleton. They encompass several subgroups, the most prominent being the isoflavones, the rotenoids, the pterocarpans and the coumestans (**Figure 1**). The multiplicity of isoflavonoid structures is accounted by different oxidation levels of the backbone, the presence of additional heterocyclic rings and the diversity of substituents. These substances occur

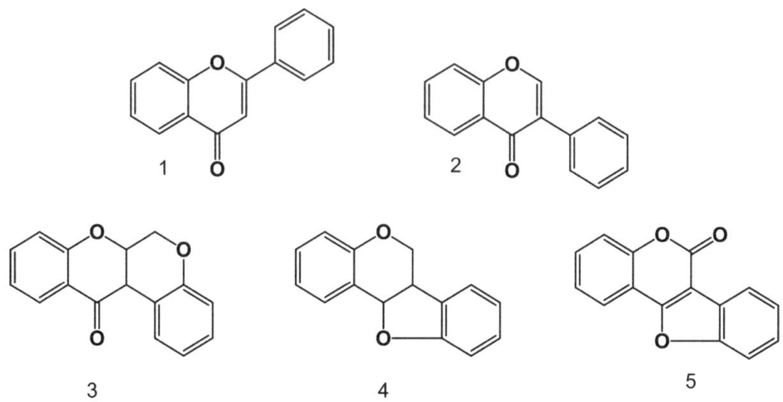

Figure 1. Flavonoid and isoflavonoid backbones. 1: Flavones, 2: Isoflavones, 3: Rotenoids, 4: Pterocarpans and 5: Coumestans.

mostly in free forms as aglycones (**Figure 2**); glycosides are formed with glucose, rhamnose or apiose as the sugar moiety and are mainly O-glycosides.

Isoflavonoids are a major biochemical marker of the Fabaceae, especially of the subfamily Faboideae [2]. As much as 95% of the isoflavonoid aglycones and about 90% of isoflavones were reported in this family [3]. In recent years, the distribution of isoflavonoids has reliably been proven in a variety of nonlegumes, including bryophytes, gymnosperms and angiosperms. In flowering plants, they occur in over 50 families, and their distribution pattern is unrelated with the degree of phylogenetic closeness. Such families include among monocots such as the Iridaceae, Liliaceae, Asphodelaceae, Poaceae, Zingiberaceae and Cyperaceae as well as a large number of dicots such as Asteraceae, Apiaceae, Malvaceae, Rosaceae, Rutaceae or Solanaceae [1, 4]. Among plant organs, isoflavonoids are mainly present in underground parts, wood and bark as compared to flowers and leaves [1]. *In planta*, they act as anti-microbial compounds synthesized in response to the attack of pathogens, representing the first identified type of phytoalexins. Isoflavonoids may either be preexistent in plant tissues before microbial attacks or be produced only upon exposure to infection or environmental stressors. The prominent distribution of these secondary metabolites

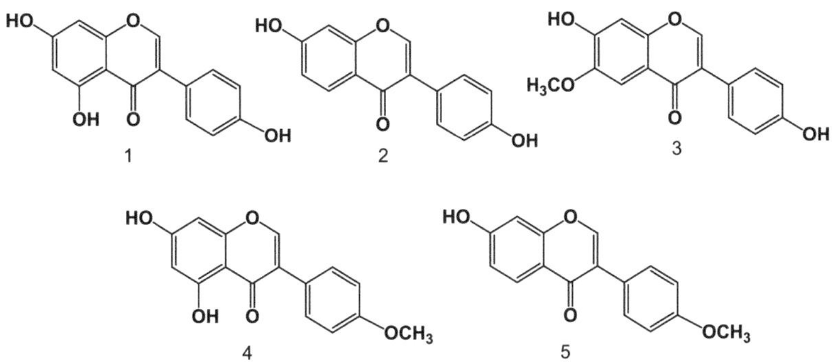

Figure 2. Main isoflavone aglycones: 1: Genistein, 2: Daidzein, 3: Glycitein, 4: Biochanin A and 5: Formononetin.

in legumes is related to their physiological role in nodulation. The isoflavonoids present in root exudates are associated with the attraction of symbiotic *Rhizobium* bacteria and promotion of their growth within root nodules [5], leading to an improved nitrogen uptake.

2. Biosynthesis and regulation

The biosynthesis of isoflavonoids occurs as a branch of the phenylpropanoid pathway, involved in the biologic obtainment of all flavonoids [6]. The departing point is represented by the amino acid phenylalanine, sequentially converted to cinnamate, p-coumarate and p-coumaroyl-CoA by the relevant enzymes of each step (phenylalanine ammonia-lyase, cinnamic acid-4-hydroxylase, and 4-coumarate-CoA ligase, respectively). Subsequently, chalcone synthase is involved in the creation of the 15-carbon flavonoid backbone from p-coumaroyl-CoA; several derivatives may be produced via different branch pathways. The crucial enzyme for isoflavone biosynthesis is isoflavone synthase, a cytochrome P450 monooxygenase, which catalyzes the migration of the aryl moiety and transforms flavanones to isoflavones [7]. The reaction involves keto-enol tautomerism of flavanones and epoxidation, followed by dehydration [8].

Relevant metabolic pathways for the biosynthesis of the major isoflavonoids genistein (precursor naringenin) and daidzein (precursor liquiritigenin) have been well-studied in soybean (*Glycine max* (L.) Merr.). Their understanding opened possibilities of metabolic engineering, leading to the development of soybean lines with an increased content of isoflavones in comparison to wild-type seeds. On the other hand, the health benefits of isoflavones prompted strategies to induce the synthesis of these compounds by non-legume plants (broccoli and tomatoes). This approach could be achieved in plants with an active phenylpropanoid pathway [9]. In fact, the flavanone naringenin is not only the substrate of isoflavone synthase, but as well that of flavanone-3-hydroxylase, involved in the biosynthesis of flavonols, anthocyanins and condensed tannins. The introduction of the isoflavone synthase gene in plants that do not express this enzyme was able to trigger genistein production in corn [7] and the model cruciferous *Arabidopsis thaliana* [10]. The disadvantage was represented by low isoflavone content, related to the competition between the flavonoid and the isoflavone pathways. Blocking the alternative flavonoid/anthocyanin branch of the phenylpropanoid pathway while upregulating the synthesis of isoflavones by introducing foreign transcription factors yielded very favorable results. Yu et al. could increase up to fourfold the level of isoflavones in soy by introducing the transcription factors C1 and R from corn in soybean (involved in the regulation of the phenylpropanoid pathway genes), and by co-suppressing flavanone-3-hydroxylase, hence diverting the whole substrate to the isoflavonoid pathway [11]. An additional regulation of isoflavone biosynthesis involves the enzymes catalyzing the glycosylation of aglycones, as the majority of isoflavonoids accumulate in conjugated form. In soybean, the malonylglycosides are predominant, followed by glucosides. These conjugates are obtained upon catalysis by malonyl transferase and glycosyl transferase, respectively, in specific positions. The engineered obtainment of isoflavonoids in nonlegumes is crucially related to the presence of isoflavonoid-specific malonyl transferases and glycosyl transferases [12]. In soy, the content and composition of isoflavonoids are subjected to polygenic regulation and highly variable in response to drought, temperature, fertilization, carbon dioxide content and genetic factors [13]. The level of isoflavonoids is higher in wild-growing populations than in cultivated soybean; this situation is thought to be a consequence of domestication [14].

3. Isolation

The isoflavonoid aglycones such as genistein and daidzein are compounds with a low polarity and hence practically insoluble in water. The polarity is lowered by methylation, as in formononetin and biochanin A. After glycosylation, the water solubility increases; glucosides have a higher solubility in water than their malonylated and acetylated derivatives. The glycosidic bond may be hydrolyzed under acidic or basic conditions [15]. Early extraction of isoflavones was performed by refluxing alcohol, but had the disadvantage of converting malonyl- and acetyl-glucosides into glucosides and aglycones [16]. A mixture of methanol 80% was as well proposed [17]. An optimized extraction method was developed by Griffith and Collison, using acetonitrile/water, without the addition of an acid [18]. Acetonitrile is considered to yield higher extraction ratios than solvents such as acetone, ethanol and methanol, during the analysis of 12 main soy isoflavones from foods; the organic solvent (53%) is mixed with water [19]. The preparative isolation of isoflavones could be achieved by high-speed countercurrent chromatography (HSCCC). In one setting of HSCCC, acid-free solvents were employed; the isolation of malonylglucosides was performed with the aid of a solvent *tert*-butyl-methyl ether/*n*-butanol/acetonitrile/water in a ratio of 1/3/1/5 [20]. Monoglucosylated and acetylated isoflavones were obtained more recently by HSCCC after a cleaning-up step on Amberlite XAD-7 material [21].

Quantification of isoflavones is usually performed by HPLC-DAD, using reversed-phase columns and eluents containing 95% acetonitrile with 0.1% trifluoroacetic acid. Validated methods with good peak resolution are available [18]. Detection may be performed at 262 nm. For the quantification in biologic samples (urine, saliva and blood), HPLC-MS/MS spectrometry is employed after solid-phase extraction (SPE) of isoflavones; the glycosides are hydrolyzed enzymatically prior to SPE [22].

4. Bioavailability and metabolism

Bioavailability of isoflavones is based on data from absorbtion, metabolism, distribution and excretion studies. After the intake of pure compounds, isoflavone-rich extracts or foods containing high levels of isoflavones, the parent compound and their metabolite can be found in plasma and urine of human volunteers. Following ingestion, soy isoflavones attain maximal plasma concentration within 4–8 h and then eliminate from the body through the bile and kidneys with an elimination half-life that is on average 8 h [23, 24]. Aglycones are well absorbed due to their low water solubility and small molecular weight [25]. After ingestion, isoflavone glycosides are hydrolyzed by intestinal glucosidases, which partially release the aglycones daidzein,

http://dx.doi.org/10.5772/67896

genistein and glycitein [24]. These may be absorbed or converted to a number of metabolites including equol and p-ethyl phenol [24].

It was proved that intestinal microflora plays an important role in the metabolism and bioavailability of isoflavones [26]. It is considered that about 50% of Asians and 25% of non-Asians host the intestinal bacteria that convert daidzein into the isoflavonoid equol [27]. Variation in individual metabolism of phytoestrogens due to differences in gut microflora might influence the serum concentration of phytoestrogens. It was found that the capacity to produce equol is higher among Japanese and Korean men than among American men [28]. After an intake of 50 mg isoflavone, the urinary excretion was 42% for daidzein and 16% for genistein [29]. Fermented soy products, or supplements in which the soy extract has been hydrolyzed, contain mostly the aglycone forms of isoflavones; however, following ingestion, the plasma profile of isoflavone metabolites is the same, no matter the form ingested [30]. Free aglycones, released after hydrolysis, are absorbed by passive diffusion across the intestin [31].

The extensive metabolization of the aglycones is evident in the extremely low content of the free form in body fluids, of less than 1% in human plasma and urine [32]. The metabolization of genistein has intensively been studied and thoroughly reviewed [25]. Following the faith of most xenobiotics, it undergoes detoxification, being conjugated to glucuronides and sulfates. Among them, monoglucuronides are present in the highest proportion (62.5%), followed by diglucuronides, sulfoglucuronides, disulfates and monosulfates in human urine after dietary intake of soy products [33]. The percentage of sulfates was reported to be slightly higher in blood (20%) than in urine (13%) [32]. In humans, the plasma level of genistein (all forms) is situated in the micromolar range, while the level of free genistein is in the higher nanomolar range. The rate of metabolization after oral ingestion is less than half an hour, with conjugations occurring in the intestine, the liver but as well in the kidneys, heart and lungs. The tissue distribution of genistein is highest in the gastro-intestinal tract and liver, consistent with its enterohepatic recycling [34]. While the differences between individuals are very high, the oral bioavailability of genistein is low due to extensive metabolization and high expression level of efflux transporters (such as breast cancer resistance protein (BCRP)) [25].

5. Dietary intake and sources of isoflavones

As already mentioned above, isoflavones are flavonoid compounds that are biogenetically produced by plants belonging mainly to the Fabaceae family. Main sources of dietary isoflavones are soybeans (*Glycine max*) [35, 36] and red clover (*Trifolium pratense*) [37]. Other plants with a high content of isoflavones are: mung beans (*Vigna radiata*) [38], kudzu (*Pueraria lobata*), lupine (*Lupinus spp.*), fava bean (*Vicia faba*), psoralea (*Psoralea corylifolia*) [39], poinciana (*Caesalpinia pulcherrima*) [40] and alfalfa (*Medicago sativa*) [41].

Soybeans are widely employed for the preparation of food and dietary supplements. They contain both isoflavone aglycones: genistein, daidzein, glycitein, and glucosides: 7-*O*-glucosides: genistin, daidzin, glycitin, and three 6″-*O*-acetyl glucosides: 6″-*O*-acetyl-genistin, 6″-*O*-acetyldaidzin, and 6″-*O*-acetyl-glycitin, and three 6″-*O*-malonyl-glucosides: 6″-*O*-malonyl-genistin, 6″-*O*-malonyldaidzin, and 6″-*O*-malonylglycitin (**Figure 3**) [42, 43]. The highest percentage of soy isoflavones in soybeans is represented by genistein glucosides, approximately 50%. Daidzein glucosides are about 40% and glycitein glucosides 5–10% [15]. More exactly, the concentration of daidzein and daidzin in soy extract is 10.4 and, respectively, 244.5 mg/g soy extract corresponding to 1.7 and 39.6%. Concerning the concentration of genistein and genistin, it is 5.3 mg/g soy extract and, respectively, 319.6 mg/g soy extract representing 0.9 and 51.8% [44].

Comparing the isoflavones from soybeans with those extracted from red clover, it is remarkable that there are four main isoflavones (daidzein, genistein, formononetin and biochanin A) from red clover and only three in soybeans [15].

In China, the first reference to soybeans dates from 2853 BC [45]. In contrast, Western cultures have adopted these products only lately. A variety of soy foods is currently available throughout the world, produced with modern processing techniques or using the traditional methods. They can be classified into fermented (tofu, okara, yuba, fresh greed soybeans, whole dry soybeans, soy nuts, whole-fat soy flour, soy sprouts, soymilk and soymilk products) and nonfermented (tempeh, natto, miso, fermented tofu and soymilk products and soy sauces) products [46]. Epidemiological studies among the Japanese population suggest that fermented soybean products have better effects than nonfermented soy products, probably because of a higher bioavailability of isoflavone aglycones [47].

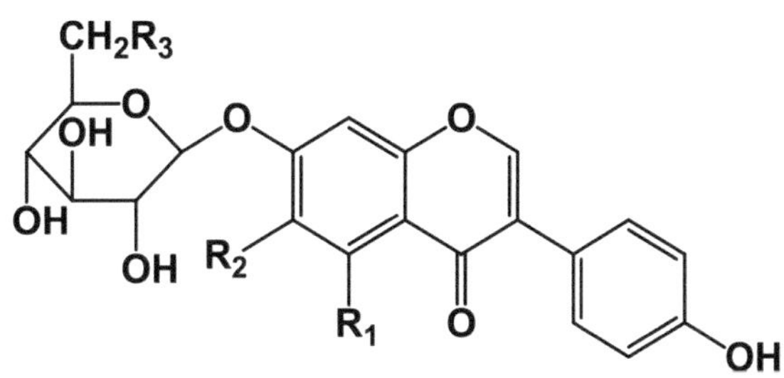

Name	R_1	R_2	R_3
Daidzin	H	H	H
Acetydaidzin	H	H	$COCH_3$
Malonyldaidzin	H	H	$COCH_2COOH$
Genistin	OH	H	H
Acetylgenistin	OH	H	$COCH_3$
Malonylgenistin	OH	H	$COCH_2COOH$
Glycitin	H	OCH_3	H
Acetylglycitin	H	OCH_3	$COCH_3$
Malonylglycitin	H	OCH_3	$COCH_2COOH$

Figure 3. Structure of main isoflavone glycosides.

http://dx.doi.org/10.5772/67896

During the processing of raw soybeans, the composition of isoflavones is altered. The loss of isoflavones in the water used to soak raw soybeans, whey and the okara, was 4, 18 and 31%, respectively [42]. The isoflavone loss during coagulation in tofu processing was 44% and during alkaline extraction in soy protein isolate production was 53% [48]. The recoveries of isoflavones in tofu and in soy beverage comparing to their initial concentration in raw soybeans were found to be 36 and 54%, respectively [42].

Several reviews are available on the content of isoflavones in soy foods, including Japanese foods [49], foods used in the US [50, 51] and Europe [52]. A comparison of the most frequently used foods containing soy is performed in **Table 1**.

Food	Daidzein	Genistein	Daidzein	Genistein	Daidzein	Genistein
Country	Japanese foods		Food in the US		Food in Europe	
	μg/g wet food		μg/g		μg/g	
Soybean (raw)	1006.3	1437.7	613.3	863.3	580	840
Soybean (boiled)	135.8	472.5	74.1	70.6	150	320
Soybean sprout	49.6	79.3	50	67		
Mung bean sprout	3.4	2.4	0.6	0.8	39	680
Soymilk	78.2	156.6	48.84	60.7	–	–
Okara—bean curd residue	33.1	57.1	36.2	44.7	–	–
Tofu (momen type)	166.2	269.2	–	–	–	–
Tofu (silken type)	130.1	206.4	91.5	84.2	–	–
Tofu (packed type)	168.6	280.7	–	–	–	–
Tofu (baked type)	166.8	291.2	102.6	104.3	–	–
Tofu (dried type)	168.2	556.7	–	–	–	–
Deep fried tofu	–	–	138.0	184.3	–	–
Deep fried tofu (thick type)	148.7	257.4	–	–	–	–
Deep fried tofu (thin type)	84.2	179.1	–	–	–	–
Miso	–	–	164.3	232.4	590	670
Tempeh—fermented soybean from Indonesia	525.9	1326.2	226.6	361.5	–	–
Navy beans (haricot)	–	–	0.137	4.08	0.130	0.110
Broccoli	–	–	0.06	0.08	0	0
Carrots	–	–	0.016	0.017	0	0
Strawberries	–	–	trace	trace	0.046	–
Spinach	–	–	0	0	0	0
Onion	–	–	0	0	0	0

Table 1. Concentrations of daidzein and genistein in Japanese foods [49], food in the US, [50, 51] and food in EU [52].

Soy foods, such as tofu and tempeh, are extremely rich in isoflavones compared to other type of foods [53]. Raw soybeans contain the highest level of genistein and daidzein in the Japanese foods. The soybeans in US have a higher amount of genistein (863.3 μg/g) and daidzein (613.3 μg/g) than in Europe, which have 840 μg/g genistein and 580 μg/g daidzein [51, 52]. The richest sources of isoflavones are tofu, tempeh and miso in Japanese, European and American foods. Small variation from one region to another may occur. The isoflavone level in vegetables, fruits and other types of food are extremely low, sometimes in traces or not detectable. Among these, the navy beans have a higher level of genistein 4.08 μg/g.

One gram of soy protein in soybeans and traditional soy foods contain about 3.5 mg of isoflavones. One serving of a traditional soy food (100 g of tofu or 250 mL soymilk) provides about 25 mg isoflavones. In more refined products, it is possible that 80–90% of the isoflavone content to be lost during processing [27].

5.1. Soy foods in Asia

In Asia, soybeans are used in producing traditional foods such as tofu, soymilk and fermented products, while in Western nations, soybeans are used in the form of refined soy protein ingredients that are further used in food processing [45]. In Japan, the most popular soy food is tofu, served at all meals and in dessert products. Fermented foods such as natto and miso were very popular among Japanese and today are also largely consumed. Japanese adults consume approximately 6–11 g of soy protein and 25–50 mg of isoflavones (expressed as aglycone equivalents) per day. The results were higher than in Hong Kong and Singapore [54]. Concerning genistein and daidzein, the annual report of the national nutrition survey in 1997 in Japan shows that the dietary intake of isoflavones daidzein and genistein was 64.6 and 111.6 μmol/day/capita (16.4 and 30.1 mg/day/capita). The isoflavones intake was mostly attributable to tofu, natto and miso [49].

The mean plasma concentrations of total isoflavones are estimated to be 492.7 nM for genistein and 282.5 nM for daidzein in Japanese men, and 33.2 nM for genistein and 17.9 nM for daidzein in British men [55].

In China, not only tofu, yuba, soymilk and many regional specialities are served, but also soy powder mixes are becoming popular [45]. Isoflavone intake differs very much from a region to another [54]. In Taiwan, the meat substitutes from soy (chicken-like and fish-like products) are highly appreciated and, in Indonesia, tempeh is the most popular soy food [45].

5.2. Soy foods in Europe

Meat and dairy substitutes are the most popular soyfoods in Europe [45]. Traditional soyfoods rich in isoflavones (tofu, tempeh and miso) are rarely eaten in the UK, but soya dairy alternatives (milk, cheese and yogurts) are more commonly eaten. Some commercial products (bread, biscuits and breakfast cereals) contain soy ingredients as food additives which contribute to isoflavone intake [56].

Data from the 1998 UK Total Diet Study shows that daily intake of isoflavone aglycones (daidzein, genistein and glycitein) is approximately 3 mg/day [57].

In a group of 9 omnivores and a group of 10 vegetarians, mean isoflavone intake was measured after a 7-day food diary; mean daily isoflavone intake in the omnivorous and vegetarian groups was 1.2 and 7.4 mg, respectively. Main isoflavones food sources for the omnivorous group were soya yogurts, wholemeal bread and rolls, and for the vegetarian group were soymilk (plain), meat-substitute foods with soy protein isolate, beans, raisins and wholemeal bread and rolls [58].

5.3. Soy foods in the US

The consumption of soy food in the US resembles with the one in Europe. Popular soy foods in the US are tofu, soy sauce, soymilk, miso and tempeh. There can be found some new, adapted soy foods such as tempeh burgers, veggie burgers, tofu hot dogs, tofu ice cream, soymilk yogurt, soymilk cheeses, soy flour pancake mix and myriad [46]. Isoflavone intake is less than 3 mg/day in the US [27].

The Study of Women's Health Across the Nation demonstrated that median intake of daidzein and genistein by white subjects, African American subjects, and Japanese subjects in the US were 6.2 and 3.9 μg/day, 2.7 and 1.7 μg/day, and 4676 and 7151 μg/day, respectively. Women were aged between 42 and 52 years [59].

5.4. Soy foods in Africa

In certain African countries, soy foods gained acceptance due to the high protein level and nutrition quality. In other countries, it is used because of the food aid. In South Africa, modern soy foods are used [45]. The intake of soybeans in South Africa (0.64 g/day) is comparable with the soybeans intake in Germany (0.64 g/day).

As it can be observed in **Table 2**, the soybeans intake in Asian countries is higher than in the Western countries. For example, in 2001, in Taiwan, Japan and China, per capita consumption was approximately 19.15, 7.73 and 7.31 kg/year, respectively. In contrast, in the US, Germany and South Africa, per capita consumption was 0.33, 0.24 and 0.23 kg/year,

Country	Soybeans	
	kg/year	g/day
Taiwan	19.15	52.46
Japan	7.73	21.19
China	7.31	20.03
US	0.33	0.89
Germany	0.24	0.66
South Africa	0.23	0.64
World average	2.39	6.54

Source: Ref. [45].

Table 2. Annual per capita consumption of soybeans (2001).

respectively. Concerning isoflavone intake among Japanese adults, it is ranged from about 30–50 mg/day but is less than 3 mg/day in the US and Europe [27]. In Asian countries, the mean isoflavone consumption is 25–50 mg/day, whereas in Western countries, 1–2 mg/day is typical [60]. However, in the last decade, in Western countries, production and intake of soy foods have increased due to its important health benefits (relief of menopausal symptoms, improvement in bone health and reduced risk of certain types of cancers) [61].

6. Isoflavones: new insights regarding the therapeutic effect

An increased consumption of soy products has been remarqued in the last decade of the twentieth century and was associated with the awareness concerning the health benefits of these products, besides their role as a source of proteins [62]. The therapeutic activities of soy isoflavones were also evaluated in numerous studies for their estrogenic [63], lipid-lowering [64], anti-diabetic [65], anti-inflammatory [66], cardioprotective [67, 68] or anticancer effects [69]. *In vitro* and *in vivo* studies were performed in order to evaluate the therapeutic potential and the mechanisms of action of these compounds [70, 71].

Isoflavones are known as phytoestrogens due to their ability to bind the estrogen receptors and to exhibit estrogenic-like properties. Dietary supplements containing soy isoflavones are used to alleviate menopause disturbances as an alternative to hormone therapy [72]. Isoflavones attracted the attention of researchers after the observation of fertility problems in sheep grazing on a clover type rich in isoflavones [62]. The decreased risk of breast, prostate and colon cancer in Asian countries was associated with a higher intake of soy products in these population compared to western countries [70].

The possible benefits of isoflavones thus generated several studies to establish their therapeutic properties. Isoflavones exhibit both estrogenic and anti-estrogenic activity, binding to both α and β subtypes of estrogen receptor (ER) [73]. It was stated that isoflavones act as estrogen antagonists or agonists depending on estrogen concentration. Therefore, they are estrogen antagonists in a high estrogen environment, but when the estrogen quantity is reduced, they act as agonists [74]. The affinity for the estrogen receptors is different; genistein presents a 20–30 times higher affinity for ERβ than for ERα, while daidzein presents a weak affinity for both receptors, still higher for ERβ [70, 75]. Administration of 80 mg/day red clover isoflavones (containing genistein, daidzein, formononetin and biochanin A) for 90 days reduced hot flushes and night sweats, important vasomotor symptoms commonly found in postmenopausal women [76]. Low doses of isoflavones (25 mg/day containing 51.8% genistein, 43.3% daidzein and 4.9% glycitein) can reduce depression and insomnia in postmenopausal women [72].

Soy isoflavones present beneficial effects on lipid metabolism. The lipid-lowering activity of isoflavones has been observed in studies performed in animals. Soy isoflavones reduced the levels of triglycerides (TG) and low-density lipoprotein (LDL) in obese rats, and also exhibited benefits on obesity. The mechanisms involved are the suppression of mechanistic target of rapamycin complex 1 (mTORC1) activity that determines a reduction of

AKT phosphorylation [77]. A decrease of total cholesterol, LDL cholesterol levels and an improvement of ApoA1/ApoB ratio were noticed after the administration of 435 mg isoflavones/day for 2 months in women with type 2 diabetes [64].

The benefits of isoflavones in glucose and lipid metabolism have been previously reported, but the mechanisms of action are not yet fully understood [78]. Potential benefits in obesity were also observed after intake of isoflavones. Dietary modifications in early stages of this condition are important to prevent cardiovascular and metabolic disorders [79]. Some of the mechanisms involved in the anti-diabetic properties of isoflavone genistein are the enhancing of β-cell proliferation and the regulation of insulin secretion [65]. Isoflavones intake might reduce the risk of type 2 diabetes [53]. However, the beneficial role of isoflavones in this disease was not supported by other studies [78, 80]. A decreased risk of type 2 diabetes was noticed for women with high concentrations of genistein in plasma, but this correlation was not observed in men [81]. After the administration of a diet supplemented with 0.02% (w/w) genistein for 8 weeks (10–12 mg genistein/day) in Zucker diabetic fatty (ZDF) rats, there were no beneficial effects on glucose homeostasis or on skeletal muscle oxidative stress [78]. The conflicting results regarding the benefits of isoflavones in diabetes led to the assumption that other soy compounds might be responsible for this activity [65].

Genistein not only reduces weight gain in female obese *ob/ob* mice after the administration of 600 mg/kg diet for 4 weeks, but also promotes oxidative stress in the vasculature and inflammation in the perivascular adipose tissue [82]. Daidzin and glycitin (0.06% in diet) decrease blood glucose, insulin and HbA1c levels in mice with obesity and diabetes induced by a high-fat diet [79].

The effects of isoflavones on components of the metabolic syndrome were also evaluated. An extract from the roots of *Pueraria lobata*, administered 0.2% in the diet of female stroke-prone spontaneously hypertensive rats (SP-SHR) improved blood glucose levels, decreased plasma total cholesterol levels and reduced blood pressure, thus indicating beneficial effects on risk factors that led to the development of metabolic syndrome. The major isoflavones in the extract were puerarin (25.3%), daidzin (7.1%) and daidzein (0.8%) [83].

Even though the effects on obesity and diabetes were not sustained by other studies, an increase in bone mass was noticed in obese mice treated with 600 mg genistein/kg for 4 weeks [84]. The bone protective properties were also observed for other compounds. Formononetin, an isoflavone found mainly in the roots of *Astragalus membranaceus* and *Astragalus mongholicus*, improved the mechanical properties of the bones and exerted beneficial effects in rats with osteoporosis induced by ovariectomy after the administration of 10 mg/kg formononetin for 4 weeks [85]. The results of a meta-analysis investigating trials that evaluated the effects of isoflavone in osteoporosis, revealed that the effects of these compounds depend on dose and duration of administration. An increase in mineral density of the bone by 54% and a reduction in the fracture risk were noticed in women included in these studies [86]. A clinical trial regarding the effects of soy isoflavones on bone loss and menopausal symptoms did not reveal benefits in this condition. Women included in the study received tablets containing 200 mg soy isoflavones (genistein and daidzein)/day for 2 years [87].

The cardiovascular diseases are more frequent in women after menopause, due to the modifications in the production of estrogen [88]. Compounds with estrogenic properties, such as the isoflavones found in soy, were evaluated for their potentially cardioprotective activity [89, 90]. The administration of 80 mg/day isoflavones extracted from soybeans for 12 weeks in patients with primary or recurrent ischemic stroke determined a decrease in high sensitivity, C-reactive protein and improved vascular endothelial function [68]. A meta-analysis evaluating the results of nine trials concluded that isoflavone supplementation improves endothelial function in postmenopausal women that presents low flow-mediated dilatation (FMD) levels, but not in ones with high baseline FMD levels [89]. Nevertheless, the benefits of isoflavones regarding the protective effects in ischemic stroke are questionable. Yu et al. associated the high intake of soy isoflavones (53.6 mg isoflavones/day) with an increase in the risk of ischemic stroke in women [91].

The protective effect against the inflammatory vascular disease of isoflavones is due to their anti-inflammatory activity. The inhibition of monocyte adhesion to endothelial cells that involves the activation of PPARγ is related to this effect [66]. The anti-inflammatory properties are due to inhibition of interleukin-6 (IL-6), interleukin-12 (IL-12), interleukin-1β (IL-1β), and tumor necrosis factor-α (TNF-α) production, NF-κB regulation and their antioxidant activity [92]. Pro-oxidant effects at high doses were also noticed for these compounds. The supplementation with 640 mg/kg daidzein in pigs revealed antioxidant properties in muscle, but were accompanied by pro-oxidant effects in fat and liver tissues [93].

Administration of soy isoflavones (daidzein, genistein and glycetin) in capsules, 20 mg twice daily in female patients improved irritable bowel disease and association with vitamin D can determine a synergistic effect [94]. The estrogen-like effects of isoflavones genistein and daidzein seem to be involved in their beneficial effects on sleep status. A high intake of isoflavones was related to an optimal sleep duration and an increased quality of sleep in Japanese adults [95].

Age-related skin modifications in women emerging with a decline in estrogen production can be reduced by isoflavones. Genistein improves skin appearance and is used to reduce wrinkles and skin dryness in cosmetic preparations. It increases the skin resistance and contributes to skin reparation [96].

Anti-bacterial effects were also reported for isoflavones. For instance, biochanin A, a methylated isoflavone from red clover, inhibits the growth of *Chlamydia trachomatis* and *Chlamydia pneumoniae* [97]. Isoflavones act as anti-viral agents against several types of viruses including herpes simplex virus and human immunodeficiency virus (HIV). Genistein, a tyrosine kinase inhibitor, is one of the compounds most studied for these properties and revealed positive effects in inhibiting HIV-1 infection, especially when applied in entry and early post-entry stages [98].

7. Isoflavonoids: new insights regarding cancer chemoprevention

It is very well known that dietary components are, nowadays, considered important therapeutic agents used to prevent various chronic disorders, especially cancer, cardiovascular pathologies and inflammation processes [99]. For example, according to the statistics, the incidence of prostate cancer or breast cancer in Asian countries has been lower than in Western countries, mainly because of their high consumption of soy products [100]. Furthermore, the incidence of breast cancer in Asian women who had immigrated to Western countries proved to be similar to that of Western women [101]. In addition, soy isoflavones, with the main representative genistein, proved to be promising phytotherapeutic drugs with chemopreventive effects in chronic disorders caused by exposure to solar UVB radiations, including nonmelanoma skin cancer [102, 103].

Genistein has been extensively studied in different types of cancers cells and cancer animal models [104]. Unfortunately, the low oral bioavailability of genistein [105], due to its high lipophilicity and its extensive metabolism by the phase II enzymes [106], has limited its use in clinical trials [107].

The data regarding the chemopreventive effects of daidzein are limited, more studies being focused on the curative effect of daidzein in cancer and not in its prophylactic properties [108].

Previous *in vitro* evidences have shown that genistein may act not only as an agonist, but also as an antagonist on cancer cells depending on both its concentration and the type of cancer cells on which was tested [109]. In this regard, genistein because of its biphasic effect may be involved not only in preventing, but also in promoting cancer [110].

The *in vitro* chemopreventive effect of genistein might be related to its involvement in epigenetic regulations of gene expression, having a direct effect on histone modifications and DNA methylation [111]. Genistein is also a strong inhibitor of the tyrosine kinase [112] and topoisomerase activities [113]. The *in vitro* apoptotic effect of genistein may be related to the inactivation of NF-kB and Akt signaling pathways [114], although its specific mechanisms of inducing apoptosis have not being fully understood [115]. According to a previous *in vitro* study on pancreatic cancer cells, genistein suppressed the ovarian cancer cell growth and migration through the inhibition of mRNA [116]. The chemopreventive properties of genistein have also been proved on MCF-7 breast cancer cells, in which genistein has inhibited the cell proliferation by inactivating the IGF-1R-PI3 K/Akt pathway and decreasing the Bcl-2/Bax mRNA and protein expressions [115]. Another mechanism responsible for the chemopreventive effects of genistein might consist in the suppression of the microsomal CYP1a1 gene expression in Hepa-1c1c7 liver cancer cells [117].

The association between genistein and daidzein has proved antiproliferative effects on human colon adenocarcinoma grade II cell line (HT-29) [118], the anticancer effect of daidzein being related to copper-dependent pathway and redox cycle [108].

Among the soy isoflavones, GLY has been the most potent activator of extracellular signal-regulated kinase (ERK1/2), exhibiting a significant antiproliferative effect on RWPE-1 nontumorigenic prostate epithelial cells [119]. Thus, according to another study on the same type of cells, GLY has also shown to decrease the expression of cytokeratin 18 and prostate-specific antigen (PSA) [120].

Furthermore, *in vivo* data have shown that oral consumption of genistein during the early prepubertal period decreased the susceptibility of developing breast cancer later in life [121]. Genistein, as a phytoestrogen, may interact with the estrogen receptors. In this regard, a previous study on HER2 overexpressing mice has shown that genistein mimicked the estradiol effects, in the presence of estrogen receptor alpha [112], while in postmenopausal women, in the absence of estradiol, genistein directly reduced the anticancer activity of cisplatin, a cytostatic drug commonly used in breast cancer [122]. For instance, according to Tonetti et al. *in vitro* and *in vivo* studies, the concomitant administration of tamoxifen with genistein or daidzein might not be safe because this association has produced bigger size tumors than tamoxifen alone [123].

Soy isoflavones exhibited *in vivo* protective effects against skin chronic disorders including cancer by reducing pro-inflammatory cytokines and oxidative stress and through activation of NF-kB [124]. For example, genistein has suppressed UV-induced skin carcinogenesis in mice through its moderate inhibitory effect on ornithine decarboxylase activity [125]. Moreover, 7,3′,4′-trihydroxyisoflavone, a major metabolite of daidzein, has reduced UVB-induced skin cancer in mice through inhibition of cyclooxygenase-2 (COX-2) expression by suppressing NF-kB transcription activity [103]. In this regard, genistein loaded-PLA nanocapsules indicated to be a promising formulation with chemopreventive effects against skin cancer in porcine ear skin not only by increasing the penetration of genistein in skin deeper layers, but also by limiting its degradation in time [107].

According to Ghaemi et al. study on human papillomavirus (HPV) associated-cervical cancer in mice, genistein has also indicated immunomodulatory effects through increment of interferon-gamma (IFN-gamma) level, lymphocyte proliferation and lactate dehydrogenase (LDH) release [126].

Consequently, the isoflavones from soy products may be considered promising alternative therapies to prevent various types of cancer, more experimental and clinical studies being necessary for establishing the safe dose that can be used especially in patients susceptible to hormone-dependent tumors.

8. Anti-inflammatory and chemopreventive activity of isoflavonoid genistein alone and incorporated in modern pharmaceutical formulations

From the main isoflavones reported above, we have studied genistein. One of the most important lines in our research group on this topic involves the analysis of the chemopreventive effect of the phytoestrogen genistein against malignant melanoma. The isoflavonoid genistein (4′,5,7-trihydroxyisoflavone) is the aglycone of heteroside genistin. It is the most studied compound from the class of isoflavones together with daidzein, glycitein, formononetin, equol and biochanin A. It is the major active compound from soy seeds, *Glycine max* (L.) Merr., family *Fabaceae* [127–129]. Regarding the biological activities, recent papers report that: genistein induces apoptosis, inhibits cell proliferation, modulates cell cycle progression on different cancerous cell lines, inhibits angiogenesis, suppress lymphocyte activation and proliferation, stabilizes mast cells and presents mild anti-inflammatory properties [130, 131]. The phytoestrogen also inhibits the production of reactive oxygen species (ROS) which is directly correlated with DNA modification and tissue damage. Production of ROS, especially by activated cells of the immune system, has been postulated to play an important role in carcinogenesis, particularly in tumor promotion [132, 133].

In a recent complex study employing, the B164A5 and B16F10 murine melanoma cell lines, we have shown testing a wide range of concentrations (150, 100, 50, 30, 15, 5 and 1 μM) that this phytocompound is an active antiproliferative and pro-apoptotic agent on this two cell lines. MTT (3-(4,5-Dimethylthiazol-2-yl)-2,5-diphenyltetrazolium bromide) assay has shown an IC_{50} of 41.1 μM genistein for B164A5 cells and 61.4 μM genistein for B16F10 cells. The carboxyfluorescein diacetate succinimidyl ester (CFSE) assay has shown that after 24 h of incubation, proliferation of B16 cells was decreased when treated with 30 μM genistein. Genistein at 100 μM was able to cause G2/M arrest in the cell cycle of these murine melanoma cell lines [134]. DAPI staining was performed in order to detect first signs of apoptosis. When B16 cells were incubated with 100 μM genistein, this phenomenon could be detected and translated by a reduction of the cell number and increase in nuclear fragmentation compared to the control group [134]. Western blot analysis was furtherer conduced, for four important proteins involved in the process of apoptosis, namely caspase 3, poly(ADP-ribose) polymerase (PARP), Bax and Bcl-2. Incubation with 100 μM genistein conduced for both B16 cell lines to cleaved caspase-3 as well as cleaved PARP as responsible for the mechanism of apoptotic events. In another study using all concentrations previously tested (150, 100, 50, 30, 15, 5 and 1 μM genistein) and after a period of incubation of 72 h, we have shown that the phytoestrogen does not induce caspase-2 activation *in vitro* on B16 melanoma cell lines [135]. In order to get the full picture about apoptosis, namely to detect early and late apoptotic cells, annexin V-FITC/7AAD staining was performed by fluorescence cytometry in parallel for the B16 melanoma cells as well as for the bone marrow-derived dendritic cells (BMDCs). At its highest concentration, genistein induced slightly more apoptotic events in the B16 cells than in the BMDCs [134]. The aim of the above-mentioned study was to find a drug that "kills" the cancerous cells and stimulate the immune activity. On this purpose, the potential immune stimulatory activity of genistein was tested by measuring the anti-tumorigenic cytokine IL-12p70 released by murine primary BMDCs. The phytoestrogen, at the concentration of 5 μM decreased the level of IL-12p70 of the LPS-stimulated DCs. These findings were in line with the observation that also the IL-12p35 mRNA levels were downregulated. T cell activity was further screened by analyzing the concentration of IFN-γ and IL-2 cytokine in the supernatant of spleen cells of OT I mice expressing the ovalbumin-specific transgenic T cell receptors. Results have shown that genistein had no effect on the concentration of IFN-γ and IL-2 cytokine [134].

Besides the chemopreventive activity for murine melanoma, our research group have investigated genistein alone and incorporated in randomly methylated β-cyclodextrin (RAMEB), hydroxypropyl-β-cyclodextrin (HPBCD) and hydroxypropyl-γ-cyclodextrin (HPGCD) in a molar ratio 1:1 for a range of biological activities. This approach was chosen in order to increase the water solubility of this lipophilic compound. CDs are cyclo-oligosaccharides presenting a hydrophilic outside and hydrophobic inner side with the ability to form host-guest inclusion complexes with an increased number of chemical structures [136]. Firstly, quantum chemical calculations were performed analyzing the behavior in gas phase, in water and in dimethyl sulfoxide, the solvent used for the solubilization of active agents for all the mentioned assays. Additionally, it was proofed that incorporation of genistein in the above-mentioned CDs took place by a series of consecrated techniques such as phase solubility studies, differential scanning calorimetry (DSC), X-ray diffraction and scanning electron microscopy (SEM) assays [136]. Genistein and its inclusion complexes were studied *in vitro* on four types of cancer cells lines such as HeLa (cervical adenocarcinoma), MCF-7 (breast adenocarcinoma), A2780 (human ovarian carcinoma) and A431 (skin epidermoid carcinoma) cell lines using the following concentrations: 1, 3, 10, 30, 60 and 90 μM and a period of incubation of 72 h. A2780 human ovarian carcinoma cell line proofed to be the most sensitive to genistein, followed by HeLa. Proliferation was not significantly affected for the other two cell lines. After incorporation in the above-mentioned CDs, changes in the antiproliferative action occurred with respect to the tested cell line. Cervical adenocarcinoma HeLa cell line was more sensitive for all three inclusion complexes when compared to pure genistein. The same behavior was found for A2780 cell line, except for the complex with RAMEB. Complexation with RAMEB conduced to an increased IC_{50} also for MCF-7 and A431 cell lines. For this two cell lines, complexation of genistein with HPBCD seemed to be the best option [136]. In the same study, genistein and its CD complexes were analyzed by the agar disk-diffusion method and the dilution method against several bacterial strains: *Bacillus subtilis, Enterococcus faecalis, Escherichia coli, Salmonella typhimurium, Shigella sonnei, Pseudomonas aeruginosa* and *Staphylococcus aureus*. Tested compounds at the concentration of 10 mM presented antibacterial activity only for *B. subtilis* [137]. The last line of this study was drawn toward tests for antiangiogenic effects employing the chorioallantoic membrane of the chicken embryo. Pure genistein presented antiangiogenic effects and the HPBCD complex showed superior activity. Also for the other two complexes, namely HPGCD and RAMEB could be described an antiangiogenic effect but a decreased one as evaluated by applying the 0–5 score [136].

Another attempt to increase the bioavailability of this lipophilic phytoestrogen was directed toward the synthesis and analysis of a genistein ester derivative with myristic acid and complexed with beta cyclodextrin. The successful synthesis of the new compound as well as the successful inclusion in beta cyclodextrin was determined using consecrated assays such as TLC analysis, HPLC analysis, FTIR spectroscopy, MS spectroscopy, differential scanning calorimetry (DSC) and scanning electron microscopy (SEM). Samples were tested *in vitro*, using the MTT proliferation assay on three human cell lines: HeLa—cervix adenocarcinoma, A2780—ovary carcinoma and A431—skin epidermoid carcinoma. Results have shown that, after a period of incubation of 72 h at the concentrations of 10 and 30 μM, respectively, genistein is an active agent on HeLa (cervix adenocarcinoma) and A2780 (ovary carcinoma) cell lines. The new formulations did not decrease the viability of the cancerous cells. This behavior may be explained by the increased stability of the complex within the *in vitro* environment [138].

A new modern formulation explored by the pharmaceutical industry, but not only, is the polyurethane microstructures (PM). What determined us to focus on these compounds? Depending on the structure of the particle, such approach can offer: the possibility of amending the lipo- or water-solubility of inclusion structures, protection from external agents such as UV radiation, strong acidic or alkaline environments, drug delivery toward a specific receptor or retard activity of the biologically active compound due to the use of transport vehicles with low speed of degradation [139]. Based on these hypotheses, we have synthesized PM with a yield of encapsulation of 68.3% genistein (w/w). The formulation was tested *in vitro* using the

MTT proliferation assay on three human breast cancer cell lines MCF7, MDA-MB-231 and T47D-human breast adenocarcinoma cell lines. Tests were performed also for the antimicrobial and antifungal activity against the following strains: *S. aureus, E. coli, P. aeruginosa, Salmonella enteritidis, B. subtilis, Bacillus cereus* and *Candida albicans* employing the dilution method. Results made us to conclude that the PM are a bad in *vitro* carrier partner for genistein [137].

In vivo assays were also performed in order to test the chemopreventive effects of genistein against murine melanoma. During the research in our group, we have observed in a murine model of melanoma, obtained by subcutaneous injection of 0.1 ml of $1*10^5$ B164A5 cells/mouse that, genistein after a period of 15 days at a dose of 15 mg/kg, body weight decreased tumor volume and weigh with about 30% and reduced distance tumors. Noninvasive measurements using the Multiprobe Adapter System (MPA5) from Courage-Khazaka, Germany, Mexameter® MX 18 showed that genistein reduced the quantity of melanin and the degree of erythema directly correlated with the number of days of treatment [140]. Being very well known, the link between inflammation and cancer, we have analyzed the effect of genistein in an animal model of ear inflammation alone and after incorporation in hydroxypropyl-beta-cyclodextrin (HPBCD) and randomly-methylated-beta-cyclodextrin (RAMEB). Cyclodextrins (CDs) are well known agents used to increase the hydro solubility of active lipophilic agents [139]. The study concluded that the phytoestrogen, at a dose of 2 mg can be reconsidered as an active anti-inflammatory natural compound on C57BL/6 J animal model of inflammation. Additionally, complexation of genistein with the above-mentioned CDs was done and led to a stronger anti-inflammatory effect [141]. In a recent study, in order to attempt to increase the bioavailability of this phytoestrogen, we have adopted a new strategy that combines two elements: the formulation and the modality of administration. The formulation was lamellar lyotropic liquid crystal in which genistein was incorporated at the concentration 3% and the formulation was applied local, with or without electroporation (EP), using the Mezoforte Duo Mez 120905-D device on C57BL6J. Results have shown that tumors appeared later when electroporation was applied. During the 21 days of the experiment, genistein incorporated in the new modern formulation, applied topically classic decreased the tumor volume, the degree of erythema and amount of melanin for mice bearing B16 murine melanoma tumors. When the formulation was applied by electroporation, the prognosis was even better. However, the new approach had no effect in terms of serum concentrations of the protein S100B and serum neuron-specific enolase (NSE), or the tissue expression of the platelet-derived growth factor receptor β (PDGFRβ) antibody [142]. Also, soy total extract incorporated in the new modern lamellar lyotropic liquid crystal formulation was tested *in vitro* on the B164A5 mouse melanoma cell line for its pro-apoptotic potential, employing two consecrated assays: 4′,6-diamidino-2-phenylindole (DAPI) and annexin-FITC-7AAD double staining. 200 μg/ml of soy extract, respectively 200 μg/ml of soy extract incorporated into the lamellar lyotropic liquid crystalline formulation were incubated for 72 h together with this murine melanoma cell line. Results have shown that soy extract has pro-apoptotic properties and incorporation in the new formulation does not affect in a negative manner this effect thus being a suitable excipient for *in vivo* studies [143]. In a recent study, diffusion and penetration of genistein, respectively, genistein incorporated in lamellar lyotropic liquid crystalline formulation through different membranes (a synthetic membrane *in vitro*, chick chorioallantoic membrane (CAM) *ex ovo*, and excised human epidermis *ex vivo*) were also investigated by conventional treatment without EP, and also with the mediation of EP by the help of a Franz diffusion cell system. *In vivo* ATR-FTIR and *ex vivo* Raman spectroscopy were applied in order to analyze the effect on mice skin [144]. Results have shown that the new formulation is a suitable carrier for the lipophilic genistein. The formulation with the active agent penetrated the skin, but when electroporation was applied, the transdermal drug transport was more rapid and effective. This observation was validated by both ATR-FTIR and Raman spectroscopy [144].

The research of our group on the phytoestrogen genistein points toward the clear conclusion that this phytocompound is an active chemopreventive agent against malignant melanoma both *in vitro* and *in vivo*. A series of attempts were made in order to increase the bioavailability of this lipophilic compound. We cannot say that we have found the optimal formulation, but we have managed to improve results compared to pure substance. Further studies will be conducted on this matter.

Author details

Corina Danciu[1]*, Diana Simona Antal[2], Florina Ardelean[2], Aimée Rodica Chiş[2], Codruţa Şoica[3], Florina Andrica[3] and Cristina Dehelean[4]

*Address all correspondence to: corina.danciu@umft.ro

1 Department of Pharmacognosy, Faculty of Pharmacy, University of Medicine and Pharmacy "Victor Babeş", Timişoara, România

2 Department of Pharmaceutical Botany, Faculty of Pharmacy, University of Medicine and Pharmacy "Victor Babeş", Timişoara, România

3 Department of Pharmaceutical Chemistry, Faculty of Pharmacy, University of Medicine and Pharmacy "Victor Babeş", Timişoara, România

4 Department of Toxicology, Faculty of Pharmacy, University of Medicine and Pharmacy "Victor Babeş", Timişoara, România

References

[1] Reynaud J, Guilet D, Terreux R, Lussignol M, Walchshofer N. Isoflavonoids in non-leguminous families: an update. Nat Prod Rep. 2005; **22(4)**: 504-15.

[2] Veitch NC. Isoflavonoids of the leguminosae. Nat Prod Rep. 2013; **30(7)**: 988-1027. DOI: 10.1039/c3np70024k.

[3] Hegnauer R, Grayer-Barkmeijer RJ. Relevance of seed polysaccharides and flavonoids for the classification of the leguminosae: a chemotaxonomic approach. Phytochemistry. 1993; **34**: 3-16.

[4] Mackova Z, Koblovska R, Lapcik O. Distribution of isoflavonoids in non-leguminous taxa – an update. Phytochemistry. 2006; **67**: 849-55.

[5] Dakora FD. Defining new roles for plant and rhizobial molecules in sole and mixed plant cultures involving symbiotic legumes. New Phytologist. 2003; **158(1)**: 39-49.

[6] Subramanian S, Stacey G, Yu O. Distinct, crucial roles of flavonoids during legume nodulation. Trends Plant Sci. 2007; **12(7)**: 282-5. DOI: 10.1016/j.tplants.2007.06.006.

[7] Jung W, Yu O, Lau SM, O'Keefe DP, Odell J, Fader G, McGonigle B. Identification and expression of isoflavone synthase, the key enzyme for biosynthesis of isoflavones in legumes. Nat Biotechnol. 2000; **18(2)**: 208-12; erratum Jung W, Yu O, Lau S-MC, O'Keefe DP, Odell J, Fader G, McGonigle B. Nat Biotechnol. 2000; 18: 559.

[8] Hagmann M, Grisebach H. Enzymatic rearrangement of flavanone to isoflavone. FEBS Letters. 1984; **175(2)**: 199-202.

[9] Dhaubhadel S. Regulation of isoflavonoid biosynthesis in Soybean Seeds, Soybean-Biochemistry, Chemistry and Physiology, Prof. Tzi-Bun Ng (Ed.), 2011; ISBN: 978-953-307-219-7, InTech, Available from: http://www.intechopen.com/books/soybean-biochemistry-chemistry-and-physiology/regulation-ofisoflavonoid-biosynthesis-in-soybean-seeds.

[10] Yu O, Jung W, Shi J, Croes RA, Fader GM, McGonigle B, Odell JT. Production of the isoflavones genistein and daidzein in non-legume dicot and monocot tissues. Plant Physiology. 2000; **124(2)**: 781-93.

[11] Yu O, Shi J, Hession AO, Maxwell AA, McGonigle B, Odell JT. Metabolic engineering to increase isoflavone biosynthesis in soybean seeds. Phytochemistry. 2003; **63(7)**: 753-63.

[12] Liu CJ, Blount JW, Steele CL, Dixon RA. Bottlenecks for metabolic engineering of isoflavone glycoconjugates in Arabidopsis. Proc Nat Acad Sci USA. 2002; **99(22)**: 14578-83.

[13] Dastmalchi M, Dhaubhadel S. Soybean seed isoflavonoids: biosynthesis and regulation. Phytochemicals – biosynthesis, function and application. Recent Adv Phytochem, 2014; **44**: 1-21.

[14] Li M-W, Muñoz Nacira B., Wong C-F, Wong F-L, Wong K-S, Wong JW-H, Xinpeng Q, Li K-P, Ng M-S, Lam H-M. QTLs regulating the contents of antioxidants, phenolics, and flavonoids in soybean seeds share a common genomic region. Front Plant Sci. 2016; **7**: 854. DOI: 10.3389/fpls.2016.00854.

[15] Albulescu M, Popovici M. Isoflavones-biochemistry, pharmacology and therapeutic use. Rev Roum Chim. 2007; **52(6)**: 537-50.

[16] Nguyenle T, Wang E, Cheung AP. An investigation on the extraction and concentration of isoflavones in soy-based products. J Pharm Biomed Anal. 1995; **14(1-2)**: 221-32.

[17] Coward L, Smith M, Kirk M, Barnes S. Chemical modification of isoflavones in soyfoods during cooking and processing. Am J Clin Nutr. 1998; **68(6)**: 1486S–91.

[18] Griffith AP, Collison MW. Improved methods for the extraction and analysis of isoflavones from soy-containing foods and nutritional supplements by reversed-phase

high-performance liquid chromatography and liquid chromatography – mass spectrometry. J Chromatography A. 2001; **913**: 397-413.

[19] Murphy P, Barua K, Hauck CC. Solvent extraction selection in the determination of isoflavones in soy foods. J Chromatogr B. 2002; **777(1-2)**: 129-38.

[20] Degenhardt A, Winterhalter P. Isolation and purification of isoflavones from soy flour by high-speed countercurrent chromatography. Eur Food Res Technol. 2001; **213(4)**: 277-80.

[21] Stürtz M, Lander V, Schmid W, Winterhalter P. Preparative isolation of isoflavones from soy and red clover. Mol Nutr Food Res. 2006; **50(4-5)**: 356-61.

[22] Cao Y, Calafat AM, Doerge DR, Umbach D, Bernbaum JC, Twaddle NC, Ye X, Rogan WJ. Isoflavones in urine, saliva, and blood of infants: data from a pilot study on the estrogenic activity of soy formula. J Expo Sci Environ Epidemiol. 2009; **19(2)**: 223-34. DOI: 10.1038/jes.2008.44.

[23] Setchell KD, Brown NM, Desai P, Zimmer-Nechemias L, Wolfe BE, Brashear WT, Kirschner AS, Cassidy A, Heubi JE. Bioavailability of pure isoflavones in healthy humans and analysis of commercial soy isoflavone supplements. J Nutr. 2001; **131**: 1362-75.

[24] Cassidy A. Factors affecting the bioavailability of soy isoflavones in humans. J AOAC Int. 2006; **89**: 1182-8.

[25] Yang Z, Kulkarni K, Zhu W, Hu M. Bioavailability and pharmacokinetics of genistein: mechanistic studies on its ADME. Anticancer Agents Med Chem. 2012; **12(10)**: 1264-80.

[26] Setchell KD, Borriello SP, Hulme P, Kirk DN, Axelson M. Nonsteroidal estrogens of dietary origin: possible roles in hormone-dependent disease. Am J Clin Nutr. 1984; **40**: 569-78.

[27] Messina M. Soy and health update: evaluation of the clinical and epidemiologic literature. Nutrients. 2016; **8**: 754. DOI: 10.3390/nu8120754.

[28] He J, Wang S, Zhou M, Yu W, Zhang Y, He X. Phytoestrogens and risk of prostate cancer: a meta-analysis of observational studies. World J Surg Oncol. 2015; **13**: 231. DOI: 10.1186/s12957-015-0648-9.

[29] Manach C, Williamson G, Morand C, Scalbert A, Rémésy C. Bioavailability and bioefficacy of polyphenols in humans. I. Review of 97 bioavailability studies. Am J Clin Nutr. 2005; **81**: 230-42.

[30] Setchell KD, Brown NM, Desai PB, Zimmer-Nechimias L, Wolfe B, Jakate AS, Creutzinger V, Heubi JE. Bioavailability, disposition, and dose-response effects of soy isoflavones when consumed by healthy women at physiologically typical dietary intakes. J Nutr. 2003; **133**: 1027-35.

[31] Scalbert A, Williamson G. Dietary intake and bioavailability of polyphenols. J Nutr. 2000;**130**: 2073-208.

[32] Gu L, House SE, Prior RL, Fang N, Ronis MJ, Clarkson TB, Wilson ME, Badger TM. Metabolic phenotype of isoflavones differ among female rats, pigs, monkeys, and women. J Nutr. 2006; **136(5)**: 1215-21.

[33] Adlercreutz H, van der Wildt, Kinzel J, Attalla H, Wähälä K, Mäkelä T, Hase T, Fotsis T. Lignan and isoflavonoid conjugates in human urine. J Steroid Biochem Mol Biol. 1995; **52(1)**: 97-103.

[34] Coldham NG, Sauer MJ. Pharmacokinetics of [(14)C] genistein in the rat: gender-related differences, potential mechanisms of biological action, and implications for human health. Toxicol Appl Pharmacol. 2000; **164(2)**: 206-15.

[35] Messina MJ. Legumes and soybeans: overview of their nutritional profiles and health effects. Am J Clin Nutr. 1999; **70**: 439-50.

[36] Liu ZM, Ho SC, Chen YM, Ho S, To K, Tomlinson B, Woo J. Whole soy, but not purified daidzein, had a favorable effect on improvement of cardiovascular risks: a 6-month randomized, double-blind, and placebo-controlled trial in equol-producing postmenopausal women. Mol Nutr Food Res. 2014; **58**: 709-17. DOI: 10.1038/ejcn.2015.24.

[37] Pakalapati G, Li L, Gretz N, Koch E, Wink M. Influence of red clover (Trifolium pratense) isoflavones on gene and protein expression profiles in liver of ovariectomized rats. Phytomedicine. 2009; **16**: 845-55. DOI: 10.1016/j.phymed.2009.03.003.

[38] Zaheer K, Akhtar MH. An updated review of dietary isoflavones: nutrition, processing, bioavailability and impacts on human health. Crit Rev Food Sci Nutr. 2017; **57(6)**: 1280-1293. DOI: 10.1080/10408398.2014.989958.

[39] Kaufman PB, Duke JA, Brielmann H, Boik J, Hoyt JE. A comparative survey of leguminous plants as sources of the isoflavones, genistein and daidzein: implications for human nutrition and health. J Altern Complement Med. 1997; **3**: 7-12. DOI: 10.1089/acm.1997.3.7.

[40] Kumar MP, Sankeshi V, Naik RR, Thirupathi P, Das B, Raju TN. The inhibitory effect of Isoflavones isolated from Caesalpinia pulcherrima on aldose reductase in STZ induced diabetic rats. Chem Biol Interact. 2015; **237**: 18-24. DOI: 10.1016/j.cbi.2015.05.010.

[41] Soto-Zarazúa MG, Rodrigues F, Pimentel FB, Bah MM, Oliveira MB. The isoflavone content of two new alfalfa-derived products for instant beverage preparation. Food Funct. 2016; **7**: 364-71. DOI: 10.1039/c5fo01115a.

[42] Jackson CJ, Dini JP, Lavandier C, Rupasinghe HP, Faulkner H, Poysa V, Buzzell D, DeGrandis S. Effects of processing on the content and composition of isoflavones during manufacturing of soy beverage and tofu. Process Biochem. 2002; **37**: 1117-23. DOI: 10.1016/S0032-9592(01)00323-5.

[43] Zhang J, Ge Y, Han F, Li B, Yan S, Sun J, Wang L. Isoflavone content of soybean cultivars from maturity group 0 to VI grown in northern and southern China. J Am Oil Chem Soc. 2014; **91**: 1019-28. DOI: 10.1007/s11746-014-2440-3.

[44] Soukup ST, Helppi J, Müller DR, Zierau O, Watzl B, Vollmer G, Diel P, Bub A, Kulling SE. Phase II metabolism of the soy isoflavones genistein and daidzein in humans, rats and mice: a cross-species and sex comparison. Arch Toxicol. 2016; **90**: 1335-47. DOI: 10.1007/s00204-016-1663-5.

[45] Golbitz P, Jordan J. Soyfoods: Market and Products. Soy Applications in Food. Taylor & Francis; Boca Raton, Florida, USA. 2006. pp. 2-20.

[46] Golbitz P. Traditional soyfoods: processing and products. J Nutr. 1995; **125**: 570-2.

[47] Nagino T, Mitsuyoshi KA, Masuoka N, Chiaki KA, Michitoshi AN, Miyazaki K, Kamachi K, Isozaki M, Suzuki C, Kasuga C, Tanaka A. Intake of a fermented soymilk beverage containing moderate levels of isoflavone aglycones enhances bioavailability of isoflavones in healthy premenopausal Japanese women: a double-blind, placebo-controlled, single-dose, crossover trial. Biosci Microbiota Food Health. 2016; **35**: 9-17. DOI: 10.12938/bmfh.2015-011.

[48] Wang HJ, Murphy PA. Mass balance study of isoflavones during soybean processing. J. Agric Food Chem. 1996; **44**: 2377-83. DOI: 10.1021/jf950535p.

[49] Arai Y, Watanabe S, Kimira M, Shimoi K, Mochizuki R, Kinae N. Dietary intakes of flavonols, flavones and isoflavones by Japanese women and the inverse correlation between quercetin intake and plasma LDL cholesterol concentration. J Nutr. 2000; **130**: 2243-50.

[50] de Kleijn MJ, van der Schouw YT, Wilson PW, Adlercreutz H, Mazur W, Grobbee DE, Jacques PF. Intake of dietary phytoestrogens is low in postmenopausal women in the United States: the Framingham study1-4. J Nutr. 2001; **131**: 1826-32.

[51] Bhagwat S, Haytowitz DB, Holden JM. USDA Database for the Isoflavone Content of Selected Foods Release 2.0. Maryland: US Department of Agriculture. 2008; 15.

[52] Liggins J, Bluck LJ, Runswick S, Atkinson C, Coward WA, Bingham SA. Daidzein and genistein contents of vegetables. Br J Nutr. 2000; **84**: 717-25. DOI: 10.1017/S0007114500002075.

[53] Ding M, Pan A, Manson JE, Willett WC, Malik V, Rosner B, Giovannucci E, Hu FB, Sun Q. Consumption of soy foods and isoflavones and risk of type 2 diabetes: a pooled analysis of three US cohorts. Eur J Clin Nutr. 2016; **70**: 1381-7. DOI: 10.1038/ejcn.2016.117.

[54] Messina M, Nagata C, Wu AH. Estimated Asian adult soy protein and isoflavone intakes. Nutr Cancer. 2006; **55**: 1-2. DOI: 10.1207/s15327914nc5501_1.

[55] Masilamani M, Wei J, Sampson HA. Regulation of the immune response by soybean isoflavones. Immunol Res. 2012; **54**: 95-110. DOI: 10.1007/s12026-012-8331-5.

[56] Mulligan AA, Welch AA, McTaggart AA, Bhaniani A, Bingham SA. Intakes and sources of soya foods and isoflavones in a UK population cohort study (EPIC-Norfolk). Eur J Clin Nutr. 2007; **61**: 248-54. DOI: 10.1038/sj.ejcn.1602509.

[57] Clarke DB, Lloyd AS. Dietary exposure estimates of isoflavones from the 1998 UK Total Diet Study. Food Addi Contam. 2004; **21**: 305-16. DOI: 10.1080/02652030410001668781.

[58] Ritchie MR, Cummings JH, Morton MS, Steel CM, Bolton-Smith C, Riches AC. A newly constructed and validated isoflavone database for the assessment of total genistein and daidzein intake. Br J Nutr. 2006; **95**: 204-13. DOI: 10.1079/BJN20051603.

[59] Greendale GA, FitzGerald G, Huang MH, Sternfeld B, Gold E, Seeman T, Sherman S, Sowers M. Dietary soy isoflavones and bone mineral density: results from the study of women's health across the nation. Am J Epidemiol. 2002; **155**: 746-54. DOI: 10.1093/aje/155.8.746.

[60] Reverri EJ, LaSalle CD, Franke AA, Steinberg FM. Soy provides modest benefits on endothelial function without affecting inflammatory biomarkers in adults at cardiometabolic risk. Mol Nutr Food Res. 2015; **59**: 323-33. DOI: 10.1002/mnfr.201400270.

[61] Sathyapalan T, Rigby AS, Bhasin S, Thatcher NJ, Kilpatrick ES, Atkin SL. Effect of soy in men with type 2 diabetes mellitus and subclinical hypogonadism – a randomized controlled study. J Clin Endocrinol Metab. 2017 Feb 1;**102**(2):425-433. DOI: 10.1210/jc. 2016-2875.

[62] Wang Q, Ge X, Tian X, Zhang Y, Zhang J, Zhang P. Soy isoflavone: the multipurpose phytochemical (review). Biomed Rep. 2013; **1(5)**: 697-701.

[63] Vitale DC, Piazza C, Melilli B, Drago F, Salomone S. Isoflavones: estrogenic activity, biological effect and bioavailability. Eur J Drug Metab Pharmacokinet. 2013; **38(1)**: 15-25.

[64] Chi XX, Zhang T, Zhang DJ, Yu W, Wang QY, Zhen JL. Effects of isoflavones on lipid and apolipoprotein levels in patients with type 2 diabetes in Heilongjiang Province in China. J Clin Biochem Nutr. 2016; **59(2)**: 134-8.

[65] Gilbert ER, Liu D. Anti-diabetic functions of soy isoflavone genistein: mechanisms underlying its effects on pancreatic β-cell function. Food Funct. 2013; **4(2)**: 200-12. DOI: 10.1039/c2fo30199g.

[66] Chacko BK, Chandler RT, D'Alessandro TL, Mundhekar A, Khoo NK, Botting N, Barnes S, Patel RP. Anti-inflammatory effects of isoflavones are dependent on flow and human endothelial cell PPARgamma. J Nutr. 2007; **137(2)**: 351-6.

[67] Gil-Izquierdo A, Penalvo JL, Gil JI, Medina S, Horcajada MN, Lafay S, Silberberg M, Llorach R, Zafrilla P, Garcia-Mora P, Ferreres F. Soy isoflavones and cardiovascular disease epidemiological, clinical and –omics perspectives. Curr Pharm Biotechnol. 2012; **13(5)**: 624-31.

[68] Chan YH, Lau KK, Yiu KH, Li SW, Chan HT, Fong DY, Tam S, Lau CP, Tse HF. Reduction of C-reactive protein with isoflavone supplement reverses endothelial dysfunction in patients with ischaemic stroke. Eur Heart J. 2008; **29(22)**: 2800-7. DOI: 10.1093/eurheartj/ehn409.

[69] Zhang YF, Kang HB, Li BL, Zhang RM. Positive effects of soy isoflavone food on survival of breast cancer patients in China. Asian Pac J Cancer Prev. 2012; **13(2)**: 479-82.

[70] Pilšáková L, Riečanský I, Jagla F. The physiological actions of isoflavone phytoestrogens. Physiol Res. 2010; **59(5)**: 651-64.

[71] Kalaiselvan V, Kalaivani M, Vijayakumar A, Sureshkumar K, Venkateskumar K. Current knowledge and future direction of research on soy isoflavones as a therapeutic agents. Pharmacogn Rev. 2010; **4(8)**: 111-7. DOI: 10.4103/0973-7847.70900.

[72] Hirose A, Terauchi M, Akiyoshi M, Owa Y, Kato K, Kubota T. Low-dose isoflavone aglycone alleviates psychological symptoms of menopause in Japanese women: a randomized, double-blind, placebo-controlled study. Arch Gynecol Obstet. 2016; **293(3)**: 609-15. DOI: 10.1007/s00404-015-3849-0.

[73] Möller FJ, Diel P, Zierau O, Hertrampf T, Maass J, Vollmer G. Long-term dietary isoflavone exposure enhances estrogen sensitivity of rat uterine responsiveness mediated through estrogen receptor alpha. Toxicol Lett. 2010; **196(3)**: 142-53. DOI: 10.1016/j.toxlet.2010.03.1117.

[74] Hwang CS, Kwak HS, Lim HJ, Lee SH, Kang YS, Choe TB, Hur HG, Han KO. Isoflavone metabolites and their in vitro dual functions: they can act as an estrogenic agonist or antagonist depending on the estrogen concentration. J Steroid Biochem Mol Biol. 2006; **101(4-5)**: 246-53.

[75] Gencel VB, Benjamin MM, Bahou SN, Khalil RA. Vascular effects of phytoestrogens and alternative menopausal hormone therapy in cardiovascular disease. Mini Rev Med Chem. 2012; **12(2)**: 149-74.

[76] Lipovac M, Chedraui P, Gruenhut C, Gocan A, Kurz C, Neuber B, Imhof M. The effect of red clover isoflavone supplementation over vasomotor and menopausal symptoms in postmenopausal women. Gynecol Endocrinol. 2012; **28(3)**: 203-7. DOI: 10.3109/09513590.2011.593671.

[77] Huang C, Pang D, Luo Q, Chen X, Gao Q, Shi L, Liu W, Zou Y, Li L, Chen Z. Soy isoflavones regulate lipid metabolism through an AKT/mTORC1 pathway in diet-induced obesity (DIO) male rats. Molecules. 2016; **21(5)**. pii: E586. DOI: 10.3390/molecules21050586.

[78] van Bree BW, Lenaers E, Nabben M, Briedé JJ, Jörgensen JA, Schaart G, Schrauwen P, Hoeks J, Hesselink MK. A genistein-enriched diet neither improves skeletal muscle oxidative capacity nor prevents the transition towards advanced insulin resistance in ZDF rats. Sci Rep. 2016; **6**: 22854. DOI: 10.1038/srep22854.

[79] Zang Y, Igarashi K, Yu C. Anti-obese and anti-diabetic effects of a mixture of daidzin and glycitin on C57BL/6J mice fed with a high-fat diet. Biosci Biotechnol Biochem. 2015; **79(1)**: 117-23.DOI: 10.1080/09168451.2014.955453.

[80] Nanri A, Mizoue T, Takahashi Y, Kirii K, Inoue M, Noda M, Tsugane S. Soy product and isoflavone intakes are associated with a lower risk of type 2 diabetes in overweight Japanese women. J Nutr. 2010; **140(3)**: 580-6. DOI: 10.3945/jn.109.116020.

[81] Ko KP, Kim CS, Ahn Y, Park SJ, Kim YJ, Park JK, Lim YK, Yoo KY, Kim SS. Plasma isoflavone concentration is associated with decreased risk of type 2 diabetes in Korean

women but not men: results from the Korean genome and epidemiology study. Diabetologia. 2015; **58(4)**: 726-35. DOI: 10.1007/s00125-014-3463-x.

[82] Simperova A, Al-Nakkash L, Faust JJ, Sweazea KL. Genistein supplementation prevents weight gain but promotes oxidative stress and inflammation in the vasculature of female obese ob/ob mice. Nutr Res. 2016; **36(8)**: 789-97. DOI: 10.1016/j.nutres.2016.03.011.

[83] Peng N, Prasain JK, Dai Y, Moore R, Arabshahi A, Barnes S, Carlson S, Wyss JM. Chronic dietary kudzu isoflavones improve components of metabolic syndrome in stroke-prone spontaneously hypertensive rats. J Agric Food Chem. 2009; **57(16)**: 7268-73. DOI: 10.1021/jf901169y.

[84] Michelin RM, Al-Nakkash L, Broderick TL, Plochocki JH. Genistein treatment increases bone mass in obese, hyperglycemic mice. Diabetes Metab Syndr Obes. 2016; **9**: 63-70. DOI: 10.2147/DMSO.S97600.

[85] Kaczmarczyk-Sedlak I, Wojnar W, Zych M, Ozimina-Kamińska E, Taranowicz J, Siwek A. Effect of formononetin on mechanical properties and chemical composition of bones in rats with ovariectomy-induced osteoporosis. Evid Based Complement Alternat Med. 2013; **2013**: 457052. DOI: 10.1155/2013/457052.

[86] Wei P, Liu M, Chen Y, Chen DC. Systematic review of soy isoflavone supplements on osteoporosis in women. Asian Pac J Trop Med. 2012; **5(3)**: 243-8. DOI: 10.1016/S1995-7645(12)60033-9.

[87] Levis S, Strickman-Stein N, Ganjei-Azar P, Xu P, Doerge DR, Krischer J. Soy isoflavones in the prevention of menopausal bone loss and menopausal symptoms: a randomized, double-blind trial. Arch Intern Med. 2011; **171(15)**: 1363-9. DOI: 10.1001/archinternmed.2011.330.

[88] Yang XP, Reckelhoff JF. Estrogen, hormonal replacement therapy and cardiovascular disease. Curr Opin Nephrol Hypertens. 2011; **20(2)**: 133-8. DOI: 10.1097/MNH.0b013e3283431921.

[89] Li SH, Liu XX, Bai YY, Wang XJ, Sun K, Chen JZ, Hui RT. Effect of oral isoflavone supplementation on vascular endothelial function in postmenopausal women: a meta-analysis of randomized placebo-controlled trials. Am J Clin Nutr. 2010; **91(2)**: 480-6. DOI: 10.3945/ajcn.2009.28203.

[90] Goodman-Gruen D, Kritz-Silverstein D. Usual dietary isoflavone intake is associated with cardiovascular disease risk factors in postmenopausal women. J Nutr. 2001; **131(4)**: 1202-6.

[91] Yu D, Shu XO, Li H, Yang G, Cai Q, Xiang YB, Ji BT, Franke AA, Gao YT, Zheng W, Zhang X. Dietary isoflavones, urinary isoflavonoids, and risk of ischemic stroke in women. Am J Clin Nutr. 2015; **102(3)**: 680-6. DOI: 10.3945/ajcn.115.111591.

[92] Yu J, Bi X, Yu B, Chen D. Isoflavones: anti-inflammatory benefit and possible caveats. Nutrients. 2016; **8(6)**. pii: E361. DOI: 10.3390/nu8060361.

[93] Chen W, Ma X, Lin Y, Xiong Y, Zheng C, Hu Y, Yu D, Jiang Z. Dietary supplementation with a high dose of daidzein enhances the antioxidant capacity in swine muscle but

experts pro-oxidant function in liver and fat tissues. J Anim Sci Biotechnol. 2016; **7**: 43. DOI: 10.1186/s40104-016-0102-z.

[94] Jalili M, Hekmatdoost A, Vahedi H, Poustchi H, Khademi B, Saadi M, Zemestani M, Janani L. Co-administration of soy isoflavones and Vitamin D in management of irritable bowel disease. PLoS One. 2016; **11(8)**: e0158545. DOI: 10.1371/journal.pone.0158545.

[95] Cui Y, Niu K, Huang C, Momma H, Guan L, Kobayashi Y, Guo H, Chujo M, Otomo A, Nagatomi R. Relationship between daily isoflavone intake and sleep in Japanese adults: a cross-sectional study. Nutr J. 2015; **14**: 127. DOI: 10.1186/s12937-015-0117-x.

[96] Polito F, Marini H, Bitto A, Irrera N, Vaccaro M, Adamo EB, Micali A, Squadrito F, Minutoli L, Altavilla D. Genistein aglycone, a soy-derived isoflavone, improves skin changes induced by ovariectomy in rats. Br J Pharmacol. 2012; **165(4)**: 994-1005. DOI: 10.1111/j.1476-5381.2011.01619.x.

[97] Hanski L, Genina N, Uvell H, Malinovskaja K, Gylfe Å, Laaksonen T, Kolakovic R, Mäkilä E, Salonen J, Hirvonen J, Elofsson M, Sandler N, Vuorela PM. Inhibitory activity of the isoflavone biochanin A on intracellular bacteria of genus *Chlamydia* and initial development of a buccal formulation. PLoS One. 2014; **9(12)**: e115115. DOI: 10.1371/journal.pone.0115115.

[98] Stantchev TS, Markovic I, Telford WG, Clouse KA, Broder CC. The tyrosine kinase inhibitor genistein blocks HIV-1 infection in primary human macrophages. Virus Res. 2007; **123(2)**: 178-89.

[99] Nabavi SF, Daglia M, Tundis R, Loizzo MR, Sobarzo-Sanchez E, Orhan IE, et al. Genistein: a boon for mitigating ischemic stroke. Curr Top Med Chem. 2015;**15**(17): 1714-21.

[100] Dong X, Xu W, Sikes RA, Wu C. Combination of low dose of genistein and daidzein has synergistic preventive effects on isogenic human prostate cancer cells when compared with individual soy isoflavone. Food Chem. 2013; **141(3)**: 1923-33. DOI: 10.1016/j.foodchem.2013.04.109.

[101] Yu D, Shin HS, Lee YS, Lee D, Kim S, Lee YC. Genistein attenuates cancer stem cell characteristics in gastric cancer through the downregulation of Gli1. Oncol Rep. 2014; **31(2)**: 673-8. DOI: 10.3892/or.2013.2893.

[102] Brownlow B, Nagaraj VJ, Nayel A, Joshi M, Elbayoumi T. Development and in vitro evaluation of vitamin E-enriched nanoemulsion vehicles loaded with genistein for chemoprevention against UVB-induced skin damage. J Pharm Sci. 2015; **104(10)**: 3510-23. DOI: 10.1002/jps.24547.

[103] Lee DE, Lee KW, Byun S, Jung SK, Song N, Lim SH, et al. 7,3′,4′-Trihydroxyisoflavone, a metabolite of the soy isoflavone daidzein, suppresses ultraviolet B-induced skin cancer by targeting Cot and MKK4. J Biol Chem. 2011; **286(16)**: 14246-56. DOI: 10.1074/jbc.M110.147348.

[104] Ahn JC, Biswas R, Chung PS. Combination with genistein enhances the efficacy of photodynamic therapy against human anaplastic thyroid cancer cells. Lasers Surg Med. 2012; **44(10)**: 840-9. DOI: 10.1002/lsm.22095.

[105] Xiong P, Wang R, Zhang X, DeLa Torre E, Leon F, Zhang Q, et al. Design, synthesis, and evaluation of genistein analogues as anti-cancer agents. Anticancer Agents Med Chem. 2015; **15(9)**: 1197-203.

[106] Wiegand H, Wagner AE, Boesch-Saadatmandi C, Kruse HP, Kulling S, Rimbach G. Effect of dietary genistein on phase II and antioxidant enzymes in rat liver. Cancer Genomics Proteomics. 2009; **6(2)**: 85-92.

[107] Zampieri AL, Ferreira FS, Resende EC, Gaeti MP, Diniz DG, Taveira SF, et al. Biodegradable polymeric nanocapsules based on poly(DL-lactide) for genistein topical delivery: obtention, characterization and skin permeation studies. J Biomed Nanotechnol. 2013; **9(3)**: 527-34.

[108] Ullah MF, Ahmad A, Bhat SH, Khan HY, Zubair H, Sarkar FH, et al. Simulating hypoxia-induced acidic environment in cancer cells facilitates mobilization and redox-cycling of genomic copper by daidzein leading to pro-oxidant cell death: implications for the sensitization of resistant hypoxic cancer cells to therapeutic challenges. Biometals. 2016; **29(2)**: 299-310. DOI: 10.1007/s10534-016-9916-6.

[109] Kim SH, Kim CW, Jeon SY, Go RE, Hwang KA, Choi KC. Chemopreventive and chemotherapeutic effects of genistein, a soy isoflavone, upon cancer development and progression in preclinical animal models. Lab Anim Res. 2014; **30(4)**: 143-50. DOI: 10.5625/lar.2014.30.4.143.

[110] Liu Y, Santillo MF, Flynn TJ, Ferguson MS. Sex hormone modulation of both induction and inhibition of CYP1A by genistein in HepG2/C3A cells. In Vitro Cell Dev Biol Anim. 2015; **51(4)**: 426-31. DOI: 10.1007/s11626-014-9848-9.

[111] Zhang Y, Chen H. Genistein, an epigenome modifier during cancer prevention. Epigenetics. 2011; **6(7)**: 888-91.

[112] Sakla MS, Shenouda NS, Ansell PJ, Macdonald RS, Lubahn DB. Genistein affects HER2 protein concentration, activation, and promoter regulation in BT-474 human breast cancer cells. Endocrine. 2007; **32(1)**: 69-78. DOI: 10.1007/s12020-007-9006-1.

[113] Zhang Z, Wang CZ, Du GJ, Qi LW, Calway T, He TC, et al. Genistein induces G2/M cell cycle arrest and apoptosis via ATM/p53-dependent pathway in human colon cancer cells. Int J Oncol. 2013; **43(1)**: 289-96. DOI: 10.3892/ijo.2013.1946.

[114] Li HQ, Luo Y, Qiao CH. The mechanisms of anticancer agents by genistein and synthetic derivatives of isoflavone. Mini Rev Med Chem. 2012; **12(4)**: 350-62.

[115] Chen J, Duan Y, Zhang X, Ye Y, Ge B, Chen J. Genistein induces apoptosis by the inactivation of the IGF-1R/p-Akt signaling pathway in MCF-7 human breast cancer cells. Food Funct. 2015; **6(3)**: 995-1000. DOI: 10.1039/c4fo01141d.

[116] Xu L, Xiang J, Shen J, Zou X, Zhai S, Yin Y, et al. Oncogenic MicroRNA-27a is a target for genistein in ovarian cancer cells. Anticancer Agents Med Chem. 2013; **13(7)**: 1126-32.

[117] Froyen EB, Steinberg FM. Genistein decreases basal hepatic cytochrome P450 1A1 protein expression and activity in Swiss Webster mice. Nutr Res. 2016; **36(5)**: 430-9. DOI: 10.1016/j.nutres.2016.01.001.

[118] Lepri SR, Zanelatto LC, da Silva PB, Sartori D, Ribeiro LR, Mantovani MS. Effects of genistein and daidzein on cell proliferation kinetics in HT29 colon cancer cells: the expression of CTNNBIP1 (beta-catenin), APC (adenomatous polyposis coli) and BIRC5 (survivin). Hum Cell. 2014; **27(2)**: 78-84. DOI: 10.1007/s13577-012-0051-6.

[119] Clubbs EA, Bomser JA. Glycitein activates extracellular signal-regulated kinase via vascular endothelial growth factor receptor signaling in nontumorigenic (RWPE-1) prostate epithelial cells. J Nutr Biochem. 2007; **18(8)**: 525-32. DOI: 10.1016/j.jnutbio.2006.09.005.

[120] Clubbs EA, Bomser JA. Basal cell induced differentiation of noncancerous prostate epithelial cells (RWPE-1) by glycitein. Nutr Cancer. 2009; **61(3)**: 390-6. DOI: 10.1080/01635580802582728.

[121] Wang J, Jenkins S, Lamartiniere CA. Cell proliferation and apoptosis in rat mammary glands following combinational exposure to bisphenol A and genistein. BMC Cancer. 2014; **14**: 379. DOI: 10.1186/1471-2407-14-379.

[122] Hu XJ, Xie MY, Kluxen FM, Diel P. Genistein modulates the anti-tumor activity of cisplatin in MCF-7 breast and HT-29 colon cancer cells. Arch Toxicol. 2014; **88(3)**: 625-35. DOI: 10.1007/s00204-013-1184-4.

[123] Tonetti DA, Zhang Y, Zhao H, Lim SB, Constantinou AI. The effect of the phytoestrogens genistein, daidzein, and equol on the growth of tamoxifen-resistant T47D/PKC alpha. Nutr Cancer. 2007; **58(2)**: 222-9. DOI: 10.1080/01635580701328545.

[124] Khan AQ, Khan R, Rehman MU, Lateef A, Tahir M, Ali F, et al. Soy isoflavones (daidzein & genistein) inhibit 12-O-tetradecanoylphorbol-13-acetate (TPA)-induced cutaneous inflammation via modulation of COX-2 and NF-kappaB in Swiss albino mice. Toxicology. 2012; **302(2-3)**: 266-74. DOI: 10.1016/j.tox.2012.08.008.

[125] Rahman Mazumder MA, Hongsprabhas P. Genistein as antioxidant and antibrowning agents in in vivo and in vitro: a review. Biomed Pharmacother. 2016; **82**: 379-92. DOI: 10.1016/j.biopha.2016.05.023.

[126] Ghaemi A, Soleimanjahi H, Razeghi S, Gorji A, Tabaraei A, Moradi A, et al. Genistein induces a protective immunomodulatory effect in a mouse model of cervical cancer. Iran J Immunol. 2012; **9(2)**: 119-27. DOI: IJIv9i2A5.

[127] Jiang W, Hu M. Mutual interactions between flavonoids and enzymatic and transporter elements responsible for flavonoid disposition via phase II metabolic pathways. RSC Adv. 2012; **2(21)**: 7948-63. DOI: 10.1039/C2RA01369J.

[128] Poluzzi E, Piccinni C, Raschi E, Rampa A, Recanatini M, Ponti FD. Phytoestrogens in postmenopause: the state of the art from a chemical, pharmacological and regulatory perspective. Curr Med Chem. 2014; **21(4)**: 417-36. DOI: 10.2174/09298673113206660297.

[129] Taylor CK, Levy RM, Elliott JC, Burnett BP. The effect of genistein aglycone on cancer and cancer risk: a review of in vitro, preclinical, and clinical studies. Nutr Rev. 2009; **67(7)**: 398-415. DOI: 10.1111/j.1753-4887.2009.00213.x.

[130] Liao CY, Lee CC, Tsai C, Hsueh CW, Wang CC, Chen IH, et al. Novel Investigations of flavonoids as chemopreventive agents for hepatocellular carcinoma. BioMed Research International. 2015; **2015**: e840542. DOI: org/10.1155/2015/840542.

[131] Polkowski K, Mazurek AP. Biological properties of genistein. A review of in vitro and in vivo data. Acta Pol Pharm. Mar-Apr 2000; **57(2)**: 135-55.

[132] Peterson G. Evaluation of the biochemical targets of genistein in tumor cells. J Nutr. 1995; **125(3 Suppl)**: 784S–9.

[133] Grivennikov SI, Greten FR, Karin M. Immunity, inflammation, and cancer. Cell. 2010; **140(6)**: 883-99. DOI: 10.1016/j.cell.2010.01.025.

[134] Pfarr K, Danciu C, Arlt O, Neske C, Dehelean C, Pfeilschifter JM, et al. Simultaneous and dose dependent melanoma cytotoxic and immune stimulatory activity of betulin. PLoS One. 2015; **10(3)**: e0118802. DOI: 10.1371/journal.pone.0118802. eCollection 2015.

[135] Danciu C, Caraba A, Bojin F, Soica C, Simu G, Ciurlea S,Peev C, Citu IM, Panzaru I.Genistein does not induce caspase 2 activation in vitro on B16 melanoma cell lines. Farmacia. 2014; **62(4)**: 753-60.

[136] Danciu C, Soica C, Oltean M, Avram S, Borcan F, Csanyi E, et al. Genistein in 1:1 inclusion complexes with ramified cyclodextrins: theoretical, physicochemical and biological evaluation. Int J Mol Sci. 2014; **15(2)**: 1962-82. DOI: 10.3390/ijms15021962.

[137] Danciu C, Borcan F, Soica C, Zupko I, Csányi E, Ambrus R, et al. Polyurethane microstructures--a good or bad in vitro partner for the isoflavone genistein? Nat Prod Commun. 2015; **10(6)**: 951-4.

[138] Danciu C, Soica C, Dehelean C, Zupko I, Csanyi E, Pinzaru I, et al. Preliminary in vitro evaluation of genistein chemopreventive capacity as a result of esterification and cyclodextrin encapsulation. Anal Cell Pathol. 2015; **2015(2015)**: e262930. DOI: org/10.1155/2015/262930.

[139] Loftsson T, Brewster ME. Pharmaceutical applications of cyclodextrins. 1. Drug solubilization and stabilization. J Pharm Sci. 1996; **85(10)**: 1017-25.

[140] Danciu C, Borcan F, Bojin F, Zupko I, Dehelean C. Effect of the isoflavone genistein on tumor size, metastasis potential and melanization in a B16 mouse model of murine melanoma. Nat Prod Commun. 2013; **8(3)**: 343-6.

[141] Danciu C, Soica C, Csanyi E, Ambrus R, Feflea S, Peev C, et al. Changes in the anti-inflammatory activity of soy isoflavonoid genistein versus genistein incorporated in two types of cyclodextrin derivatives. Chem Cent J. 2012; **6(1)**: 58. DOI: 10.1186/1752-153X-6-58.

[142] Danciu C, Berkó S, Varju G, Balázs B, Kemény L, Németh IB, et al. The effect of electroporation of a lyotropic liquid crystal genistein-based formulation in the recovery of murine melanoma lesions. Int J Mol Sci. 2015; **16(7)**: 15425-41. DOI: 10.3390/ijms160715425.

[143] Danciu C, Biris M, Boglárka B, Csanyi E, Pavel IZ, Pop G, Soica C, Ceuta L, Nita L, Morgovan C,Stoian D.Pro-apoptotic effect of soy total extract incorporated in lyotropic liquid crystals formulation. Revista de Chimie. 2015; **66(7)**: 1038-41.

[144] Balázs B, Sipos P, Danciu C, Avram S, Soica C, Dehelean C, Varju G, Eros E, Szucs MB, Berkó S, and Csányi E. ATR-FTIR and Raman spectroscopic investigation of the electroporation-mediated transdermal delivery of a nanocarrier system containing an antitumor drug. Biomed Opt Express. 2016; **7(1)**: 67-78. DOI: 10.1364/BOE.7.000067.

Chapter 13

Flavonoids: Anticancer Properties

Duraipandiyan Veeramuthu,
William Raja Tharsius Raja,
Naif Abdullah Al-Dhabi and
Ignacimuthu Savarimuthu

Additional information is available at the end of the chapter

http://dx.doi.org/ 10.5772/68095

Abstract

Flavonoids are plant secondary metabolites. They are mainly classified into four major groups, such as flavanols, flavones, anthocyanidins, and isoflavonoids. Furthermore, they are divided into some subclasses. They are available in dietary foods and they cure various diseases. Certain plants and spices contain flavonoids, which have been commonly used for thousands of years in traditional medicine. Some of the flavonoids have been clinically used in many countries. Baicalein and its glycosides are one among them to have been experimented clinically. Flavonoids have the capability to regulate cell division and proliferation in an important pathway. They have medicinal activities including anticancer properties. The isoflavone analog rotenone is one of the flavonoid compounds, which has been revealed to be actual anticancer agent. *Scutellaria* species having flavones retain cytotoxic activities against many human cancer cell lines. At the same time, they do not harm the myeloid cells, normal peripheral and normal epithelial blood cells. Epidemiological studies also confirmed that the intake of dietary flavonoids reduces a risk condition in cancer.

Keywords: flavonoids, cancer, dietary foods, pathway, epidemiological study

1. Introduction

Flavonoids are plant-based secondary metabolites. The intake of flavonoids is always safe and without adverse effects. Recent studies also suggest that the consumption of different fruits and vegetables has the capability to fight against cancers and decrease the cancer risk level at least by 20%. Based on this, the scientific community has focused its attention on plant-based compounds in order to control cancers. Many compounds, such as flavonoids, were isolated from plants and shown to have anticancer activity notably. This was confirmed through *in vitro* and *in vivo* studies [1]. Our dietary foods contain different types of flavonoids in various food additives. Grains and herbs have flavones. Fruits and vegetables hold flavonols and their glycosides. Citrus juices, legumes, and tea contain flavanones, isoflavones, and catechins, respectively. Some flavonoids are able to fight against breast cancer [2]. The health benefits of flavonoids may be helpful to find new drug discoveries. Such compounds are listed with their specific subclasses. Apigenin, baicalein, luteolin, and chrysin belong to the subclass of flavones; kaempferol, myricetin, and quercetin are closer to the subclass of flavonols; hesperetin is flavanone compound; genistein and daidzein go with the subclass of isoflavones; baicalin, catechin, and rutin fit with flavone glycosides, flavan-3-ols, and flavonol glycosides, respectively. There are different types of tumors which can be organized and categorized as oral (pharyngeal, laryngeal), gastrointestinal (esophageal, gastric, pancreatic), colorectal, liver, reproductive (ovarian, endometrial, prostate), breast, and lung cancer. The various diseases including cancers are controlled by the intake of flavonoids. Cytotoxicity in cancer cell line is shown mainly because of flavonoid compounds which do not affect normal cells. This was proved by cytotoxicity assay. Apigenin and luteolin come under the flavonoid subclass, flavones which have the ability to regulate macrophage function in cancer cell elimination and act as a potential inhibitor of cell proliferation. Many *in vitro* and *in vivo* studies confirmed that flavonoids have good activity against various cancer cell lines. Flavonoids have the ability to perform antiproliferation and cytotoxicity in cancer cell lines. They are used for human clinical trial which was conducted on flavone acetic acid.

In 2011, a database of U.S Department of Agriculture explains to us the flavonoid content in 500 foods in which isoflavone, proanthocyanidin, and other compounds are identified [3]. This definitely helps us calculate the flavonoid intake and its cancer-preventive properties. The amount of intake and the time of exposure have considerable say in the anticancer response to flavonoid-rich diets. Some intervention trials of flavonoids have shown their capacity to prevent cancer. They have the ability to block cell cycle followed by apoptosis. In recent years, they have been used for the treatment of prostate, pancreatic, breast, cervical, and ovarian cancers. Several protein kinases, epidermal growth factor receptors (EGFRs), platelet-derived growth factor receptors (PDGFRs), vascular endothelial growth factor receptors (VEGFRs), and cyclin-dependent kinases (CDKs) [4] play important roles in cancer pathology. COX (cyclooxygenase), LOX (lipoxygenase), and xanthine oxidase enzymes are also responsible for cancer pathologies. Flavonoids have the power to decrease and sometimes control all these pathogenic factors completely.

2. Anticancer properties of flavonoids

Major classes of flavonoids possess anticancer properties. The sources of flavonoids are also explained in this context. Flavanols are present in strawberries, apple, chocolate, cocoa, beans, cherry, green, and black tea. They have the potential to fight against human oral, rectal, and prostate cancer. The major sources of anthocyanidins are blueberries, blackberries, blackcurrant, and aubergine. These natural resources are used to treat colorectal cancer. The major sources of flavones are Siberian larch tree, onion, milk thistle, acai palm, lemon juice, orange juice, grape juice, kale, cherries, leek, Brussel sprouts, pepper, broccoli, capsicum, parsley, and celery. They have the ability to fight against breast cancer, lung cancer, leukemia, thyroid, stomach, laryngeal, colon, and oral cancer. Sources of isoflavonoids are soybeans, soy flour, soy milk, beer, and tempeh. They fight against prostate cancer, breast cancer, colon, kidney, and thyroid cancer [5].

2.1. Different groups of flavonoids

Flavonoids are mainly classified into four major groups: flavanols, flavones, anthocyanidins, and isoflavonoids. The major groups of these flavonoids are displayed in the subsequent text (**Figure 1**). A chemical structure of compound is drawn for each flavonoid group (**Figure 2**). Compounds from various subclasses of flavonoids are put together in their respective flavonoid groups. The major classification of flavones and anthocyanidins is displayed in **Figure 3**. Furthermore, they are divided into some subclasses. Among these subclasses, flavan-3-ols contain catechin, gallocatechin, catechin-3-gallate, epicatechin, and epigallocatechin (EGC). Kaempferol, myricetin, quercetin, and rutin belong to the subclass of flavonol [5]. Some other compounds are also classified under the specific subclasses of flavonoids (**Figure 3**).

2.2. Epidemiological information for flavonoids

Many studies on the distribution of diseases prove that flavonoids have positive effects in curbing cancer. It has been evidenced by various studies that the possibility of developing cancer could be reduced if more amount of flavonoid is administered [6, 7].

There was a case-control type study on 250 breast cancer-positive individuals based on population in Shanghai from 1996 to 1998. It was revealed in the corresponding controls; Dai et al. [8] noted that the number of breast cancer-positive individuals had less of isoflavonoids as well as lignans compared to the controls (urine sample of the cancer-positive individual was taken prior to administering therapy). The middle discharge rate of aggregate isoflavonoids

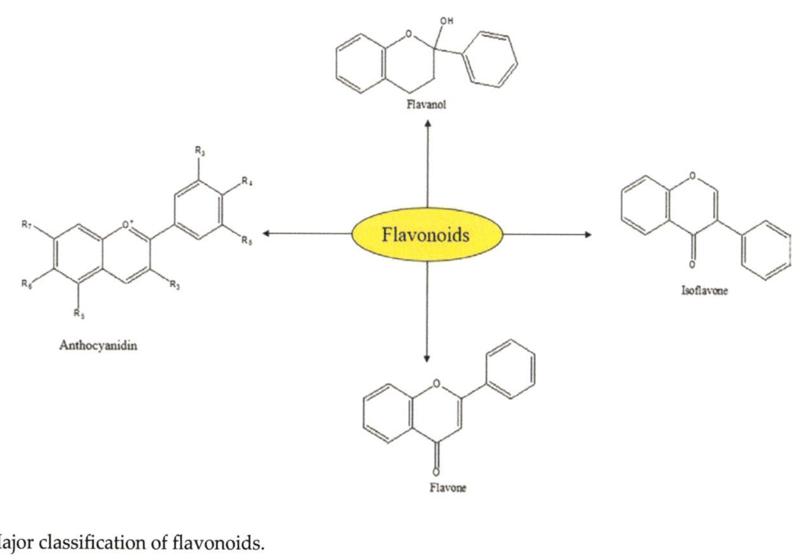

Figure 1. Major classification of flavonoids.

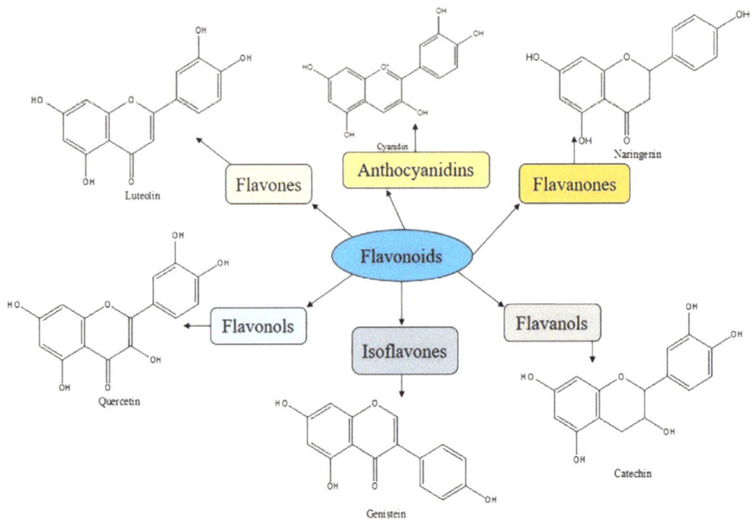

Figure 2. Different classes of flavonoids and their compound chemical structures.

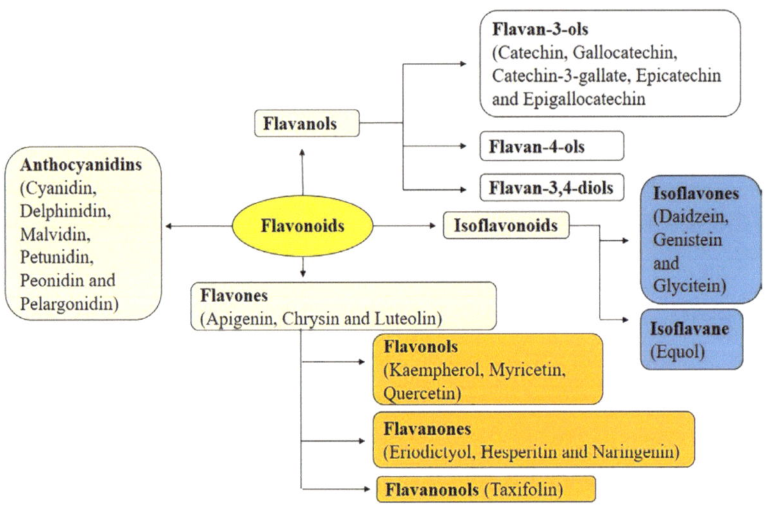

Figure 3. Different groups of flavonoids and their respective compounds.

was 13.97 nmol/mg creatinine in cases and 23.09 in controls (P = 0.01), and that of aggregate lignans was 1.77 in cases and 4.16 in controls (P < 0.01). Thus, it was recommended that flavonoids are capable of averting breast cancer.

Another lung cancer study was done on the observation of individuals beyond the age of 25. A total number of 9959 lung cancer-positive Finnish men and women between the ages of 25–99 showed reduced lung cancer after administering flavonoids through diet. The inference was made based on vitamin E, vitamin C, beta-carotene, or total calories consumption. There was a study on 10,054 individuals of both men and women by Knekt and coworkers [9] on the amount of flavonoid consumption in Finnish diet. The study revealed a lesser possibility for lung cancer with the higher consumption of quercetin and the lesser possibility of prostate cancer with more consumption of myricetin. Thus, flavonoids were proved to play a vital role in preventing cancer occurrence.

There was also a case-control work done based on population in Hawaii in order to study in detail the relation between the probability of lung cancer and the consumption of flavonoids through diet. For the study, they took 582 individuals who were lung cancer-positive and the same number of controls of matching age, sex, and ethnicity. The consumption of flavonoids such as onion, white grapefruits, apples, and quercetin was reversely related to the probability of suffering lung cancer [10]. The outcome of the above study is found to be similar to the previous study done in Uruguay on 541 lung cancer-positive individuals and 540 controls but fewer incidents of lung cancer due to vitamin E and beta-carotene.

Besides, it was also found in a case-control work carried out by the group of researchers in Uruguay between 1996 January and 1997 December that reduced incidents of an esophagus, oral cavity, larynx, and pharynx cancer by 70% achieved by flavonoids. Flavonoids like kaempferol and quercetin are also found to be preventing gastric cancer unlike carotenoids like alpha-carotene, lutein, beta-carotene, and lycopene in yet another case-control study carried out in Spain which consisted of 354 gastric cancer-positive individuals and 354 controls. An observation was done on 34,651 women free from postmenopausal cancer between the ages of 55 and 69 during 1986 and 1998. In modification with prospective confounders, the consumption of catechin was reversely related to only the rectal cancer occurrence [11]. These prove the potential ability of flavonoids for a cancer cure.

In this way, the administering of flavonoids is effective in preventing cancer in most if not in all studies. Reports [12] also show that flavonoids are ineffective. It is mainly because of the uneven availability of the same. However, it should not be fully neglected without detailed study.

2.3. Case-control study in cancer

Two case-control studies were conducted in six counties in New Jersey (205 cases of ovarian cancer and 390 controls) [13] and in the North-East United States (1231 cases and 1244 controls). These revealed that there was no link between total flavonoid consumption and ovarian cancer [14]. Some of the cancer case studies have been discussed in the subsequent text.

2.3.1. Gastrointestinal cancers

A case study showed that there is an inverse association between flavanone intake and esophageal cancer, and this could reduce by the intake of citrus fruits. An increased risk of gastric cancer is found among smoking men. The intake of epigallocatechin (EGC) plays an important role to slow down the disease.

2.3.2. Pancreatic cancer

Researchers analyzed the intake of flavonoids and the risk of pancreatic cancer during the study. The results reported that flavonoid-rich diets can decline pancreatic cancer risk in male smokers. Inverse relationships were also found among current smokers between a risk of pancreatic cancer and the intake of total flavonols, quercetin, kaempferol, and myricetin.

2.3.3. Colorectal cancer

Isoflavone intake was inversely related to colorectal cancer risk in men and postmenopausal women. Cases were analyzed in Japan, Netherlands, and in the UK in both men and women regarding the intake of isoflavone and its inverse effect on colorectal cancer. Total catechin, (β)-catechin, myricetin and (−)-epicatechin and kaempferol were effective against colorectal cancer. These results may have associations for the use of dietary flavonoids in the prevention of rectal cancer.

2.4. Inhibition of pro-oxidant enzymes

NADPH oxidase I (NOX 1) enzyme produces superoxide, which is overexpressed in colon and prostate cancer cell lines [15]. Superoxide is one of the reactive oxygen species (ROS). Superoxide dismutase (SOD) is one of the antioxidants which can inhibit a pro-oxidant enzyme (**Figure 4**). Generally, flavonoids have the ability to inhibit DNA damaging, mutagenic signaling, cell proliferation, and proto-oncogenes (cFOS, cJUN, and cMyc). Diagrams are drawn using Microsoft PowerPoint 2013 and converted to JPEG format.

2.5. Flavonoids from *Scutellaria* species

Wogonin and baicalein from *Scutellaria* species have been tested in a mouse for anticancer activity. *S. baicalensis* has an O-methylated flavone called wogonin and a flavone called baicalein, which were isolated from the roots of the same plant as well as from *S. lateriflora*. A flavone glycoside called baicalin is also found in *Scutellaria* species. Oral administering of 20 mg/kg baicalein was able to inhibit prostate cancer nearly 55%. Both the compounds have therapeutic potential against cancer. The identified flavonoids from *Scutellaria* species are about 60. The reported minor flavonoids from the same species are Apigenin, Luteolin [16], and Chrysin. They possess antitumor activities. *Scutellaria* alone or in combination with other herbs has the cytostatic effect on several cancer cell lines in vitro and in vivo mouse model [17]. One of the anticancer drugs is wogonin. It comes under flavonoids. It is considered as chemotherapeutic agent to decrease their side effects. It has a hepatoprotective effect

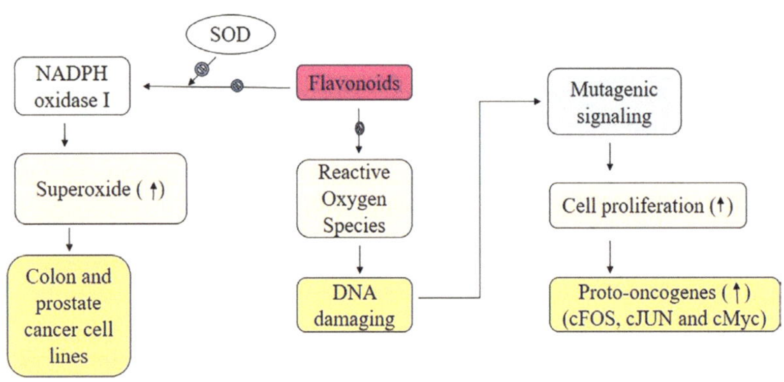

Figure 4. Inhibition of pro-oxidant enzymes.

and prompts apoptosis in caspase 3 pathway. It alternates p21 protein expression. Wogonin and its derivatives possess anticancer activity. Wogonin induced apoptosis in lung cancer. It was experimented and proved in the nude mouse model [18–20]. It goes through multiple apoptosis pathways such as ROS (Reactive Oxygen Species)-mediated and ER stress-dependent pathway (**Figure 5**).

2.6. Flavonoid compounds for cancer treatment

2.6.1. Apigenin

Apigenin has anti-mutagenic properties. It inhibits benzo[a]pyrene- and 2-aminoanthracene-induced bacterial mutagenesis. It scavenges free radicals and promotes metal chelation in *in vivo* tumor models [21]. It affords protective effect in murine skin and colon cancer models [22]. It would suppress this enzyme effectively. It also increases glutathione concentration and enhances the endogenous defense against oxidative stress [23]. It was experimented against skin carcinogenesis model. It inhibits dimethylbenzanthracene-induced skin tumors. It has been administered against UV-light-induced cancers. The result showed that it could diminish the occurrence of UV light-induced cancers and was able to increase tumor-free cells. Apigenin plays an effective role to inhibit casein kinase (CK)-2 expression in both prostate and breast cancers [24]. It inhibits HIF-1α and VEGF expression via PI3K/Akt/p70S6K1 and HDM2/p53 pathways in human ovarian cancer cells [25].

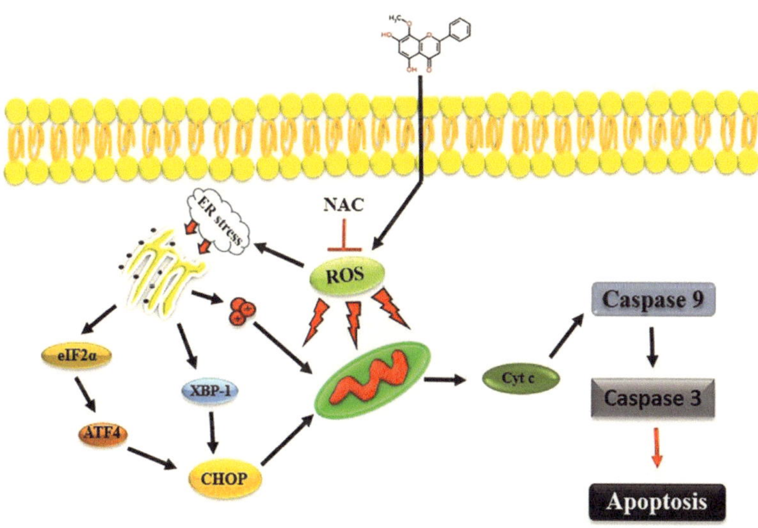

Figure 5. Mechanism of action of wogonin-induced apoptosis in human lung cancer cells. Wogonin induces apoptosis with extrinsic apoptotic pathway and ROS-intervened ER stress-dependent pathway. NAC (N-acetyl-L-cysteine) is used to identify and test ROS. In mammalian cells, the major ER stress sensors such as pancreatic ER kinase (PERK), activating transcription factor-4 (ATF4), ionizing radiation, eIF2α, and CHOP will carry the signal from the ER lumen to cytoplasm and nucleus in order to recruit ER stress and also to develop tumor progression. Wogonin goes through this pathway and generates apoptosis at the end.

2.6.2. *Kaempferol*

Kaempferol has anticancer effects and acts as a chemopreventive agent. It was found to be curbing the growth of various carcinomas such as glioblastoma (LN229, U87MG, and T98G), leukemia (HL-60 and Jurkat), lung cancer (H460 and A549), breast adenocarcinoma (MCF- 7, BT-549, and MDA-MB-231), osteosarcoma (U-2 OS), prostate cancer (LNCaP, PC-3, and DU145), colorectal carcinoma (Caco-2, HCT-116, DLD-1, and Lovo), and pancreatic cancer (MIA PaCa-2, Panc 1). It is used to arrest the cell cycle in cancer cells. It has been used as antiapoptotic agent on cancer cells. Kaempferol is very effective against metastasis and angiogenesis [26].

2.6.3. *Quercetin and diosmin*

Quercetin is one of the dietary flavonoids, which suppresses tumor growth by inhibiting protein tyrosine kinase (PTK). About 10 μM of this compound confirmed antiproliferative activity against colon cancer cells, Caco-2, and HT-29. Diosmin is one of the important *Citrus* flavonoids, which showed antiproliferative activity in the same cancer cell line. The proliferation of MCF-7 human breast cancer cell line was controlled by the intake of citrus flavones.

2.6.4. *Tangeretin*

Among these phenolic compounds (gallic acid, baicalein, myricetin, 7,3′ dimethylhesperetin, quercetin, and luteolin), flavone tangeretin showed better anticancer activity against B16F10, SK-MEL-1 and SK-MEL-5 melanoma cell lines [40, 41], human hepatoma HepG2, Hep3B, and PLC/PRF/5 cell lines [42], HL-60 leukemia cell line [43], and human lung DMS-14 cell line, breast MCF-7 and MDA-MB-435 cell lines, colon HT-29 cell line, and prostate DU-145 cell line [41]. The in vitro studies confirmed that the compound was more effective against various cancer cell lines.

2.7. Anticancer activity of flavonoids

Fruits and vegetables are having an enormous amount of flavonoids, which have been used as cancer chemopreventive agents. Flavonol quercetin is contained in dietary fruits and vegetables especially onion and apple. Quercetin flavonol is used to treat prostate, lung, stomach, and breast cancers [27]. Many biological properties in flavonoids and isoflavonoids are sometimes proved to be cancer chemopreventive. The mechanism of action of flavonoids in the molecular study is cell cycle arrest, heat-shock protein inhibition, tyrosine kinase inhibition, downregulation of p53 protein, estrogen receptor-binding capacity, inhibition of Ras protein, and expression of Ras protein. The most genetic abnormalities in human cancers are based upon p53-mutated proteins. The protein may be downregulated because of flavonoid intake. The flavonoid expression on p53 proteins may lead to arrest cancer cells in G2 and mobile phase of cell cycle. Tyrosine kinases are proteins. They are considered as growth factor signals for the nucleus. The expression of the protein is involved in oncogenesis. The anticancer drug is able to inhibit tyrosine kinase activity. Quercetin has been used in human phase I clinical trial against tyrosine kinase activity. It is proved that it could be considered as antitumor agent without the cytotoxic side effects [28]. It does arrest cell cycle in proliferating lymphoid cells. Flavonoids inhibit heat-shock proteins in several malignant cell lines, comprising leukemia, colon cancer, and breast cancer [29].

2.8. Antitumor effects

Reactive oxygen species (ROS) can harm DNA and lead to mutations. It is involved in cell signaling and cell growth. It increases the DNA exposure to mutagens. Stefani et al. reported that flavonoids can have inhibition effect against carcinogenesis. Apigenin, fisetin, and luteolin flavonoids have been used to inhibit cell proliferation effectively. A variety of endogenous angiogenic and angiostatic factors have the responsibility for regulating angiogenesis. Flavonoids have the power to fight against angiogenesis. Lumen formation, endothelial cells migration, and their proliferation are the important steps in angiogenesis. Angiogenesis inhibitors can interfere with these steps. Flavonoids play an essential role among the known angiogenesis inhibitors. The inhibition of protein kinases is the possible mechanism for the treatment of angiogenesis. These enzymes are involved in the process of signal transduction against angiogenesis.

2.9. Cancer chemoprevention

Carcinogenesis, the multistep process of tumor development, primarily involves the acquisition of the hallmark capabilities of cancer namely sustaining proliferative signaling, shirking growth suppressors, fighting cell death, triggering invasion and metastasis, and inducing angiogenesis by the incipient cells. Aberrations in multiple intracellular signaling cascades and progressive accumulation of mutations during carcinogenesis present considerable opportunities for the development of clinical interventions in preventing cancer initiation, treating neoplasms during premalignant stages, and inhibiting tumor progression. Natural agents that can target the hallmarks of cancer have attracted the attention of several researchers due to their chemical diversity, structural complexity, inherent biologic activity, affordability, easy availability, and lack of substantial toxic effects. The potential targets of chemopreventive agents include multiple signaling pathways such as ROS generation and signaling, cyclooxygenase-2 (COX-2) and lipoxygenase (LOX) pathways, and numerous cellular molecules like XMEs, transcription factors, proteins involved in cell cycle, apoptosis, invasion and angiogenesis, and enzymes involved in epigenetic modifications.

2.10. Mechanism of action on flavonoids

Flavonoids are proved to be effective chemopreventive agents. The chemopreventive functions of flavonoids are estrogenic/antiestrogenic activity, antiproliferation or apoptosis, prevention of oxidation, induction of cell cycle arrest, regulation of the host immune system, induction of detoxification enzymes, anti-inflammatory activity, and changes in cellular signaling [30]. The research study suggests that the medicinal plant, *Glycyrrhiza inflata* has anticancer activity and also does the mechanism of action on flavonoids. Licorice is the root of *G. inflata* which contains more anticancer properties. Licorice total flavonoids (LTFs) are used effectively against cancer [31].

2.11. Dietary flavonoids on apoptotic pathway

Flavonoids enter through the outer membrane. Bad, Bax, and Bak are the proapoptotic regulators. Bcl-2 and Bcl-x are the apoptosis regulator proteins. Proapoptotic regulators and apoptosis regulator proteins release cytochrome c in the mitochondria (**Figure 6**). Apaf1, dATP, and procaspase-9 are bound with cytochrome c to form the apoptosome. Caspase is activated because of the cleavage of procaspase-9. At the same time, death receptors can interrelate with procaspase-8 to create its active form. A bid can control programmed cell death and can also release cytochrome c. At the end, apoptosis is performed [32].

2.12. Role of intrinsic and extrinsic signaling pathways

The intrinsic and extrinsic signaling pathways are involved in apoptosis. Cellular stress factors are involved in the intrinsic apoptotic pathway. They include ROS generation, endoplasmic reticulum (ER) stress, growth factor deprivation, and ionizing radiation. All these cellular stress factors are responsible for releasing cytochrome c from mitochondria. Apoptosome is the formation of a cytosolic multiprotein complex. It contains the adapter protein apoptotic protease-activating factor 1 (Apaf-1), cytochrome c, and pro-caspase-9.

In the place of apoptosome, caspase-9 begins and activates caspase-3 which cleaves target proteins leading to apoptosis. Pro-apoptotic (e.g., Bax, Bad, Bid, and Bak) and anti-apoptotic (e.g., Bcl-2, Mcl-1, and Bcl-xL) Bcl-2 family proteins have control over this death pathway. The extrinsic pathway is a process whereby the involvement of ligation of a ligand occurs with corresponding receptors. Ligands, such as CD95L, CD95, and TNF, are bound to the corresponding receptors. CD95L [CD95 (Fas/APO-1)-ligand] arbitrates apoptosis. This ligand

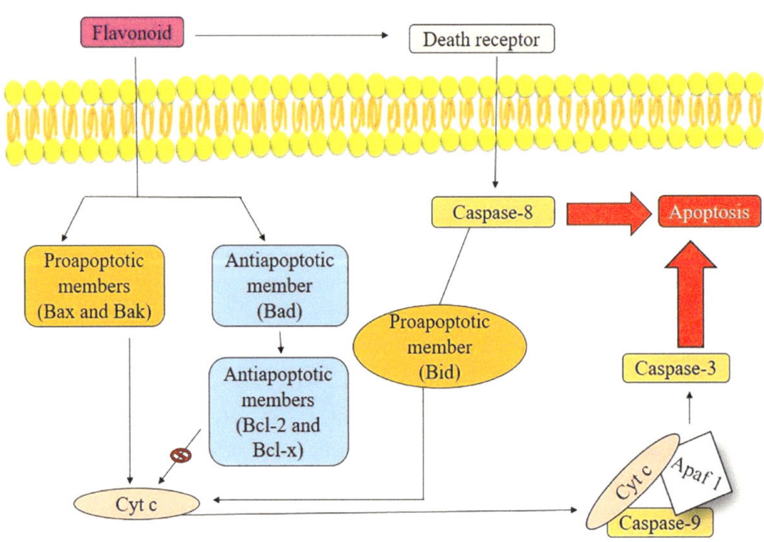

Figure 6. Flavonoids on apoptotic pathway.

http://dx.doi.org/ 10.5772/68095

binds to the corresponding receptor, CD95 [CD95 (APO-1/Fas)], on the surface of sensitive cells. The corresponding receptor is a prototype death receptor. Fas associated via death domain (FADD), pro-caspase 8, and FLICE-inhibitory protein (FLIP) are collectively called as DISC (death-inducing-signaling-complex). DISC activates caspase-8 which can further activate caspase-3 and leads to apoptosis. One of the other ligands is TNF (tumor necrosis factor). The corresponding receptor is TNF-R. Complex I contains receptor-interacting protein 1 (RIP 1), TNF receptor-associated death domain (TRADD), and telomeric repeat-binding factor 2 (trf 2). It is attached to the receptor itself. Complex II holds RIP 1, TRADD, FADD, and pro-caspase 8. It can be recruited from complex I. The instigation of pro-caspase-8, in turn, activates caspase-3. Mitochondria produce numerous death signals which are needed by the extrinsic death pathway. Caspase 8 activates the extrinsic pathway. It is able to link with an intrinsic pathway. It can also activate the apoptotic gene, Bid. The intrinsic pathway is connected with the apoptotic genes such as Bax and Bak. The above apoptotic gene formation results in cytochrome c. Finally, apoptosis occurs (**Figure 7**).

2.13. *In vivo* and *in vitro* studies on cancer

Quercetin and apigenin can inhibit melanoma cell growth. These compounds have potential to fight against invasive and metastatic cancers. This study has been conducted and proved with mice [33]. *In vitro* studies have confirmed that some flavonoids could inhibit the cell growth of colon, prostate, liver, and breast cancer [34]. Flavonoids can suppress carcinogenesis and also prevent cancer. Thus, these studies confirm the effectiveness of flavonoids in preventing cancer [35].

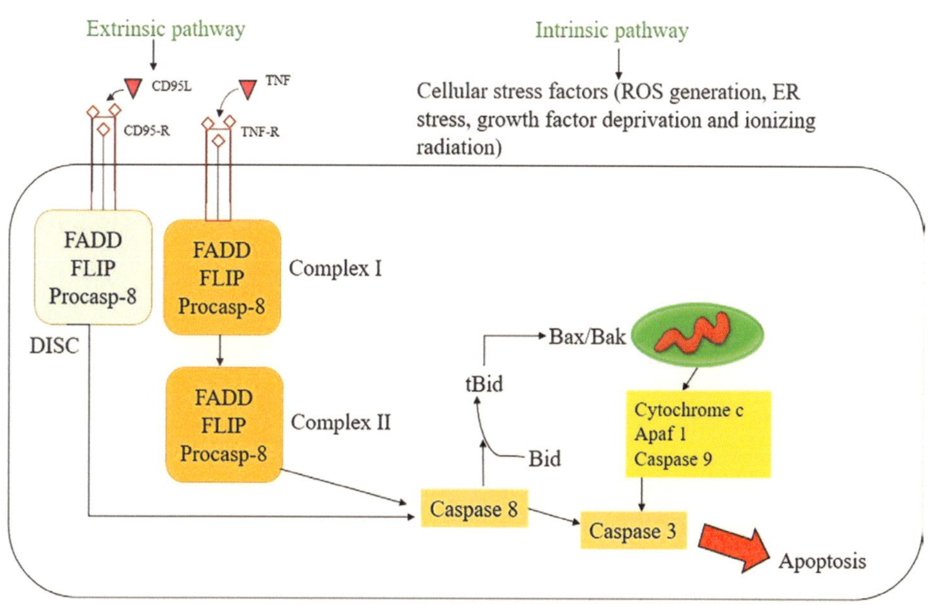

Figure 7. Intrinsic and extrinsic signaling pathways.

2.14. Flavonoids in cancer treatment

Oral cancer was developed chemically and was treated with flavonoids in the rat using 4-nitroquinoline 1-oxide-induced model. It was found later that flavonoid inhibited oral cancer. Kawaii et al. studied about some citrus flavonoids and found that they inhibited the proliferation of cancer cells such as lung carcinoma A549 and gastric TGBC11TKB cancer cell lines. It did not affect the human normal cell lines.

2.15. Cancer process and cancer therapy

Cancer is considered as a genetic illness caused by mutated genes. It is implicated in cell proliferation and cell death. DNA damage may lead to cell death. Three groups of genes are mainly involved in the cancer process. They are oncogenes (damaged proto-oncogenes), the tumor suppressor genes, and the DNA repair genes. Mutated proto-oncogenes lead to oncogenes. They are the responsible genes to proliferate the cells. Tumor suppressor genes code for proteins especially protein p53 and act as checkpoints to cell proliferation or cell death. They can persuade cell cycle arrest in a damaged cell. DNA repair genes can be mutated and lead to a failure in DNA repair [36].

Chemotherapy, radiotherapy, surgery, and some other therapies are available in order to control the risk level of the various cancers and to give a complete cure to the disease. When cancer cells are spread in a human body, chemotherapy is preferred to kill the cancer cells mainly [36].

2.16. Effects of ASMq on TGF-β1 and TNF-α protein expression

Abnormal Savda Munziq (ASMq), a traditional Uyghur medicine, has anticancer activities. TGF-β1 and TNF-α protein expression studies are conducted using Western blot. U27 tumor mice model is used for this study. Based on this study, CTX group showed a decreased level of TGF-β1 and TNF-α proteins. ASMq groups with different dosages expressed decreased TGF-β1 protein and were increased in TNF-α proteins. Compared to CTX group, TGF-β1 protein expression of ASMq groups was decreased and protein level was increased in TNF-α [37].

2.17. Definition of cancer prevention

The time period between 2000 and 2006 has witnessed 1.3% of cancer decline among men and the same time period (from 1998 to 2006) has seen 0.5% decline among women. Twenty-five percent of death has also occurred due to the consequence of this disease [38]. The advent of modern technology and its advances have not considerably reduced mortality caused by cancer; it still remains a major threat. As once quipped by Benjamin Franklin, "An ounce of prevention is worth a pound of cure." It is clear that the prevention of the disease is better than the cure. Sporn in 1976 defined cancer chemoprevention as a method "to arrest or reverse premalignant cells." Cancer-chemopreventive capacity of flavonoid is characterized by inhibiting inflammation, scavenging various free radicals, adhesion, suppressing cell proliferation, cell cycle arrest, and apoptosis [39].

2.18. Flavonoids in cancer prevention

A study was conducted on 9959 men and women with regard to consumption of flavonoids and its anticancer activity. It was found that the association between the two was inverse. After some observations, they found that the highest quartile of flavonoid intake reduced the lung cancer up to 50%. Flavonoids can prevent cancers and cure this disease too. Cell cycle arrest happens at G1/S phase, at G2/M, and both phases of G1/S and G2/M phases, and also oxidative radical damages on DNA can be rectified by the dietary flavonoids.

Flavonoids and isoflavonoids are highly antiproliferative, and their compounds come to be handy in curbing the cell cycle or induce apoptosis. They are found to be effective in stopping both G1/S and G2/M of the cell cycle in cultured cancer cell lines. For example, some studies have found that quercetin (30–100 mM) stops the cell cycle at G1/S in human colonic COLO320 DM cells and leukemic T-cells and prompts apoptosis.

2.19. Overview on flavonoids

Flavonoids are commonly nontoxic compounds. They can be used along with synthetic drugs which may have little toxic substances and side effects. The effect of toxic substances in marketed drugs may be decreased due to flavonoid content in the combinational drugs. Therefore, synergistic studies are more effective. Some flavonoids have a chemopreventive effect on nitrosamine-induced carcinogenesis. Many flavonoids protect the genome from chemical carcinogens. Rich anticancer properties are found in various flavonoids which are used to decrease about 20% risk level in cancers. They prevent cancers and also are able to cure the disease. This was proved by *in vitro* and *in vivo* studies.

Acknowledgements

The authors like to extend their sincere appreciation to the Deanship of Scientific Research at King Saud University for its funding of this research through the Research Group project No RGP-213.

Author details

Duraipandiyan Veeramuthu[1,2*], William Raja Tharsius Raja[2], Naif Abdullah Al-Dhabi[1] and Ignacimuthu Savarimuthu[2]

*Address all correspondence to: avdpandiyan@yahoo.co.in

1 Addiriyah Researcher for Environmental Studies, Department of Botany and Microbiology, College 16 of Science, King Saud University, Riyadh, Kingdom of Saudi Arabia

2 Division of Ethnopharmacology, Entomology Research Institute, Loyola College, Chennai, Tamil Nadu, India

References

[1] Havsteen BH. The biochemistry and medical significance of the flavonoids. Pharmacology & Therapeutics. 2002;**96**:67-202

[2] Iwase Y, Takemura Y, Ju-ichi M, Mukainaka T, Ichiishi E, Ito C, Furukawa H, Yano M, Tokuda H, Nishino H. Inhibitory effect of flavonoid derivatives on Epstein-Barr virus activation and two-stage carcinogenesis of skin tumors. Cancer Letters. 2001; **173**:105-109

[3] Neveu V, Perez-Jimenez J, Vos F, Crespy V, du Chaffaut L, Mennen L, Knox C, Eisner R, Cruz J, Wishart D, Scalbert A. Phenol-explorer: An online comprehensive database on polyphenol contents in foods. Database. 2010;**2010**:bap024

[4] Singh RP, Agarwal R. Natural flavonoids targeting deregulated cell cycle progression in cancer cells. Current Drug Targets. 2006;**7**:345-354

[5] Batra P, Sharma AK. Anti-cancer potential of flavonoids: Recent trends and future perspectives. 3 Biotech. 2013;**3**:439-459

[6] Le Marchand L. Cancer preventive effects of flavonoids—A review. Biomedicine & Pharmacotherapy. 2002;**56**:296-301

[7] Chun HS, Chang HJ, Choi EH, Kim HJ, Ku KH. Molecular and absorption properties of 12 soy isoflavones and their structure–activity relationship with selected biological activities. Biotechnology Letters. 2005;**27**:1105-1111

[8] Dai Q, Franke AA, Jin F, Shu XO, Hebert JR, Custer LJ, Cheng J, Gao YT, Zheng W. Urinary excretion of phytoestrogens and risk of breast cancer among Chinese women in Shanghai. Cancer Epidemiology and Prevention Biomarkers. 2002;**11**:815-821

[9] Knekt P, Kumpulainen J, Järvinen R, Rissanen H, Heliövaara M, Reunanen A, Hakulinen T, Aromaa A. Flavonoid intake and risk of chronic diseases. The American Journal of Clinical Nutrition. 2002;**76**:560-568

[10] Le Marchand L, Murphy SP, Hankin JH, Wilkens LR, Kolonel LN. Intake of flavonoids and lung cancer. Journal of the National Cancer Institute. 2000;**92**:154-160

[11] Arts IC, Jacobs DR, Gross M, Harnack LJ, Folsom AR. Dietary catechins and cancer incidence among postmenopausal women: The Iowa Women's Health Study (United States). Cancer Causes and Control. 2002;**13**:373-382

[12] Arts IC, Hollman PC, Bueno de Mesquita HB, Feskens EJ, Kromhout D. Dietary catechins and epithelial cancer incidence: The Zutphen elderly study. International Journal of Cancer. 2001;**92**:298-302

[13] Bandera EV, King M, Chandran U, Paddock LE, Rodriguez-Rodriguez L, Olson SH. Phytoestrogen consumption from foods and supplements and epithelial ovarian cancer risk: A population-based case control study. BMC Women's Health. 2011;**11**:40

[14] Gates MA, Vitonis AF, Tworoger SS, Rosner B, Titus-Ernstoff L, Hankinson SE, Cramer DW. Flavonoid intake and ovarian cancer risk in a population-based case-control study. International Journal of Cancer. 2009;**124**:1918-1925

[15] Park HS, Lim JH, Kim HJ, Choi HJ, Lee IS. Antioxidant flavone glycosides from the leaves of *Sasa borealis*. Archives of Pharmacal Research. 2007;**30**:161-166

[16] Bonham M, Posakony J, Coleman I, Montgomery B, Simon J, Nelson PS. Characterization of chemical constituents in *Scutellaria baicalensis* with anti-androgenic and growth-inhibitory activities toward prostate carcinoma. Clinical Cancer Research. 2005;**11**:3905-3914

[17] Yu J, Liu H, Lei J, Tan W, Hu X, Zou G. Antitumor activity of chloroform fraction of *Scutellaria barbata* and its active constituents. Phytotherapy Research. 2007;**21**:817-822

[18] Chen YC, Shen SC, Lee WR, Lin HY, Ko CH, Shih CM, Yang LL. Wogonin and fisetin induction of apoptosis through activation of caspase 3 cascade and alternative expression of p21 protein in hepatocellular carcinoma cells SK-HEP-1. Archives of Toxicology. 2002;**76**:351-359

[19] Wang W, Guo Q, You Q, Zhang K, Yang Y, Yu J, Liu W, Zhao L, Gu H, Hu Y, Tan Z. Involvement of bax/bcl-2 in wogonin-induced apoptosis of human hepatoma cell line SMMC-7721. Anti-Cancer Drugs. 2006;**17**:797-805

[20] Wei L, Lu N, Dai Q, Rong J, Chen Y, Li Z, You Q, Guo Q. Different apoptotic effects of wogonin via induction of H_2O_2 generation and Ca^{2+} overload in malignant hepatoma and normal hepatic cells. Journal of Cellular Biochemistry. 2010;**111**:1629-1641

[21] Middleton E, Kandaswami C, Theoharides TC. The effects of plant flavonoids on mammalian cells: Implications for inflammation, heart disease, and cancer. Pharmacological Reviews. 2000;**52**:673-751

[22] Van Dross R, Xue Y, Knudson A, Pelling JC. The chemopreventive bioflavonoid apigenin modulates signal transduction pathways in keratinocyte and colon carcinoma cell lines. The Journal of Nutrition. 2003;**133**:3800S–3804S

[23] Myhrstad MC, Carlsen H, Nordström O, Blomhoff R, Moskaug JØ. Flavonoids increase the intracellular glutathione level by transactivation of the γ-glutamylcysteine synthetase catalytical subunit promoter. Free Radical Biology and Medicine. 2002;**32**:386-393

[24] Hessenauer A, Montenarh M, Gotz C. Inhibition of CK2 activity provokes different responses in hormone-sensitive and hormone-refractory prostate cancer cells. International Journal of Oncology. 2003;**22**:1263-1270

[25] Fang J, Xia C, Cao Z, Zheng JZ, Reed E, Jiang BH. Apigenin inhibits VEGF and HIF-1 expression via PI3K/AKT/p70S6K1 and HDM2/p53 pathways. The FASEB Journal. 2005;**19**:342-353

[26] Rajendran P, Rengarajan T, Nandakumar N, Palaniswami R, Nishigaki Y, Nishigaki I. Kaempferol, a potential cytostatic and cure for inflammatory disorders. European Journal of Medicinal Chemistry. 2014;**86**:103-112

[27] Brusselmans K, Vrolix R, Verhoeven G, Swinnen JV. Induction of cancer cell apoptosis by flavonoids is associated with their ability to inhibit fatty acid synthase activity. Journal of Biological Chemistry. 2005;**280**:5636-5645

[28] Duthie GG, Duthie SJ, Kyle JA. Plant polyphenols in cancer and heart disease: Implications as nutritional antioxidants. Nutrition Research Reviews. 2000;**13**:79-106

[29] Davis W, Lamson MS, Matthew S, Brignall ND. Antioxidants and cancer III: Quercetin. Alternative Medicine Review. 2000;**5**:196-208

[30] García-Lafuente A, Guillamón E, Villares A, Rostagno MA, Martínez JA. Flavonoids as anti-inflammatory agents: Implications in cancer and cardiovascular disease. Inflammation Research. 2009;**58**:537-552

[31] Xia GU, Matsidik R, Ablise M, Lei SH, Abudula A. Anti-cancer activity of flavonoids from Xinjiang *Glycyrrhiza inflata* Licorice on proliferation, cytotoxicity and apoptosis in cervical carcinoma cells. Journal of Medicinal Plants Research. 2013;**7**:173-178

[32] Ramos S. Effects of dietary flavonoids on apoptotic pathways related to cancer chemoprevention. The Journal of Nutritional Biochemistry. 2007;**18**:427-442

[33] Caltagirone S, Rossi C, Poggi A, Ranelletti FO, Natali PG, Brunetti M, Aiello FB, Piantelli M. Flavonoids apigenin and quercetin inhibit melanoma growth and metastatic potential. International Journal of Cancer. 2000;**87**:595-600

[34] Kilani-Jaziri S, Frachet V, Bhouri W, Ghedira K, Chekir-Ghedira L, Ronot X. Flavones inhibit the proliferation of human tumor cancer cell lines by inducing apoptosis. Drug and Chemical Toxicology. 2012;**35**:1-10

[35] Pollard M, Suckow MA. Dietary prevention of hormone refractory prostate cancer in Lobund-Wistar rats: A review of studies in a relevant animal model. Comparative Medicine. 2006;**56**:461-467

[36] Lopez-Lazaro M. Flavonoids as anticancer agents: Structure-activity relationship study. Current Medicinal Chemistry-Anti-Cancer Agents. 2002;**2**:691-714

[37] Omarniyaz Z, Yu Y, Yang T, Shan L, Miao W, Reyimu R, Upur H, Aikemu A. Anti-tumor effects of abnormal Savda Munziq on the transplanted cervical cancer (U27) mouse model. BMC Complementary and Alternative Medicine. 2016;**16**:477

[38] Jemal A, Siegel R, Xu J, Ward E. Cancer statistics, 2010. CA: A Cancer Journal for Clinicians. 2010;**60**:277-300

[39] Huang WY, Cai YZ, Zhang Y. Natural phenolic compounds from medicinal herbs and dietary plants: Potential use for cancer prevention. Nutrition and Cancer. 2009;**62**:1-20

[40] Rodriguez J, Yanez J, Vicente V, Alcaraz M, Benavente-Garcia O, Castillo J, Lorente J, Lozano JA. Effects of several flavonoids on the growth of B16F10 and SK-MEL-1 melanoma cell lines: Relationship between structure and activity. Melanoma Research. 2002;**12**:99-107

[41] Manthey JA, Guthrie N. Antiproliferative activities of citrus flavonoids against six human cancer cell lines. Journal of Agricultural and Food Chemistry. 2002;**50**:5837-5843

[42] Chiang LC, Ng LT, Lin IC, Kuo PL, Lin CC. Anti-proliferative effect of apigenin and its apoptotic induction in human HepG2 cells. Cancer Letters. 2006;**237**:207-214

[43] Hirano T, Abe K, Gotoh M, Oka K. Citrus flavone tangeretin inhibits leukaemic HL-60 cell growth partially through induction of apoptosis with less cytotoxicity on normal lymphocytes. British Journal of Cancer. 1995;**72**:1380

Chapter 14

Flavonoid Complexes as Promising Anticancer Metallodrugs

Valentina Uivarosi and
Alexandra-Cristina Munteanu

Additional information is available at the end of the chapter

http://dx.doi.org/10.5772/67879

Abstract

Flavonoid metal complexes commonly exhibit an improvement of biological activity compared to the parent ligands. This chapter is focused on the antioxidant and anticancer properties of flavonoid metal complexes, in correlation with their binding ability to vital macromolecules such as nucleic acids and serum proteins. Perspectives for an adequate formulation of these complexes were also discussed.

Keywords: flavonoids, metal complexes, anticancer properties, nucleic acid binding, protein binding, formulation

1. Introduction

The urgency to overcome the biophysical and biomedical drawbacks of current chemotherapeutic treatments led the scientists to consider flavonoid metal complexes as viable options in cancer therapy. Both *in vitro* and *in vivo* studies report that flavonoids and their metal ion complexes exert pleiotropic effects on tumor promotion and progression.

Our work is an attempt to help the design of novel flavonoid-metal ion complexes with improved pharmacological activity and a broader range of antitumor mechanisms of action. In this chapter, we have analyzed relevant data available in literature on the antitumor activity of the flavonoid-metal ion complexes, regarding their cellular targets, their role in cancer cell death, growth and proliferation, and structure-activity analysis.

A novel metal-based compound with antitumor activity and promising clinical efficacy should meet the following criteria: (1) possess good intrinsic properties, molecular stability, allowing the drug to arrive intact at the target cells; (2) exert efficient interaction with transport proteins in blood and membranes; (3) show good DNA-binding properties; (4) have selective activity against cancerous cells over normal cells; and (5) preferably have activity against tumor cells that are resistant to cisplatin and derivatives. These aspects will be discussed in the following pages.

2. Flavonoids: general information, main classes, and chelating properties

Flavonoids (from the Latin "flavus," yellow) are secondary plant metabolites naturally occurring in seeds, fruit skin, peel, and bark of plants [1]. Flavonoids are important components of the human diet, the major sources of flavonoids being apples, red fruits, onions, citrus fruits, nuts, and beverages such as tea, beer, and wine [2]. Although they are not considered nutrients, due to the variety of pharmacological activities in the mammalian body, flavonoids are more correctly referred to as "nutraceuticals" [3].

These compounds possess a common flavane (2-phenyl-benzo-γ-piran) nucleus, consisting of an aromatic A-ring fused to a heterocyclic C-ring, attached through a single carbon-carbon bond to a benzene B-ring (**Figure 1**).

According to the oxidation degree of the C-ring, the hydroxylation pattern of the nucleus, and the C^3 substituent, the flavonoids can be categorized into seven subclasses: flavones, flavonols, flavanones, flavanols (catechins), flavanonols, isoflavones, anthocyanins, and anthocyanidins [4–6]. Thus, the total number of polyphenolic compounds exceeds 4000. In all these subclasses, rings B and C are linked at C^2, with the sole exception of the isoflavones (linked at C^3). Many flavonoids occur naturally as glycosides; the carbohydrate substituents include D-glucose, L-rhamnose, glucorhamnose, galactose, and arabinose [7]. **Table 1** lists the subclasses of flavonoids, and a set of representatives, to which we will refer to in this chapter.

Flavonoids possess three possible metal-chelating sites that can bind metal ions: (i) the 3-hydroxy-4-ketone groups in the C-ring, (ii) the 5-hydroxy group in the A-ring and 4-carbonyl group in the C–ring, and (iii) 3′,4′-dihydroxy groups, located on the B-ring (**Figure 2**).

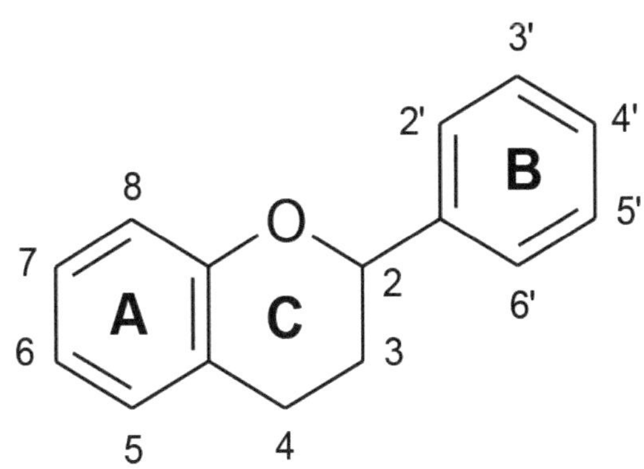

Figure 1. Basic flavonoid structure.

http://dx.doi.org/10.5772/67879

Flavonoid subclass	General structure	Representatives
Flavones	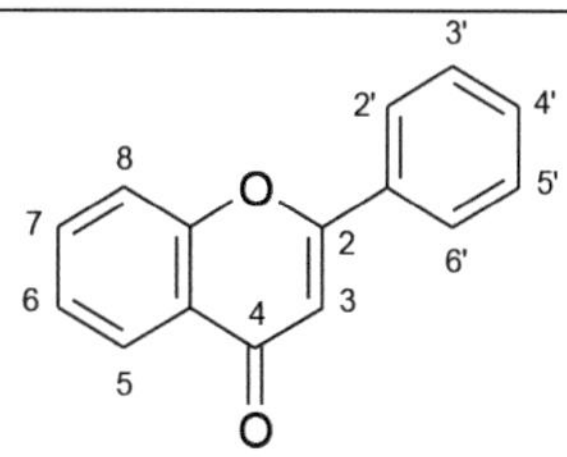	Chrysin: 5=7=OH Apigenin: 5=7=4′=OH Luteolin: 5=7=3′=4′=OH Diosmetin: 5=7=3′=OH, 4′=OCH_3 Diosmin: Diosmetin-7-rutinoside
Flavonols	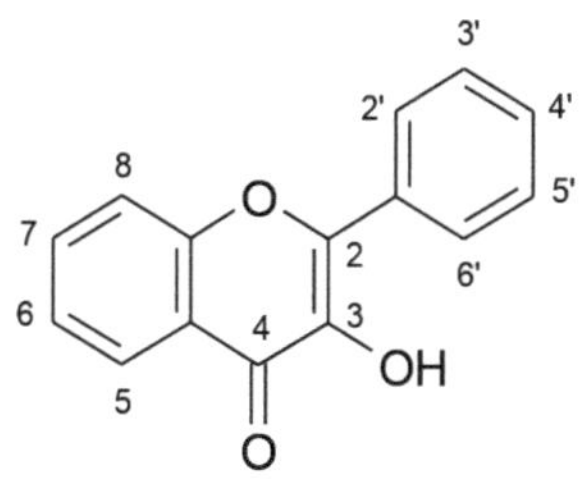	Fisetin:7=3′=4′=OH Kaempferol: 5=7=4′=OH Quercetin: 5=7=3′=4′=OH Morin: 5=7=2′=4′=OH Myricetin: 5=7=3′=4′=5′=OH Galangin: 3=5=7=OH Rutin: Quercetin-3-rutinoside
Flavanones	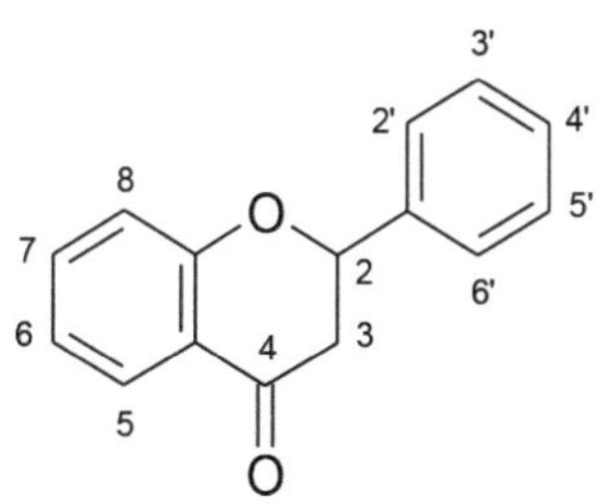	Naringenin: 5=7=4′=OH Hesperetin : 5=7=3′=OH, 4′=OCH3 Naringin: Naringenin-7-neohesperidoside Hesperidin: Hesperitin-7-rutinoside Dio
Flavanols	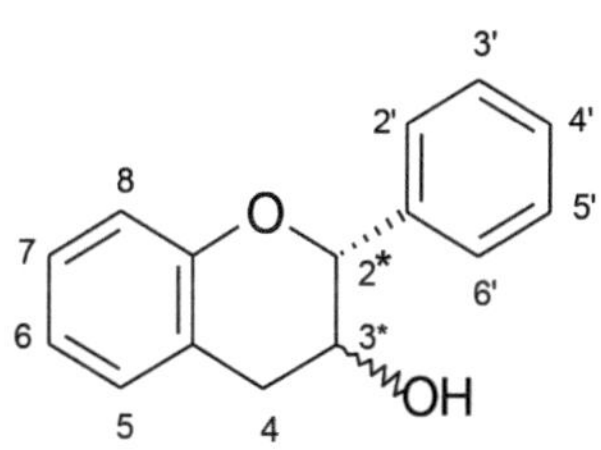	Catechin (2*R, 3*S): 5=7=3′=4′=OH Epicatechin (2*R, 3*R): 5=7=3′=4′=OH Epigallocatechin (2*R, 3*R): 5=7=3′=4′=5′=OH Epicatechin gallate (2*R, 3*R): 5=7=3′=4′=OH, 3-gallic acid ester Epigallocatechin gallate (2*R, 3*R): 5=7=3′=4′=5′=OH, 3-gallic acid ester
Flavanonols	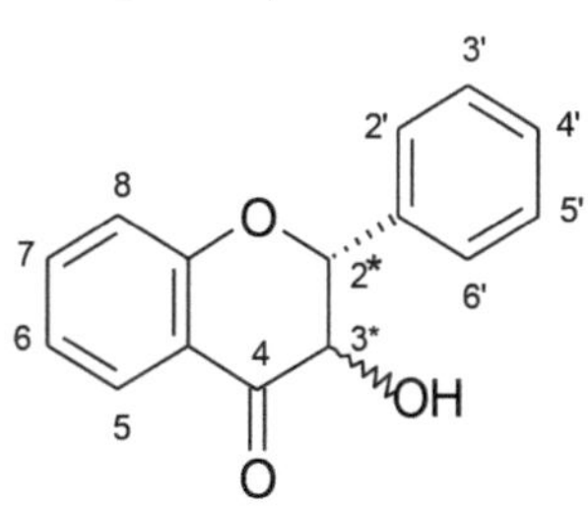	Taxifolin (2*R, 3*R): 5=7=3′=4′=OH
Isoflavones	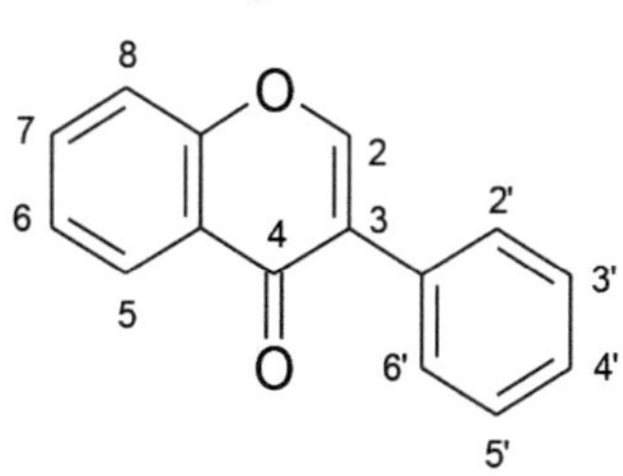	Daidzein : 7=4′=OH Genistein: 5=7=4′=OH

Flavonoid subclass	General structure	Representatives
Anthocyanidins, anthocyanins	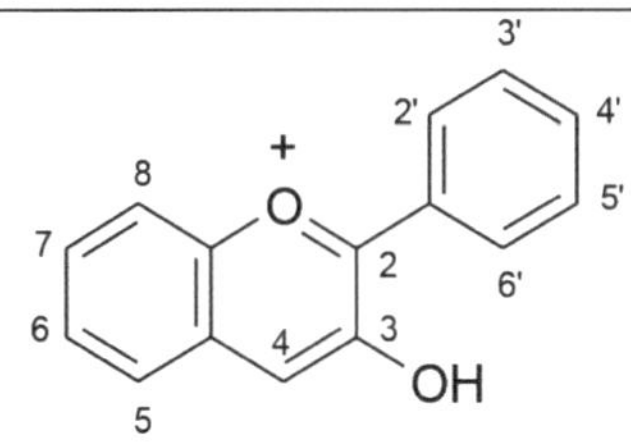	**Anthocyanidins** Pelargonidin: 5=7=4′=OH Cyanidin: 5=7=3′=4′=OH Delphinidin: 5=7=3′=4′=5′=OH Malvidin: 5=7=4′=OH, 3′=5′=OCH **Anthocyanins** Cyanidin 3-glucoside Cyanidin 3-galactoside Cyanidin 3-rutinoside Malvidin 3-glucoside

Table 1. Main flavonoid [21] subclasses.

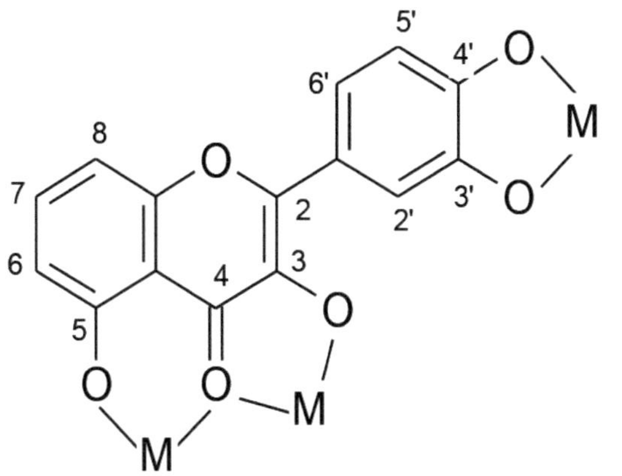

Figure 2. Typical chelation sites in forming the flavonoid complexes [2].

Cornard and Merlin [8] have reported that in acidic conditions, the 3-hydroxy-4-ketone or the 5-hydroxy-4-keto groups of quercetin (Q) are involved in coordination, whereas in alkaline milieu, the second chelating site, 3′,4′-dihydroxy group, located on the B-ring, is also involved.

3. Antioxidant and prooxidant activity of flavonoids and their metal complexes

Flavonoids have been reported to possess various biological effects: anticarcinogenic [9–11], antiviral [12], anti-inflammatory, immune stimulation [13, 14], antiallergic [15], and reducing the risk of cardiovascular disease [16–18].

Many of these beneficial health effects, including anticancer activity, arise from the antioxidant properties of these polyphenolic compounds [19], which are based on the following mechanisms:

(1) Direct radical scavenging: the flavonoid molecules are oxidized by free radicals ($R^{\bullet}$) resulting in more stable, less reactive radicals [20]:

$$FlOH + R\bullet \rightarrow FlO\bullet + RH \quad (1)$$

where $R^{\bullet}$ refers to either reactive oxygen species, ROS (hydroxyl, superoxide, peroxyl, alkyl peroxyl), or reactive nitrogen species, RNS (nitric oxide, peroxynitrite).

(2) Chelation of metal ions; the property of flavonoids to form chelates can be considered from two points of view:

- flavonoids bind metal ions like Fe(II), Fe(III), and Cu(II), thus preventing free radical generation in Fenton reactions:

$$M^{n+} + H_2O_2 \rightarrow M^{(n+1)+} + OH^- + OH\bullet \quad (2)$$

- flavonoids form metal complexes that are generally more powerful antioxidants than the free ligands.

(3) Inhibition of various enzymes that are responsible for ROS/RNS formation (xanthine oxidase, reduced nicotinamide adenine dinucleotide phosphate (NADPH) oxidase).

(4) Upregulation of enzymatic and non-enzymatic systems involved in the elimination and detoxification processes of oxidant species, such as reduced glutathione (GSH), GSH peroxidase, superoxide dismutase, and so on.

(5) Regeneration of the antioxidant species such as ascorbate and α-tocopherol by electron transfer reactions [5].

Many studies have correlated the antioxidant activity of flavonoids with the following structural features (**Figure 3**):

1. *ortho*-dihydroxy substitution in the ring B;
2. the presence of a 2,3 double bond and of a 4-oxo function;
3. a 3-hydroxy-4-keto and/or 5-hydroxy-4-keto conformation [21].

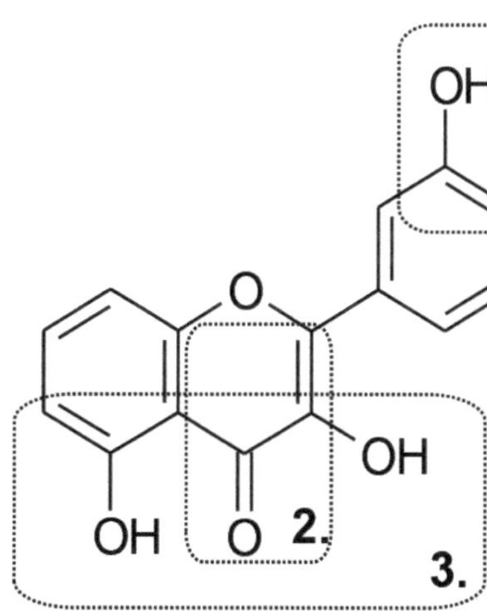

Figure 3. Main structural features required for antioxidant activity of flavonoids.

Among these main structural properties, it was observed that the antioxidant activity increases with the number of hydroxyl substituents and that the aglycones are more potent antioxidants than their corresponding glycosides [22].

Apart from the antioxidant activity, flavonoids can also exhibit prooxidant activity. This behavior is associated with the low one-electron reduction potentials; some of them may autoxidize slowly under neutral conditions and faster in alkaline media, generating ROS [23]. Another mechanism of prooxidant activity is correlated with their ability to reduce Cu^{2+} and Fe^{3+}, thus generating Fe^{2+} and Cu^{+} which in turn reduces hydrogen peroxide to hydroxyl radicals [24]. Prooxidant activity is thought to be directly proportional to the total number of hydroxyl groups [25], especially in the B-ring [26], and the concentration of the flavonoid [27].

Commonly, the prooxidant activity of flavonoids has been associated with various toxic effects, but it is also possible for it to generate beneficial properties. For example, the apoptotic and bactericidal activity of epigallocatechin gallate is based on its ability to reduce O_2 to yield H_2O_2 [28]. Likewise, the capacity of flavonoids to stimulate the activity of the detoxifying enzymes is a major mechanism for the chemopreventive properties of flavonoids, thus protecting the organism against mutagens and carcinogens [29].

A great number of metal complexes of flavonoid derivatives have been obtained, in order to improve the antioxidant activity of the parent flavonoid (**Table 2**).

Complex	Molar ratio metaion:ligand	Effect on the flavonoid antioxidant activity	Ref.
Flavonoid subclass: flavones			
VO(II)-luteolin	1:2	Increase	[30]
Ge(IV)-chrysin	1:1	Increase	[31]
VO(II)-chrysin	1:2	Increase	[32], [33]
Ce(IV)-chrysin	1:2	Increase	[34]
VO(II)-diosmin	1:1	Increase the scavenging activity only for ROO•, not for $O_2^{\bullet -}$, DPPH•, OH•	[35]
Flavonoid subclass: flavonols			
Mg(II)-quercetin	1:1	Increase	[36]
Al(III)-quercetin	2:1	Increase	[37]
Ge(IV)-quercetin	2:1	?	[38]
Sn(II)-quercetin	Not determined	Slightly decrease	[39]
Fe(II)-quercetin	1:2	Increase	[40]
Co(II)-quercetin	2:1	Increase	[41]
Cu(II)-quercetin	1:1	Increase	[42]
Cu(II)-quercetin	2:1	Increase	[43]
Cd(II)-quercetin	1:1	Decrease	[44]

Complex	Molar ratio metaion:ligand	Effect on the flavonoid antioxidant activity	Ref.
M(II)-quercetin M = Mn, Co, Ni, Cu, Zn	1:2	Increase	[45]
M(II)/(III)-quercetin M(II) = Fe, Cu, Zn M(III) = Al	2:1	Increase	[46]
Ln(III)-quercetin Ln = La, Nd, Eu, Gd, Tb, Dy, Tm, Y	1:3	Increase	[47]
VO(II)-morin	1:2	Increase the scavenging activity for $OH^•$ and O_2^- Decrease scavenging activity for $ROO^•$.	[48]
Cr(III)-morin	1:2	Increase	[49]
Cu(II)-morin	1:1	Increase	[50]
M(II)/(III)-galangin M(II) = Fe, Cu, Zn M(III) = Al	1:1	Increase	[46]
Sn(II)-rutin	1:2	Decrease	[51]
Sn(II)-rutin	3:2	Decrease	[52]
Cr(III)-rutin	2:1	Slightly increase	[53]
Cr(III)-rutin	3:1	Increase	[54]
VO(II)-rutin	1:2	Increase	[55]
Ni(II)-rutin	1:2	Increase	[56]
Cu(II)-rutin	1:1	Increase	[57]
Zn(II)-rutin	2:1	Increase	[58]
Flavonoid subclass: flavanones			
Cu(II)-naringenin-phenanthroline	1:1:1	Increase	[59]
M(II)-naringenin M=Cu, Zn, Ni	1:2	Increase	[60]
Ce(IV)-naringenin	1:2	Increase	[34]
Cu(II)-hesperetin-phenanthroline	1:1:1	Increase	[59]
Cu(II)-naringin	1:1	Increase	[61]
VO(II)-hesperidin	1:1	Increase	[62]
Flavonoid subclass: flavanols			
M(II)/(III)-catechin M(II) = Fe, Cu, Zn M(III) = Al	1:1	Increase	[46]
Al(III)-catechin	1:1	Decrease	[37]
Flavonoid subclass: isoflavones			
Cu(II)-genistein Fe(III)-genistein Cu(II)-biochanin A Fe(III)-biochanin A	1:2	Increase for Cu(II) Prooxidant activity for Fe(III)	[63]
Ce(IV)-daidzein	2:1	Increase	[34]

Table 2. Metal complexes of flavonoids investigated for their antioxidant activity.

By analyzing the data from **Table 2** and the scientific literature, some observations appear to be relevant:

- most metal complexes behave as more powerful antioxidants compared to the parent flavonoids;
- the higher antioxidant activity of the metal complexes was generally explained by acquisition of an additional radical-scavenging metal center by the complexes, probably a superoxide-dismutase-mimicking center;
- in some cases, the metal (e.g., Sn^{2+}, Cd^{2+}) complexes show lower antioxidant activity than the parent flavonoids; the lower antioxidant activity can be explained by the fact that chelation of these metal ions significantly changes the chemical properties of the flavonoid, or might increase the oxidation potentials relative to those of the free flavonoid;
- some metal complexes, especially those of Fe(III) ions, display prooxidant activity.

4. Anticancer activity of flavonoids metal complexes

The cytotoxic activity of flavonoids involves the inhibition of several molecular targets and pathways: DNA topoisomerases I and II [64], cyclin-dependent kinases CDK2 and/or CDK1 [65], androgen receptor signaling [66], actin polymerization [67, 68], activation of p53, and inhibition of NFkB pathways [69]. Flavonoids activate the caspase-mediated signal transduction pathways, consecutively stimulating the tumor-suppressor protein p53, which consequently triggers cell apoptosis [69].

In many cases, the antitumor activity of the flavonoid metal complexes has been reported to be greater than that of the free corresponding flavonoids [47]. This may be mediated through the regulation of important cell-cycle events, alterations in the DNA structure, prooxidant effects, or interactions with the phospholipid bilayer. Flavonoid metal complexes have been reported to be active against gastric cancer cells, human hepatocellular carcinoma cells [70], human cervical carcinoma cells [71], leukemia cells [61], human colon adenocarcinoma cells, human hepatoma cells, and osteoblast cancer cells [62]. A very important and promising feature displayed by these complexes is that a number of them have proven to be selective toward cancerous cells over normal cells [72]. **Table 3** contains a selection of metal complexes of flavonoids which have shown antitumoral activity.

As it has been shown in **Table 5**, complexes contain diverse structures, derived from the variety of flavonoids and metal ions used in the drug design. Regarding the flavonoid structures in metal complexes, the type of the substituents on either ring in the ligand structure, whether electron-withdrawing or electron donating, seems to have minor importance. However, their position appears to be crucial for the cytotoxic activity. *Ortho*-substitution of the B-ring appears to be unfavorable, while *meta*- and *para*-substitution augments the anticancer activity. This may be due to a structural effect, as the B (phenyl)-ring is more twisted in the *ortho*-substituted compounds than in the *meta*- and *para*-derivatives, which may increase the interaction with biological targets. The purpose of combining these ligands with metal centers is to obtain anticancer agents with extended mechanistic range, facilitating single-molecule multi-target anticancer therapy.

Complex	Comments	Ref.
4'-Methoxy-5,7-dihydroxy-isoflavone Complexes with Mn(II)*, Cu(II), Zn(II), Co(II), Ni(II) Metal ion:isoflavone molar ratio 1:2	Cell lines: *HeLa* (cervical cancer), *MDA-MB-435* (breast carcinoma), *SW620* (colon carcinoma), *HepG2* (liver cancer), *A549* (non-small-cell lung carcinoma). *was most efficient, especially on line SW620 (IC_{50} < cisplatin). Flow cytometry analysis showed that * acts in the G2/M phase of the cell cycle, causing early apoptosis.	[73]
Rutin complex with Zn(II) (Znrut) Metal ion:isoflavone molar ratio 1:1	*KG1* (leukemia): IC_{50} = 91.4 μM; *K562* (leukemia): IC_{50} = 173.2 μM; *Jurkat* (leukemia): IC_{50} = 150.2 μM; *RPMI8226* (multiple myeloma): IC_{50} = 196.6 μM; *B16F10* (melanoma): IC_{50} = 160.7 μM; *SK-Mel-28* (melanoma): IC_{50} = 194.0 μM. In the Ehrlich ascites carcinoma model, Znrut exerts modulating effect of mitochondrial membrane potential and expression of genes involved in cell cycle control, apoptosis and angiogenesis.	[74]
Diosmin complex with oxovanadyl (IV) $[VO(dios)(OH)_3]Na_5.6H_2O$ (VOdios)	*A549*: dios- IC_{50} > 100 μM, at 100 μM, VOdios decreases cell viability by 31%; *T47D* (breast tumor): dios exerts no antiproliferative effect, VOdios – IC_{50} = 23.3 μM; *SKBR3* (breast adenocarcinoma): dios – IC_{50} > 100 μM, VOdios – IC_{50} = 46.4 μM; *MDAMB231* (breast adenocarcinoma): dios – IC_{50} = 70 μM, VOdios – IC_{50} = 11.6 μM. An antiproliferative non-apoptotic mechanism, caspase 3/7 and oxidative stress independent, has been proposed.	[35]
Luteolin complex with oxovanadyl (IV) $[VO(lut)(H_2O)_2]Na{\cdot}3H_2O$ (VOlut)	*MDAMB231*: IC_{50} = 17 μM (lut: IC_{50} = 88.3 μM); *A549*: IC_{50} = 60.5 μM (lut: IC_{50} = 66.3 μM); Both Lut and VOlut generate ROS and mitotic arrest; VOlute causes nuclear and cytoplasmic membrane damage.	[75]
Naringenin complex with oxovanadyl (IV) $[VO(nar)_2]{\cdot}2H_2O$ (VOnar)	*A549*: nar shows no antiproliferative effect, VOnar: IC_{50} > 100 μM (at 100-μM cell viability decreases by 35%); *SKBR3*: nar shows no antiproliferative effect, VOnar: IC_{50} = 73 μM; *MDAMB231*: nar: IC_{50} > 100 μM (at 100-μM cell viability decreases by 30%), VOnar: IC_{50} = 20 μM. VOnar antiproliferative effect was accompanied by ROS generation, cell membrane damage and DNA degradation, cell cycle arrest, caspase 3/7 activation, and mitochondrial membrane potential decrease.	[76]
Chrysin complex with oxovanadyl (IV) $[VO(chrys)_2EtOH]_2$ (VOchrys)	*MC3T3E1* (normal osteoblast cell line), *UMR106* (osteosarcoma). Both the ligand and the complex exert similar inhibitory effect on normal cells and cancer cells (60% inhibition at 100 μM for osteosarcoma cells). Complexation does not improve the antitumor properties of the free flavonoid. *In vitro* and *in vivo* studies were performed on 3D human osteosarcoma and xenograft osteosarcoma mice models. The pharmacological results show that VOchrys had an inhibitory effect on cell viability, affecting the shape and volume of the spheroids and suppressed MG-63 tumor growth in the mice models without toxic and side effects.	[32, 33, 77]
Quercetin (Quer) complex with oxovanadyl (IV) $[VO(Quer)_2EtOH]_n$ (VOQuer)	*MC3T3E1* (normal osteoblast cell line), *UMR106* (osteosarcoma). The complexation of quercetin does not improve its potential anticarcinogenic properties. The complex was more cytotoxic toward normal osteoblasts. VOQuer interacts with two markers of osteoblastic differentiation; it stimulates type I collagen production and inhibits bovine ALP-specific activity; QuerVO stimulated the phosphorylation of ERKs in a dose-response manner, showing that the mitogenic effect of the complex in osteoblasts may involve the ERK pathways. VOQuer shows potential for promoting osteoblast differentiation.	[78]
cis-$[Pt(PPh_3)_2(3\text{-}Hfl)]Cl$ **(1)** *cis*-$[Pt(PPh_3)_2(etga)]$ **(2)** PPh = triphenylphosphine, 3-Hfl = 3-hydroxyflavone monoanion, etga = etilgalat dianion	*U87* (glioblastoma): IC_{50}(μM) values: 3-Hfl: 27.5 ± 2.3; **(1)**: 26.3 ± 2.1; etga: 97.7 ± 2.6; **(2)**: 123.7 ± 3.8; cisplatin: 1.76 ± 0.22; *MCF-7* (breast tumor): IC_{50}(μM) values: 3-Hfl: 108.1 ± 3.5; **(1)**: 55.5 ± 1.7; etga: > 200; **(2)**: >200; cisplatin: 14 ± 3.	[79]

Complex	Comments	Ref.
Ru(II) *cis*-dichlorobis(3-imino-2-methoxy-flavanone) **(1)** Ru(II) *cis*-dichlorobis(3-imino-2-ethoxy-flavanone) **(2)** Ru(II) *cis*-dichloro(3-nitrozoflavone)(3-hydroxyiminoflavanone)**(3)**	*A549*: IC_{50}(μM) values: cisplatin: 3.8 ± 0.8; **(1)**: ~100; **(2)**: 5.3 ± 0.4; **(3)**: 3.8 ± 0.3; *A2780* (ovarian cancer): IC_{50}(μM) values: cisplatin: 0.2 ± 0.1; **(1)**: 39.6 ± 3; **(2)**: 2.5 ± 0.2; **(3)**: 1.8 ± 0.1; *A2780cis* (cisplatin-resistant subline): IC_{50}(μM) values: cisplatin: 16.8 ± 1.5; (1): 62.5 ± 16; (2): 4.6 ± 0.2; (3): 1.5 ± 0.04; *Toledo* (diffuse large B-cell lymphoma): IC_{50}(μM) values: cisplatin: 0.5 ± 0.07; **(1)**: 13.9 ± 1.2; **(2)**: 0.5 ± 0.1; **(3)**: 0.6 ± 0.04; *Toledo-cis* (cisplatin-resistant subline): IC_{50}(μM) values: cisplatin: 8.3 ± 0.6; **(1)**: 56.6 ± 2.8; **(2)**: 2.9 ± 0.2; **(3)**: 2.8 ± 0.1. *Lymphocytes*: IC_{50} (μM) values: cisplatin: 0.2 ± 0.1; **(1)**: 19.4 ± 3; **(2)**: 3.4 ± 0.3; **(3)**: 1.6 ± 0.2. The complexes had minor effects on hemostasis or on the red blood cell lysis *in vitro*, at cytotoxic concentrations and are therefore unlikely to cause hematologic disorders *in vivo*.	[80]
Ru(II) complex with η^6-*p*-cymene and 3-hydroxyflavone derivatives	*CH1 (ovarian carcinoma), SW480 (colon carcinoma, A549 (non-small-cell lung carcinoma), 5637 (human urinary bladder), LCLC-103H (human large-cell lung), DAN-G (human pancreatic carcinoma cell lines).* The IC_{50} values were in the low micromolar range. The chemosensitive CH1 cell line was very sensitive to the complexes with IC_{50} values < 7.9 μM; for SW480, the IC_{50} values: 3.4–26 μM. The lowest cytotoxic potency was found for *A549* (IC_{50}: 8.6–51 μM). Some of the complexes have an influence on the cell cycle distribution; at concentrations around the IC_{50} values of the complexes, an increase in the cell fraction in G0/G1 phase was observed. Most of the complexes strongly inhibit CDK2, inhibit topoisomerase IIα to a greater extent than the corresponding free ligands (the activity on topo-IIα correlates with the cytotoxic effect of the complexes).	[81]
Quercetin complexes with Zn(II), Mn(II) $[Zn(que)_2(H_2O)_2]$ **(1)** $[Mn(que)_2(H_2O)_2]$ **(2)**	*HepG2*: IC_{50}(μM) values: $(que)_2$: 13.3 ± 1.0; **(1)**: 5.46 ± 0.36; **(2)**: 8.0 ± 0.3; *SMMC-7221* (hepatocarcinom): IC_{50}(μM) values: $(que)_2$: 13.3 ± 1.0; **(1)**: 7.66 ± 0.30; **(2)**: 8.0 ± 0.3; *A549*: IC_{50}(μM) values: $(que)_2$:13.3 ± 1.0; **(1)**: 10.0 ± 0.45; **(2)**: 8.0 ± 0.3. The antitumor activity of **(2)** may be partially attributed to the interaction with the GC-rich DNA sequences and to DNA protein-binding interactions. These processes lead to downregulation of survivin gene expression, caspase activation, and induction of apoptosis. HepG2 cells had undergone morphological changes typical of apoptosis, characterized by nuclear shrinkage, chromatin condensation, and fragmentation subsequent to exposure to complex **(1)**. The cytotoxic effect is partially attributed to intercalation between DNA base pairs.	[82, 83]
Chrysin **(1)**, apigenin **(2)**, genistein **(3)** homoleptic Cu(II) Complexes **(4)–(6)**, respectively	*518A2* (melanoma): IC_{50}(μM) values: **(1)**: > 50; **(2)**: > 50; **(3)** > 50; **(4)**: 10.9 ± 1.6; **(5)**: 13.2 ± 1.1; **(6)**: 10.9 ± 1.4; *HCT-116* (colon carcinoma): IC_{50}(μM) values: **(1)**: > 50; **(2)**: > 50; **(3)** > 50; **(4)**: 14.8 ± 2.7; **(5)**: 15.8 ± 2.5; **(6)**: 10.4 ± 1.6; *KB-V1/Vbl* (cervix carcinoma): IC_{50}(μM) values: **(1)**: 12.5 ± 2.1; **(2)**: 20.6 ± 1.7; **(3)** 28.8 ± 3.0; **(4)**: 4.6 ± 0.5; **(5)**: 8.0 ± 1.2; **(6)**: 6.7 ± 0.5; *MCF-7/Topo* (breast carcinoma): IC_{50}(μM) values: **(1)**: 28.2 ± 6.1; **(2)**: 31.2 ± 0.9; **(3)** 37.0 ± 3.0; **(4)**: 7.8 ± 1.5; **(5)**: 11.8 ± 0.7; **(6)**: 8.5 ± 1.4; The cytotoxic activity of the complexes was correlated with an arrest of the cell cycle of *518A2* cells at the G2/M transition. The complexes gave better results than the free flavonoids in decreasing the migration propensity of *518A2* cells in wound-healing assays. The antimetastatic effects of complex **(6)** derive from the remodeling of the actin cytoskeleton and the increase in cadherin-catenin complex formation, factors that favor cell-cell adhesion. Complex (**6**) decreased the expression and secretion of the metastasis-related matrix metalloproteinases MMP-2 and MMP-9. Thus, the coordination of apigenin and genistein to Cu(II) enhances the antitumoral properties of the free flavonoids and expands their mechanistic range.	[84]

Complex	Comments	Ref.
Chiral L/D-valine (val) quercetin (Q) organotin(IV) complexes $[(CH_3)_2Sn(Q)(val)]$ **1**$_{L/D}$ $[(C_6H_5)_2Sn(Q)(val)]$ **2**$_{L/D}$	*HeLa*: IC_{50} (μg mL^{-1}) values: < 10; *MCF-7*: IC_{50} (μg mL^{-1}) values: < 10; *HepG2*: IC_{50} (μg mL^{-1}) values: **1**$_L$: 34.6; **2**$_L$: < 10; *MIA-Pa-Ca-2* (pancreas cancer): IC_{50} (μg mL^{-1}) values: < 10; The mechanistic activity of the complexes appears to involve interaction with topoisomerase I and DNA.	[85]

ALP: alkaline phosphatase; ERK: extracellular-regulated kinase; CDK2: cyclin A protein kinase.

Table 3. Metal complexes of flavonoids with antitumor activity.

It should be kept in mind that with the failure of ASA404/vadimezan (Antisoma/Novartis) in a phase III trial for advanced non-small-cell line cancer when given in combination with carboplatin and taxol [86], there are no more benzopyran-4-one derivatives strong candidates for anticancer drugs. The research regarding the antitumor properties of flavonoid metal complexes offers promising results and these should be further improved in order for the complexes to enter clinical trials.

5. Interactions of flavonoids metal complexes with proteins

Protein binding can influence the blood levels and the pharmacokinetic behavior of a drug and possibly its pharmacologic and toxicologic profiles. Human serum albumin (HSA) is quantitatively ~55% of total serum proteins. Analyses of the crystal structure of the protein indicate that the main binding sites in HSA are located in the hydrophobic cavities of subdomains II and III of site A. These binding sites are known as Sudlow I and II, respectively. The remaining single tryptophan in HSA structure is located within the site's Sudlow I (Trp-214) [87]. Serum albumin (SA) is involved in blood transport for many compounds and metal ions. The interactions of a drug with SA play a major role in drug efficacy. Flavonoids and their metal complexes interact in the microenvironment surrounding the tryptophan residue of SA.

Transferrin (Tf) is a monomeric glycoprotein, containing two main metal ion binding sites: the N-terminal lobe and the C-terminal lobe, similar in structure, each binding one Fe(III) ion. It has been found that iron-binding sites in the serum Tf are only saturated to an extent of 39% of Fe(III), meaning that the free lobes can bind other metal ions [88]. Cancer cells, as active and rapidly proliferating cells, express high levels of transferrin receptors. Consequently, transferrin has been explored as a potential drug carrier for targeted delivery into tumor cells [89].

Moreover, human topoisomerase II-α (topo II-α) is currently a target for anticancer chemotherapy. Topo II-α is involved in DNA transcription, replication, and chromosome segregation. Although these biological functions are vital for insuring genomic integrity, the ability to inhibit topo II-α and generate enzyme-mediated DNA damage in cancer cells is an effective strategy in antitumor therapy [90]. Electrostatic forces, hydrogen bonds, and van der Waals interactions are involved in the interactions of flavonoid metal complexes and proteins [72].

Flavonoids and their metal complexes bind bovine serum albumin (BSA) or human (HSA), in vitro, with different affinities (**Table 4**).

Complex	Comments	Ref.
Diosmin-oxovanadyl (IV) complex $[VO(dios)(OH)_3]\ Na_5 6H_2O$ (VOdios)	Diosmin and VOdios interactions with BSA follow a static mechanism, a ratio of 1: 1 compound: protein; the interaction is enabled via hydrogen bonds and van der Waals forces. At 298 K: $K_{b\ dios} = 1.84 \pm 0.32 \times 10^4\ M^{-1}$ $K_{b\ VOdios} = 55.3 \pm 0.07 \times 10^5\ M^{-1}$	[35]
Luteolin- oxovanadyl (IV) complex $[VO(lut)(H_2O)_2]\ Na{\cdot}3H_2O$ (VOlut)	Luteolin interacts with the microenvironment around tryptophan in BSA, by electrostatic forces; the complex interacts with the protein through hydrogen bonds and van der Waals forces. The interaction takes place in a ratio of 1: 1 compound: protein through a static mechanism. At 298 K: $K_{b\ lut} = 65.10 \pm 0.90 \times 10^6\ M^{-1}$ $K_{b\ VOdios} = 79.43 \pm 0.56 \times 10^6\ M^{-1}$	[75]
Naringenin-oxovanadyl (IV) complex $[VO(nar)_2]{\cdot}2H_2O$ (VOnar)	The interaction takes place in a ratio of 1:1 compound:protein through a static mechanism. Complexation lowered the binding affinity to BSA of naringenin, probably due to steric hindrance. At 298 K: $K_{b\ nar} = 10.20 \pm 0.30 \times 10^4\ M^{-1}$ $K_{b\ VOnar} = 0.31 \pm 0.01 \times 10^4\ M^{-1}$	[76]
Quercetin and morin and their sulfonic derivatives-oxovanadyl (IV) complexes $VO(que)_2$ $VO(mor)_2$ $VO(que^S)_2$ $VO(mor^S)_2$	In the systems with apo-human transferrin (apo-HTF) and albumin (HSA), $VO(que)_2$ and $VO(Que^S)_2$ are stable, while VO $(mor)_2$ and $VO(mor^S)_2$ undergo displacement reactions, when the species $(VO)_x$(HSA) and (VO)(apo-HTF)/ $(VO)_2$(apo-HTF) and VO-apo-HTF-mor^S and mor^S-VO-HAS are formed. The complexes interact strongly with the proteins by the formation of hydrogen bonds with polar groups on the protein surface. $VO(que)_2/VO(Que^S)_2$ does not interact with hemoglobin, while $VO(mor)_2/VO(mor^S)$ forms adducts with hemoglobin (Hb).	[91]
Baicalein-Al(III) complex (ALBC)	The results of the competitive binding and molecular-docking studies indicate that BC binds to HSA at site I (subdomain IIA), while ALBC binds mainly at site II (subdomain IIIA). BC binding had a greater influence than ALBC on the secondary structure of the protein. At 297 K: $K_{b\ BC} = 1.73 \times 10^5\ M^{-1}$ $K_{b\ ALBC} = 1.67 \times 10^5\ M^{-1}$	[92]
Dihydromyricetin Mn(II), Cu(II) and Zn(II) complexes MnDMY CuDMY ZnDMY	DMY-BSA interaction is achieved by van der Waals forces and hydrogen bonds, while complexes bind through hydrophobic and hydrogen-bonding forces (conclusions based on thermodynamic values for parameters). At 300 K: $K_{b\ DMY} = 1.30 \times 10^5\ M^{-1}$ $K_{b\ MnDMY} = 2.13 \times 10^5\ M^{-1}$ $K_{b\ CuDMY} = 2.95 \times 10^5\ M^{-1}$ $K_{b\ ZnDMY} = 7.76 \times 10^4\ M^{-1}$	[93]
Morin-La(III) complex (La-Mo)	La-mo-HSA interaction was studied by means of spectrofluorometry and circular dichroism. It has been found that La-mo is an efficient interaction with HSA hydrogen bonding and van der Waals forces. The thermodynamic parameters (ΔG, ΔH, ΔS) that characterize the interaction had negative values, implying that the binding is thermodynamically favorable and the degree of reversibility is modest. Circular dichroism spectra show a reduction in the α-helix-type structures from 60.0 to 56.9% and an increase in the β-chain-type structures from 6.0 to 7.1% in HSA. Molecular-docking studies show that La-mo competes with warfarin site Sudlow I of subdomain II in HSA structure. At 299 K: $K_{b\ La\text{-}Mo} = 1.5752 \pm 0.007 \times 10^5\ M^{-1}$	[94]

Complex	Comments	Ref.
Ru(II)-η^6-*p*-cymene complex 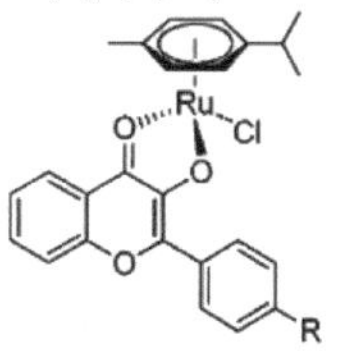 $R=CH_3$, F or Cl	The complex interacts with topoisomerase II-α and the interaction depends on the substituent located at the *para*-position of the phenyl ring. The complex where R=Cl is the most potent topoisomerase inhibitor. The complexes showed better activity than the ligands.	[95]
Quercetin-Cu^{II}-Sn_2^{IV} heterobimetallic complex	The complex is a potent inhibitor of topoisomerase I at 30 μM.	[96]

K_b: binding constant; K_i: inhibition constant.

Table 4. Selection of flavonoid metal complexes-protein interactions studies.

6. Interactions of flavonoids metal complexes with nucleic acids

The ability of small molecules to interact with DNA ranks among the most important mechanisms of action enabling antitumor activity, since intercalation between adjacent base pairs inhibits DNA replication. Most of the flavonoids and their metal complexes show affinity toward nucleic acids. Flavonoids bind DNA as a result of electrostatic interactions, as is the case of quercetin [97] and morin [98 and references therein]. Their complexes, on the other hand, are bulkier, and display more diversified mechanisms of interaction with DNA, including "major" or "minor groove" binding and/or intercalation. Due to their structural planarity, flavonoid complexes are prone to act as intercalators [99]. Moreover, the emergence of electrostatic interactions between the metal cation and anionic phosphate groups of DNA structure stabilizes the adducts formed between the complexes and DNA. It is considered that the DNA base pairs remove the flavonoid molecules in the metal complex, since the binding affinity between the negatively charged phosphate groups and the positively charged metal ions is stronger than that between the flavonoid molecule and the metal center [100, 101]. In most cases, active compounds possess quasi-planar structures, with a medium-sized planar area and hydrophobic character [82]. A selection of the metal complexes of flavonoids that interact with DNA is presented in **Table 5**.

Another aspect regarding DNA interaction refers to the complexes' cleavage activity. This property can benefit their antitumor activity, but can also cause oxidative DNA damage (and consequently cell death) in normal cells [22]. Therefore, efforts toward increasing the complexes' selectivity against cancer cells are of prior importance.

The development of electrochemical DNA biosensors has been of high interest in this field, since metal flavonoid complexes show promising results in DNA recognition. Flavonoid complex-based biosensors can be useful in several domains, such as transducing DNA hybridization, drug design, and diagnosis [102, 103, 105].

Complex	Comments	Ref.
Fe(II)-quercetin complex Fe(quer)$_2$	The interaction of the complex with DNA occurs in a moderate intercalative manner. Fe(quer)$_2$ shows cleavage activity on plasmid DNA (pBR322) under physiological conditions, via oxidative pathway.	[40]
Ni(II)-rutin (R) complex metal:ligand molar ratio 1:2 (NiR)	The interaction of NiR with DNA was studied using fluorescence spectra and agarose gel electrophoresis. The complex can intercalate moderately between DNA base pairs and shows significant, dose-dependent cleavage activity on pBR322 plasmid DNA from the SC form to the NC form.	[52]
Fe(III) chlorobis(flavonolato) (methanol) complex metal:ligand molar ratio 1:2 Fe(III)-(3hf)	The complex shows cleavage activity toward CT-DNA via an oxidative mechanism with higher efficacy in the presence of reducing agents (ascorbate/hydrogen peroxide).	[101]
Cd(II)-morin complex Cd(mor)$_2$	The interaction with salmon sperm dsDNA was studied by means of electrochemical methods. Cd(mor)$_2$ can intercalate into the double-helix DNA; according to the Hill model for cooperative binding, the equilibrium dissociation constant and the binding stoichiometry were calculated to be $K = 2.5 \times 10^{-5}$ M and $m = 1.761$, respectively.	[102]
Co(II)-morin complex Co(mor)$_2$	Competitive experiments, viscosity, and electrochemical studies indicate that the complex binds to DNA via a weak partial intercalation. $K_b = 2 \times 10^3$ M^{-1} at 20°C.	[103]
Morin **(1)** Cu(II)-morin complex **(2)** Morin-β-CD **(3)**	Intercalation was proposed as the mode of binding of the ligand, the complex with salmon sperm dsDNA via cyclic and square wave voltammetry, UV-vis spectroscopy techniques. K_b values: **(1)**: 1.58, **(2)**: 2.29, **(3)**: 3.20 × 10^5 M^{-1} at 298 K. Other studies indicate that morin interacts with DNA in a non-intercalative manner [98].	[104]
Cu(II)-rutin (R) complex Cu$_2$R$_3$	The interaction of Cu$_2$R$_3$ with salmon sperm dsDNA was studied by means of cyclic voltammetry and fluorescence spectroscopy. The complex interacts with DNA via intercalation and nonspecific electrostatic interaction. Cu$_2$R$_3$ was used for the construction of an electrochemical DNA biosensor for DNA hybridization detection, showing relatively good sensitivity and selectivity.	[105]
Cu(II)-hesperetin complex	$K_b = 1.5 \times 10^6$ M^{-1} The UV-vis, fluorescence and CD spectral measurements revealed that both the complex and the ligand interact with CT-DNA via intercalation. The binding affinity of the complex is stronger than that of free ligand.	[59, 70]
Cu(II)-quercetin complex metal:ligand molar ratio 1:2	By means of UV-vis spectrophotometry, cyclic voltammetry, and synchronous fluorescence spectroscopy, an intercalative binding mode was proposed. $K_b = 1.82 \pm 0.2 \times 10^5$ M^{-1} The equilibrium constant of the exchange process in the intercalation reaction was found to be approximately 5 × 10^{-1}; 35% of the bound complex was not involved in intercalation.	[106]
Tb(III)/Eu(III)-quercetin system	A sensitive method for the determination of CT-DNA is proposed based on the quenching effect of DNA on the RLS intensity of Tb(III)/Eu(III)-quercetin system.	[100]

Complex	Comments	Ref.
Chiral L/D-valine (val) quercetin (Q) diorganotin(IV) complexes $[(CH_3)_2Sn(Q)(val)]$ **1**$_{L/D}$ $[(C_6H_5)_2Sn(Q)(val)]$ **2**$_{L/D}$	UV-vis, fluorescence titrations, thermal denaturation, and circular dichroism studies revealed the electrostatic mode of DNA binding of the complexes, although partial intercalation into the DNA helix and hydrophobic interactions cannot be excluded. K_b values (: **1**$_L$: 2.17×10^5 M^{-1}; **1**$_D$: 1.00×10^4 M^{-1}; **2**$_L$: 4.74 $\times 10^5$ M^{-1}; **2**$_D$: 1.40×10^4 M^{-1}) revealed that the L-enantiomers of exhibited exceptionally high-binding propensity as compared to their D-enantiomers. **1**$_L$ and **2**$_L$ exhibit supercoiled DNA cleavage ability mediated by single strand breaks.	[85]
La(III)-chrysin complex	$K_{b\ chrysin} = 1.29 \times 10^6$ M^{-1}; $K_{b\ complex} = 5.44 \times 10^5$ M^{-1} Spectrophotometric methods and viscosity measurements indicated that La(III) complex and chrysin can both bind to DNA by intercalation modes.	[107]
Eu(III)-quercetin complex $Eu(quer)_3$	The CT-DNA interactions were studied using cyclic voltammetry and double potential step chronocoulometry at glass carbon electrode for the surface method. Quercetin binds to DNA mainly by electrostatic interaction and the complex binds to DNA by means of intercalation and electrostatic forces. For the dsDNA-modified GCE systems, the following parameters were obtained: $\Gamma_{s\ quercetin} = 2.28 \pm 0.2 \times 10^{-10}$ mol/cm^3 $n_{quercetin} = 1.2$ $\Gamma_{s\ Eu(quer)3} = 1.65 \pm 0.2 \times 10^{-10}$ mol/cm^3 $n_{Eu(quer)3} = 1.8$	[97]
Mn(II)-quercetin complex metal:ligand molar ratio 1:2	UV/VIS, fluorescence spectroscopy, and viscosity measurements were carried out. The results indicate that the complex binds preferentially in the GC-rich regions via an intercalative mode.	[83]
Zn(II)-quercetin complex metal:ligand molar ratio 1:2	The complex promotes plasmid DNA cleavage, producing single- and double-DNA strand breaks. The amount of conversion of the SC form to the NC form depends on the concentration of the complex and the duration of incubation with DNA. The rate of conversion of SC to NC is 1.68×10^{-4} s^{-1} at pH 7.2 (100 μM complex).	[108]
Cu^{II}-$\mathbf{Sn}_2^{IV}$-quercetin Zn^{II}-$\mathbf{Sn}_2^{IV}$-quercetin	The heterobimetallic-type complexes form electrostatic interactions with DNA (between Sn^{IV} and the phosphate groups) and covalent bond between the metal center and nitrogenous bases. The complex Cu^{II}-$\mathbf{Sn}_2^{IV}$-quercetin exhibits nuclease and SOD-mimetic activity.	[96]

ds: double-stranded; CT-DNA: calf thymus DNA; RLS: resonance light scattering; Γ_s: saturation coverage value; GC: guanine-cytosine; SC: supercoiled; NC: nicked circular form.

Table 5. Selection of flavonoid metal complexes-DNA interaction studies.

Regarding the interaction of the complexes with RNA, a few studies have been cited in the literature. La(III)-quercetin complex enhances binding to plant viral satellite dsRNA [109]. Both quercetin and the complex interact with dsDNA, dsRNA, and ssRNA. The affinities of La(III)-quercetin for dsDNA and dsRNA were significantly higher compared to the free ligand, revealing significant impact of La(III) in binding to polynucleotides, most likely due to the electrostatic interactions between La(III) and the phosphate groups surrounding binding sites. Similar results were observed for interactions of the La(III)-quercetin complex with ssRNA [110].

Undisputedly, there is a consistent amount of experimental evidence regarding the interaction of flavonoids and their metal complexes with nucleic acids. However, some details with respect to the binding sites in the DNA structure need further investigations. There appears that the complexes possess higher affinity toward GC-rich sequences in DNA [111], but this assumption needs to be backed up by more data.

7. Perspectives in the formulation of flavonoids metal complexes

In order to enhance the water solubility and to control the release of flavonoids metal complexes, many efforts have been focused on the preparation of cyclodextrin complexes, or of novel micro- and nano-carriers, such as liposomes and organic compounds-inorganic particles hybrid materials (HMs). In the following subchapter, the most important systems developed for the inclusion of flavonoids metal complexes with antioxidant or antitumor activity will be reviewed, as well as the systems that include flavonoids with antitumor activity based on the complexation process between flavonoid and metal ions.

7.1. Liposomal systems

Flavonoids' ability to penetrate into the hydrophobic regions of lipid bilayers in biological membranes is a key factor to prevent peroxidation of unsaturated double bonds. In this regard, the lipophilicity of flavonoids is essential for an adequate penetration. According to Tweedy's chelation theory, the polarity of the metal ion is reduced, as a consequence of chelation, mainly because of the partial sharing of its positive charge with the ligand's donor groups and possible electron delocalization over the entire ring. Consequently, the lipophilic character of the chelates increases, favoring their permeation through the lipid layers of biological membranes [112]. For two complexes of quercetin and taxifolin with Fe(II), it was assumed that the oxygen charges are generally decreases, while the main negative charge is localized on the iron atom. Thus, after chelating the metal ion, the polarity of flavonoid molecules generally decreased, while the iron atom becomes the most polar and hence hydrophilic part of the molecule. Overall, the lipophilicity of the complexes is considerably larger than that of the corresponding free flavonoids [113, 114].

On the other side, the hydrophobicity of flavonoids chelates dramatically reduces the water solubility, which restricts their medical applications. Integration of these complexes into liposomes may increase their bioavailability and improve the therapeutic effect.

The interaction study of quercetin-iron complexes with dimyristoylphosphatidylcholine (DMPC) or palmitoyl-oleoyl phosphatidylethanolamine (POPE) multilamellar liposomes revealed that, during preparation, quercetin should be added first to the suspension of liposomes [115]. It was presumed that quercetin can increase the permeability of lipid bilayers to iron cations, showing ionophore activity toward iron cations.

7.2. Cyclodextrins

The inclusion into cyclodextrins is a convenient alternative to solve the problems related to the low solubility of hydrophobic drugs, such as flavonoid complexes. Inclusion of Cu(II) and

Cr(III) complexes of flavonoids morin, quercetin, and 6-hydroxyflavone into β-cyclodextrin led to an enhancement of aqueous solubility [116]. The anticancer activity of metal-flavonoid complexes was evaluated in terms of dsDNA binding in the environment of beta-cyclodextrin, and it was revealed that DNA could bind Cu-flavonoid-βCD through intercalation and Cr-flavonoid-βCD via an electrostatic-binding mode.

7.3. Organic/inorganic hybrid nanosystems

Based on the property of flavonoids to reduce metal ions such as Ag(I) and Au(III), some hybrid systems of ***flavone/metal nanoparticles*** have been developed. It was suggested that flavanones can be adsorbed on the surface of metallic nanoparticles through the interaction of the metals with the carbonyl groups or π electrons in the flavonoid structures [117].

The effect of 3-hydroxyflavone (3-HF) in a silver nanoparticles (SNPs) complex on the cell viability and on the cell morphology of L929 mouse fibroblast cells was studied *in vitro*. The contribution of the carrier protein BSA to 3-HF properties has also been investigated. Determination of the cell viability using MTT assay revealed that 3-HF in BSA/SNPs systems presented no cytotoxic effect in L929 mouse fibroblast cells at any of the tested concentrations [118].

In order to enhance the interaction efficacy with biomacromolecules and therefore increase its therapeutic potential, morin was conjugated with citrate-coated Au nanoparticles (M-C-AuNPs). Interactions of M-C-AuNPs and C-AuNPs with BSA were studied in order to compare the efficiency of M-C-AuNPs and C-AuNPs in biological systems. It was found that the binding affinity toward BSA of M-C-AuNPs is significantly higher than that of C-AuNPs', indicating that M-C-AuNPs might show better BSA interaction efficiency, better biocompatibility, and chemical stability than C-AuNPs [119].

Taking into account the potential biomedical applications, for example, targeted drug delivery, several flavonoids with antioxidant and antitumor activities have been conjugated with ***magnetic nanoparticles***, mainly Fe_3O_4. Flavonoid molecules can bind to Fe_3O_4 via the hydroxyl substituents in their deprotonated form. Most of the biomedical applications of magnetic nanoparticles require surface modification for drug loading or anchoring linkers in support of sustained drug release. Surface functionalization with PEG (polyethylene glycol), PVP (polyvinylpyrrolidone), PVA (polyvinyl acetate), peptides, carbohydrates, proteins, and so on facilitates the drug loading and controlled release, and also controls the stability of the system [120].

Quercetin-conjugated superparamagnetic Fe_3O_4 nanoparticles were investigated for the *in vitro* cytotoxic activity on breast cancer cell lines. The MTT assay revealed that the dextran-coated Fe_3O_4 nanoparticles did not exhibit notable toxicity against MCF7 cells, whereas the cytotoxicity of quercetin-conjugated Fe_3O_4 nanoparticles increased significantly in comparison with the free quercetin [120]. The results sustain that the quercetin-conjugated Fe_3O_4 nanoparticles are promising anticancer agents for targeted drug delivery.

A drug delivery methodology was proposed to study a new quercetin release system in the form of magnetite-quercetin-copolymer (MQC), as perspective of targeting specific organs within the body. The quercetin-magnetite nanoparticles (Fe_3O_4) system was incorporated into a triblock copolymer of ethylene oxide and oxyphenylethylene, as a model of drug carrier system for anticancer agents [121].

Furthermore, quercetin loading on ***mesoporous carriers*** was performed in order to enable the sustained delivery of the bioactive compound. Mesoporous nanosized silicas are widely used as carriers for drug delivery. However, the appropriate chemical surface modification of the mesoporous matrix is essential, taking into account that the silanol groups of the silica surface do not possess sufficient selectivity to adsorb drug molecules with different functionalities [122]. Quercetin was loaded on the pure silica and Zn-modified mesoporous MCM-41 and SBA-16 supports, and the formation of complexes between quercetin and pure siliceous or Zn-modified MCM-41 and SBA-16 mesoporous silica was determined. Quercetin has a higher binding affinity for the Zn^{2+} cation than to the silanol groups. Therefore, the release of quercetin is easier from the silicate samples containing only superficial silanol groups than from the Zn-modified samples. The obtained mesoporous delivery systems with Zn-quercetin complex showed promising results for further use in dermal formulations [123].

Natural zeolite of the clinoptilolite type (CT, particle size of up to 200 μm) and its modified forms with different concentrations (0.06–5%) of the pharmaceutically active compounds quercetin and quercetin dihydrate (QD) have been investigated for their anticancer activity. Carcinoma cell lines Jurkat, CEM, HeLa, MCF7, A549, and MDA were treated with various amounts of natural clinoptilolite and their modified forms CTQ and CTQD. Incorporation of the flavonoids quercetin and quercetin dihydrate with antiproliferative activity had no synergic effect on the zeolite cytotoxicity, but the protective effect of cancer cells. The tumor cell lines studied after the application of modified zeolite CTQ or CTQD had lower antiproliferative activity in comparison with the natural zeolite of the clinoptilolite-type CT. The modified zeolite CTQD had greater antiproliferative effects than modified zeolite CTQ [124].

8. Conclusions and future perspectives

There is an impressive number of studies with reference to the antitumor activity of flavonoids. However, even though the metal complexation of flavonoids generally results in enhanced antitumor activity, the number of studies regarding flavonoid metal complexes is much smaller. Moreover, none of the flavonoid metal complexes that have shown great antitumor activity has been included in clinical trials so far. This may happen due to the challenges implied by the study of the multiple possible mechanisms induced by the presence of the metal ion. Stability of the complexes in physiological context may also raise some important issues. Therefore, further studies are required to understand their mechanisms of action and their biotransformation in the human body.

Author details

Valentina Uivarosi* and Alexandra-Cristina Munteanu

*Address all correspondence to: valentina_uivarosi@yahoo.com

Department of General and Inorganic Chemistry, Faculty of Pharmacy, "Carol Davila" University of Medicine and Pharmacy, Bucharest, Romania

References

[1] Kumar P, Kumar S, Tripathi MK, Mehta N, Ranjan R, Bhat ZF, Sing PK. Flavonoids in the development of functional meat products: A review. Veterinary World. 2013;**6**:573-578. DOI: 10.5455/vetworld.2013.573-578

[2] Symonowicz M, Kolanek M. Flavonoids and their properties to form chelate complexes. Biotechnology and Food Sciences. 2012;**76**:35-41.

[3] Tapas AR, Sakarkar DM, Kakde RB. Flavonoids as nutraceuticals: A review. Tropical Journal of Pharmaceutical Research. 2008;**7**:1089-1099. DOI: 10.4314/tjpr.v7i3.14693

[4] Beecher GR. Overview of dietary flavonoids: nomenclature, occurrence and intake. Journal of Nutrition. 2003;**133**:3248S–3254S.

[5] Leonarduzzi G, Testa G, Sottero B, Gamba P, Poli G. Design and development of nanovehicle-based delivery systems for preventive or therapeutic supplementation with flavonoids. Current Medicinal Chemistry. 2010;**17**:74-95. DOI: 10.2174/092986710789957760

[6] Laguerre M, Lecomte J, Villeneuve P. Evaluation of the ability of antioxidants to counteract lipid oxidation: Existing methods, new trends and challenges. Progress in Lipid Research. 2007;**46**:244-282. DOI:10.1016/j.plipres.2007.05.002

[7] Leopoldini M, Russo N, Toscano M. The molecular basis of working mechanism of natural polyphenolic antioxidants. Food Chemistry. 2011;**125**:288-306. DOI: 10.1016/j.foodchem.2010.08.012

[8] Cornard JP, Merlin JC. Comparison of the chelating power of hydroxyflavones. Journal of Molecular Structure 2003, **651-653**:381-387. DOI: 10.1016/S0022-2860(02)00655-5

[9] Block G. The data support a role for antioxidants in reducing cancer risk. Nutrition Reviews. 1992;**50**:207-213. DOI: 10.1111/j.1753-4887.1992.tb01329.x

[10] Elangovan V, Sekar N, Govindasamy S. Chemoprotective potential of dietary bioflavonoids against 20-methylcholanthrene-induced tumorigenesis. Cancer Letters. 1994;**87**:107-113. DOI: 10.1016/0304-3835(94)90416-2

[11] Le Marchand L. Cancer preventive effects of flavonoids—a review. Biomedicine and Pharmacotherapy. 2002;**56**:296-301. DOI: 10.1016/S0753-3322(02)00186-5

[12] Selway JW. Antiviral activity of flavones and flavans. Progress in Clinical and Biological Research. 1986;**213**:521-536.

[13] Middleton E Jr. Effect of plant flavonoids on immune and inflammatory cell function. Advances in Experimental Medicine and Biology. 1998;**439**:175-182.

[14] Gomes A, Fernandes E, Lima JL, Mira L, Corvo ML. Molecular mechanisms of anti-inflammatory activity mediated by flavonoids. Current Medicinal Chemistry. 2008;**15**:1586-1605. DOI: **10**.2174/092986708784911579

[15] Gabor M. Anti-inflammatory and anti-allergic properties of flavonoids. Progress in Clinical and Biological Research. 1986;**213**:471-480.

[16] Facino RM, Carini M, Aldini G, Berti F, Rossoni G, Bombardelli E, Morazzoni P. Diet enriched with procyanidins enhances antioxidant activity and reduces myocardial post-ischemic damage in rats. Life Sciences. 1999;**64**:943-949. DOI: 10.1016/S0024-3205(98)00605-5

[17] Herteg MG., Feskens EJM, Hollmen PCH, Katan MB, Kromhout D. Dietary flavonoids and risk of coronary heart disease. Lancet.1993;**342**:1007-1011

[18] Hodgson JM. Tea flavonoids and cardiovascular disease. Asia Pacific Journal of Clinical Nutrition. 2008;**17**:288-290

[19] Kumar S, Pandey AK. Chemistry and biological activities of flavonoids: An overview. The Scientific World Journal. 2013;Article ID 162750:16 pages. DOI: 10.1155/2013/162750

[20] Nijveldt RJ, van Nood E, van Hoorn DE, Boelens PG, van Norren K, van Leeuwen PA. Flavonoids: A review of probable mechanisms of action and potential applications. American Journal of Clinical Nutrition. 2001;**74**:418-25.

[21] Heim KE, Tagliaferro AR, Bobilya DJ. Flavonoid antioxidants: chemistry, metabolism and structure-activity relationships. Journal of Nutritional Biochemistry. 2002;**13**:572-84.

[22] Samsonowicz M, Regulska E. Spectroscopic study of molecular structure, antioxidant activity and biological effects of metal hydroxyflavonol complexes. Spectrochimica Acta Part A: Molecular and Biomolecular Spectroscopy. 2017;**173**:757-771. DOI: 10.1016/j.saa.2016.10.031

[23] Galati G, O'Brien PJ. Potential toxicity of flavonoids and other dietary phenolics: significance for their chemopreventive and anticancer properties. Free Radical Biology & Medicine. 2004;**37**:287-303. DOI: 10.1016/j.freeradbiomed.2004.04.034

[24] Procházková D, Boušová I, Wilhelmová N. Antioxidant and prooxidant properties of flavonoids. Fitoterapia. 2011;**82**:513-523. DOI:10.1016/j.fitote.2011.01.018

[25] Cao G, Sofic E, Prior RL. Antioxidant and prooxidant behavior of flavonoids: Structure-activity relationships. Free Radical Biology & Medicine. 1997;**22**:749-760. DOI: 10.1016/S0891-5849(96)00351-6

[26] Hanasaki Y, Ogawa S, Fukui S. The correlation between active oxygens scavenging and antioxidative effects of flavonoids. Free Radical Biology & Medicine. 1994;**16**:845-850. DOI: 10.1016/0891-5849(94)90202-X

[27] Cherrak SA, Mokhtari-Soulimane N, Berroukeche F, Bensenane B, Cherbonnel A, Merzouk H, Elhabiri M. *In vitro* antioxidant *versus* metal ion chelating properties of flavonoids: A structure-activity investigation. PLoS One. 2016;**11**:e0165575 DOI: 10.1371/journal.pone.0165575

[28] Nakagawa HK, Hasumi JT, Nagai K, Wachi M. Generation of hydroxide peroxide primarily contributes to the induction of Fe(II)-dependent apoptosis in Jurkat cells by (−)-epigallocatechin-3-gallate. Carcinogenesis. 2004;**25**:1567-1574. DOI: 10.1093/carcin/bgh168

[29] Lotito SB, Frei B. Consumption of flavonoid-rich foods and increased plasma antioxidant capacity in humans: cause, consequence, or epiphenomenon. Free Radical Biology & Medicine. 2006;**41**:1727-1746. DOI: 10.1016/j.freeradbiomed.2006.04.033

[30] Roy S, Mallick S, Chakraborty T, Ghosh N, Singh AK, Manna S, Majumdar S. Synthesis, characterisation and antioxidant activity of luteolin–vanadium(II) complex. Food Chemistry. 2015;**173**:1172-1178. DOI: 10.1016/j.foodchem.2014.10.141

[31] Jiang J, Yao S, Cai H-H, Yang P-H, Cai J. Synthesis and synergetic effects of chrysin–organogermanium (IV) complex as potential anti-oxidant. Bioorganic & Medicinal Chemistry Letters. 2013;**23**:5727-5732. DOI: 10.1016/j.bmcl.2013.07.073

[32] Naso, L, Ferrer EG, Leyama L, Rojo T, Etcheverry SB, Williams P. Role of oxidative stress in the antitumoral action of a new vanadyl(IV) complex with the flavonoid chrysin in two osteoblast cell lines: relationship with the radical scavenger activity. Journal of Biological Inorganic Chemistry. 2010;**15**:889-902. DOI 10.1007/s00775-010-0652-z

[33] León IE J. Cadavid-Vargas F, Tiscornia I, Porro V, Castelli S, Katkar P, Desideri A, Bollati-Fogolin M, Etcheverry SB. Oxidovanadium(IV) complexes with chrysin and silibinin: Anticancer activity and mechanisms of action in a human colon adenocarcinoma model. Journal of Biological Inorganic Chemistry. 2015;**20**:1175-1191. DOI: 10.1007/s00775-015-1298-7

[34] Jabbari M, Gharib F. Solvent dependence on antioxidant activity of some water-insoluble flavonoids and their cerium(IV) complexes. Journal of Molecular Liquids. 2012;**168**:36-41. DOI:10.1016/j.molliq.2012.02.001

[35] Naso L, Martínez VR, Lezama L, Salado C, Valcarcel M, Ferrer EG, Williams PA. Antioxidant, anticancer activities and mechanistic studies of the flavone glycoside diosmin and its oxidovanadium(IV) complex. Interactions with bovine serum albumin. Bioorganic & Medicinal Chemistry. 2016;**24**:4108-4119. DOI: 10.1016/j.bmc.2016.06.053

[36] Ghosh N, Chakraborty T, Mallick S, Mana S, Singha D, Ghosh B, Roy S. Synthesis, characterization and study of antioxidant activity of quercetin–magnesium complex. Spectrochimica Acta Part A: Molecular and Biomolecular Spectroscopy. 2015;**151**:807-813. DOI: 10.1016/j.saa.2015.07.050

[37] Dias K, Nikolaou S, De Giovani WF. The in vitro antioxidant properties of the Al-quercetin/βCD and Al-catechin/βCD inclusion compounds, rationalized in terms of their electrochemical behaviour. Medicinal Chemistry Research. 2012;**21**:2920-2925. DOI: 10.1007/s00044-011-9812-0

[38] Li S-P, Xie W-L, Cai H-H, Cai J-Y, Yang P-H. Hydroxyl radical scavenging mechanism of human erythrocytes by quercetin–germanium (IV) complex. European Journal of Pharmaceutical Sciences. 2012;**47**:28-34. DOI: 10.1016/j.ejps.2012.04.019

[39] Dehghan G, Khoshkam Z. Tin(II)–quercetin complex: Synthesis, spectral characterisation and antioxidant activity. Food Chemistry. 2012;**131**:422-426. DOI:10.1016/j.foodchem.2011.08.074

[40] Raza A, Xu X, Xia L, Xia C, Tang J, Ouyang Z. Quercetin-iron complex: Synthesis, characterization, antioxidant, DNA binding, DNA cleavage, and antibacterial activity studies. Journal of Fluorescence. 2016;**26:**2023-2031. DOI: 10.1007/s10895-016-1896-y

[41] Bukhari SB, Memon S, Tahir MM, Bhanger MI. Synthesis, characterization and investigation of antioxidant activity of cobalt–quercetin complex. Journal of Molecular Structure. 2008;**892**:39-46. DOI:10.1016/j.molstruc.2008.04.050

[42] Pękal A, Biesaga M, Pyrzynska K. Interaction of quercetin with copper ions: Complexation, oxidation and reactivity towards radicals. Biometals. 2011;**24**:41-49. DOI: 10.1007/s10534-010-9372-7

[43] Bukhari SB, Memon S, Mahroof-Tahir M, Bhanger MI. Synthesis, characterization and antioxidant activity copper–quercetin complex. Spectrochimica Acta Part A. 2009;**71**:1901-1906. DOI :10.1016/j.saa.2008.07.030

[44] Ravichandran R, Rajendran M, Devapiriam D. Antioxidant study of quercetin and their metal complex and determination of stability constant by spectrophotometry method. Food Chemistry. 2014;**146**:472-478. DOI: 10.1016/j.foodchem.2013.09.080

[45] Zhou J, Wang L, Wang J, Tang N. Antioxidative and anti-tumour activities of solid quercetin metal(II) complexes. Transition Metal Chemistry. 2001;**26**:57-63. DOI: 10.1023/A:1007152927167

[46] de Souza RFV, De Giovani WF. Antioxidant properties of complexes of flavonoids with metal ions. Redox Report. 2004;**9**:97-104. DOI: 10.1179/135100004225003897

[47] Zhou J, Wang L, Wang J, Tang N. Synthesis, characterization, antioxidative and antitumor activities of solid quercetin rare earth(III) complexes. Journal of Inorganic Biochemistry. 2001;**83**:41-48. DOI: 10.1016/S0162-0134(00)00128-8

[48] Naso LG, Lezama L, Rojo T, Etcheverry SB, Valcarcel M, Roura M, Salado C, Ferrer EG, Williams PA. Biological evaluation of morin and its new oxovanadium(IV) complex as antioxidant and specific anti-cancer agents. Chemico-Biological Interactions. 2013;**206**:289-301. DOI: 10.1016/j.cbi.2013.10.006

[49] Panhwar QK, Memon S. Synthesis of Cr(III)-morin complex: Characterization and antioxidant study. The Scientific World Journal. 2014; Article ID 845208, 8 pages. DOI: 10.1155/2014/845208

[50] Panhwar QK, Memon S, Bhanger MI. Synthesis, characterization, spectroscopic and antioxidation studies of Cu(II)–morin complex. Journal of Molecular Structure. 2010;**967**:47-53. DOI: 10.1016/j.molstruc.2009.12.037

[51] Roy S, Majumdar S, Singh AK, Ghosh B, Ghosh N, MannaS, Chakraborty T, Mallik S. Synthesis, characterization, antioxidant status, and toxicity study of vanadium–rutin complex in balb/c mice. Biological Trace Element Research. 2015;**166**:183-200. DOI: 10.1007/s12011-015-0270-2

[52] Raza A, Bano S, Xu X, Zhang RX, Khalid H, Iqbal FM, Xia C, Tang J, Ouyang Z. Rutin–nickel complex: Synthesis, characterization, antioxidant, DNA binding, and DNA cleavage activities. Biological Trace Element Research. DOI: 10.1007/s12011-016-0909-7

[53] Afanas'ev IB, Ostrachovich EA, Korkina LG. Effect of rutin and its copper complex on superoxide formation and lipid peroxidation in rat liver microsomes. FEBS Letters. 1998;**425**:256-258. DOI: 10.1016/S0014-5793(98)00244-0

[54] Ikeda NE, Novak EM, Maria DA, Velosa AS, Pereira RM. Synthesis, characterization and biological evaluation of Rutin–zinc(II) flavonoid-metal complex. Chemico-Biological Interactions. 2015;**239**:184-191. DOI: 10.1016/j.cbi.2015.06.011

[55] Zhai G, Qu W, Yan Z, Zhu W, Duan Y, Wang J. Synthesis, spectral and antioxidant properties of tin(II)-rutin complex. Chemistry of Natural Compounds. 2014;**50**:624-628. DOI: 10.1007/s10600-014-1039-0

[56] Panhwar QK, Memon S. Synthesis, characterization and antioxidant study of Tin(II)–rutin complex: Exploration of tin packaging hazards. Inorganica Chimica Acta. 2013;**407**:252-260. DOI: 10.1016/j.ica.2013.08.001

[57] Panhwar QK, Memon S. Synthesis, characterization, and antioxidant study of Cr(III)–rutin complex. Chemical Papers. 2014;**68**:614-623. DOI: 10.2478/s11696-013-0494-6

[58] Qi Y, Jiang M, Cui Y-L, Zhao L, Liu S. Novel reduction of Cr(VI) from wastewater using a naturally derived microcapsule loaded with rutin–Cr(III) complex. Journal of Hazardous Materials. 2015;**285**:336-345. DOI: 10.1016/j.jhazmat.2014.12.008

[59] Tamayo LV, Gouvea LR, Sousa AC, Albuquerque RM, Teixeira SF, de Azevedo RA, Louro SRW, Ferreira AK, Beraldo H. Copper(II) complexes with naringenin and hesperetin: cytotoxic activity against A 549 human lung adenocarcinoma cells and investigation on the mode of action. Biometals. 2016;**29**:39-52. DOI: 10.1007/s10534-015-9894-0

[60] Wang H-L, Yang Z-Y, Wang B. Synthesis, characterization and the antioxidative activity of copper(II), zinc(II) and nickel(II) complexes with naringenin. Transition Metal Chemistry. 2006;**31**:470-474. DOI: 10.1007/s11243-006-0015-3

[61] Pereira RMS, Andrades NED, Paulino N, Sawaya ACHF, Eberlin MN, Marcucci MC, Favero GM, Novak EM, Bydlowski SP. Synthesis and characterization of a metal complex containing naringin and Cu, and its antioxidant, antimicrobial, antiinflammatory and tumor cell cytotoxicity. Molecules. 2007;**12**:1352-1366. DOI: 10.3390/12071352

[62] Etcheverry SB, Ferrer EG, Naso L, Rivadeneira J, Salinas V, Williams PA. Antioxidant effects of the VO(IV) hesperidin complex and its role in cancer chemoprevention. Journal of Biological Inorganic Chemistry. 2008;**13**:435-447. DOI: 10.1007/s00775-007-0332-9

[63] Dowling S, Regan F, Hughes H. The characterisation of structural and antioxidant properties of isoflavone metal chelates. Journal of Inorganic Biochemistry. 2010;**104**:1091-1098. DOI: 10.1016/j.jinorgbio.2010.06.007

[64] López-Lázaro M, Willmore E, Austin CA. The dietary flavonoids myricetin and fisetin act as dual inhibitors of DNA topoisomerases I and II in cells. Mutation Research/Genetic Toxicology and Environmental Mutagenesis. 2010;**696**:41-47. DOI: 10.1016/j.mrgentox.2009.12.010

[65] Casagrande F, Darbon JM. Effects of structurally related flavonoids on cell cycle progression of human melanoma cells: Regulation of cyclin-dependent kinases CDK2 and CDK1. Biochemical Pharmacology. 2001;**61**:1205-1215. DOI: 10.1016/S0006-2952(01)00583-4

[66] Khan N, Asim M, Afaq F, Abu Zaid M, Mukhtar H. A novel dietary flavonoid fisetin inhibits androgen receptor signaling and tumor growth in athymic nude mice. Cancer Research. 2008;**68:**8555-8563. DOI: 10.1158/0008-5472

[67] Bohl M, Tietze S, Sokoll A, Madathil S, Pfennig F, Apostolakis J, Fahmy K, Gutzeit HO. Flavonoids affect actin functions in cytoplasm and nucleus. Biophysical Journal. 2007;**93**:2767-2780. DOI:10.1529/biophysj.107.107813

[68] Sinha S, Amin H, Nayak D, Bhatnagar M, Kacker P, Chakraborty S, Ghosal S. Assessment of microtubule depolymerization property of flavonoids isolated from *Tanacetum gracile* in breast cancer cells by biochemical and molecular docking approach. Chemico-Biological Interactions. 2015;**239**:1-11. DOI: 10.1016/j.cbi.2015.06.034

[69] Erdogan S, Doganlar O, Doganlar ZB, Serttas R, Turkekul K, Dibirdik I, Bilir A. The flavonoid apigenin reduces prostate cancer CD44+ stem cell survival and migration through PI3K/Akt/NF-κB signaling. Life Sciences. 2016;**162**:77-86. DOI: 10.1016/j.lfs.2016.08.019

[70] Tan M, Zhu J, Pan Y, Chen Z, Liang H, Liu H, Wang H. Synthesis, cytotoxic activity, and DNA binding properties of copper (II) complexes with hesperetin, naringenin, and apigenin. Bioinorganic Chemistry and Applications. 2009;Article ID 347872, 9 pages. DOI: 10.1155/2009/347872

[71] Durgo K, Halec I, Sola I, Franekic J. Cytotoxic and genotoxic effects of the quercetin/lanthanum complex on human cervical carcinoma cells in vitro. Archives of Industrial Hygiene and Toxicology. 2011;**62**:221-227. DOI: 10.2478/10004-1254-62-2011-2122

[72] Kasprzak MM, Erxleben A, Ochocki J. Properties and applications of flavonoid metal complexes. RSC Advances. 2015;**5**:45853-45877. DOI: 10.1039/C5RA05069C

[73] Chen X, Tang LJ, Sun YN, Qiu PH, Liang G. Syntheses, characterization and antitumor activities of transition metal complexes with isoflavone. Journal of Inorganic Biochemistry. 2010;**104**:379-384. DOI: 10.1016/j.jinorgbio.2009.11.008

[74] Andrades Ikeda NE, Novak EM, Maria DA, Velosa AS, Pereira RM. Synthesis, characterization and biological evaluation of Rutin-zinc(II) flavonoid -metal complex. Chemico-Biological Interactions. 2015;**239**:184-191. DOI: 10.1016/j.cbi.2015.06.011

[75] Naso LG, Lezama L, Valcarcel M, Salado C, Villacé P, Kortazar D, Ferrer EG, Williams PA. Bovine serum albumin binding, antioxidant and anticancer properties of an oxidovanadium(IV) complex with luteolin. Journal of Inorganic Biochemistry 2016;**157:**80-93. DOI: 10.1016/j.jinorgbio.2016.01.021

[76] Islas MS, Naso LG, Lezama L, Valcarcel M, Salado C, Roura-Ferrer M, Ferrer EG, Williams PA. Insights into the mechanisms underlying the antitumor activity of an oxidovanadium (IV) compound with the antioxidant naringenin. Albumin binding studies. Journal of Inorganic Biochemistry. 2015;**149**:12-24. DOI: 10.1016/j.jinorgbio.2015.04.011

[77] León IE, Cadavid-Vargas JF, Resasco A, Maschi F, Ayala MA, Carbone C, Etcheverry S B. *In vitro* and *in vivo* antitumor effects of the VO-chrysin complex on a new three-dimensional osteosarcoma spheroids model and a xenograft tumor in mice. Journal of Biological Inorganic Chemistry. 2016;**21**:1009-1020. DOI: 10.1007/s00775-016-1397-0

[78] Ferrer EG, Salinas MV, Correa MJ, Naso L, Barrio DA, Etcheverry SB, Lezama L, Rojo T, Williams PA. Synthesis, characterization, antitumoral and osteogenic activities of quercetin vanadyl(IV) complexes. Journal of Biological Inorganic Chemistry. 2006;**11**:791-801. DOI: 10.1007/s00775-006-0122-9

[79] Dell'Anna MM, Censi V, Carrozzini B, Caliandro R, Denora N, Franco M, Veclani D, Melchior A, Tolazzi M, Mastrorilli P. Triphenylphosphane Pt(II) complexes containing biologically active natural polyphenols: Synthesis, crystal structure, molecular modeling and cytotoxic studies. Journal of Inorganic Biochemistry. 2016;**163**:346-361. DOI: 10.1016/j.jinorgbio.2016.08.006

[80] Kasprzak M, Fabijańska M, Chęcińska L, Studzian K, Markowicz-Piasecka M, Sikora J, Mikiciuk-Olasik E, Ochocki J. Small differences in structure, a large difference in activity—Comparing a new Ru(II)-3-hydroxyiminoflavanone complex with analogous Ru(II) compounds. Inorganica Chimica Acta. 2016;**457**:69-80. DOI: 10.1016/j.ica.2016.11.021

[81] Kurzwernhart A, Kandioller W, Bächler S, Bartel C, Martic S, Buczkowska M, Mühlgassner G, Jakupec MA, Kraatz HB, Bednarski PJ, Arion VB, Marko D, Keppler BK, Hartinger CG. Structure-activity relationships of targeted RuII(η^6-p-Cymene) anticancer complexes with flavonol-derived ligands. Journal of Medicinal Chemistry. 2012;**55**: 10512-10522. DOI: 10.1021/jm301376a

[82] Tan J, Zhu L, Wang B. DNA binding, cytotoxicity, apoptotic inducing activity, and molecular modeling study of quercetin zinc(II) complex. Bioorganic Medicinal Chemistry. 2009;**17**:614-620. DOI: 10.1016/j.bmc.2008.11.063

[83] Tan J, Zhu L, Wang B. GC (Guanine-cytosine)-selective DNA-binding and antitumor activity of a quercetin-manganese (II) complex. Chemistry & Biodiversity. 2011;**8**:1550-1559. DOI: 10.1002/cbdv.201000313

[84] Spoerlein C, Mahal K, Schmidt H, Schobert R. Effects of chrysin, apigenin, genistein and their homoleptic copper(II) complexes on the growth and metastatic potential of cancer cells. Journal of Inorganic Biochemistry. 2013;**127**:107-115. DOI: 10.1016/j.jinorgbio.2013.07.038

[85] Parveen S, Tabassum S, Arjmand F. Human topoisomerase I mediated cytotoxicity profile of l-valine-quercetin diorganotin(IV) antitumor drug entities. Journal of Organometallic Chemistry. 2016;**823**:23-33. DOI: 10.1016/j.jorganchem.2016.09.015

[86] Lara PNJ, Douillard J-Y, Nakagawa K, von Pawel J, McKeage MJ, Albert I, Losonczy G, Reck M, Heo DS, Fan X, Fandi A, Scagliotti G. Randomized phase III placebo-controlled trial of carboplatin and paclitaxel with or without the vascular disrupting agent vadimezan (ASA404) in advanced non-small-cell lung cancer. Journal of Clinical Oncology: Official Journal of the American Society of Clinical Oncology. 2011;**29**:2965-2971. DOI: 10.1200/JCO.2011.35.0660

[87] Abou-Zied OK, Al-Shihi OI. Characterization of subdomain IIA binding site of human serum albumin in its native, unfolded, and refolded states using small molecular probes. Journal of American Chemical Society. 2008;**130**:10793-801. DOI: 10.1021/ja8031289

[88] Sarzehi S, Chamani J. Investigation on the interaction between tamoxifen and human holo-transferrin: Determination of the binding mechanism by fluorescence quenching, resonance light scattering and circular dichroism methods. International Journal of Biological Macromolecules. 2010;**47**:558-569. DOI: 10.1016/j.ijbiomac.2010.08.002

[89] Zhang XF, Han RM, Sun XR, Li GY, Yang QF, Li Q, Gai W, Zhang M, Chen L, Yang G, Tang YL. The effect of the skeleton structure of flavanone and flavonoid on interaction with transferrin. Bioorganic & Medicinal Chemistry Letters. 2013;**23**:6677-6681. DOI: 10.1016/j.bmcl.2013.10.042

[90] Nitiss JL. Targeting DNA topoisomerase II in cancer chemotherapy. Nature reviews Cancer. 2009;**9**:338-350. doi:10.1038/nrc2607.

[91] Sanna D, Ugone V, Pisano L, Serra M, Micera G, Garribba E. Behavior of the potential antitumor $V^{IV}O$ complexes formed by flavonoid ligands. 2. Characterization of sulfonate derivatives of quercetin and morin, interaction with the bioligands of the plasma and preliminary biotransformation studies. Journal of Inorganic Biochemistry. 2015;**153**:167-177. DOI: 10.1016/j.jinorgbio.2015.07.018

[92] Wang J, Wang Q, Wu D, Yan J, Wu Y, Li H. Comparative studies on the interactions of baicalein and Al(III)-baicalein complex with human serum albumin. Luminescence 2016;**31**:54-62. DOI: 10.1002/bio.2921

[93] Guo Q, Yuan J, Zeng J. Binding of dihydromyricetin and its metal ion complexes with bovine serum albumin. Biotechnology & Biotechnological Equipment. 2014;**28**:333-341. DOI: 10.1080/13102818.2014.915077

[94] Liu E, Zhang H. Interaction of the La(III)–morin complex with human serum albumin. Journal of Solution Chemistry. 2014;**43**:1402-1413. DOI: 10.1007/s10953-014-0210-3

[95] Kurzwernhart A, Kandioller W, Bartel C, Bächler S, Trondl R, Mühlgassner G, Jakupec MA, Arion VB, Marko D, Keppler BK, Hartinger CG. Targeting the DNA-topoisomerase complex in a double-strike approach with a topoisomerase inhibiting moiety and covalent DNA binder. Chemical Communications.2012;48:4839-4841. DOI: 10.1039/C2CC31040F

[96] Tabassum S, Zaki M, Afzal M, Arjmand F. New modulated design and synthesis of quercetin-Cu(II)/Zn(II)-Sn_2(IV) scaffold as anticancer agents: in vitro DNA binding profile, DNA cleavage pathway and Topo-I activity. Dalton Transactions. 2013;**42**:10029-10041. DOI: 10.1039/c3dt50646k

[97] Kang J, Zhuo L, Lu X, Liu H, Zhang M, Wu H. Electrochemical investigation on interaction between DNA with quercetin and Eu-Qu3 complex. Journal of Inorganic Biochemistry. 2004;**98**:79-86. DOI: 10.1016/j.jinorgbio.2003.08.015

[98] Jamali AA, Tavakoli A, Nazhad Dolatabadi JE. Analytical overview of DNA interaction with Morin and its metal complexes. European Food Research and Technology. 2012 **235**:367-373. DOI: 10.1007/s00217-012-1778-8

[99] Selvaraj S, Krishnaswamy S, Devashya V, Sethuraman S, Krishnan UM. Flavonoid-metal ion complexes: a novel class of therapeutic agents. Medicinal Research Reviews, 2014;**34**:677-702. DOI: 10.1002/med.21301

[100] Li J, Kang J, Lu J, Li X, Tang J, Zhang H, Zhang Y. Determination of calf thymus DNA using resonance light-scattering quenching method based on the terbium (III) (Tb^{3+})/ europium (III) (Eu^{3+})–quercetin system. Journal of Luminescence. 2009;**129**:906-911. DOI: 10.1016/j.jlumin.2009.03.015

[101] El Amrani FB, Perelló L, Real JA, González-Alvarez M, Alzuet G, Borrás J, García-Granda S, Montejo-Bernardo J. Oxidative DNA cleavage induced by an iron(III) flavonoid complex: Synthesis, crystal structure and characterization of chlorobis(flavonolato) (methanol) iron(III) complex. Journal of Inorganic Biochemistry. 2006;100:1208-1218. DOI: 10.1016/j.jinorgbio.2006.01.036

[102] Niu S, Wu M, Bi S, Zhang S. Reaction of Cd(II)-Morin with dsDNA for biosensing of ssDNA oligomers with complementary, GCE-immobilized ssDNA. Bioelectrochemistry. 2008;73:64-69. DOI: 10.1016/j.bioelechem.2008.04.021

[103] Song Y, Yang P, Yang M, Kang J, Qin S, Lü B, Wang L. Spectroscopic and voltammetric studies of the cobalt (II) complex of Morin bound to calf thymus DNA. Transition Metal Chemistry. 2003;28:712-716. DOI: 10.1023/A:1025418332121

[104] Temerk YM, Ibrahim MS, Kotb M. Voltammetric and spectroscopic studies on binding of antitumor Morin, Morin-Cu complex and Morin-β-cyclodextrin with DNA. Spectrochimica Acta Part A: Molecular and Biomolecular Spectroscopy. 2009;71:1830-1836. DOI: 10.1016/j.saa.2008.07.001

[105] Niu SY, Wu ML, Hu LZ, Mei ZH, Liu SF. Nucleic acid biosensor for DNA hybridization detection using rutin-Cu as an electrochemical indicator. Electrochimica Acta. 2009;54:1564-1569. DOI: 10.1016/j.electacta.2008.09.038

[106] Ni Y, Du S, Kokot S. Interaction between quercetin-copper(II) complex and DNA with the use of the neutral red dye fluorophore probe. Analytica Chimica Acta, 2007;584:19-27. DOI: 10.1016/j.aca.2006.11.006

[107] Zeng YB, Yang N, Liu WS, Tang N. Synthesis, characterization and DNA-binding properties of La(III) complex of chrysin. Journal of Inorganic Biochemistry. 2003;97:258-264. DOI: 10.1016/S0162-0134(03)00313-1

[108] Tan J, Wang B, Zhu L (). Hydrolytic cleavage of DNA by quercetin zinc(II) complex. Bioorganic and Medicinal Chemistry Letters. 2007;17:1197-1199. DOI: 10.1016/j.bmcl.2006.12.023

[109] Rusak G, Piantanida I, Bretschneider S, Ludwig-Muller J. Complex formation of quercetin with lanthanum enhances binding to plant viral satellite double stranded RNA. Journal of Inorganic Biochemistry. 2009;103:1597-1601. DOI: 10.1016/j.jinorgbio.2009.08.008

[110] Marinic M, Piantanida I, Rusak G, Zinic M. Interactions of quercetin and its lanthane complex with double stranded DNA/RNA and single stranded RNA: Spectrophotometric sensing of poly G. Journal of Inorganic Biochemistry. 2006;100:288-298. DOI: 10.1016/j.jinorgbio.2005.11.013

[111] Tan J, Zhu L, Wang B. From GC-rich DNA binding to the repression of survivin gene for quercetin nickel (II) complex: Implications for cancer therapy. Biometals. 2010;23:1075-1084. DOI: 10.1007/s10534-010-9353-x

[112] Al-Amiery AA, Kadhum AAH, Mohamad AB. Antifungal and antioxidant activities of pyrrolidone thiosemicarbazone complexes. Bioinorganic Chemistry and Applications. 2012; Article ID 795812, 6 pages. DOI:10.1155/2012/795812

[113] Tarahovsky YS, Yagolnik EA, Muzafarov EN, Abdrasilov BS, Kim YA. Calcium-dependent aggregation and fusion of phosphatidylcholine liposomes induced by complexes of flavonoids with divalent iron. Biochimica et Biophysica Acta. 2012;1818:695-702. DOI:10.1016/j.bbamem.2011.12.001

[114] Kim YA, Tarahovsky YS, Yagolnik EA, Kuznetsova SM, Muzafarov EN. Lipophilicity of flavonoid complexes with iron(II) and their interaction with liposomes. Biochemical and Biophysical Research Communications. 2013;431:680-685. DOI: 10.1016/j.bbrc.2013.01.060

[115] Kim YA, Tarahovsky YS, Yagolnik EA, Kuznetsova S. Integration of quercetin-Iron complexes into phosphatidylcholine or phosphatidylethanolamine liposomes. Applied Biochemistry and Biotechnology. 2015;176:1904-1913. DOI: 10.1007/s12010-015-1686-z

[116] Jabeen E, Janjua NK, Hameed S. β-Cyclodextrin assisted solubilization of Cu and Cr complexes of flavonoids in aqueous medium: A DNA-interaction study. Spectrochimica Acta Part A: Molecular and Biomolecular Spectroscopy. 2014;128:191-196. DOI: 10.1016/j.saa.2014.02.132

[117] Gan PP, Li SFY. Potential of plant as a biological factory to synthesize gold and silver nanoparticles and their applications. Reviews in Environmental Science and Biotechnology. 2012;11:169-206. DOI: 10.1007/s11157-012-9278-7

[118] Voicescu M, Craciunescu O, Moldovan L, Anastasescu M, Angelescu DG, Teodorescu VS. Physicochemical characterization and in vitro cytotoxic effect of 3-hydroxyflavone in a silver nanoparticles complex. Journal of Fluorescence. 2015;25:1215-1223. DOI: 10.1007/s10895-015-1608-z

[119] Yue H-L, Hu Y-J, Huang H-G, Jiang S, Tu B. Development of morin-conjugated Au nanoparticles: Exploring the interaction efficiency with BSA using spectroscopic methods. Spectrochimica Acta Part A: Molecular and Biomolecular Spectroscopy. 2014;**130**:402-410. DOI: 10.1016/j.saa.2014.04.022

[120] Kumar SR, Priyatharshni S, Babu VN, Mangalaraj D, Viswanathan C, Kannan S, Ponpandian N. Quercetin conjugated superparamagnetic magnetite nanoparticles for in-vitro analysis of breast cancer cell lines for chemotherapy applications. Journal of Colloid and Interface Science. 2014;436:234-242. DOI: 10.1016/j.jcis.2014.08.064

[121] Barreto ACH, Santiago VR, Mazzetto SE, Denardin JC, Lavín R, Mele G, Ribeiro MENP, Vieira IGP, Gonçalves T, Ricardo NMPS, Fechine PBA. Magnetic nanoparticles for a new drug delivery system to control quercetin releasing for cancer chemotherapy. Journal of Nanoparticles Research. 2011;13:6545-6553. DOI: 10.1007/s11051-011-0559-9

[122] Horcajada P, Ramila A, Ferey G, Vallet-Regí M. Influence of superficial organic modification on MCM-41 matrices on drug delivery rate. Solid State Sciences. 2006;8:1243-1249. DOI: 10.1016/j.solidstatesciences.2006.04.016

[123] Popova M, Trendafilova I, Szegedi A, Mihály J, Németh P, Marinova SG, Aleksandrov HA, Vayssilov GN. Experimental and theoretical study of quercetin complexes formed on pure silica and Zn-modified mesoporous MCM-41 and SBA-16 materials. Microporous and Mesoporous Materials. 2016;228:256-265. DOI: 10.1016/j.micromeso.2016.04.001

[124] Tomečkova V, Rehakova M, Mojžišova G, Magura J, Wadsten T, Zelenakova K. Modified natural clinoptilolite with quercetin and quercetin dihydrate and the study of their anticancer activity. Microporous and Mesoporous Materials. 2012;**147**:59-67. DOI: 10.1016/j.micromeso.2011.05.031

Chapter 15

Flavonoid Actions on Receptors for the Inhibitory Neurotransmitter GABA

Tina Hinton, Jane R. Hanrahan and
Graham A.R. Johnston

Additional information is available at the end of the chapter

http://dx.doi.org/10.5772/67971

Abstract

Flavonoids, both naturally occurring and synthetic, are known to have multiple effects on the activation of ionotropic receptors for γ-aminobutyric acid (GABA), the major inhibitory neurotransmitter in our brains. They can act as positive or negative allosteric modulators, enhancing or reducing the effect of GABA. They can elicit a direct activation of the receptors. They can also act to modulate the action of other modulators. This ability to influence function via their actions on GABA receptors permits a range of effects of flavonoids, including relief of anxiety, anticonvulsant, analgesic and sedative actions.

Keywords: apigenin, hispidulin, luteolin, EGCG, synthetic flavonoids, synergism

1. Introduction

Flavonoids have shown a range of effects, such as anxiolytic, sedative, anticonvulsant and analgesic properties, via their actions on the central nervous system (CNS). These effects occur through a variety of interactions with different receptors and signalling systems, including γ-aminobutyric acid (GABA) receptors. γ-Aminobutyric acid (GABA) is the primary inhibitory neurotransmitter in the mammalian brain, released by up to 40% of neurons [1]. GABA acts on two classes of receptors—ionotropic and metabotropic [2]. Ionotropic receptors for GABA are ligand-gated chloride channels located in the neuronal membrane. When activated by GABA, these channels permit the passage of chloride ions down their electrochemical gradient. This usually results in the inward flow of chloride ions and the inhibition of neuronal firing. Metabotropic receptors for GABA are G-protein-coupled receptors that modulate neuronal activity via a variety of second messengers. While an extensive literature on the interactions of flavonoids with ionotropic GABA receptors exists [3], there are no examples of flavonoids acting on metabotropic GABA receptors, though they are known to act on other G-protein-coupled receptors such as adrenergic receptors [4].

This overview highlights the effects of some representative flavonoids on ionotropic GABA receptors acting as positive or negative allosteric modulators, increasing or decreasing the effect of GABA, as directly acting allosteric agonists, and as second-order modulators influencing the action of other modulators. Of particular interest are flavonoids that show subtype selectivity on GABA receptors. This overview also highlights the pre-clinical evidence for these representative flavonoids as anxiolytics, sedatives and anticonvulsants through their interactions with the GABAergic system. Further, the synergistic actions of flavonoids are reviewed.

2. Ionotropic GABA receptors

There are two classes of ionotropic GABA receptors: $GABA_A$ and $GABA_C$ receptors. $GABA_A$ receptors are relatively complex proteins, while $GABA_C$ receptors are relatively simple [2]. These receptors are pharmacologically distinguished by selective antagonists—$GABA_A$ receptors are antagonized by the convulsant alkaloid bicuculline and are insensitive to (1,2,5,6-tetrahydropyridin-4-yl)methylphosphinic acid (TPMPA), whereas $GABA_C$ receptors are insensitive to bicuculline and are selectively antagonized by TPMPA. $GABA_B$ receptors are metabotropic receptors selectively activated by the GABA analogue baclofen, and insensitive to bicuculline and TPMPA. $GABA_A$, $GABA_B$ and $GABA_C$ receptors differ in their physiology, pharmacology and molecular biology.

Ionotropic GABA receptors are part of a superfamily of ligand-gated ion channels comprising excitatory, cation-selective channels such as nicotinic acetylcholine receptors, $5\text{-}HT_3$ receptors and zinc-activated channels, as well as inhibitory, anion-selective channels such as $GABA_A$ and $GABA_C$ receptors, strychnine-sensitive glycine receptors and invertebrate glutamate-gated chloride channels [5]. Receptors of this superfamily require five subunits to assemble a single ion channel. The ion channel may be homomeric formed by five identical subunits as is the case of $GABA_C$ receptors, or heteromeric, consisting of a combination of at least two different subunits, for example, the $GABA_A$ receptors [6]. This superfamily of ligand-gated ion channels is referred to as the cys-loop receptors due to a conserved characteristic cys-cys disulphide bond forming a loop of 13 amino acids in the N-terminal extracellular domain that contains the orthosteric agonist-binding site for the transmitter. The cys-loop is believed to be important for both cell surface expression of the receptor and cooperative interaction between agonist-binding sites and the channel gate.

$GABA_A$ receptors are heteromeric pentamers, composed of a variety of protein subunits. In humans, there are 19 isoforms of $GABA_A$ subunits, that is, six α, three β, three γ and one of δ, ε, π, θ, known to form heteromeric $GABA_A$ receptors. The most widely distributed complex in the brain is composed of two α, two β and one γ subunit, but many other combinations are known to be found in specific brain areas. Theoretically, many thousands of $GABA_A$ receptors

http://dx.doi.org/10.5772/67971

could exist, made up of different combinations of subunits. Specific subunit combinations are thought to be associated with selective actions [7]. Using transgenic mice, it was demonstrated that receptors containing $\alpha1$-subunits serve as targets for sedative-hypnotics-mediating sedation, while $\alpha2$- and/or $\alpha3$-containing receptors mediate anxiolysis, and $\alpha5$-containing receptors are involved in memory [8].

The action of GABA on $GABA_A$ receptors can be modulated by many well-known agents. These include benzodiazepines, such as diazepam, barbiturates, anaesthetic agents, ethanol, neurosteroids and flavonoids. Consequently, modulators of $GABA_A$ receptors are important targets for drug development, particularly modulators that are selective for $GABA_A$ receptors made up of specific subunit combinations [7].

$GABA_C$ receptors are relatively simple compared to $GABA_A$ receptors. Three subunits, ρ1, ρ2 and ρ3, have been cloned from retina and restricted brain regions [9]. These subunits usually express as pentameric homomeric $GABA_C$ receptors that are activated by lower concentrations of GABA than $GABA_A$ receptors. The amino acid sequence similarity between $GABA_A$ and $GABA_C$ subunits is 35–45% and is as high as 75% in the transmembrane region. The genes coding for the three $GABA_C$ subunits are found on different chromosomes to those coding for $GABA_A$ subunits. Modulators of $GABA_C$ receptors are rare (zinc, lanthanides and some neurosteroids), although some flavonoids have been show to inhibit $GABA_C$ receptors, for example, apigenin and quercetin [10, 11]. $GABA_A$ and $GABA_C$ receptors have distinct pharmacological profiles, with some agents having opposite effects (agonist/antagonist) on the two classes of receptors. As $GABA_C$ receptors are much less widely distributed in the brain than $GABA_A$ receptors, they are considered to be important targets for drug development [9].

2.1. Flavonoids and benzodiazepines

Flavonoids were first linked to benzodiazepines when *S*-(−)-equol, 4′-hydroxy-7-methoxyisoflavone and 3′,7-dihydroxyisoflavone, isolated from bovine urine, were shown to inhibit benzodiazepine binding to brain membranes [12]. These flavonoids most likely originated from plant sources in the bovine diet. The pioneering studies on naturally occurring and synthetic flavonoids carried out by the research groups of Marder, Medina, Paladini in Argentina during the 1990s drew attention to flavonoids as initially as 'a new family of benzodiazepine receptor ligands [13, 14].

At the time of initial research into flavonoids at GABA receptors, benzodiazepines were amongst the most widely prescribed pharmaceuticals, and numerous flavonoids of the various classes were investigated both *in vitro* and *in vivo* as potential leads for novel benzodiazepine site ligands. The major therapeutic actions of benzodiazepines are now known to result from their action as positive allosteric modulators of GABA at $GABA_A$ receptors, that is, they enhance the action of GABA on these receptors by acting at another site on $GABA_A$ receptors distinct from site that interacts with GABA. This positive modulatory action of benzodiazepines can be antagonized by flumazenil, a neutralizing modulator used therapeutically to reverse benzodiazepine effects. Benzodiazepines have also been shown to act at high concentrations via a flumazenil-insensitive low-affinity-binding site, separate to the high-affinity, flumazenil-sensitive-binding site [15]. Thus, benzodiazepines act via 'two distinct and separable mechanisms' on $GABA_A$ receptors.

Flavonoids have been shown to modulate $GABA_A$ receptors at low concentrations in either a flumazenil-sensitive or flumazenil-insensitive manner [3]. Thus, flavonoids can influence $GABA_A$ receptors via the classical benzodiazepine-binding site, as well as independently of the classical benzodiazepine-binding site [3, 16]. Many flavonoids elicit biphasic responses, enhancing GABA actions at low concentrations and inhibiting at high concentrations. Additionally, some flavonoids act as agonists particularly at high concentrations and directly gate the receptor in the absence of GABA [17]. Clearly then, flavonoids interact with at least two, and possibly more specific active sites on $GABA_A$ receptors.

2.2. Types of flavonoids

Flavonoids form a class of molecules that consist of a benzopyran moiety (A and C rings) with a phenyl substituent (B ring), as shown in **Figure 1**. The degree of oxidation of the C ring, the hydroxylation pattern of the C ring structure and the substitution in the 3-position demarcate the different subgroups of flavonoids [18]. The predominant subgroups of naturally occurring flavonoids include flavonols (e.g. quercetin [10]), flavones (e.g. apigenin and luteolin [19, 20]), isoflavones (e.g. genistein [21]), flavanones (e.g. naringenin [22]) and flavanols (e.g. epigallocatechin gallate (EGCG) [11]). Each of the flavonoids listed is known to influence $GABA_A$ receptors and to produce CNS effects.

2.3. Apigenin, hispidulin and luteolin

The flavones apigenin, hispidulin and luteolin are closely related structurally, as shown in **Figure 2**. Compared with apigenin, hispidulin has an extra methoxy group in the 6-position and luteolin has an extra hydroxyl group in the 4′-position (**Figure 2**). These small structural differences significantly impact the effects of these flavones on experimental animal behaviours and on $GABA_A$ receptors.

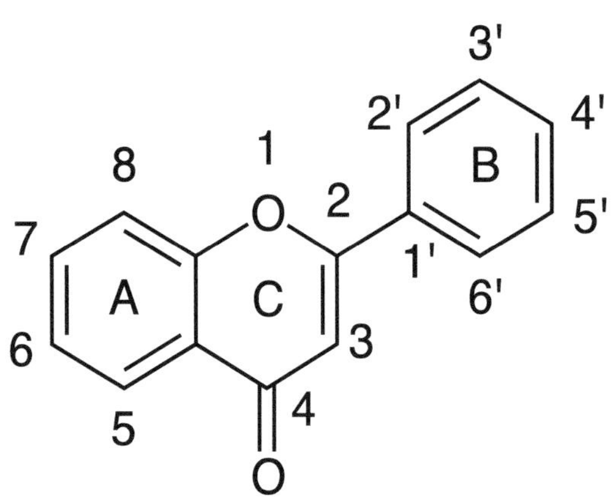

Figure 1. Structure of flavone, showing the generic structure of flavonoids with numbering system and ring designation.

http://dx.doi.org/10.5772/67971

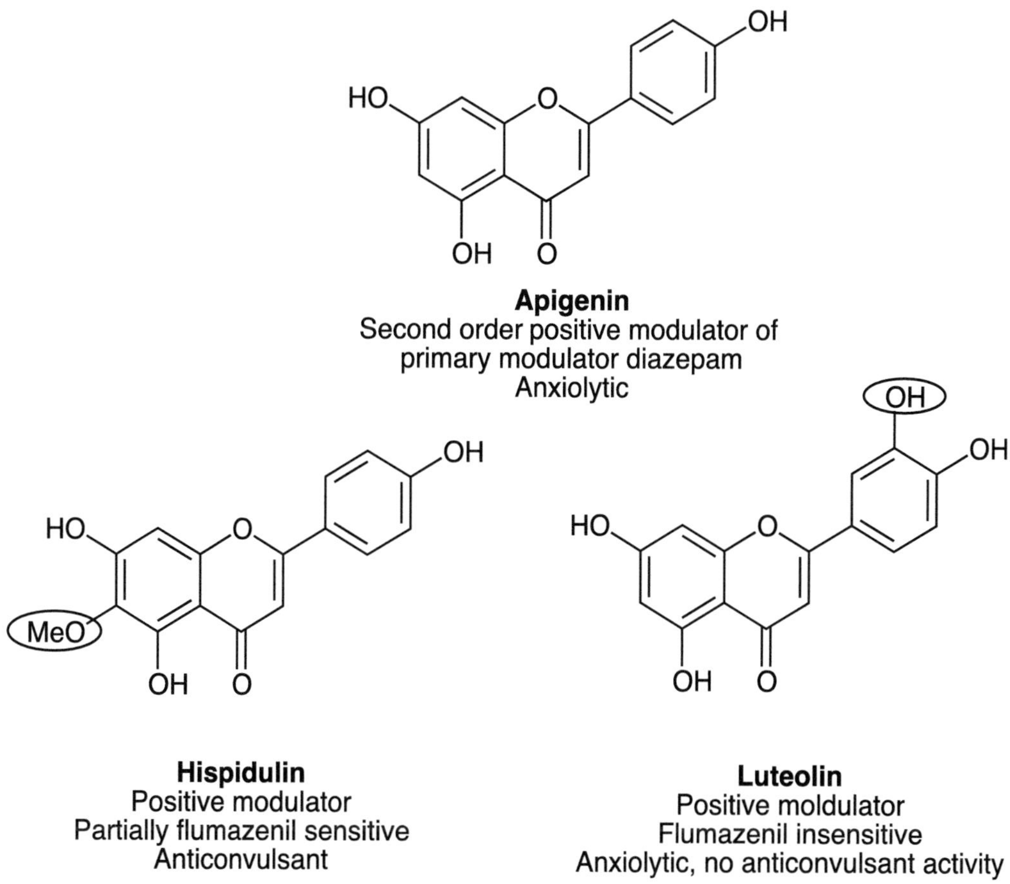

Figure 2. Structures of the flavanones apigenin, hispidulin and luteolin with the differences from apigenin circled.

Apigenin (**Figure 2**), the major flavonoid found in chamomile (*Matricaria chamomilla*), has complex modulatory actions on $GABA_A$ receptors. In cultured cerebellar granule cells, apigenin is described as a negative modulator of GABA action, and it is a weak inhibitor of flumazenil binding to cerebellar membranes [19]. The inhibition of the GABA response at $\alpha1\beta2\gamma2S$ $GABA_A$ receptors expressed in oocytes in the presence of flumazenil (0.1–1 mM) was the first definitive report of the flumazenil-insensitive actions of flavonoids on recombinant $GABA_A$ receptors [10]. Similar flumazenil-insensitive negative modulatory actions of apigenin on recombinant $GABA_A$ receptors were subsequently reported by other investigators [11, 23].

Functional electrophysiological studies have also demonstrated that apigenin can act as a second-order modulator of the first-order modulation of $GABA_A$ receptors by benzodiazepines [11]. This effect of apigenin was observed at concentrations where apigenin alone had no detectable modulatory effects on GABA responses. The second-order positive modulation of the diazepam-enhanced response was observed at the maximal flumazenil-sensitive concentration of diazepam. It is unlikely that apigenin acts by enhancing diazepam binding as it is known to inhibit such binding. Furthermore, apigenin does not influence the binding of muscimol, a potent $GABA_A$ agonist. The observed second-order modulation may result from alterations in the coupling of the flumazenil-sensitive benzodiazepine allosteric sites with the orthosteric GABA sites on $GABA_A$ receptors. This action was selective for diazepam modulation and was not observed for pentobarbitone or allopregnanolone enhancement. In order for this to be a mechanism for the inhibition of locomotor activity by apigenin, there would have to be primary modulation by endogenous benzodiazepines, the so-called endozepines [24].

The second-order modulation (or metamodulation) has also been noted in other systems [25, 26] and may represent an obscure novel mechanism of drug action deserving further investigation, with the potential to lead to decreased therapeutic doses. It is not easy to study as it involves the dose-dependent interactions between three ligands, requiring the study of a number of dose combinations, and thus may be difficult to observe. Synergistic interactions have been described between other flavonoids on GABA receptor-related behaviours [27–29] and at glycine receptors between strychnine and flavonoids [30]. Complex tertiary interactions between flavonoids and other substances may be a subtle feature of cys-loop ligand-gated ion channels.

Clear anxiolytic effects for apigenin have been shown using the elevated plus maze model of anxiety in mice at doses that did not cause myorelaxation or sedation [29, 31], and in rats (5 mg/kg) [32], as did apigenin 7-glucoside (2.5 and 5 mg/kg) [32]. One study using rats was unable to demonstrate an anxiolytic effect of apigenin on the light-dark avoidance model of anxiety at doses of 1–25 mg/kg [33]. On the other hand, 25 mg/kg apigenin was shown to inhibit locomotor activity in a flumazenil-insensitive manner, since flumazenil pre-treatment did not inhibit this effect [33]. It was concluded that the sedative action of apigenin seen in this study 'cannot be ascribed to an interaction with GABA-benzodiazepine receptor, since it is not counteracted by the benzodiazepine antagonist flumazenil' [33]. Nevertheless, this study was undertaken prior to the discovery of the flumazenil-insensitive action of flavonoids on $GABA_A$ receptors. Other possible mechanisms for the action of apigenin on locomotor activity include reduced activation by L-glutamate of NMDA receptors [34] and inhibition of L-glutamate release via reduction of calcium ion entry [35].

The strongest evidence for flavonoid modulation of $GABA_A$ receptors in the brain comes from the anticonvulsant hispidulin, found in the sage plant (*Salvia officinalis*) (**Figure 2**) [36]. Structurally, hispidulin differs from apigenin only by the addition of a 6-methoxy substituent. In functional studies on recombinant $\alpha1\beta2\gamma2$ $GABA_A$ receptors expressed in *Xenopus* oocytes, hispidulin was inactive when applied alone, and at concentrations up to 10 μM was found to positively modulate the effect of 4 μM GABA [23]. This positive modulation was partially blocked by flumazenil. However, hispidulin at 10 μM was inactive at $\alpha1\beta2$ receptors, which lack the flumazenil-sensitive benzodiazepine site. Hispidulin was further shown to exhibit a biphasic action and to be approximately equipotent at each of six different α-subunit containing $GABA_A$ receptors—$\alpha1,2,3,5,6\beta2\gamma2S$, enhancing at low concentrations (EC_{50} 0.8–5 mM) and inhibiting at higher concentrations (>30 mM). The fact that hispidulin is inactive at $\alpha_1\beta_2$ receptors but is active at $\alpha6$-containing receptors that are insensitive to benzodiazepines suggests that hispidulin may act via more than one binding site on $GABA_A$ receptors, at least one of which may represent a novel site. Interestingly, previous studies indicated that a range of natural and synthetic flavones had no affinity for recombinant $\alpha6\beta3\gamma$ receptors [37]. Thus, hispidulin appears to show a different profile of activity to apigenin at $GABA_A$ receptor subtypes.

Hispidulin was also shown to displace ^{3}H-flumazenil binding in human frontal cortex crude synaptic membrane preparations [38]. Using ^{14}C-hispidulin, the flavone was found to pass the blood-brain barrier using a rat perfusion model [23]. Further, hispidulin exhibited a flumazenil-sensitive anticonvulsant action, similar to diazepam, in seizure-prone Mongolian gerbils used as an animal model of epilepsy [23].

Luteolin (**Figure 2**), found in many plants including celery and green pepper, differs from apigenin by an additional 3′-hydroxyl group. Following acute administration in mice, luteolin at doses of 1–50 mg/kg failed to demonstrate anticonvulsant or myorelaxation effects, or to have any impact on locomotor activity [39]. On the other hand, 5 mg/kg luteolin increased open-arm entries in the elevated plus maze, suggesting an anxiolytic effect, and reduced haloperidol-induced catalepsy [39]. Both of these effects, however, disappeared at higher doses tested. Another study testing 50 mg/kg luteolin in rats in the elevated plus maze also failed to demonstrate any anxiolytic action [40], though this may be partly explained by the finding that the same dose reduced locomotor activity, suggesting a sedative effect that may have masked any anxiolytic action. Although it has not been tested *in vitro*, it may also be possible that, like some other flavonoids, luteolin possesses positive allosteric modulatory actions at low doses and negative allosteric modulation at higher doses. A combination of flumazenil with luteolin also failed to show any significant difference to the control group, and the researchers concluded that 'luteolin does not produce anxiolysis by modulation of the $GABA_A$ receptor' [40]. Given that we now know of flumazenil-insensitive actions of flavonoids on $GABA_A$ receptors, this may be an incorrect conclusion.

Following chronic administration (14 days), luteolin showed antidepressant activity in the forced swim test, significantly reducing latency to immobilization and increasing total immobilization to the same extent as diazepam, interpreted as increased adaptation (rather than increased helplessness) in the model used [39]. Further evidence of luteolin's antidepressant activity was shown using the forced swim test [20]. A dose-dependent reduction in the duration of immobility was observed in mice following acute administration of doses (5 and 10 mg/kg) that did not alter locomotor activity when tested using the open field.

In a rat neuropathic pain model, luteolin produced analgesia in a bicuculline-sensitive, flumazenil-insensitive manner [23]. Luteolin (10 mg/kg) also improved spatial memory in rats in a scopolamine-induced amnesia model in the Y maze [41], although the proposed mechanisms relate to brain-derived neurotrophic factor, acetylcholine and lipid peroxidation. Further studies are required to demonstrate any involvement of the GABAergic system in this observed memory enhancement. Finally, a study of luteolin in four mouse seizure models showed no anticonvulsant activity [42].

Evidence that some of these behavioural effects of luteolin may be mediated by binding to $GABA_A$ receptors exists. Luteolin displaced ^{3}H-flunitrazepam (1.5 nM) binding on rat cortical crude synaptic membrane preparations, with a K_i of 60 μM, suggesting weak binding to the benzodiazepine-binding site on $GABA_A$ receptors [39]. Further, luteolin was shown to promote GABA-mediated chloride influx in human neuroblastoma cells, which was attenuated by the $GABA_A$ receptor antagonist bicuculline [20].

The studies reviewed here reveal that apigenin, hispidulin and luteolin appear to show differing profiles of activity at $GABA_A$ receptor subtypes and differing effects *in vivo*, demonstrating that small differences in chemical structure have profound effects on the biological properties of these flavonoids.

2.4. (−)-Epigallocatechin gallate (EGCG)

EGCG (**Figure 3**) is a flavanol, and the major polyphenol in green tea (*Camellia sinensis*) [43]. At low concentrations, EGCG (0.1 mM) has a potent second-order modulatory action on the first-order modulation by diazepam at $\alpha1\beta2\gamma2L$ $GABA_A$ receptors but inhibits the action of GABA at higher concentrations (>1 mM) [11]. As a second-order modulator, EGCG is an order of magnitude more potent than apigenin [11]. In addition, it has been found that EGCG, at concentrations that have no influence on the activation of $GABA_A$ receptors by GABA, was able to reverse β-carboline (a negative modulator of $GABA_A$ receptors)-mediated inhibition of GABA currents in cultured hippocampal neurons [44]. Also, up to 100 μM EGCG significantly increased chloride influx in primary cultured cerebellar cells [45]. This indicates that EGCG may act as a second-order modulator with respect to the first-order modulation by both positive and negative modulators that act via benzodiazepine-binding sites on $GABA_A$ receptors.

EGCG demonstrates dose-dependent stress-reducing, anxiolytic, sedative and hypnotic properties in a number of animal models [44–46], with evidence that these activities are mediated at least in part by $GABA_A$ receptors [44, 46]. EGCG has effects on learning and memory [47] that may be useful in the treatment of Alzheimer's disease [48]. It has also been suggested for Parkinson's disease therapy [49] while some novel EGCG derivatives may be useful for neuropathic pain [50]. It is possible that activity at $GABA_A$ receptors underlies many of the reported actions of EGCG.

2.5. Synthetic flavonoids

Classical structure activity-based design led to the development of synthetic flavonoid ligands with high affinity for the classical flumazenil-sensitive benzodiazepine-binding site on $GABA_A$ receptors [51]. Numerous synthetic flavonoids have been shown to influence $GABA_A$ receptors. Of particular interest are a series of 6-substituted flavones that show the full repertoire of effects on $GABA_A$ receptors: positive, neutralizing and negative modulation and direct activation [3, 52], at both the flumazenil-sensitive benzodiazepine site and flumazenil-insensitive site(s).

6-Bromoflavone was shown to be a flumazenil-sensitive positive allosteric modulator at $GABA_A$ receptors, whereas 6-fluoro- and 6-chloroflavone were demonstrated to act as neutralizing modulators [52]. On the other hand, 6,2′-dihydroxyflavone was found to be a negative modulator. By contrast, 6-methylflavone has been shown to act as a flumazenil-insensitive modulator of $GABA_A$ receptors [53].

http://dx.doi.org/10.5772/67971

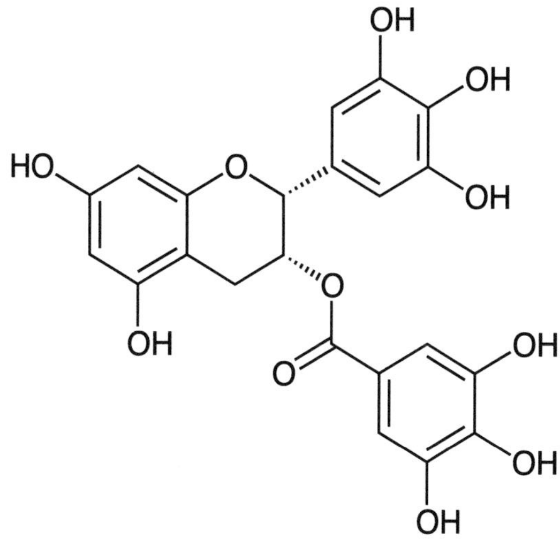

(-)-Epigallocatechin gallate (EGCG)
Second order positive modulator of primary modulator diazepam
Counteracts negative modulation by methyl β-carboline-3-caroxylate
Anxiolytic, sedative hypnotic

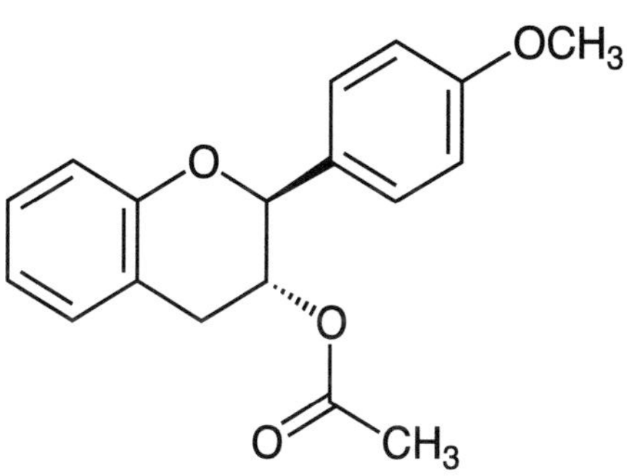

Fa131
Positive modulator
Flumazenil-insensitive
Selective for α2 over α1, α3, α5
Non-sedating anxiolytic

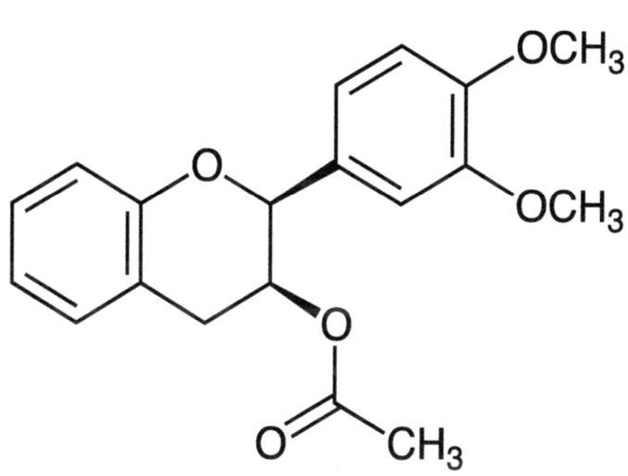

Fa173
Blocks Fa131, etomidate, loreclezole and
low but not high affinity diazepam.
Inactive against propofol, thiopental
and allopregnanolone

Figure 3. Structures of the flavanols EGCG, Fa131 and Fa173.

2′-Methoxy-6-methylflavone has demonstrated anxiolytic effects in mice at 1 and 10 mg/kg using the elevated plus maze model of anxiety [54]. It was also found to act as a positive modulator at $\alpha2\beta1\gamma2L$ and all $\alpha1$-containing $GABA_A$ receptor subtypes, demonstrated via recombinant $GABA_A$ receptors expressed in *Xenopus* oocytes [54]. By contrast, it directly activated $\alpha2\beta2/3\gamma2L$ receptors without potentiating GABA [54]. Activation by 2′-methoxy-6-methylflavone was attenuated by bicuculline and gabazine but not flumazenil, indicating a novel site of action. This suggests that there is a further flavonoid site on $GABA_A$ receptors that mediate opening of the chloride channel in the absence of GABA.

Quantitative structure-efficacy relationships have shown that flavone analogues differing only at position 6 show significantly different pharmacological properties at $GABA_A$ receptors [52]. This study clearly shows the importance of the 6-position as a determination of activity. However, further studies on 6-substitued flavones are needed to study the complex nature of the activation and modulation of $GABA_A$ receptor subtypes and to explore the unique therapeutic potential of these synthetic flavones.

Another interesting series of synthetic flavonoids are the flavan-3-ol esters, analogues of EGCG, a naturally occurring flavanol-3-ester. Fa131 (trans-(2S,3R)-3-acetoxy-4′-methoxyflavan, **Figure 3**) is a non-sedating anxiolytic and a selective positive modulator of α2-containing $GABA_A$ receptors, shown on the basis of efficacy [55, 56]. The efficacy of 2100% enhancement exceeds the highest efficacy previously recorded, which was 1250% by (+)-borneol at these receptors [57].

Fa173 (cis-(2S,3S)-3-acetoxy-3′,4′-dimethoxyflavan, **Figure 3**), a diastereo-isomeric flavan-3-ol ester with additional methoxy in the 3′ position, was shown to block the modulatory actions of Fa131 [58]. Additionally, Fa173 also blocked the enhancement of the GABA response by the anaesthetic etomidate, the sedative anticonvulsant loreclezole, and selectively blocked the low-affinity effect of diazepam (100 μM) at $\alpha1\beta2\gamma2L$ and $\alpha1\beta2$ $GABA_A$ receptors, but not the high-affinity effect of diazepam (100 nM). Fa173 was found not to inhibit the positive modulation of GABA by the anaesthetic propofol, barbiturate thiopental, or neuroactive steroid allopregnanolone. This suggests that Fa131, etomidate, loreclezole and high (non-flumazenil-sensitive) doses of benzodiazepine all exert their positive modulatory effects via a common or overlapping binding site that can be blocked by the neutralizing modulator Fa173. Of these agents, Fa131 alone shows selectivity for α2-containing $GABA_A$ recombinant receptors. Fa131 is the first positive modulator to distinguish between α2- and α3-containing $GABA_A$ receptors, highlighting the potential of targeting flumazenil-insensitive allosteric sites in the search for new anxio-selective drugs.

2.6. Synergism between flavonoids

As flavonoids are significant constituents of our diet, it is important that we understand how natural flavonoids might influence brain function. Except when consumed as dietary supplements, flavonoids are generally consumed as a mixture of different flavonoids from one or more foodstuffs [59]. The effects of mixtures of flavonoids and other modulators on $GABA_A$ receptors need to be more thoroughly investigated. Synergies have been noted between individual flavonoids [29, 60, 61], and between flavonoids and benzodiazepines [27, 28].

Hesperidin, a glycosylated flavonone isolated from Valerian species, has shown synergistic effects in mice. The combination of hesperidin (2 mg/kg) with apigenin (1 mg/kg), 6,3′-dinitroflavone (0.02 mg/kg) or diazepam (0.3 mg/kg) enhanced the barbiturate-induced sleeping time in mice [27, 29]. Both hesperidin and diazepam administered separately showed a dose-dependent reduction in exploratory parameters (number of head dips,

http://dx.doi.org/10.5772/67971

time spent head-dipping and rearing behaviour), indicative of increased sedation, in mice on the holeboard assay [27]. Isobolar analysis revealed synergism between diazepam and hesperidin when administered together. For all exploratory parameters measured, a 1.3 to 2-fold increase in potency was observed compared to the administration of either drug alone. Further, these synergistic actions could not be explained by any influence of either drug on plasma concentrations of the other [27]. Loscalzo et al. [28] further demonstrated a potentiation of sedation in mice when hesperidin was administered together with either bromazepam, alprazolam, flunitrazepam or midazolam, through a reduction in exploratory parameters and locomotor activity using the holeboard assay and open-field test, respectively.

Using human mammary epithelial carcinoma cells (MCF-7), Choi and colleagues [60] showed that isoflavones daidzein (derived from soybean) and baicalein (from *Scutellaria baicalensis*) stimulated oestrogen receptor phosphorylation and transcriptional activation of oestrogen-responsive element (ERE). When tested together, the observed effects on oestrogen receptor phosphorylation and transcription of the ERE were further increased in comparison to when the drugs were tested alone. A combination of baicalein and daidzein was shown to produce a synergistic effect in stimulating oestrogenic activity in MCF-7 cells, calculated using the median-effect principle. Further, using an Aβ-aggregation assay to model cellular pathology of Alzheimer's disease, daidzein and baicalein were demonstrated to reduce Aβ-aggregation. As oestrogen is neuroprotective, this synergistic action of the isoflavones on oestrogen receptors, as well as in reducing Aβ-aggregation, suggests that the combination of flavonoids could provide valuable neuroprotective effects and prevention of neurodegenerative disease [60].

Another study examining the synergistic effects of flavonoids found that the flavonol EGCG and the flavone luteolin synergistically inhibited TGF-β-induced myofibroblast phenotypes in prostate fibroblast cell lines [61]. Myofibroblasts are converted from fibroblasts by TGF-β and IL-6 in the tumour microenvironment. These cells play a role in cell proliferation, migration and invasion. TGF-β-induced fibronectin expression was used as a marker of myofibroblast expression. Both EGCG (20–40 μM) and luteolin (20 μM) were shown to reduce TGF-β-induced fibronectin expression alone. In combination, these compounds showed greater efficacy in reducing fibronectin expression at concentrations that were less effective when administered alone. The authors concluded that a combination of EGCG and luteolin could prove effective in reducing the toxic effects of either compound by requiring lower doses, and in preventing advancement of tumour growth by reducing the myofibroblast phenotype.

3. Conclusion

Since flavonoids were first linked to benzodiazepine-binding sites on $GABA_A$ receptors many years ago, recent studies have clearly demonstrated that the actions of flavonoids on these receptors are far more complex than a single action at a single site. In addition to the now relatively well-characterized flumazenil-sensitive benzodiazepine-binding sites, there is significant interest in flumazenil-insensitive, non-benzodiazepine-binding sites for flavonoids. This overview has sought to highlight the action of representative flavonoids on $GABA_A$ receptors to illustrate the range of activities.

Recent studies have identified the presence of multiple sites on ionotropic GABA receptors at which flavonoids can act, modulating the effect of GABA. The sites may include ones that are insensitive to the classical benzodiazepine-binding site antagonist (neutralizing modulator) flumazenil and described as low-affinity benzodiazepine sites [15]. Perhaps, these would be more appropriately described as flavonoid sites as they appear to be activated by many naturally occurring and synthetic flavonoids.

Author details

Tina Hinton[1], Jane R. Hanrahan[2] and Graham A.R. Johnston[1*]

*Address all correspondence to: graham.johnston@sydney.edu.au

1 Discipline of Pharmacology, School of Medical Sciences, The University of Sydney, NSW, Australia

2 Faculty of Pharmacy, The University of Sydney, NSW, Australia

References

[1] Bowery NG, Smart TG. GABA and glycine as neurotransmitters: a brief history. Br J Pharmacol. 2006;**147 Suppl 1**:S109-19.

[2] Chebib M, Johnston GAR. GABA-activated ligand gated ion channels: Medicinal chemistry and molecular biology. J Med Chem. 2000;**43**:1427-47.

[3] Hanrahan JR, Chebib M, Johnston GAR. Interactions of flavonoids with ionotropic GABA receptors. Adv Pharmacol. 2015;**72**:189-200.

[4] Li W, Du L, Li M. Alkaloids and flavonoids as alpha(1)-adrenergic receptor antagonists. Curr Med Chem. 2011;**18**:4923-32.

[5] Thompson AJ, Lester HA, Lummis SC. The structural basis of function in Cys-loop receptors. Q Rev Biophys. 2010;**43**:449-99.

[6] Olsen RW, Sieghart W. $GABA_A$ receptors: subtypes provide diversity of function and pharmacology. Neuropharmacology. 2009;**56**:141-8.

[7] Rudolph U, Mohler H. $GABA_A$ receptor subtypes: Therapeutic potential in Down syndrome, affective disorders, schizophrenia, and autism. Annu Rev Pharmacol Toxicol. 2014;**54**:483-507.

[8] Rudolph U, Mohler H. GABA-based therapeutic approaches: $GABA_A$ receptor subtype functions. Curr Opin Pharmacol. 2006;**6**:18-23.

[9] Johnston GAR, Chebib M, Hanrahan JR, Mewett KN. $GABA_C$ receptors as drug targets. Curr Drug Targets - CNS & Neurol Dis. 2003;**2**:260-8.

[10] Goutman JD, Waxemberg MD, Donate-Oliver F, Pomata PE, Calvo DJ. Flavonoid modulation of ionic currents mediated by $GABA_A$ and $GABA_C$ receptors. Eur J Pharmacol. 2003;**461**:79-87.

[11] Campbell EL, Chebib M, Johnston GAR. The dietary flavonoids apigenin and (−)-epigallocatechin gallate enhance the positive modulation by diazepam of the activation by GABA of recombinant $GABA_A$ receptors. Biochem Pharmacol. 2004;**68**:1631-8.

[12] Luk KC, Stern L, Weigele M, O'Brien RA, Spirst N. Isolation and identification of "diazepam-like" compounds in bovine brain. J Nat Prod. 1983;**46**:852-61.

[13] Medina JH, Viola H, Wolfman C, Marder M, Wasowski C, Calvo D, et al. Overview—flavonoids—a new family of benzodiazepine receptor ligands. Neurochem Res. 1997;**22**:419-25.

[14] Paladini AC, Marder M, Viola H, Wolfman C, Wasowski C, Medina JH. Flavonoids and the central nervous system: from forgotten factors to potent anxiolytic compounds. J Pharm Pharmacol. 1999;**51**:519-26.

[15] Walters RJ, Hadley SH, Morris KDW, Amin J. Benzodiazepines act on $GABA_A$ receptors via two distinct and separable mechanisms. Nat Neurosci. 2000;**3**:1274-81.

[16] Hanrahan JR, Chebib M, Johnston GAR. Flavonoid modulation of $GABA_A$ receptors. Br J Pharmacol. 2011;**163**:234-45.

[17] Chua HC, Absalom NL, Hanrahan JR, Viswas R, Chebib M. The direct actions of GABA, 2′-methoxy-6-methylflavone and general anaesthetics at beta3gamma2L $GABA_A$ receptors: evidence for receptors with different subunit stoichiometries. PLoS ONE [Electronic Resource]. 2015;**10**(10):e0141359.

[18] Spencer JP, Vauzour D, Rendeiro C. Flavonoids and cognition: the molecular mechanisms underlying their behavioural effects. Arch Biochem Biophys. 2009;**492**:1-9.

[19] Avallone R, Zanoli P, Puia G, Kleinschnitz M, Schreier P, Baraldi M. Pharmacological profile of apigenin, a flavonoid isolated from Matricaria chamomilla. Biochem Pharmacol. 2000;**59**:1387-94.

[20] de la Pena JB, Kim CA, Lee HL, Yoon SY, Kim HJ, Hong EY, et al. Luteolin mediates the antidepressant-like effects of Cirsium japonicum in mice, possibly through modulation of the $GABA_A$ receptor. Arch Pharm Res. 2014;**37**:263-9.

[21] Dunne EL, Moss SJ, Smart TG. Inhibition of $GABA_A$ receptor function by tyrosine kinase inhibitors and their inactive analogues. Mol Cell Neurosci. 1998;**12**:300-10.

[22] Jager AK, Almquist JP, Vangsoe SAK, Stafford GI, Adsersen A, van Staden J. Compounds from Mentha aquatica with affinity to the GABA-benzodiazepine receptor. S Afr J Bot. 2007;**73**:518-21.

[23] Kavvadias D, Sand P, Youdim KA, Qaiser MZ, Rice-Evans C, Baur R, et al. The flavone hispidulin, a benzodiazepine receptor ligand with positive allosteric properties, traverses the blood–brain barrier and exhibits anti-convulsive effects. Br J Pharmacol. 2004;**142**:811-20.

[24] Baraldi M, Avallone R, Corsi L, Venturini I, Baraldi C, Zeneroli ML. Natural endogenous ligands for benzodiazepine receptors in hepatic encephalopathy. Metab Brain Dis. 2009;**24**:81-93.

[25] Mesce KA. Metamodulation of the biogenic amines: Second-order modulation by steroid hormones and amine cocktails. Brain Behav Evol. 2002;**60**:339-49.

[26] Ribeiro JA, Sebastiao AM. Modulation and metamodulation of synapses by adenosine. Acta Physiol. 2010;**199**:161-9.

[27] Fernandez SP, Wasowski C, Paladini AC, Marder M. Synergistic interaction between hesperidin, a natural flavonoid, and diazepam. Eur J Pharmacol. 2005;**512**:189-98.

[28] Loscalzo LM, Wasowski C, Paladini AC, Marder M. Opioid receptors are involved in the sedative and antinociceptive effects of hesperidin as well as in its potentiation with benzodiazepines. Eur J Pharmacol. 2008;**580**:306-13.

[29] Marder M, Viola H, Wasowski C, Fernandez S, Medina JH, Paladini AC. 6-Methylapigenin and hesperidin: new valeriana flavonoids with activity on the CNS. Pharmacol Biochem Behav. 2003;**75**:537-45.

[30] Raafat K, Breitinger U, Mahran L, Ayoub N, Breitinger HG. Synergistic inhibition of glycinergic transmission in vitro and in vivo by flavonoids and strychnine. Toxicol Sci. 2010;**118**:171-82.

[31] Viola H, Wasowski C, Levi de Stein M, Wolfman C, Silvera R, Medina AE, et al. Apigenin, a component of *Matricaria recutita* flowers, is a central benzodiazepine receptors-ligand with anxiolytic effects. Planta Med. 1995;**61**:213-6.

[32] Kumar D, Bhat ZA. Apigenin 7-glucoside from Stachys tibetica Vatke and its anxiolytic effect in rats. Phytomedicine. 2014;**21**:1010-4.

[33] Zanoli P, Avallone R, Baraldi M. Behavioural characterisation of the flavonoids apigenin and chrysin. Fitoterapia. 2000;**71**:S117–S23.

[34] Losi G, Puia G, Garzon G, de Vuono MC, Baraldi M. Apigenin modulates GABAergic and glutamatergic transmission in cultured cortical neurons. Eur J Pharmacol. 2004;**502**:41-6.

[35] Chang CY, Lin TY, Lu CW, Wang CC, Wang YC, Chou SSP, et al. Apigenin, a natural flavonoid, inhibits glutamate release in the rat hippocampus. Eur J Pharmacol. 2015;**762**:72-81.

[36] Atif M, Ali I, Hussain A, Hyder S, Khan FA, Maalik A, et al. Pharmacological assessment of hispidulin—a natural bioactive flavone. Acta Pol Pharm. 2016;**73**:565-78.

[37] Marder M, Estiu G, Blanch LB, Viola H, Wasowski C, Medina JH, et al. Molecular modelling and QSAR analysis of the interaction of flavone derivatives with the benzodiazepine site of $GABA_A$ receptor complex. Bioorg Med Chem. 2001;**9**:323-35.

[38] Kavvadias D, Monschein V, Sand P, Riederer P, Schreier P. Constituents of sage (*Salvia officinalis L.*) with *in vitro* affinity to human brain benzodiazepine receptor. Planta Med. 2003;**69**:113-7.

[39] Coleta M, Campos MG, Cotrim MD, Lima TC, Cunha AP. Assessment of luteolin (3′,4′,5,7-tetrahydroxyflavone) neuropharmacological activity. Behav Brain Res. 2008;**189**: 75-82.

[40] Raines T, Jones P, Moe N, Duncan R, McCall S, Ceremuga TE. Investigation of the anxiolytic effects of luteolin, a lemon balm flavonoid in the male Sprague–Dawley rat. AANA J. 2009;**77**:33-6.

[41] Yoo DY, Choi JH, Kim W, Nam SM, Jung HY, Kim JH, et al. Effects of luteolin on spatial memory, cell proliferation, and neuroblast differentiation in the hippocampal dentate gyrus in a scopolamine-induced amnesia model. Neurol Res. 2013;**35**:813-20.

[42] Shaikh MF, Tan KN, Borges K. Anticonvulsant screening of luteolin in four mouse seizure models. Neurosci Lett. 2013;**550**:195-9.

[43] Afzal M, Safer AM, Menon M. Green tea polyphenols and their potential role in health and disease. Inflammopharmacology. 2015;**23**:151-61.

[44] Vignes M, Maurice T, Lante F, Nedjar M, Thethi K, Guiramand J, et al. Anxiolytic properties of green tea polyphenol (−)-epigallocatechin gallate (EGCG). Brain Res. 2006;**1110**:102-15.

[45] Park KS, Han JY, Moon DC, Hong JT, Oh KW. (−)-epigallocatechin-3-O-gallate augments pentobarbital-induced sleeping behaviors through Cl− channel activation. J Med Food. 2011;**14**:1456-62.

[46] Adachi N, Tomonaga S, Tachibana T, Denbow DM, Furuse M. (−)-Epigallocatechin gallate attenuates acute stress responses through GABAergic system in the brain. Eur J Pharmacol. 2006;**531**:171-5.

[47] Gu HF, Nie YX, Tong QZ, Tang YL, Zeng Y, Jing KQ, et al. Epigallocatechin-3-gallate attenuates impairment of learning and memory in chronic unpredictable mild stress-treated rats by restoring hippocampal autophagic flux.[Erratum appears in PLoS One. 2015;10(2):e0117649; PMID: 25658954]. PLoS ONE [Electronic Resource]. 2014;**9**.

[48] Chang X, Rong C, Chen Y, Yang C, Hu Q, Mo Y, et al. (−)-Epigallocatechin-3-gallate attenuates cognitive deterioration in Alzheimer's disease model mice by upregulating neprilysin expression. Exp Cell Res. 2015;**334**:136-45.

[49] Renaud J, Nabavi SF, Daglia M, Nabavi SM, Martinoli MG. Epigallocatechin-3-gallate, a promising molecule for Parkinson's disease? Rejuvenation Res. 2015;**18**:257-69.

[50] Xifro X, Vidal-Sancho L, Boadas-Vaello P, Turrado C, Alberch J, Puig T, et al. Novel epigallocatechin-3-gallate (EGCG) derivative as a new therapeutic strategy for reducing neuropathic pain after chronic constriction nerve injury in mice. PLoS ONE [Electronic Resource]. 2015;**10**(4):e0123122.

[51] Nilsson J, Sterner O. Modulation of $GABA_A$ receptors by natural products and the development of novel synthetic ligands for the benzodiazepine binding site. Curr Drug Targets. 2011;**12**:1674-88.

[52] Ren L, Chan WM, Wang F, Xu Z, Zhao C, Mat WK, et al. Effects of flavone 6-substitutions on $GABA_A$ receptors efficacy. Eur J Pharmacol. 2011;**670**:121-9.

[53] Hall BJ, Chebib M, Hanrahan JR, Johnston GAR. Flumazenil-independent positive modulation of γ-aminobutyric acid by 6-methylflavone at human recombinant $\alpha1\beta2\gamma2L$ and $\alpha1\beta2$ $GABA_A$ receptors. Eur J Pharmacol. 2004;**491**:1-8.

[54] Karim N, Curmi J, Gavande N, Johnston GAR, Hanrahan JR, Tierney ML, et al. 2′-Methoxy-6-methylflavone: a novel anxiolytic and sedative with subtype selective activating and modulating actions at $GABA_A$ receptors. Br J Pharmacol. 2012;**165**:880-96.

[55] Fernandez SP, Mewett KN, Hanrahan JR, Chebib M, Johnston GA. Flavan-3-ol derivatives are positive modulators of $GABA_A$ receptors with higher efficacy for the alpha(2) subtype and anxiolytic action in mice. Neuropharmacology. 2008;**55**:900-7.

[56] Mewett KN, Fernandez SP, Pasricha AK, Pong A, Devenish SO, Hibbs DE, et al. Synthesis and biological evaluation of flavan-3-ol derivatives as positive modulators of $GABA_A$ receptors. Bioorg Med Chem. 2009;**17**:7156-73.

[57] Granger RE, Campbell EL, Johnston GAR. (+)- And (−)-borneol: efficacious positive modulators of GABA action at human recombinant alpha1beta2gamma2L $GABA_A$ receptors. Biochem Pharmacol. 2005;**69**:1101-11.

[58] Fernandez SP, Karim N, Mewett KN, Chebib M, Johnston GAR, Hanrahan JR. Flavan-3-ol esters: new agents for exploring modulatory sites on $GABA_A$ receptors. Br J Pharmacol. 2012;**165**:965-77.

[59] Johnston GAR. Flavonoid nutraceuticals and ionotropic receptors for the inhibitory neurotransmitter GABA. Neurochem Int. 2015;**89**:120-5.

[60] Choi RCY, Zhu JTT, Yung AWY, Lee PSC, Xu SL, Guo AJY, et al. Synergistic action of flavonoids, baicalein, and daidzein in estrogenic and neuroprotective effects: a development of potential health products and therapeutic drugs against Alzheimer's disease. Evid Based Compl Altern Med. 2013;**2013**:635694-Article ID

[61] Gray AL, Stephens CA, Bigelow RL, Coleman DT, Cardelli JA. The polyphenols (−)-epigallocatechin-3-gallate and luteolin synergistically inhibit TGF-beta-induced myofibroblast phenotypes through RhoA and ERK inhibition. PLoS ONE [Electronic Resource]. 2014;**9**(10):e109208.

Section 4

Flavonoids in Food

Chapter 16

Flavonoids: Important Biocompounds in Food

Saul Ruiz-Cruz, Saraí Chaparro-Hernández,
Karen L. Hernández-Ruiz, Luis A. Cira-Chávez,
María I. Estrada-Alvarado, Laura E. Gassos Ortega,
José de Jesús Ornelas-Paz and Marco A. Lopez Mata

Additional information is available at the end of the chapter

http://dx.doi.org/10.5772/67864

Abstract

Flavonoids are secondary metabolites in plants that show some desirable characteristics. These compounds can be grouped in different classes on the basis of their basic structure. It has been reported that flavonoids are important for human health because of their antioxidant, antibacterial, antiviral, and anti-inflammatory activities and because they act as free radical scavengers as they are potential reducing agents that protect from oxidative damage, which are conferred by the content of hydroxyl groups. In recent years, flavonoids have been investigated based on their ability to reduce the incidence of many diseases, to inhibit cell damage, to repair DNA process and to reduce oxidative stress. Besides, flavonoids have been demonstrated to have cardioprotective effects, have potential to improve coronary vasodilatation and prevent LDLs from oxidizing and also showed potential neuroprotective effects. Moreover, flavonoids have been used in the food industry due to their ability to preserve foods, to provide colour and flavour and to make dietary supplements, among other important industrial applications.

Keywords: biocompound, flavonoid, antioxidant

1. Introduction

Plant metabolites can be divided into two groups: primary metabolites, which are involved in the nutrition and the essential metabolic processes (e.g. carbohydrates, lipids and proteins); and secondary ones, which have an important function in the interaction between a plant and its environment, such as pigments or defensive compounds. Secondary metabolites included a group of compounds known as phenolic; in this group, we can find the flavonoids.

Flavonoids arc a large group of natural substances with variable structures present almost in all growing parts of the plants, being reported as the most abundant plant pigment along with chlorophyll and carotenoids, also providing fragrance and taste to fruits, flowers and seeds, which makes them attractants for other organisms [1, 2]. These compounds are also one of the largest groups of secondary metabolites [3]. Besides their relevance in plants, flavonoids are important for human health because of their high pharmacological activities. Recent interest in these substances has been stimulated by the potential health benefits arising from the antioxidant activities of these polyphenolic compounds.

In this chapter, we focused in the antioxidant, antibacterial, antiviral and anti-inflammatory activities of flavonoids, which according to many studies are conferred mainly by the content of hydroxyl groups attached to base structures of these compounds. Biochemical actions of these compounds depend primarily on the presence and position of their substituent groups, which can affect the metabolism of each one [4]. One of the most important characteristics of flavonoids is that they often occur in the glycosidic form, which possibly let them take place in the gastrointestinal tract [5].

However, they attract attention due to their antioxidant activity and reduce free radical formation and also scavenge free radicals [6]. Flavonoids have other important biological activities such as protect skin from UV light exposure, protect DNA from damage, strengthening of capillaries, anti-inflammatory effect and protective action against radiation, moistening, softening, soothing, antiseptic and other. Due these properties, flavonoids can be used as ingredients in the production of cosmetics and pharmaceutical products [7, 8].

Nowadays, flavonoids are major bioactive compounds known for their potential health benefits, which have been used against many chronic diseases such as cancer, antiviral, inflammation, cardiovascular and neurodegenerative disorders; it is widely assumed that active dietary constituents are antioxidant nutrients present in fruits and vegetables [9].

2. Classification

In general, all flavonoids have a general structural compound with a chromane-type skeleton composed of three phenolic rings referred to as A, B and C rings (**Figure 1**) with phenyl

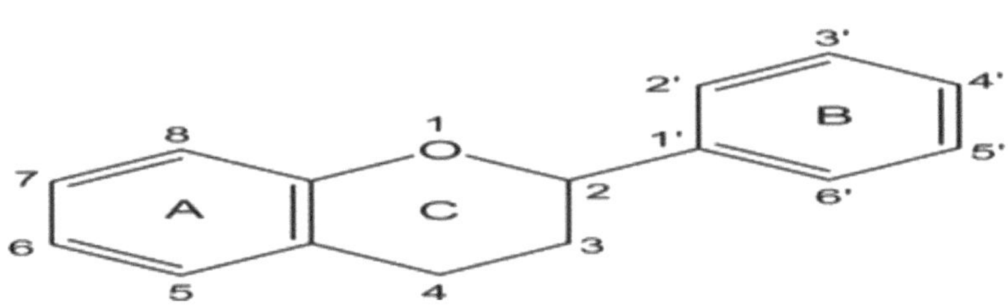

Figure 1. Flavonoid basic structure [13].

substituent in the C2 or C3 position in B ring. Their functions depend on their structural class, degree of hydroxylation and conjugation and degree of polymerization. They vary in the structure around the heterocyclic oxygen ring, but all have the characteristic C6-C3-C6 carbon skeleton [3, 10–12].

They can be grouped into different classes with respect to the basic structure, which allows a wide substitution patron and variations in C ring leading to their classification [11].

2.1. Flavones

In the classification, we can find that in the main structure of *flavones* a double bond between carbons 2 and 3 of the central ring is present (**Table 1**). Also, the structure of flavones has a carbonyl group in position 4 in C ring. These types of flavonoids are pale yellow in colour, and their representative forms are apigenin, chrysin, myricetin, rutin, sibelin, luteolin, diosmetin and quercetin [5, 14–17]

2.2. Flavonols

As flavones, *flavonols'* structure exhibits a double bond between carbons 2 and 3, and a carbonyl group in carbon 4, also a hydroxyl group in carbon 3 of C ring, for example, quercetin, rhamnetin and kaempferol.

	Structure	Position	3	5	7	3′	4′	5′
Flavones	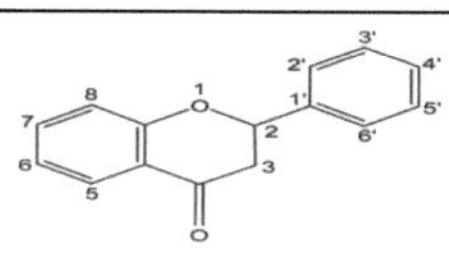	Apigenin	-	OH	OH	-	OH	-
		Luteolin	-	OH	OH	OH	OH	-
		Chrysin	-	OH	OH	-	-	-
Flavonones	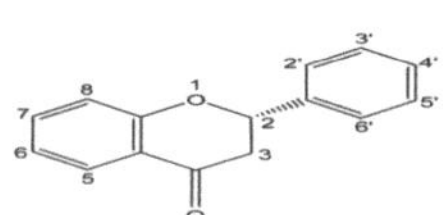	Naringenin	-	OH	OH	OH	OH	-
		Hesperetin	-	OH	OH	-	OH	-
		Neohesperetin	-	OH	OH	-	-	-
Flavonols	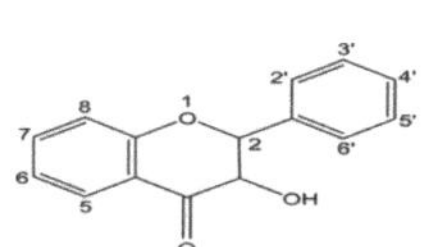	Quercetin	-	OH	OH	OH	OH	-
		Kaempferol	-	OH	OH	-	OH	-
		Galangin	-	OH	OH	-	-	-
Isoflavones	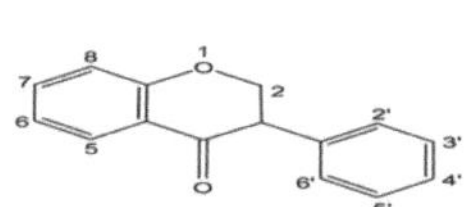	Genistein	-	OH	OH	-	OH	-
		Daidzein	-	OH	OH	-	-	-
		Glycitein	-	OH	OH	-	-	-
Flavanols	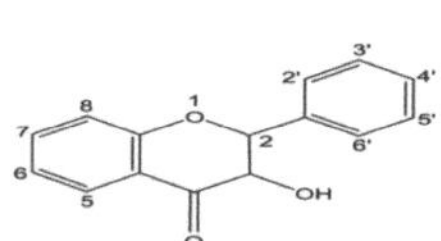	(+) – catechin	βOH	OH	OH	OH	OH	-
		(–) + epicatechin	αOH	OH	OH	OH	OH	-
		(–) – epigallocatechin	αOH	OH	OH	OH	OH	OH

Table 1. Flavonoid subgroup structures and their different substitution patterns [13, 23].

2.3. Flavanones

The basic structural model of flavanones has 2-phenylbenzopiran-4-one-skeleton. These compounds are of great interest due to the fact that they play an important role in the metabolic pathway of the other flavonoids. Their metabolic precursors are chalcones, flavones, dihydroflavonols; isoflavones are biosynthesized from the flavanones [18]. Their chemical structures have a hydroxyl group in 3 position of C ring [19]. Physically, some flavanones are colourless (Hesperidin), pale (naringenin, eriodictyol) and others yellow (neohesperidin) [15, 16].

2.4. Flavanols

Flavanols, well known as flavan-3-ols or catechin, have a hydroxyl group in C3 of C ring, with no carbonyl group. These flavonoids are colourless (catechins, gallocatechin, epicatechin, epigallocatechin, gallate) and some of them show yellow colouration (procyanidin, theaflavines).

2.5. Anthocyanidines

This is the other subgroup, which structurally has a hydroxyl group in 3 position and also a double bond between carbons 3 and 4 of C ring. These molecules are water soluble, and are the pigments responsible for the red to purple color in plants.

2.6. Isoflavones

This is a distinctive subclass of flavonoids. They structurally have a 3-phenylchromone skeleton with same three phenolic rings referred to as A, B and C rings. But also, in some of their derivative compounds, they can form an additional heterocyclic ring (D ring) (rotenoid, coumestane, pterocarpan) [20].

3. Flavonoids in foods

Flavonoids are synthesized in all parts of the plant. They provide colour, fragrance and taste to the fruits, flowers and seeds, which make them attractants for insects, birds and mammals, which aid in pollen or seed transmission [2]. Flavonoid distribution in plants is affected by different factors, including variation and exposure degree of light. The production of high oxidized flavonoids is accelerated by light [21]. Major classes of flavonoids based on abundance in food are flavonols, flavones, isoflavones, flavanones, flavandiols, anthocyanins, proanthocyanidins and cathechins. Other classes of flavonoids include flavan-3-ols, anthocyanidins, chalcones, and other biosynthetic intermediates of the flavonoid biosynthesis are aurones, bioflavonoids and dihydrochalcones [22].

The content of flavonoids in food varies, which depends on the source, vegetable, fruits or seeds, as well as processed food. The content of some important flavonoids in foods such as cabbage, spinach, carrots, peas, mushrooms, peaches, strawberries, orange juice, white wine or brewed coffee is low or <10 mg/kg or L. In some other vegetables and fruits, such as lettuce, beans, red pepper, tomato, grapes, cherries, red wine and tea, their content is < 50 mg/kg or L. On the other hand, their content in broccoli, kale, French beans, celery and cranberries is high, over 50 mg/kg or L. Apple contains flavonoids such as quercetin (184 mg) and epicatechin (180 mg) in approximate 120 g of flesh with 80 g of skin [24]. Citrus fruit such as oranges and grapefruits contain flavonoids such as naringin and hesperidin. Naringin is mainly present in grapefruits and sour oranges, while hesperidin is present in sweat oranges, mandarins and lemons. In addition, 500 ml of orange juice contains 292 mg of hesperidin [25, 26]. Berry fruits (strawberries, blueberries and cranberries) contain high levels of anthocyanins in their skin and flesh; strawberries contain 33.63 mg per 100 g, whereas blueberries contain 13.52 mg [27]. Onions contain up to 22 mg of quercetin per 100 g [28]. **Table 2** shows various classes of flavonoids and some of their food sources.

Flavonoid subgroup	Source	Specific flavonoid	Reference
	Erica andevalensis, Garcinia kola (seeds), Achyrocline satureioides, Silydum marianum, Scutellaria baicalensis		[32]
Anthocyanins	Fruits and flowers, blueberries, bananas, strawberries, red wine, cranberries, blackcurrants	Anthocyanidin, cyanidin, delphinidin, malvidin, pelargonidin, petunidin, peonidin	[21, 27, 33, 34]
Flavanols (flavan-3-ols)	Apples, Hops, green tea, Beer, kiwi, cocoa products	Catechins, gallactocatechins, epicatechin, epigallocatechin gallate	[24, 33–36]
	Wine, fruit juice	Procyanidin	
	Black tea	Theaflavins	
Flavanones	Citrus fruits, curcumin, oranges, grapefruits, peppermint, lemon	Hesperidin, Naringenin Neohesperidin. eriodictyol	[9, 25, 33, 37]
	Tomato		
Flavones	Herbs, cereals, fruits, parsley, thyme, bee pollen, passiflora plant	Apigenin, chrysin, luteolin	[33, 34, 38, 49]
	Vegetables, flowers	Diosmetin, luteolin	
Flavonols	Onions, cherries, apples, broccoli, kale, tomatoes, berries, black tea, red wine, tartary buckwheat, edible part of plants, leeks, bee pollen, citrus fruits	Isorhamnetin, kaempferol, quercetin, myricetin, rutin, pinocembrin	[24, 28, 38–40]
	Yellow wax and ancho peppers, canned kale		
Flavanonols	Lemon, aurantium	Taxifolin	[21]
Isoflavones	Legumes (soybeans), soy products, miso tofu	Daizein, genistein, glycitein, formononetin, biochanin A	[33, 37]

Table 2. Food sources of flavonoids.

4. Antioxidant activity

The best described property of almost all groups of flavonoids is their powerful antioxidants, which act as free radical scavengers as they are potential reducing agents and protect from oxidative reactions taking place inside the body [22]. The antioxidant activity of flavonoids is due a combination of their iron chelating and free radical scavenger properties, also the inhibition of oxidases enzymes such as lipoxygenase, cyclooxygenase, myeloperoxidase, NADPH oxidase and xanthine oxidase; avoiding thereby the formation of reactive oxygen species and organic hydroperoxides [19]. Moreover, flavonoids can inhibit enzymes by indirectly involving in the oxidative process, such as phospholipase A2, at the same flavonoids can stimulate other enzymes with antioxidant activities like catalyse and superoxide dismutase.

This feature depends of their molecular structure. This free radical scavenging activity is primarily attributed to the high reactivity of hydroxyl substituents. For example, by scavenging free radicals, the hydroxyl group on the B-ring donates hydrogen and an electron to hydroxyl, peroxyl and peroxynitrite free radicals, stabilizing them, because of the high reactivity of the hydroxyl group of the flavonoids, radicals are made inactive according to Eq. (1), where R* is a free radical and O* is the oxygen free radical) [17, 29–31].

$$\text{F-OH} + \text{R}\bullet \rightarrow \text{F-O}\bullet + \text{RH} \quad (1)$$

The significance of other hydroxyl configuration is less clear, but beyond increasing total number of hydroxyl groups, A-ring substitution correlates little with antioxidant activity. Similarly, the difference in antioxidant activity between polyhydroxylated and polymethoxylated flavonoids is most likely due to differences in both hydrophobicity and molecular planarity [30].

A lot of studies have realized the antioxidant activity of flavonoids. Majewska et al. [41] evaluated different flavonoids and compared their antioxidant activity with DPPH radical. The flavonoids such as quercetin, rhamnetin, isorhamnetin, luteolin and apigenin were tested and the results showed that quercetin has the highest antioxidant potential at lower concentration (0.1–5 μg/sample). In other studies, Choi et al. [42] tested the inhibitory activity of xanthine oxidase, linoleic acid peroxidation and scavenging capacity of DPPH with some flavonoids, and obtained the following results at a concentration of 100 μg/ml: for xanthine oxidase, catechin, morin, naringenin and quercetin showed 100% of inhibition, and rutin only 43%; for linoleic acid peroxidation, quercetin showed the highest inhibition (82%), followed by catechin (71%), rutin (63%), morin (56%), naringenin (53%); and finally for DPPH catechin, morin and quercetin obtained a high activity (100%) than rutin (95%), naringenin did not show any data.

On the other hand, using an aqueous emulsion of linoleic acid/β-carotene at 50°C, and compared with synthetic antioxidant BHT, D,L-α-tocopherol and its acetate, Burda and Oleszk [43] examined the antioxidant activity of some flavonoids and observed the highest values were obtained by synthetic ones (88–95%) and slightly lower, but still high antioxidant activity was shown by a group that included fisetin (61%), kaempferol (65%), galangin (64%), quercetin (63%), robinetin (61%), morin (63%) and kaempferide (60%), all of these are flavonols with a free hydroxyl group at the C-3 position.

5. Health benefits

Flavonoids are a widely distributed group of polyphenols and comprise the most studied bioactive compounds for about 50 years due their potential health benefits; they have a complex and unknown biologic function, which are based on their antioxidant activity [9, 39, 44, 45]. Furthermore, because of their remarkable activities and potential health benefits, they are used in the food and cosmetics industries. Their functions can be divided in two (i) general or nonspecific activity (related to the presence of phenolic compounds) and (ii) specific activity (depends on the particular chemical and structural characteristics of the active compound) [46], which are presented in **Table 3**.

The decrease in consumption of fruits and vegetables has been associated with several diseases. Flavonoids attract attention due to their ability to reduce the incidence of many diseases. Its recommended intake or supplementation of flavonoids with a combination of another antioxidants, vitamins and minerals can reduce incidence of cancer, cardiovascular mortality and ischemic vascular disease [39, 49–53]. According to many studies, in flavonoid-rich foods the daily intake of these compounds can be range between 50 and 800 mg/day, mainly with the consumption of fruits and vegetables [6, 54]. However, is also known that flavonoids inhibit proliferation of viruses like herpes simplex, polio and pseudorabies [47]. On the other hand, some results from other different studies have demonstrated that in very high doses, flavonoids can act as pro-oxidant; so, it is important to consider the recommended daily intake of these compounds to promote their functions on the organism [55].

For many years, plants have been used like traditional Mexican medicine due to their medicinal potential of crude extracts (mainly rich in flavonoids) to the treatment of diarrhoea and stomach diseases [55]. This potential can be attributed for many physiological functions, such as their antioxidant, antibacterial and antigiardial activities to reduce the incidence of this kind of diseases [54].

Nonspecific activity	Specific activity
Absorption of ultraviolet radiation	Affinity to estrogenic receptors
Reactive oxygen species neutralization	Anti-inflammatory activity
Inhibition of radical reactions	Impact on cardiovascular system
Chelating trace metals	Influence on regulatory systems and tissue signal transmission
Inhibition of enzymes	Microorganism growth inhibition
Activation of antioxidant enzymes	Interaction with enzymes
Interactions with membranes	Interaction with transcription factors
Antioxidant properties	Interaction with receptors

Table 3. Biological functions of flavonoids on humans [46–48].

In recent years, flavonoids have gained interest in health as protective agents against many diseases, which involve radical damage [56]. Likewise, many epidemiological studies have been developed to show the association between consumption of vegetables and fruits and the risk of disease such as cancer, neurological and cardiovascular diseases [57]. Some authors show beneficial results of flavonoids against diseases; studies about the dietary called French paradox revealed that flavonoid-rich diet is directly correlated with decrease incidence of cardiovascular diseases and increase longevity on Mediterranean population, which is associated with consumption of red wine and low-fat diets [58].

5.1. Anticancer effect of flavonoids

Cancer is a multifactorial heterogeneous disease, which is the main cause of death in some countries along with cardiovascular diseases, and it is expected to increase about 70% cases in the next 20 years. These have been related to different factors of behavioural and dietary risks, among which are the reduced consumption of vegetables, limited physical activity and consumption of alcohol and tobacco, which may cause negative impact on health [59, 60].

According to some studies, flavonoids have some pharmacological properties that inhibit cell damage [61]. Herrera et al. [51] conducted a study and evaluated the efficiency of supplementation with correct antioxidant content to reduce cancer incidence; these trials concluded that antioxidant supplementation may inhibit cancer incidence only in healthy subjects. But according with Hollman and Katan [62], bioactive compounds may inhibit various stages in cancer process.

Some studies in animals with some types of chemical carcinogen-induced or transplanted tumours have been shown good results about the prevention development and growth of various types of cancer. Some mechanisms of biologically active substances, including their antioxidant activity, may act on stages of carcinogenesis such as initiation, promotion and progression. Flavonoids can inactive the carcinogen, inhibit cell proliferation, repair DNA processes and reduce oxidative stress (in initiation and promotion stages); in the progression stage, they may exhibit antioxidant activity, induce apoptosis and develop cytotoxic action against cancer cells [4, 63]. Because inflammation is closely related to tumour promotion, bioactive compounds like flavonoids are expected to exert chemopreventive effects against carcinogenesis, especially in promotion and progression stages, and are estimated to reduce risk of cancer [64].

5.2. Flavonoids on the cardiovascular system

In the last decade, cardiovascular diseases increased the number of death around the world; this is due to the decrease in consumption of fruits and vegetables, and it is estimated that these diseases will continue affecting the population [65]. Antioxidant properties of flavonoids are interesting due their potential role in prevention of cardiovascular diseases. Other authors suggested that these compounds decrease the risk of coronary diseases, but nevertheless, their antioxidant and chelating properties are the major mechanisms to inhibit or inactivate reactive oxygen species, which play an important role to affect cardiovascular system [9, 63].

According to some authors, the mechanistic pathways of flavonoids on cardioprotection include some actions, such as improving coronary vasodilatation, decreasing the ability of platelets in the blood to clot, preventing LDLs from oxidizing, inhibition of inflammation propagation, antiapoptotic, antinecrotic, free radical scavenging abilities and its crucial role in the impact on capillary blood vessel [9, 47, 57].

Rice-Evans et al. [5] studied the correlation with the intake of flavonoids and risk of death from coronary heart disease on a group of Dutch population; these epidemiological study showed that intake natural flavonoids can reduce 2.4 times cardiovascular diseases compared people with lower intake of these bioactive compounds. They established that antioxidant and antithrombotic properties contribute to this protection and may provide a better lifestyle.

5.3. Flavonoids on the nervous system

In recent years, experimental and clinical studies have been reported that flavonoids not only have benefits against cancer and cardiovascular diseases. The interest in consumption of flavonoids has been addressed due their potential neuroprective effects to prevent different diseases related to nervous system such as Alzheimer, Parkinson and slow down cognitive decline, which can generate dementia [45]. Diets rich in these substances (at low concentration) were shown beneficial effects to maintain human cognitive functions to promote improvements in memory, protect vulnerable neurons, stimulate neuronal regeneration and prevent oxidative neuronal damages [66].

Reactive oxygen, nitrogen species, inflammatory responses and their mediators are involved in the development in the pathogenesis of various neurodegenerative diseases [61]. Recent works have revealed that flavonoid-rich food intake (for 8 weeks) provides positive effects on neurocognitive functions due to its antioxidant activity and is effective in reversing age-related deficits in brain. Previous studies in rats showed that the application of natural extracts rich in polyphenolic compounds represent an important advance to use of these extracts to the generation of nutraceuticals products to improve health in humans [44]. According some authors, the neurobiological actions of these compounds may occur in two major ways: the first is regulation of the neuronal signal cellular cascades, which is caused by neurotoxic substances and may damage neurogenesis, neuronal function and brain connectivity. Second, flavonoids seem to improve blood-flow toward brain and sensory systems and exert beneficial effects on the peripheral and central nervous [4, 63].

6. Applications in food

In recent years, many polyphenols have been used as natural antioxidants in oils and fats to prevent lipid oxidation, protect food and beverages from light exposure, to prolong shelf-life of food, supplement for animal feeds to improve their health, to protect animal products, like antimicrobial agent in foodstuffs and health functional ingredient in foods and dietary supplements; these applications can be attributed mainly to antioxidant and antimicrobial activities [47, 67, 68]. The use of synthetic additives in the food industry has been declining in recent years; this is mainly because man seeks to reduce risk in suffering from diseases which have been associated with synthetic products consumption. Flavonoids are phytochemicals that cannot be synthesized by humans; however, these compounds may be used as food additives to improve health-beneficial effects and increase their amount in humans [69, 70].

Flavonoids are present in significant amounts in many fruits and vegetables; natural antioxidants and flavonoids have been reported as two of the most important micronutrients, which can be used in industry to reduce the use of synthetic compounds on foods and improve health in humans due their potential to decrease several diseases. These bioactive compounds can be used to prolong shelf-life and preserve many foods due to their antimicrobial and antioxidant properties [44]. Moreover, in recent years interest in acquiring ever more natural products has grown; this has been caused by restriction over the use of synthetic antioxidants in food (such as BHA and BHT) and increase the use of natural compounds such as antioxidants or food components, which have relatively low toxicity compared to synthetic products [57, 65].

Flavonoids are suitable compounds that may be used as food preservative due to their beneficial effects that have been demonstrated before: prevention of fat and oils oxidation, supplement for animal feeds, protection of vitamins and enzymes, inhibition of microbial growth in foodstuffs and health functional ingredient in foods and dietary supplements [68]. The most studied applications of flavonoids in foods are in red meats and poultry to inhibit lipid oxidation, use of natural extracts in foods as functional ingredients and retard spoilage microorganism growth in meats. Some studies have indicated that these compounds have protective effect on meat quality and other products [67].

Recently, flavonoids have gained attention due their antioxidant properties, which can act against oxygen free radical and lipid peroxidation [71]. Many studies have mentioned that *in vitro* antioxidant activity of these compounds depends on the arrangement of functional groups on its structures; however, the mechanism of their antioxidant activity is substantially influenced by their configuration and total number of hydroxyl groups presents in the flavonoid molecule and the glycosylation of flavonoid molecule, but could be increased by polymerization of flavonoid monomers [58, 72, 73].

The protective effect of flavonoids in biological systems is attributed to their antioxidant activity due to their capability to donate a hydrogen atom [58]. However, it is important to consider the antioxidant effect *in vivo*, which can be different comparing with *in vitro* assays: (i) low concentration on systemic circulation compared with endogenous and exogenous antioxidant compounds, (ii) high level of biotransformation that flavonoids suffer during their absorption and distribution in the body, which can decrease their antioxidant activity and (iii) large doses of flavonoids may decrease bioavailability of trace elements, such vitamins, folic acid or other antioxidants [4, 74].

7. Other applications

Plant bioactive compounds (primarily flavonoids) have been used in dermatology and cosmetic preparations for a long time due to several associated properties such as antioxidant, antimicrobial, anti-inflammatory and therapeutic properties [8, 75]. Flavonoids are gaining popularity as cosmetic ingredients because they can protect skin from UV light and cure the skin from oxidative stress, so they can improve skin appearance [76]. The activity of flavonoids on the skin is associated with their antiradical properties due to the presence of phenolic groups, but these activities on the skin are poorly investigated. Many studies have been demonstrated that the scavenging of flavonoids depends mainly to a high degree on their structures and physicochemical properties [47].

In order to protect the skin against UV radiation, the use of herbal and biodegradable products has been increased; cosmetic products are being recognized for their benefits in the prevention of various pathologies and in the improvement of several dermatologic conditions, which represent a viable alternative for industrial application of natural extracts [70, 76]. Recent dermatological studies have been demonstrated that herbal extracts were added to cosmetic preparations due to antioxidant properties of flavonoids in addition to impart UV protection and inhibit metal chelating properties to protect skin [75].

The protective action of flavonoids on the skin is manifested through their anti-inflammatory activity. This activity is given by the action of bioactive compounds on the enzymes and other factors that promote inflammation stages. The other mechanism of action is the inhibition of enzymes linked to cellular activation that promotes skin deterioration and the secretion of regulatory substances for their propagation. The affinity for protein structures and with estrogen receptors is another property of flavonoids, as well as anti-irradiating activities; because of these features, they are used in the cosmetic industry to reduce congestion problems, with the aim of reducing inflammatory symptoms that damage humans [47].

Author details

Saul Ruiz-Cruz[1]*, Saraí Chaparro-Hernández[1], Karen L. Hernández-Ruiz[1], Luis A. Cira-Chávez[1], María I. Estrada-Alvarado[1], Laura E. Gassos Ortega[1], José de Jesús Ornelas-Paz[2] and Marco A. Lopez Mata[3]

*Address all correspondence to: sruiz@itson.edu.mx

1 Department of Biotechnology and Food Sciences, Institute Technological of Sonora, Sonora, México

2 Center for Research in Food and Development, Chihuahua, México

3 Department of Health Sciences, University of Sonora, Campus Cajeme, Sonora, México

References

[1] Stalikas C. D. Extraction, separation, and detection methods for phenolic acids and flavonoids. Journal of Separation Science. 2007;**30**(18):3268-3295. DOI: 10.1002/jssc.200700262

[2] Koes R. E. & Quattrocchio F. The flavonoid biosynthetic pathway in plants: Function and evolution. BioEssays. 1994;**16**(2):123-132. DOI: 10.1002/bies.950160209

[3] de Rijke E., Out P., Niessen W. M., Ariese F., Goojier C. & Udo A. T. Analytical separation and detection methods for flavonoids. Journal of Chromatography A. 2006;**1112**(1):31-63. DOI: http://dx.doi.org/10.1016/j.chroma.2006.01.019

[4] Kozlowska A. & Szostak-Wegierek D. Flavonoids-food sources and health benefits. Roczniki Panstwowego Zakladu Higieny. 2014;**65**(2). DOI: PMID: 25272572

[5] Rice-Evans C. A., Miller N. & Paganga G. Structure-antioxidant activity relationships of flavonoids and phenolic acids. Free Radical Biology and Medicine. 1996;**20**(7):933-956. DOI: http://dx.doi.org/10.1016/0891-5849(95)02227-9

[6] Seyoum A., Asres K. & El-Fiky F. K. Structure-radical scavenging activity relationships of flavonoids. Phytochemistry. 2006;**67**(18):2058-2070. DOI: 10.1016/j.phytochem.2006.07.002

[7] Chuarienthong P., Lourith N. & Leelapornpisid P. Clinical efficacy comparison of anti-wrinkle cosmetics containing herbal flavonoids. International Journal of Cosmetic Science. 2010;**32**(2):99-106. DOI: 10.1111/j.1468-2494.2010.00522.x

[8] Malinowska P. Effect of flavonoids content on antioxidant activity of commercial cosmetic plant extracts. Herba Polonica. 2013;**59**(3):63-75. DOI: https://doi.org/10.2478/hepo-2013-0017

[9] Benavente-García O., Castillo J., Marin F.R., Ortuño A. & Del Río J.A. Uses and properties of citrus flavonoids. Journal of Agricultural and Food Chemistry. 1997;**45**(12):4505-4515. DOI: 10.1021/jf970373s

[10] Clifford A. H. & Cuppett S. L. Review: Anthocyanins-nature, occurrence and dietary burden. Journal of Science of Food and Agriculture. 2000;**80**(7):1063-1072. DOI: 10.1002/(SICI)1097-0010(20000515)80:7<1063::AID-JSFA605>3.0.CO;2-Q

[11] Martínez-Flores S., González-Gallego J., Culebras J. M. & Tuñón M. J. Los flavonoides: propiedades y acciones antioxidantes. Nutrición Hospitalaria. 2002;17(6):271-278. DOI: ISSN: 0212-1611

[12] Pal D. & Verma P. Flavonoids: A powerful and abundant source of antioxidants. International Journal of Pharmacy and Pharmaceutical Sciences. 2013;**5**(3):95-98. DOI: ISSN-0975-1491

[13] Saxena M., Saxena J. & Pradhan A. Flavonoids and phenolics acids as antioxidants in plants and human health. International Journal of Pharmaceutical Sciences Review Research. 2012;**16**(2):130-134. DOI: ISSN 097-044X

[14] Bravo L. Polyphenols: Chemistry, dietary sources, metabolism, and nutritional significance. Nutrition Reviews. 1998;**56**(11):317-333. DOI: https://doi.org/10.1111/j.1753-4887.1998.tb01670.

[15] Cook N. C. & Samman S. Review: Flavonoids-chemistry, metabolism, cardioprotective effects and dietary sources. The Journal of Nutritional Biochemistry. 1996;**7**(2):66-76. DOI: http://dx.doi.org/10.1016/S0955-2863(95)00168-9

[16] Aherne A.S. & Obrien N.M. Dietary flavonoids: Chemistry, food content and metabolism. Nutrition. 2002;**18**.

[17] Nijveldt R., Van Nood E. L. S., Van Hoorn D. E., Boelens P. G., Van Norren K. & Van Leeuwen P. A. Flavonoids: a review of probable mechanisms of action and potential applications. The American Journal of Clinical Nutrition. 2001;**74**(4):418-425. DOI: PMID: 11566638

[18] Marin F. R., Frutos M. J., Pérez-Álvarez J. A., Martínez- Sánchez F. & Del Río J. A. Flavonoids as nutraceuticals: Structural related antioxidant properties and their role on ascorbic acid preservation. Studies in Natural Products Chemistry. 2002;**26**:741-778. DOI: http://dx.doi.org/10.1016/S1572-5995(02)80018-7

[19] Escamilla C. O., Cuevas E. Y. & Guevara J. Flavonoides y sus acciones antioxidantes. Revista de la Facultad de Medicina. 2009;**52**(002):73-75.

[20] Marais J. P., Deavours B., Dixon R. A. & Ferreira D. The Dtereochemistry of Flavonoids. Springer Science. 2006;**3**. DOI: ISBN-10:0-387-28821-X

[21] Yao L. H., Jiang Y. M., Shi J., Tomás-Barberán F. A., Datta N., Singanusong R. & Chen S. S. Flavonoids in Food and their health benefits. Plant Foods for Human Nutrition. 2014;**59**:113-122. DOI: PMID: 15678717

[22] Routray W. & Orsat V. Microwave-assisted extraction of flavonoids: A review. Food and Bioprocess Technology. 2012;**5**(2):409-424. DOI: 10.1007/s11947-011-0573-z

[23] Kumar S. & Pandey A. Chemistry and Biological activities of flavonoids: An Overview. The Scientific World Journal. 2013;**2013**:1-16. DOI: 10.1155/2013/162750

[24] Bondonno C.P., Yang X. & Croft K. D. Flavonoid-rich apples and nitrate rich spinach augmnt nitric oxide status and improve endothelial function in healthy men and women a randomized controlled trial. Free Radical Biology and Medicine. 2012;**52**(1):98-102. DOI: http://dx.doi.org/10.1016/j.freeradbiomed.2011.09.028

[25] Tripoli E., La Guardia M., Giammanco S., Di Majo D. & Giammanco M. Citrus flavonoids: molecular structure, biological activity and nutritional properties: A review. Food Chemistry. 2007;**104**(2):466-479. DOI: http://dx.doi.org/10.1016/0891-5849(95)02047-0

[26] Morand C., Dubray C. & Milenkovic D. Hesperidin contributes to the vascular protective effects of orange juice: A randomized crossover study in healthy volunteers. The American Journal of Clinical Nutrition. 2011;**93**(1):73-80. DOI: 10.3945/ajcn.110.004945

[27] BasuA.,RhoneM.&LyonsT.J.Berries:Emergingimpactoncardiovascularhealth.Nutrition Reviews. 2010;**68**(3):168-177. DOI: http://doi.org/10.1111/j.1753-4887.2010.00273.x

[28] Larson A. J., Symons J. D. & Jalili T. Therapeutic potential of quercetin to decrease blood pressure: Review of efficacy and mechanisms. Advances in Nutrition: An International Review Journal. 2012;**3**(1):39-46. DOI: 10.3945/an.111.001271

[29] Cao G., Sofic E. & Prior R. L. Antioxidant and prooxidant behavior of flavonoids: structure-activity relationship. Free Radical Biology and Medicine. 1997;**22**(5):749-760. DOI: http://dx.doi.org/10.1016/S0891-5849(96)00351-6

[30] Heim K. E., Tagliaferro A. R. & Bobilya D. J. Flavonoid antioxidants: chemistry, metabolism and structure-activity relationships. Journal of Nutritional Biochemistry. 2002;**13**(10):572-584. DOI: http:dx.doi.org/10.1016/S0955-2863(02)00208-5

[31] Beecher G. R. Overview of dietary flavonoids: nomenclature, occurrence and intake. The Journal of Nutrition. 2003;**133**(10):2448S–3254S. DOI: PMID:14519822

[32] Zayachkivska O.S., Konturek S.J., Drozdowicz D., Konturek P.C., Brzozowski T. & Ghegotsky M.R. Gastroprotective effects of flavonoids in plant extracts. Journal of Physiology Pharmacology. 2005;**56**(1):219-231. DOI: PMID: 15800396

[33] Huang M. T., Osawa T., Ho C. T. & Rosen R. T. Food phytochemicals for cancer prevention. I: Fruits and vegetables. American Chemical Society. 1993;DOI: 10.1021/bk-1994-0546.fw001

[34] Ho C. T., Osawa T., Huang M. T. & Rosen R. T. Food phytochemical for cancer prevention. II: Teas, spices, and herbs. American Chemical Society. 1994.

[35] Harnly J. M., Dohert. F., Beecher G. R., Holden J. M., Hayowitz D. B., Bhagwat S. & Gebhardt S. Flavonoid content of U.S. fruits, vegetables, and nuts. Journal of Agricultural and Food Chemistry. 2006;**54**(26):9966-9977. DOI: 10.1021/jf061478a

[36] Toh J. Y., Tan V. M. H., Lim P. C. Y., Lim S. T. & Chong M. F. F. Flavonoids from Fruit and Vegetables: A focus on cardiovascular risk factors. Current Atherosclerosis Reports. 2013;**15**(12):1-7. DOI: 10.1007/s11883-013-0368-y

[37] Ho C.T., Lee C.Y. & Huang M.T. Phenolic compounds in food and their effects on health. I. Analysis, ocurrence & chemistry. Austin TX: American Chemical Society. 1992;**506**. DOI: 10.1021/bk-1992-0506

[38] Saric A., Balog T., Sobocanec S., Kusic B., Sverko V., Rusak G., Likic S., Bubalo D., Pinto B., Reali D. & Marotti T. Antioxidant effects of flavonoid from Croatin *Cystus incanus* L. rich bee pollen. Food and Chemical Toxicology. 2009;**47**(3):547-557. DOI: 10.1016/j.fct.2008.12.007

[39] Basu A., Das A. S., Majumder M. & Mukhopadhyay R. Anti-atherogenic roles of dietary flavonoids chrysin, quercetin and luteolin. Journal of Cardiovascular Pharmacology. 2016;**68**(1):89-96. DOI: 10.1097/FJC.0000000000000380

[40] Hertog M. G., Hollman P. C. & Katan M. B. Content of potentially anticarcinogenic flavonoids of 28 vegetables and 9 fruits commonly consumed in the Netherlands. Journal of Agricultural and Food Chemistry. 1992;**40**(12):2379-2383. DOI: INIST: 7332,35400003285856.0100

[41] Majewska M., Skrzycki M., Podsaid M. & Czeczot H. Evaluation of antioxidant potential of flavonoids: An in vitro study. Acta Poloniae Pharmaceutical Drug Research. 2011;**68**(4):611-615. DOI: PMID: 21796946

[42] Choi C. W., Kim S. C., Hwang S.S, Choi B. K., Ahn H. J., Lee M. Y., Park S. H. & Kim S. K. Antioxidant, activity and free radical scavenging capacity between Korean medicinal plants and flavonoids by assay guided comparison. Plant Science. 2002;**163**(6):1161-1168. DOI: http://dx.doi.org/10.1016/S0168-9452(02)00332-1

[43] Burda S. & Oleszke W. Antioxidant and antiradical activities of flavonoids. Journal of Agricultural and Food Chemistry. 2001;**48**(6):2774-2779. DOI: 10.1021/jf001413m

[44] Pandey K. B. & Rizvi S. I. Plant polyphenols as dietary antioxidants in human health and disease. Oxidative Medicine and Cellular Longevity. 2009;**2**(5):270-278. DOI: http://dx.doi.org/10.4161/oxim.2.5.9498

[45] Grassi D., Ferri C. & Desideri G. Brain protection and cognitive function: cocoa flavonoids as nutraceuticals. Current Pharmaceutical Design. 2016;**22**(2):145-151. DOI: PMID: 26561075

[46] Fraga C. G., Galleano M., Verstraeten S. V. & Oteiza P.I. Basic biochemical mechanisms behind the health benefits of polyphenols. Molecular Aspects of Medicine. 2010;**31**(6):435-445. DOI: http://dx.doi.org/10.1016/j.mam.2010.09.006

[47] Arct J. & Pytkowska K. Flavonoids as components of biologically active cosmeceuticals. Clinics in Dermatology. 2008;**26**(4):347-357. DOI: http://dx.doi.org/10.1016/j.clindermatol.2008.01.004

[48] Ravishankar D., Rajora A. K., Greco F. & Osborn H. M. Flavonoids as prospective compounds for anti-cancer therapy. The International Journal of Biochemistry & Cell Biology. 2013;**45**(12):2821-2831. DOI: http://dx.doi.org/10.1016/j.biocel.2013.10.004

[49] Brower V. Nutraceuticals: poised for a healthy slice of the healthcare market? Nature Biotechnology. 1998;**16**:728-732. DOI: 10.1038/nbt0898-728

[50] Fraser G.E. Vegetarian diets: what do we know of their effects on common chronic diseases? The American Journal of Clinical Nutrition. 2009;**89**(5):1607S–1612S. DOI: 10.3945/ajcn.2009.26736K

[51] Herrera E., Jiménez R., Aruoma O. I., Hercberg S., Sánchez-García I. & Fraga C. Aspects of antioxidants foods and supplements in health and disease. Nutrition Reviews. 2009;**67**(1):S140–S144. DOI: https://doi.org/10.1111/j.1753-4887.2009.00177.x

[52] Manzano C., Estupiñán D. & Poveda E. Efectos clínicos de los probióticos: qué dice la evidencia. Revista Chilena de Nutrición. 2012;**39**(1):98-110. DOI: http://dx.doi.org/10.4067/S0171-75182012000100010

[53] Fuchs-Tarlovsky V., Rivera M. A. C., Altamirno K. A., López-Alvarenga J.C. & Ceballos-Reyes G.M. Antioxidant supplementation has a positive effect on oxidative stress and hematological toxicity during oncology treatment in cervical cancer patients. Supportive Care in Cancer. 2013;**21**(5):1359-1363. DOI: 10.1007/s00520-012-1674-6

[54] Spencer J. P., Vauzour D. & Rendeiro C. Flavonoids and cognition: the molecular mechanisms underlying their behavioural effects. Archives of Biochemistry and Biophysics. 2009;**492**(1):1-9. DOI: 10.1016/j.abb.2009.10.003

[55] Barbosa E., Calzada F. & Campos R. In vivo antigiardial activity of three flavonoids isolated of some medicinal plants used in Mexican traditional medicine for the treatment of diarrhea. Journal of Ethnopharmacology. 2007;**109**(3):552-554. DOI: http://dx.doi.org/10.1016/j.jep.2006.09.009

[56] Van Acker S. A., Tromp M. N., Griffioen D. H., Van Bennekom W. P., Van Der Vijgh W. J. & Bast A. Structural aspects of antioxidant activity of flavonoids. Free Radical Biology and Medicine. 1996;**20**(3):331-342. DOI: http://dx.doi.org/10.1016/0891-5849(95)02047-0

[57] Soobrattee M. A., Neergheen V. S., Luximon-Ramma A., Aruoma O. I. & Bahorun T. Phenolics as potential antioxidant therapeutic agents: mechanism and actions. Mutation Research/Fundamental and Molecular Mechanisms of Mutagenesis. 2005;**579**(1):200-213. DOI: http://dx.doi.org/10.1016/j.mrfmmm.2005.03.023

[58] Procházková D., Bousová I. & Wilhelmová N. Antioxidant and prooxidant properties of flavonoids. Fitoterapia. 2011;**82**(4):513-523. DOI: http://dx.doi.org/10.1016/f.fitote.2011.01.018

[59] Lutz M. & Zureta A. Relación entre la alimentación y la salud del consumidor en aspectos nutricionales y saludables de los productos de panificación. Universidad de Velparaíso. 2009;**17**.

[60] OMS. Cáncer. Organización Mundial de la Salud. 2015;DOI: http:/www.who.int/mediacentre/factsheets/fs297/es

[61] Benavente-García O. & Castillo J. Update on uses and properties of citrus flavonoids: new findings in anticancer, cardiovascular, and anti-inflammatory activity. Journal of Agricultural and Food Chemistry. 2008;**56**(15):6185-6205. DOI: 10.1021/jf8006568

[62] Hollman P. H. & Katan M. B. Dietary flavonoids: intake, health effects and bioavailability. Food and Chemical Toxicology. 1999;**37**(9):937-942. DOI: http://dx.doi.org/10.1016/S0278-6915(99)00079-4

[63] Sokolov A. N., Pavlova M. A., Klosterhalfen S. & Enck P. Chocolate and the brain: neurobiological impact of cocoa flavnols on cognition and behavior. Neuroscience & Biobehavioral Reviews. 2013;**37**(10):2445-2453. DOI: 10.1016/j.neubiorev.2013.06.013

[64] Kang N. J., Shin S. H., Lee H. J. & Lee K. W. Polyphenols as small molecular inhibitors of signaling cascades in carcinogenesis. Pharmacology & Therapeutics. 2011;**130**(3):310-324. DOI: http://dx.doi.org/10.1016/j.pharmthera.2011.02.004

[65] OMS. Enfermedades cardiovasculares. Organización Mundial de la Salud. 2015;DOI: http://www.who.int/mediacentre/factsheets/fs317/es

[66] Huntley A. L. The health benefits of berry flavonoids for menopausal women: cardiovascular disease, cancer and cognition. Maturitas. 2009;**63**(4):297-301. DOI: http://dx.doi.org/10.1016/j.maturitas.2009.05.005

[67] Yilmaz Y. Novel uses of catechins in foods. Trends in Food Science & Technology. 2006;**17**(2):64-71. DOI: http://dx.doi.org/10.1016/j.tifs.2005.10.005

[68] Huvaere K. & Skibsted L. H. Flavonoids protecting food and beverages against light. Journal of the Science of Food and Agriculture. 2015;**95**(1):20-35. DOI: 10.1102/jsfa.6796

[69] Peterson J. & Dwyer J. Flavonoids: Dietary occurrence and biochemical activity. Nutrition Research. 1998;**18**(12):1995-2018. DOI: http://dx.doi.org/10.1016/S0271-5317(98)00169-9

[70] Rodríguez E. E., Reyes M. A., Rivera G., Espinosa L. G., Palos I. & Bocanegra V. Alimentos funcionales y compuestos bioactivos. Editores P y V México. 2015;**63**.

[71] Holiman P. C., Hertog M. G. & Katan M. B. Analysis and health effects of flavonoids. Food Chemistry. 1996;**57**(1):43-46. DOI: http://dx.doi.org/10.1016/0308-8146(96)00065-9

[72] Cai Y.Z., Sun M., Xing J., Luo Q. & Corke H. Structure-radical scavenging activity relationships of phenolic compounds from traditional Chinese medicinal plants. Life Sciences. 2006;**78**(25):2872-2888. DOI: http://dx.doi.org/10.1016/j.lfs.2005.11.004

[73] Galleano M., Verstraeten S. V., Oteiza P. I. & Fraga C. G. Antioxidants actions of flavonoids: thermodynamic and kinetic analysis. Archives of Biochemistry and Biophysics. 2010;**501**(1):23-30. DOI: http://dx.doi.org/10.1016/j.abb.2010.04.005

[74] Sies H. Polyphenols and health: update and perspectives. Archives of Biochemistry and Biophysics. 2010;**201**(1):2-5. DOI: 10.1016/j.abb.2010.04.006

[75] Kole P.L., Jadhay H. R., Thakurdesai P. & Nagappa A. N. Cosmetic potential of herbal extracts. Natural Product Radiance. 2005;**4**:315-321. DOI: http://hdl.handle.net/123456789/8117

[76] Soto M. L., Falqué E. & Domínguez H. Relevance of natural phenolics from grape and derivative products in the formulation of cosmetics. Cosmetics. 2015;**2**(3):259-276. DOI: 10.3390/cosmetics2030259

Chapter 17

An Overview of Global Flavonoid Intake and its Food Sources

Romina Escobar-Cévoli, Carlota Castro-Espín, Virginie Béraud, Genevieve Buckland and Raul Zamora-Ros

Additional information is available at the end of the chapter

http://dx.doi.org/10.5772/67655

Abstract

Dietary patterns and food availability differ greatly between regions and countries around the world. As a result, there is a large variability in the intake of total flavonoids and flavonoid subclasses, and subsequently in their major food sources. However, we need to be aware of certain methodological issues when we compare studies on flavonoid intake.

In order to evaluate the intake of flavonoids, the different potential dietary assessment methodologies (dietary questionnaires and biomarkers) will be presented. Advantages and limitations of using of the two main food composition databases on flavonoids (US Department of Agriculture and Phenol-Explorer databases) will be discussed. The intake of total flavonoid and flavonoid subclasses in the various studies around the world will be comprehensively reviewed. The major food sources of flavonoids by region/country will be described. The main determinants of the intake of flavonoids will be explained as well.

Calculating the intake of flavonoids is the first step before estimating their potential protective effects against chronic diseases and is an essential step for developing future dietary guidelines on flavonoids.

Keywords: flavonoids, intake, food sources, determinants, worldwide

1. Introduction

Flavonoids are ubiquitously distributed plant secondary metabolites, particularly in fruits, vegetables, legumes, nuts, chocolate, and derived beverages (e.g., tea, wine, and juices) [1]. They are synthesized through the phenylpropanoid pathway, converting phenylalanine into 4-coumaroyl-CoA, which then enter the flavonoid biosynthesis pathway. Finally, various enzymes modify the basic flavonoid skeleton, leading to the different flavonoid subclasses [2]. In plants, flavonoids fulfill many different functions, such as protecting against ultraviolet radiation and phytopathogens, and acting as pigments, chemical messengers, physiological regulators, and cell cycle inhibitors [2].

Over the last three decades, flavonoids have received a lot of attention in cellular and animal models due to their well-established biological properties such as antioxidant, anti-inflammatory, and anti-carcinogenic effects, especially inducing enzymes and modulating metabolic and cell signaling pathways [3]. However, the epidemiologic evidence on the reduction in the chronic disease risk is still limited and usually inconsistent [4, 5]. The strongest evidence of their health-protective effects is for cardiovascular diseases [6] and type 2 diabetes [7]. One of the potential explanations of these inconsistent results is the difficulty in accurately assessing dietary flavonoid intake and the large variability of flavonoid intake among populations/countries.

This chapter is focused on the large differences in dietary flavonoid intakes and food sources worldwide, but prior to this, it is also important to briefly summarize the complexity of flavonoid chemistry and classification, and the different possible methodologies to assess dietary flavonoid intake.

2. Chemistry and classification of flavonoids

Flavonoids are a group of natural compounds with variable phenolic structures. In 1930, a new substance was isolated from oranges. At that time, it was believed to be a member of a new class of vitamins and was named as vitamin P. Posteriorly, it appeared that this compound was the flavonoid rutin and since then more than 9000 varieties of flavonoids have been identified [8].

Chemically, flavonoids are formed by a C6–C3–C6 structure, which consists of two benzene rings (A and B) linked by a three-carbon chain that form an oxygenated heterocycle (C ring) (**Figure 1**). They can be divided into nine classes according to their chemical structures: flavanones, flavones, dihydroflavonols, flavonols, flavan-3-ols, or flavanols [including monomers, proanthocyanidins, and flavanol-derived compounds (theaflavins and thearubigins)], anthocyanins, isoflavones, chalcones, and dihydrochalcones (**Figure 2**) [4]. The various classes of flavonoids differ in the level of oxidation and pattern of substitution of the C ring, while individual compounds within a class differ in the pattern of substitution of the A and B rings. Flavonoids in nature occur mostly as glycosides, aglycones (especially flavanols), and, in few cases, as methylated derivatives [3]. The basic flavonoid structure is the aglycone. When glycosides are formed, the glycosidic linkage is normally located in positions 3 or 7 and the

http://dx.doi.org/10.5772/67655

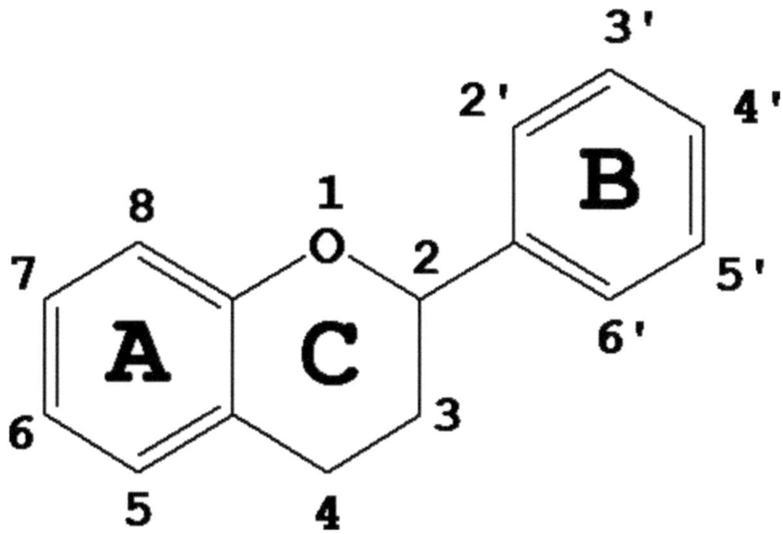

Figure 1. Basic flavonoid structure.

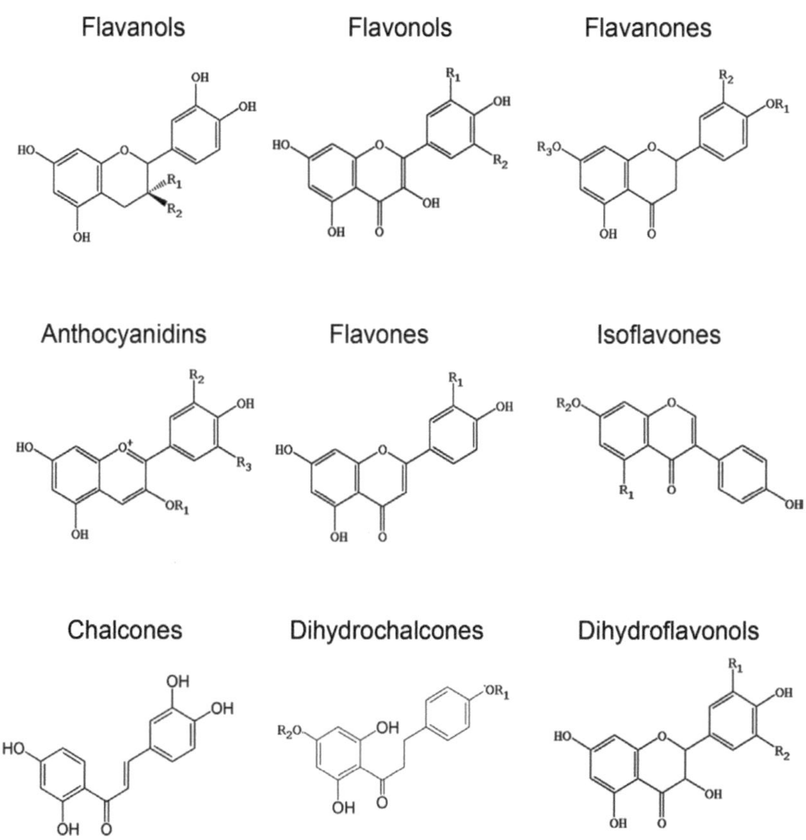

Figure 2. Flavonoid classes and their chemical structures.

sugar can be L-rhamnose, D-glucose, glucorhamnose, galactose, or arabinose [9]. This is very relevant because bioavailability differs among flavonoids, depending on the aglycone, the type of monosaccharide attached, and its position [10].

3. Dietary assessment of flavonoids

The most common method of estimating flavonoid intake in epidemiologic studies is to use dietary questionnaires, such as food frequency questionnaires (FFQ), 24-h dietary recalls, and food diaries, to record all food consumption over a known period of time dietary surveys, especially FFQ, are prone to several limitations, particularly regarding the identification of specific foods (e.g., the large variety of products available on the market, different food processing and cooking techniques, distinct ingredients, and preparation of recipes) and accurate quantification of portion sizes [11]. Despite this, FFQs are the most common method used in large epidemiological studies, mainly because they are an easy, quick, and an economical way to record the participants' habitual diet. In order to improve both quantification and quality of the data gathered, the use of a previously validated FFQ on flavonoids or flavonoid-rich foods is essential. Moreover, the use of innovative technologies and methodologies for the dietary assessment of flavonoid intake such as the collection of multiple 24-HDRs and food records and interactive computer- and camera-based technologies will certainly improve this process [12].

In order to estimate the intake of flavonoids, a food composition table or database is also needed. There are currently two main databases. The first one is the US Department of Agriculture (USDA) database on flavonoids, isoflavones, and proanthocyanidins, which was created in 2003, 1999, and 2004, respectively [13–15]. Since then, several updates have been released. The USDA databases contain worldwide food composition data on the six main flavonoid classes (flavanols, flavonols, flavanones, flavones, isoflavones, and anthocyanidins) expressed as aglycones. Phenol-Explorer is the second most common database [16]. It was developed in 2009 and contains worldwide data, but in this case, on all nine flavonoid classes (including chalcones, dihydrochalcones, and dihydroflavonols) expressed as they are found in nature (mainly as glycosides) and analyzed with chromatography without hydrolysis. Phenol-Explorer also contains data on chromatography after hydrolysis and, in this case, flavonoids are expressed as aglycones, but these are not usually used. In the studies using Phenol-Explorer using flavonoid data expressed as aglycone equivalents, flavonoid glycoside contents are converted into aglycone contents using their respective molecular weights [17, 18]. The other important difference between databases is that Phenol-Explorer does not contain data on thearubigins because the composition data quality on thearubigins is very low [19]. The non-specific spectrophotometric method used to date only provides a crude estimation of their concentrations in black teas, which are the only known food sources of thearubigins.

These differences between databases (classes of flavonoids, aglycones vs. glycosides, and inclusion of thearubigins) complicate the comparison among studies using different databases. In this book chapter, we have mostly compared studies using the USDA databases or Phenol-Explorer with data expressed as aglycones. In our comparisons, we have also made distinctions between

http://dx.doi.org/10.5772/67655

the studies including proanthocyanidins and thearubigins, which are the most controversial flavonoid compounds.

4. Geographical differences in total flavonoid intake and their food sources

The intake of flavonoids varies greatly by country/geographical region due to different dietary patterns; although differences in the flavonoid classes included, the dietary assessment methods, the food composition database use, and the method of expression (glycoses vs aglycones) applied complicate these comparisons. **Table 1** summarizes the most important descriptive studies assessing flavonoid intakes in different countries/regions. The mean intake of flavonoids worldwide ranges between 150 and 600 mg/day expressed as aglycones without thearubigins [20–26]. Taking into account thearubigins, the flavonoid intake could increase by a negligible amount in populations with low black tea intake, but would double the intake in black tea-consuming countries, such as the UK (from ~500 to >1000 mg/day) [27].

Europe is the continent with the most number of studies assessing the intake of total flavonoids, and therefore, it is possible to describe their intake quite accurately. Moreover, studies like the European Prospective Investigation into Cancer and Nutrition (EPIC) study allow us to compare the intake in 10 European countries using the same methodology [21, 26]. In Europe, an increasing south-to-north gradient is usually observed [26]. Despite the high intake of fruits, vegetables, and red wine in Mediterranean (MED) countries, the intake of total flavonoids in these countries (250–400 mg/day) is lower than in non-MED countries (350–600 mg/day), even without taking into account thearubigins. This is due to the much higher consumption of tea in non-MED compared to MED countries. These differences would be larger if thearubigins were also included [27]. The highest total flavonoid intake in Europe is in the UK, due to the traditional tea culture [21, 26]. Eastern European countries, such as Poland, also have a high intake of total flavonoids (~600 mg/day), related to their high tea consumption [32, 40]. Southern regions of France are considered as a MED region, while northern regions as a non-MED region; therefore, France has an intermediate intake of total flavonoids, with the main food sources being fruit, tea, and red wine (a combination of food sources from MED and non-MED countries) [33]. In Scandinavian countries with a low consumption of tea, such as Finland, the intake of total flavonoids (200–250 mg/day) is lower than MED countries, since the consumption of fruit is lower as well [31].

In Australia, there are various studies describing the intake of total flavonoids [20, 28, 29]. As a high tea-consuming country, Australia has a high intake of total flavonoids (650–700 mg/day), including thearubigins. In the three Australian populations documented, black tea contributes to at least 75% of total polyphenols.

In America, several studies have estimated the intake of total flavonoids, especially in the US [22, 34–36, 41], since the USDA databases were developed there [13–15]. In the US, the mean intake of total flavonoids varies from 250 to 400 mg/day, including proanthocyanidins and thearubigins [22, 34–36, 41]. Although the consumption of tea is not very high, tea is still the main food source of total flavonoids in the US, probably due to a low consumption of fruits

Study	Country	Population	Dietary assessment	FCDB	Aglycones vs. glycosides	Total flavonoid intake (mg/day)	Major class contributor	Main food sources	References
NNS95	Australia	17,326	24-HDR	USDA	Aglycones	225§	Flavanols**	Black tea	[20]
Calcium Intake Fracture Outcome Age-Related Extension Study	Australia	1136	SemiQ-FFQ	USDA	Aglycones	696	Flavanols**	Black tea	[28]
The Blue Mountains Eye	Australia	79	3 4-WFR	USDA	Aglycones	683	Flavanols**	Black tea	[29]
EPIC	MED	36,037	24-HDR	USDA, PE	Aglycones	370*	PA	Fruit	[21]
				PE	Glycosides	449*	PA	Fruit	[26]
	Non-MED			USDA, PE	Aglycones	374*	PA	Fruit	[21]
				PE	Glycosides	522*	PA	Tea	[26]
PREDIMED	Spain	7200	FFQ	PE	Glycosides	443	PA	Fruit	[27]
TOSCA.IT	Italy	2573	FFQ	USDA	Aglycones	364	PA	Fruit	[30]
SUVIMAX	France	4942	6 24-HDR	PE	Aglycones	436*	PA	Fruit, tea, red wine	[17]
					Glycosides	505*			
FINDIET 2002	Finland	2007	24-HDR	Own database	Aglycones	209*	PA	Berries, fruit	[31]
HAPIEE	Poland	10,477	FFQ	PE	Aglycones	898*	Flavanols**	Tea, cocoa	[32]
Polish National Multicenter Health Survey	Poland	6661	24-HDR	USDA, PE	Aglycones	524.6	Flavanols**	Tea, apples	[33]
					Glycosides	403*			
NIH-AARP	US	491,840	FFQ	USDA	Aglycones	203*§	Flavanols**	Tea	[34]
NHS I & NHS II & HPFS	US	156,957	FFQ	USDA	Aglycones	391	PA	Tea	[35]
NHANES	US	8809	24HDR	USDA	Aglycones	264	Flavanols**	Tea	[20, 36]
Mexican Teachers' Cohort	Mexico	106,466	FFQ	PE	Aglycones	140*	PA	Fruit and orange juice	[23]
					Glycosides	235*			
Health Survey-São Paulo	Brazil	1103	24-HDR	PE	Glycosides	54.6*	Flavanones	Citrus fruit, beans	[37]
Cross-sectional study	China	3317	SemiQ-FFQ	USDA	Aglycones	225*	Flavanols**	Soy, pome fruit	[24]

http://dx.doi.org/10.5772/67655

Study	Country	Population	Dietary assessment	FCDB	Aglycones vs. glycosides	Total flavonoid intake (mg/day)	Major class contributor	Main food sources	References
Case-control	China	66	FFQ	USDA	Aglycones	65§	Flavonols	–	[38]
KNHANES	Korea	33,581	24-HDR	USDA, PE, own database	Aglycones	318	PA	Fruit, tofu, onions	[25]
Tehran Lipid and Glucose Study	Iran	2618	FFQ	USDA	Aglycones	1652	Flavanols**	Vegetables, fruit	[39]

PE=phenol explorer; WFR=weighed food record; 24HDR=24-h dietary recall; MED=Mediterranean countries; Non-MED=non-Mediterranean countries; PA=proanthocyanidins; SemiQ-FFQ=semi-quantitative FFQ; FFQ=food frequency questionnaire; FCDB=food composition database.
*Without thearubigins.
§Without proanthocyanidins.
**Flavanols is a combination of flavan-3-ol monomers and flavanol-derived compounds.

Table 1. Intake of total flavonoids (mg/day) in different countries/regions.

and vegetables. In the US, some differences in the total flavonoid intake among ethnicities have been observed [22, 36]. Non-Hispanic whites have the highest intake of flavonoids (>300 mg/day), whereas non-Hispanic blacks, Mexican Americans, and other ethnicities consume approximately 200 mg/day. Data on other countries in America is very limited, and to the best of our knowledge, Brazil [37] and Mexico [23] are the only other countries with available data. The intake of total flavonoids is around 150 and 50 mg/day in Mexico and Brazil, respectively. These are the countries with the lowest intake of total flavonoids that have been published worldwide. Although, the studies in these countries did not take thearubigins into account, the contribution of thearubigins to total flavonoids is insignificant because Brazilians and Mexicans rarely drink tea. The main food source of total flavonoids is citrus juices, followed by fruit in Mexico and beans in Brazil [23, 37]. In these Latin-American countries, a potential relevant underestimation of total flavonoids cannot be ruled out because of missing food composition data on some tropical foods, such as fruits and vegetables (e.g., mamey, zapote, papaya, sweet potato, nopal, guava, jicama, and prickly pears) that are frequently consumed in their diets and may be good sources of flavonoids.

In Asian countries, very little complete data are available, as there is only data for East Asian countries (such as China [24, 38] and South Korea [25]). The total flavonoid intake in China ranges from 65 mg/day without proanthocyanidins [38] to 225 mg/day with proanthocyanidins [24]. In both studies, thearubigins are not included, but Chinese people drink green tea, but not black tea. Thearubigins are formed during the fermentation of green tea to black tea. In South Korea, the intake of total flavonoids is slightly higher than in China (320 mg/day) [25]. In East Asian countries, soy and its derived products (the main food sources of isoflavones) are one of the most important contributors to total flavonoids, although proanthocyanidins and flavan-3-ol monomers are the most abundant flavonoids in South Korea and China, respectively. In Japan, China, and South Korea, there are many studies focusing on isoflavones, due to their potential phytoestrogenic effects [42], but not on total flavonoids. In the Middle East, a recent study in Iran has been conducted showing a mean intake of 1650 mg/day [39]. This is the highest mean of total flavonoid intake worldwide. This high amount is because of the elevated consumption of black tea in these Middle East populations.

In summary, there is a high heterogeneity between countries in total flavonoid intake. The highest intake is in Iran [39], followed by the UK (>1000 mg/day) [27] and the lowest intake is in Brazil and Mexico (<150 mg/day) [23, 37]. The populations with a higher intake of total flavonoids are those with a high consumption of tea, especially black tea. In this case, the main contributor to total flavonoids is thearubigins, and this may be partially explained due to the inaccurate measurement of thearubigin content in black tea [19]. Populations with a lower intake of total flavonoids are those with a low consumption of tea. In these cases, the main food sources are fruit and proanthocyanidins become the main contributor to total flavonoids. In East Asia, isoflavones are also important contributors to total flavonoids due to the high consumption of soy-derived products. Further studies are warranted to estimate the intake of total flavonoids in other regions of the world with little available data, such as Latin America, Africa, and Middle East. More research on the content of flavonoids in food is also needed in order to improve the existent food composition data, adding new foods (e.g., tropical foods) and refining the thearubigin data.

http://dx.doi.org/10.5772/67655

5. Determinants of the total flavonoid intake

Total flavonoids, as with most nutrients and dietary compounds, are positively correlated to total energy intake. Therefore, subjects consuming more energy are also more likely to be those with a higher intake of total flavonoids, even though one of the most relevant food sources is tea which is a non-caloric beverage. For this reason, if we want to study the determinants of total flavonoid intake, we should adjust our models for total energy intake.

Men usually consume higher amounts of total flavonoids, but after adjusting for total energy, women actually tend to consume more total flavonoids, as reported in the US [22] and South Korean populations [25]. However, men have a higher flavonoid intake in MED countries [21] and France [33] as men tend to drink more wine, particularly red wine. In contrast, the opposite was found in non-MED countries because women tend to consume more tea than men [21].

Young adults are more likely to consume less total flavonoids than older adults [21–23, 33]. The age range with the highest intake of total flavonoids is between 55 and 70 years, probably because they consume a more traditional dietary pattern and therefore more fruit, vegetables, red wine, and tea (depending on the region). On the other hand younger adults (20–40 years) tend to follow a more Westernized diet, with less plant-based foods. Although total flavonoid intake is usually higher in the older age groups, subjects are likely to keep their dietary habits throughout adulthood. Therefore, no significant differences have been found in the estimation of total flavonoid intake after several years of follow-up [22, 41].

Fruit, tea, red wine, and vegetables are the most important food sources of total flavonoids [21, 26]. These plant-based foods are characteristic of healthy and traditional diets, and therefore, subjects with high adherence to healthy dietary and lifestyle habits are more likely to have a higher intake of total flavonoids. For example, individuals with a normal weight (BMI < 25), a high education level, and a high socioeconomic status and who are physically active and non-smokers (never and former smokers) tend to have a higher total flavonoid intake [22, 23, 26, 33].

6. Flavanols: intake and food sources

Total flavanols or flavan-3-ols is the most consumed flavonoid class by far worldwide (**Table 1**), contributing to >80% of total flavonoids. Flavanols are divided into three subclasses: flavan-3-ol monomers (including catechins and epicatechins), proanthocyanidins (including oligomers and polymers of flavanols), and flavanol-derived compounds (including thearubigins and theaflavins). In nature, flavanols are predominantly found as aglycones.

The intake of total flavanols ranges from 11 mg/day in Brazil [37] to 629 mg/day in Australia [29] (**Table 2**). These large differences among populations are due to different intakes of the main food sources (i.e., tea and fruit). Depending on the intake of tea (a beverage rich in flavan-3-ol monomers and derived compounds) and fruit (foods rich in proanthocyanidins) a rank of countries/regions can be proposed: (i) countries with a very high consumption of tea (such as Australia,

UK, and Poland), which have a mean intake between 450 and 600 mg/day [28, 29, 33, 43, 44]; (ii) countries with a moderate consumption of tea or a very high consumption of fruit (such as MED countries, Germany, and the US), which have a mean intake of 250–400 mg/day [17, 35, 43, 45]; (iii) countries with a low consumption of tea and low-moderate consumption of fruit (such as Scandinavian countries, South Korea, and Latin-American countries), with a mean intake ranging from 50 to 150 mg/day [23, 25, 31, 37]. In Latin-American countries, an underestimation of the total flavanol intake is probable, since there is little food composition data available on tropical foods, especially on fruit [13–16].

6.1. Flavan-3-ol monomers

Intake of flavan-3-ol monomers varies between 10 mg/day in Mexico [23, 58, 59] and 270 mg/day in Poland [32], contributing to approximately 15–30% of total flavanols. The main food sources of flavan-3-ol monomers are tea, fruit (e.g., apples, stone fruits), red wine, and cocoa products [1]. The intake of these foods and beverages will determine the levels of flavan-3-ol monomers in the diet. Countries with a high intake of tea have a major intake of these compounds, such as Australia, UK, and Poland (150–250 mg/day), followed by countries with a high consumption of fruit and red wine, such MED countries and France (50–100 mg/day), and finally countries with a low consumption of tea and certain types of fruit, such as Latin American and Finland (<l50;mg/day) (**Table 2**).

6.2. Proanthocyanidins

Proanthocyanidins account for >75% of total flavanols in countries with a low intake of tea (MED countries) and <40% of total flavanols in countries with a high intake of tea (non-MED countries) [40, 43]. Proanthocyanidins are characteristic flavonoids in fruit, particularly apples, stone fruits, and berries, as well as red wine, but not in tea [1]. The highest intake of proanthocyanidins is in MED countries, including France (>200 mg/day) [17, 43, 53], followed by non-MED countries and the US (150–200 mg/day) [32, 35, 36, 43, 45], and finally Latin-American and Asian countries (<150 mg/day) [23–25] (**Table 2**).

6.3. Flavanol-derived compounds

Theaflavins and thearubigins occur exclusively in tea [1]. Theaflavins are consumed in low amounts (<3 mg/day) in countries with low tea consumption and larger amounts (10–30 mg/day) in countries with an important tea culture (**Table 2**) [43]. The same pattern can be applied to thearubigins, which can only be found in black tea. The intake ranges from 1 mg/day in Spain to 530 mg/day in the UK, where it contributes to almost 50% of total flavonoids [40].

7. Flavanones: intake and food sources

Flavanones are normally the second main flavonoid class (8–10%) [21]. The variation in flavanone intakes is relatively small between most of the countries (30–40 mg/day) (**Table 2**). The highest intake is in Greece and Spain (~60 mg/day as aglycones) [27, 55], and the lowest intake

										Total Flavanols										
													Flavanol-derived compounds							
Study name	Country	N	Dietary assessment	FCDB	Glycosides vs. aglycones	Anthocyanidins	Flavones	Flavonols	Flavanones	Total	Flavan-3-olmonomers	PA	Total	Theaflavins	Thearubigins	Isoflavones	Chalcones	Dihydrochalcones	Dihydroflavonols	References
NNS95	Australia	17,326	24HDR	USDA	Aglycones	2.9	0.5	20.7	6.9		187.9		239.2	26.2	213.0					[20]
Calcium Intake Fracture Outcome Age Related Extension Study	Australia	1136	SemiQ-FFQ	USDA	Aglycones			28.0	56.0	570										[28]
Longitudinal Assessment of Ageing in Women	Australia	511	Diet history	USDA	Aglycones											4.1				[46]
The Blue Mountains Eye	Australia	79	3 4-d WFR	USDA	Aglycones	7.0	1.9	28.7	21.2	629										[29]
Membership database of Flemish Dietetic Association	Belgium	45	SemiQ-FFQ & 4DFR	based on USDA	Aglycones	6.9	4.3	19.6	37.7	107										[47]
EPIC	MED	36,037	24HDR	USDA, PE	Aglycones	37.4	5.6	24.8	33.7	282	48.5	218.4	15.0	1.5	13.5	0.4				[40, 43, 48–50]
				PE	Glycosides	48.5	14.9	32.4	29.7	316	57.7	251.7		6.5		0.6	0.004	2.3	4.6	[26]
	Non-MED			USDA, PE	Aglycones	27.3	4.1	29.5	31.8	430	93.1	180.1	157.1	7.1	150.0	0.7				[40, 43, 48–50]
				PE	Glycosides	43.9	10.1	39.2	38.0	385	134.6	224.7		25.8		0.54	0.004	2.7	3.0	[26]
EFSA database	Northern Europe	30,000	FFQ & 24HDR	USDA, PE	Aglycones					241	58.8	160.0	113.2	7.9	105.3					[51]
	Central Europe									449	99.5	114.5	235.9	17.3	218.6					
	Southern Europe									283	35.2	110.5	45.2	3.1	42.1					
PREDIMED	Spain	7200	FFQ	PE	Glycosides	38.5	41.6	80.4	132.0	147	26.7	117.0		0.3		<0.01	<0.01	3.0	2.8	[27]
Austrian institutionalized elderly population	Spain	304	FFQ	PE	Glycosides															[52]
Case-control study in Italy	Italy	4154	FFQ	USDA	Aglycones	20.0	0.5	21.6	38.3	325	57.4	267.0				0.03				[53, 54]

Study name	Country	N	Dietary assessment	FCDB	Glycosides vs. aglycones	Anthocyanidins	Flavones	Flavonols	Flavanones	Total Flavanols						Isoflavones	Chalcones	Dihydrochalcones	Dihydroflavonols	References
													Flavanol-derived compounds							
										Total	Flavan-3-olmonomers	PA	Total	Theaflavins	Thearubigins					
TOSCA.IT	Italy	2573	FFQ	USDA, PE	Aglycones	33.1	10.5	39.7	19.0	65						0.07	3.3	2.3	0.2	[30]
Case-control study in Greece	Greece	200	SemiQ-FFQ	USDA	Aglycones	12.2	3.0	9.6	58.1		10.9					0.8				[55]
SU.VI.MAX study	France	4942	6 24HDR	PE	Aglycones	35.0	18.0	34.0	13.0		87.0	227.0		9.0					5	[17]
German National Nutrition Survey II	Germany	15,371	Diet history	USDA, PE	Aglycones					386	119.8	196.4	69.7							[45]
HAPIEE study	Poland	10,477	FFQ	PE	Glycosides	29.7	15.5	106.1	103.8		270.0	176.3				1.6		11.3		[32]
Polish National Multicenter Health Survey	Poland	6661	24HDR	USDA	Aglycones	20.9	0.6	32.0	9.2	462						0.2				[33]
				PE	Glycosides	22.0	4.3	47.9	7.3	315						0.03	<0.01	7	0.07	
Leeds Women's Wellbeing Study (LWW), the Diet and Health Study (DH)	UK	246	3- 7d FR	USDA, PE	Aglycones	19.1	2.8	58.3	21.4	506						2.4		1.9		[44]
Case-control study in Scotland	Scotland	1456	SemiQ-FFQ	Own database	Aglycones		1.0	28.0	20.6		115.2	33.5								[56]
FINDIET 2002 Study	Finland	2007	24-HDR	Own database	Aglycones	47.0		5.4	27.0	128	12.0	20.0				0.9				[31]
NIH-AARP Diet and Health Study	US	491.840	FFQ	USDA	Aglycones	12.7	1.2	19.1	37.1		101.9					0.6				[34]
NHS I & NHS II & HPFS	US	156.957	FFQ	USDA	Aglycones	13.9	1.8	18.4	41.0	307	55.6	251.3								[35]
NHANES	US	8809	24HDR, FFQ	USDA	Aglycones	3.0	1.5	12.9	14.4		230.6	74.1				1.1				[20, 36]
Women's Health Initiative	US	96	FFQ	USDA	Aglycones											2.1				[57]
Mexican Teachers' Cohort	Mexico	106	SemiQ-FFQ	PE	Glycosides	27.0	1.0	12.9	60.2	121	17.4	103.9	0.0	0.0	0.0	2.1	<0.01	1.4	<0.01	[23]
Mexican case-control study	Mexico	478	FFQ	USDA	Aglycones		6.6	35.6			6.9									[58]
Cohort study in the state of Morelos	México	50	FFQ	USDA	Aglycones		7.5	29.3			10.0									[59]

										Total Flavanols			Flavanol-derived compounds							
Study name	Country	N	Dietary assessment	FCDB	Glycosides vs. aglycones	Anthocyanidins	Flavones	Flavonols	Flavanones	Total	Flavan-3-olmonomers	PA	Total	Theaflavins	Thearubigins	Isoflavones	Chalcones	Dihydrochalcones	Dihydroflavonols	References
Health Survey-São Paulo (ISA-Capital 2008)	Brazil	1103	24HDR	PE	Glycosides	6.8	3.6	14.6	16.1	11						1.5				[37]
Health conditions, nutrition, and use of medication by the elderly in Viçosa (Minas Gerais)	Brazil	620	Recall of habitual consumption	PE	Aglycones	2.2	4.7	23.7	36.4	377						1				[60]
Cross-sectional Chinese study	China	3317	SemiQ-FFQ	USDA	Aglycones	5.3	0.7	14.0	5.8		123.9	48.7				6.3				[24]
Shanghai Breast Cancer Study	China	1823	FFQ	USDA	Aglycones											68.6				[61]
Case-control study in Shanghai	China	1393	FFQ	Own database	Aglycones	27.4	10.4	123.8								3.54				[62]
Study in Suihua, northern China	China	887	FFQ	Own database	Aglycones		29.2	32.6												[63]
Case-control in China	China	560	FFQ	Own database	Aglycones											29.2				[64]
Case-control in the northeast of China	China	66	FFQ	USDA	Aglycones	17.2	2.1	22.8	5.1		13.3									[38]
KNHANES	Korea	33,581	24HDR	USDA, PE	Aglycones	37.0	1.0	64.6	35.9	122	50.6	70.8		0.1	0.6	57.5				[25]
Case-control in Japan	Japan	340	FFQ	Own database	Aglycones											39.6				[65]

PE=phenol explorer; WFR=weighed food record; 24HDR=24-h dietary recall; MED=Mediterranean countries; Non-MED=non-Mediterranean countries; PA=proanthocyanidins; SemiQ-FFQ=semi-quantitative FFQ; FFQ=food frequency questionnaire; FR=food record; FCDB=food composition databaseB_118796_SE_LE

Table 2. Intake of flavonoid classes and subclasses (mg/day) in different countries/regions.

is in China (~5mg/day) [24, 38]. Flavanone intake patterns perfectly match the consumption of their main food sources: citrus fruit and citrus juices [1].

8. Flavonols: intake and food sources

Flavonols are usually the third principal flavonoid class (7–9%) [21]. The mean intake of flavonols generally varies between 20 and 40mg/day (**Table 2**). In some countries (South Korea, China, and Poland) [25, 32, 62], a higher intake of flavonols is reported (>60mg/day as aglycones), but some of these values are probably overestimated. Flavonols are widely distributed in fruit and vegetables, particularly apples, onions, some leafy vegetables, and red wine [1].

9. Anthocyanidins: intake and food sources

Anthocyanidins are present in all red-blue-purple fruits and vegetables (e.g., berries, apples, pears, grapes, plums, and aubergine) along with red wine. They are responsible for the color of these foods. Depending on the region, anthocyanidins fluctuate from the second to the fourth most consumed flavonoid class (7–10%) [21]. The mean intake of anthocyanidins range from 20 to 40mg/day in most of the countries (**Table 2**). Countries with a high consumption of berries, such as Finland, are the top consumers of anthocyanidins (~50mg/day) [31], while Brazil and the US have a particularly low intake (<15mg/day) [20, 34, 35, 37, 60].

10. Flavones: intake and food sources

Flavones are the fifth most common flavonoid class, accounting for 1–2% of total flavonoids [21]. Their intake varies between 1 and 5mg/day using the USDA database [13] and 5–20mg/day as aglycones using the Phenol-Explorer [16] (**Table 2**). Flavones are widely distributed in plant-based foods, fruits, vegetables, juices, wine, etc. It is important to highlight that the richest sources of flavones are herbal teas, and this food item is rarely recorded in dietary questionnaires, and therefore, an underestimation of flavone intake is quite probable.

11. Isoflavones: intake and food sources

Isoflavones are the flavonoid class with the lowest intake among the typical flavonoid classes (flavanols, flavanones, flavonols, anthocyanidins, flavones, and isoflavones). The intake of flavonoids is normally <2mg/day in Westernized countries, as they mainly occur in soy products [48]. However, in the UK health conscious group of the EPIC study, including vegans and vegetarians, the mean intake rises to 18mg/day [48]. In Asian countries, where the consumption of soy is very common, the mean intake of isoflavones ranges from 30 to 70mg/day, contributing to almost 20% of total flavonoids.

http://dx.doi.org/10.5772/67655

12. Other minor flavonoids: intake and food sources

Chalcones, dihydrochalcones, and dihydroflavonols are three minor subclasses of flavonoids. Food composition data on these classes are limited and only available in Phenol-Explorer [16], but not in the USDA database [13]. Chalcones are found in beer and broad beans, and their mean intake is very low (<0.01 mg/day expressed as glycosides) (**Table 2**). Dihydrochalcones are present in apples and derived products (e.g., cider and apple juice). The mean intake is between 1.5 and 3 mg/day expressed as glycosides [17, 26], except in Poland where the intake is double (**Table 2**) [32, 33]. The principal food source of dihydroflavonols is wine, and therefore, the countries with a greater intake of wine, such as France and MED countries, have the highest consumption (~5 mg/day expressed as glycosides) (**Table 2**) [17, 26]. Countries with a low wine intake, such as Poland, have a negligible intake (<0.1 mg/day expressed as glycosides) [33].

13. Conclusions

Overall, the mean intake of total flavonoids worldwide is around 400 mg/day, ranging from 150 mg/day in Latin-American countries to 600 mg/day in Australia and the UK, without considering thearubigins. Tea (rich in flavan-3-ol monomers and thearubigins) is the most important food source of total flavonoids in countries with a tea culture, while fruits (rich in proanthocyanidins) are the main food source in the remaining countries. Flavanols is the main contributor to total flavonoids (75%), followed by far by flavanones, flavones, and anthocyanidins (7–10% each) and flavones and isoflavones (1–2% each). Chalcones, dihydrochalcones, and dihydroflavonols are also minor flavonoid subclasses. It is important to highlight that the intake of isoflavones is extremely high in East Asian countries (30–70 mg/day) compared to Western countries (<2 mg/day), since isoflavones mostly occur in soy products. Older women with healthy lifestyle and dietary habits and higher educational levels and incomes are the sector with the highest intake of total flavonoids.

Finally, more composition data are needed, particularly for tropical foods, to improve the estimation of total flavonoids, especially in some specific regions. Further studies are warranted to increase the descriptive analyses in several areas where little data are available to date, such as Latin America, Africa, and the Middle East. An accurate estimation of flavonoid intake is the first step before estimating their potential protective effects against chronic diseases and is an essential step for developing future dietary guidelines on flavonoids.

Acknowledgements

This project has been supported by the Instituto de Salud Carlos III project CP15/00100 and cofunded by FEDER funds/ European Regional Development Fund (ERDF)- a way to build Europe.

Author details

Romina Escobar-Cévoli, Carlota Castro-Espín, Virginie Béraud, Genevieve Buckland and Raul Zamora-Ros*

*Address all correspondence to: rzamora@iconcologia.net

Unit of Nutrition and Cancer, Cancer Epidemiology Research Program, Catalan Institute of Oncology, Bellvitge Biomedical Research Institute (IDIBELL), Barcelona, Spain

References

[1] Pérez-Jiménez J, Neveu V, Vos F, Scalbert A. Identification of the 100 richest dietary sources of polyphenols: an application of the Phenol-Explorer database. Eur J Clin Nutr. 2010;64,Suppl 3:S112–20.

[2] Falcone Ferreyra ML, Rius SP, Casati P. Flavonoids: biosynthesis, biological functions, and biotechnological applications. Front Plant Sci. 2012;3:222.

[3] Kumar S, Pandey AK. Chemistry and biological activities of flavonoids: an overview. Sci World J. 2013;2013:162750.

[4] Zamora-Ros R, Touillaud M, Rothwell JA, Romieu I, Scalbert A. Measuring exposure to the polyphenol metabolome in observational epidemiologic studies: current tools and applications and their limits. Am J Clin Nutr. 2014;100:11–26.

[5] van Dam RM, Naidoo N, Landberg R. Dietary flavonoids and the development of type 2 diabetes and cardiovascular diseases: review of recent findings. Curr Opin Lipidol. 2013;24:25–33.

[6] Jiang W, Wei H, He B. Dietary flavonoids intake and the risk of coronary heart disease: a dose-response meta-analysis of 15 prospective studies. Thromb Res. 2015;135(3):459–63.

[7] Liu YJ, Zhan J, Liu XL, Wang Y, Ji J, He QQ. Dietary flavonoids intake and risk of type 2 diabetes: a meta-analysis of prospective cohort studies. Clin Nutr. 2014;33(1):59–63.

[8] Xiao ZP, Peng ZY, Peng MJ, Yan WB, Ouyang YZ, Zhu HL. Flavonoids health benefits and their molecular mechanism. Mini Rev Med Chem. 2011;11:169–77.

[9] Middleton E. The flavonoids. Trends Pharm Sci. 1984;5:335–8.

[10] Manach C, Williamson G, Morand C, Scalbert A, Rémésy C. Bioavailability and bioefficacy of polyphenols in humans. I. Review of 97 bioavailability studies. Am J Clin Nutr. 2005;81(1 Suppl):230S–42S.

[11] Zamora-Ros R, Rabassa M, Llorach R, González CA, Andres-Lacueva C. Application of dietary phenolic biomarkers in epidemiology: past, present, and future. J Agric Food Chem. 2012;60:6648–57.

[12] Illner A-K, Freisling H, Boeing H, Huybrechts I, Crispim S, Slimani N. Review and evaluation of innovative technologies for measuring diet in nutritional epidemiology. Int J Epidemiol. 2012;41:1187–203.

[13] USDA. USDA database for the flavonoid content of selected foods. Beltsville, MD: USDA, 2003.

[14] USDA. USDA database for the isoflavone content of selected foods. Beltsville, MD: USDA, 1999.

[15] USDA. USDA database for the proanthocyanidin content of selected foods. Beltsville, MD: USDA, 2004.

[16] Neveu V, Pérez-Jiménez J, Vos F, Crespy V, du Chaffaut L, Mennen L, Knox C, Eisner R, Cruz J, Wishart D, et al. Phenol-Explorer: an online comprehensive database on polyphenol contents in foods. Database (Oxford) 2010;2010:bap024.

[17] Pérez-Jiménez J, Fezeu L, Touvier M, Arnault N, Manach C, Hercberg S, Galan P, Scalbert A. Dietary intake of 337 polyphenols in French adults. Am J Clin Nutr. 2011;93 (6):1220–8.

[18] Zamora-Ros R, Barupal D, Rothwell JA, Jenab M, Fedirko V, Romieu I, Aleksandrova K, Overvad K, Kyrø C, Tjønneland A, et al. Dietary flavonoid intake and colorectal cancer risk in the European Prospective Investigation into Cancer and Nutrition (EPIC) cohort. Int J Cancer. 2017;140(8):1836–1844.

[19] Lakenbrink C, Lapczynski S, Maiwald B, Engelhardt UH. Flavonoids and other polyphenols in consumer brews of tea and other caffeinated beverages. J Agric Food Chem. 2000;48:2848–52.

[20] Johannot L, Somerset S M. Age-related variations in flavonoid intake and sources in the Australian population. Public Health Nutr. 2006;9(8):1045–54.

[21] Zamora-Ros R, Knaze V, Luján-Barroso L, Romieu I, Scalbert A, Slimani N, Hjartåker A, Engeset D, Skeie G, Overvad K, et al. Differences in dietary intakes, food sources and determinants of total flavonoids between mediterranean and non-mediterranean countries participating in the European Prospective Investigation into Cancer and Nutrition (EPIC) study. Br J Nutr. 2013;109(8):1498–507.

[22] Chun OK, Chung SJ, Song, WO. Estimated dietary flavonoid intake and major food sources of US adults. J Nutr. 2007;137(5):1244–52.

[23] Zamora-Ros R, Biessy C, Rothwell JA, Monge A, Lajous M, Scalbert A, López-Ridaura R, Romieu I. Dietary polyphenol intake and their major food sources in the Mexican Teachers' Cohort. J Acad Nutr & Diet. 2017; Submitted.

[24] Zhang Z, He L, Liu Y, Liu J, Su Y, Chen Y. Association between dietary intake of flavonoid and bone mineral density in middle aged and elderly Chinese women and men. Osteoporos Int. 2014;25(10):2417–25.

[25] Jun S, Shin S, Joung H. Estimation of dietary flavonoid intake and major food sources of Korean adults. Br J Nutr. 2016;115(3):480–9.

[26] Zamora-Ros R, Knaze V, Rothwell JA, Hémon B, Moskal A, Overvad K, Tjønneland A, Kyrø C, Fagherazzi G, Boutron-Ruault MC, et al. Dietary polyphenol intake in Europe: the European Prospective Investigation into Cancer and Nutrition (EPIC) study. Eur J Nutr. 2016;55(4):1359–75.

[27] Tresserra-Rimbau A, Medina-Remón A, Pérez-Jiménez J, Martínez-González MA, Covas MI, Corella D, Salas-Salvadó J, Gómez-Gracia E, Lapetra J, Arós F, et al. Dietary intake and major food sources of polyphenols in a Spanish population at high cardiovascular risk: The PREDIMED study. Nutr Metab Cardiovasc Dis. 2013;23(10):953–9.

[28] Ivey KL, Hodgson JM, Croft KD, Lewis JR, Prince RL. Flavonoid intake and all-cause mortality. Am J Clin Nutr. 2015;101(5):1012–20.

[29] Kent K, Charlton KE, Russell J, Mitchell P, Flood VM. Estimation of flavonoid intake in older Australians: Secondary Data Analysis of the Blue Mountains Eye Study. J Nutr Gerontol Geriatr. 2015;34(4):388–98.

[30] Vitale M, Masulli M, Rivellese AA, Bonora E, Cappellini F, Nicolucci A, Squatrito S, Antenucci D, Barrea A, Bianchi C, et al. Dietary intake and major food sources of polyphenols in people with type 2 diabetes: the TOSCA.IT Study. Eur J Nutr. 2016. In press. doi: 10.1007/s00394-016-1355-1

[31] Ovaskainen ML, Törrönen R, Koponen JM, Sinkko H, Hellström J, Reinivuo H, Mattila P. Dietary intake and major food sources of polyphenols in Finnish adults. J Nutr. 2008;138 (3):562–6.

[32] Grosso G, Stepaniak U, Topor-Mądry R, Szafraniec K, Pająk A. Estimated dietary intake and major food sources of polyphenols in the Polish arm of the HAPIEE study. Nutrition. 2014;30(11–12):1398–403.

[33] Witkowska AM, Zujko ME, Waśkiewicz A, Terlikowska KM, Piotrowski W. Comparison of various databases for estimation of dietary polyphenol intake in the population of Polish adults. Nutrients. 2015;7(11):9299–308.

[34] Xiao Q, Park Y, Hollenbeck AR, Kitahara CM. Dietary flavonoid intake and thyroid cancer risk in the NIH-AARP diet and health study. Cancer Epidemiol Biomarkers Prev. 2014;23(6):1102–8.

[35] Cassidy A, O'Reilly ÉJ, Kay C, Sampson L, Franz M, Forman JP, Curhan G, Rimm EB. Habitual intake of flavonoid subclasses and incident hypertension in adults. Am J Clin Nutr. 2011;93(2):338–47.

[36] Wang Y, Chung SJ, Song WO, Chun OK. Estimation of daily proanthocyanidin intake and major food sources in the U.S. diet. J Nutr. 2011;141(3):447–52.

[37] Miranda AM, Steluti J, Fisberg RM, Marchioni DM. Dietary intake and food contributors of polyphenols in adults and elderly adults of Sao Paulo: a population-based study. Br J Nutr. 2016;115(6):1061–70.

[38] Ma Y, Gao W, Wu K, Bao Y. Flavonoid intake and the risk of age-related cataract in China's Heilongjiang Province. Food Nutr Res. 2015;59:29564.

[39] Sohrab G, Hosseinpour-Niazi S, Hejazi J, Yuzbashian E, Mirmiran P, Azizi F. Dietary polyphenols and metabolic syndrome among Iranian adults. Int J Food Sci Nutr. 2013;64 (6):661–7.

[40] Zamora-Ros R, Knaze V, Romieu I, Scalbert A, Slimani N, Clavel-Chapelon F, Touillaud M, Perquier F, Skeie G, Engeset D, et al. Impact of thearubigins on the estimation of total dietary flavonoids in the European Prospective Investigation into Cancer and Nutrition (EPIC) study. Eur J Clin Nutr. 2013;67(7):779–82.

[41] Kim K, Vance TM, Chun OK. Estimated intake and major food sources of flavonoids among US adults: changes between 1999–2002 and 2007–2010 in NHANES. Eur J Nutr. 2016;55(2):833–43.

[42] Barnes S. Soy isoflavones--phytoestrogens and what else? J Nutr. 2004;134(5):1225S–1228S.

[43] Knaze V, Zamora-Ros R, Luján-Barroso L, Romieu I, Scalbert A, Slimani N, Riboli E, van Rossum CT, Bueno-de-Mesquita HB, Trichopoulou A, et al. Intake estimation of total and individual flavan-3-ols, proanthocyanidins and theaflavins, their food sources and determinants in the European Prospective Investigation into Cancer and Nutrition (EPIC) study. Br J Nutr. 2012;108(6):1095–108.

[44] Yahya HM, Day A, Lawton C, Myrissa K, Croden F, Dye L, Williamson G. Dietary intake of 20 polyphenol subclasses in a cohort of UK women. Eur J Nutr. 2016;55(5):1839–47.

[45] Vogiatzoglou A, Heuer T, Mulligan AA, Lentjes MAH, Luben RN, Kuhnle GGC. Estimated dietary intakes and sources of flavanols in the German population (German National Nutrition Survey II). Eur J Nutr. 2014;53:635–43.

[46] Hanna KL, O'Neill S, Lyons-Wall PM. Intake of isoflavone and lignan phytoestrogens and associated demographic and lifestyle factors in older Australian women. Asia Pac J Clin Nutr. 2010;19(4):540–9.

[47] Mullie P, Clarys P, Deriemaeker P, Hebbelinck M. Estimation of daily human intake of food flavonoids. Plant Foods Hum Nutr. 2007;62(3):93–8.

[48] Zamora-Ros R, Knaze V, Luján-Barroso L, Kuhnle GG, Mulligan AA, Touillaud M, Slimani N, Romieu I, Powell N, Tumino R, Peeters PH, et al. Dietary intakes and food sources of phytoestrogens in the European Prospective Investigation into Cancer and Nutrition (EPIC) 24-hour dietary recall cohort. Eur J Clin Nutr. 2012;66(8):932–41

[49] Zamora-Ros R, Knaze V, Luján-Barroso L, Slimani N, Romieu I, Touillaud M, Kaaks R, Teucher B, Mattiello A, Grioni S, et al. Estimation of the intake of anthocyanidins and their food sources in the European Prospective Investigation into Cancer and Nutrition (EPIC) study. Br J Nutr. 2011;106(7):1090–9.

[50] Zamora-Ros R, Knaze V, Luján-Barroso L, Slimani N, Romieu I, Fedirko V, de Magistris MS, Ericson U, Amiano P, Trichopoulou A, et al. Estimated dietary intakes of flavonols,

flavanones and flavones in the European Prospective Investigation into Cancer and Nutrition (EPIC) 24 hour dietary recall cohort. Br J Nutr. 2011;106(12):1915–25.

[51] Vogiatzoglou A, Mulligan AA, Luben RN, Lentjes MA, Heiss C, Kelm M, Merx MW, Spencer JP, Schroeter H, Kuhnle GG. Assessment of the dietary intake of total flavan-3-ols, monomeric flavan-3-ols, proanthocyanidins and theaflavins in the European Union. Brit J Nutr. 2013;111:1463–73.

[52] González S, Fernández M, Cuervo A, Lasheras C. Dietary intake of polyphenols and major food sources in an institutionalised elderly population. J Hum Nutr Diet. 2014;27 (2):176–83.

[53] Rossi M, Negri E, Parpinel M, Lagiou P, Bosetti C, Talamini R, Montella M, Attilio G, Franceschi S, Vecchia C La. Proanthocyanidins and the risk of colorectal cancer in Italy. Cancer Causes Control. 2010;21:243–50.

[54] Rossi M, Negri E, Talamini R, Bosetti C, Parpinel M, Gnagnarella P, Franceschi S, Dal Maso L, Montella M, Giacosa A, et al. Flavonoids and colorectal cancer in Italy. Cancer Epidemiol Biomarkers Prev. 2006;15:1555–8.

[55] Lagiou P, Samoli E, Lagiou A, Skalkidis Y, Katsouyanni K, Petridou E, Trichopoulos D. Flavonoid classes and risk of peripheral arterial occlusive disease: a case-control study in Greece. Eur J Clin Nutr. 2006;60(2):214–9.

[56] Theodoratou E, Kyle J, Cetnarskyj R, Farrington SM, Tenesa A, Barnetson R, Porteous M, Dunlop M, Campbell H. Dietary flavonoids and the risk of colorectal cancer. Cancer Epidemiol Biomarkers Prev. 2007;16:684–93.

[57] Frankenfeld CL, Patterson RE, Horner NK, Neuhouser ML, Skor HE, Kalhorn TF, Howald WN, Lampe JW. Validation of a soy food-frequency questionnaire and evaluation of correlates of plasma isoflavone concentrations in postmenopausal women. Am J Clin Nutr. 2003;77(3):674–80.

[58] Hernández-Ramírez RU, Galván-Portillo MV, Ward MH, Agudo A, González CA, Oñate-Ocaña LF, Herrera-Goepfert R, Palma-Coca O, López-Carrillo L. Dietary intake of polyphenols, nitrate and nitrite and gastric cancer risk in Mexico City. Int J Cancer. 2009;125 (6):1424–30.

[59] Galvan-Portillo MV, Wolff MS, Torres-Sánchez LE, López-Cervantes M, López-Carrillo L. Assessing phytochemical intake in a group of Mexican women. Salud Publica Mex. 2007;49(2):126–31.

[60] Nascimento-Souza MA, de Paiva PG, Pérez-Jiménez J, do Carmo Castro Franceschini S, Ribeiro AQ. Estimated dietary intake and major food sources of polyphenols in elderly of Viçosa, Brazil: a population-based study. Eur J Nutr. 2016. In press. doi: 10.1007/s00394-016-1348-0.

[61] Frankenfeld CL, Lampe JW, Shannon J, Gao DL, Ray RM, Prunty J, Kalhorn TF, Wähälä K, Patterson RE, Thomas DB. Frequency of soy food consumption and serum isoflavone concentrations among Chinese women in Shanghai. Public Health Nutr. 2004;7(6):765–72.

[62] Li G, Zhu Y, Zhang Y, Lang J, Chen Y, Ling W. Estimated daily flavonoid and stilbene intake from fruits, vegetables, and nuts and associations with lipid profiles in Chinese adults. J Acad Nutr Diet. 2013;113(6):786–94.

[63] Sun C, Wang H, Wang D, Chen Y, Zhao Y, Xia W. Using an FFQ to assess intakes of dietary flavonols and flavones among female adolescents in the Suihua area of northern China. Public Health Nutr. 2015;18:632–9.

[64] Li L, Zhang M, Holman CDAJ, Li L, Zhang M, Holman CDAJ. Population versus hospital controls in the assessment of dietary intake of isoflavone for case-control studies on cancers in China. Nutr Cancer. 2013;65(3):390–7.

[65] Hirayama F, Lee AH, Binns CW, Hiramatsu N, Mori M. Dietary intake of isoflavones and polyunsaturated fatty acids associated with lung function, breathlessness and the prevalence of chronic obstructive pulmonary disease?: possible protective effect of traditional Japanese diet. Mol Nutr Food Res. 2010;54(7):909–17.

Chapter 18

Anthocyanin Profile of Red Maize Native from Mixteco Race and Their Antiproliferative Activity on Cell Line DU145

Mónica Herrera-Sotero, Francisco González-Cortés, Hugo García-Galindo, Enrique Juarez-Aguilar, Mauricio Rodríguez Dorantes, José Chávez-Servia, Rosa Oliart-Ros and Rosa Guzmán-Gerónimo

Additional information is available at the end of the chapter

http://dx.doi.org/10.5772/67809

Abstract

Mexico is regarded as the point of origin and biodiversity of maize, which takes the colors white, yellow, blue, or red. Red maize in particular owes its coloring to a type of polyphenolic compounds known as anthocyanins. The aim of this study was to determine the profile anthocyanin of red maize, as well as their antiproliferative activity on prostate cancer cell line DU145. Three samples of red maize were analyzed. Total polyphenols, monomeric anthocyanins, antioxidant activity by DDPH and FRAP were evaluated. The sample of red maize with the highest levels of total polyphenols and monomeric anthocyanins was selected, and its anthocyanin fraction was analyzed by HPLC-ESI-MS. Twenty compounds were detected in the anthocyanin profile, and from these, 12 anthocyanins derived of cyanidin were identified. MTT assay was used to determine the antiproliferative activity of the anthocyanin fraction from red maize at different concentrations (7–1000 μg/mL), and a significant antiproliferative activity was observed at 1000 μg/mL. Microscopy analysis showed that the anthocyanin fraction of red maize induced apoptosis in cell lines DU145. This is the first report showing that anthocyanin fraction of red maize possess antiproliferative activity in the DU145 cell line.

Keywords: red maize, phenolic compounds, anthocyanin profile, antiproliferative activity, DU145 cell line

1. Introduction

Cereals are still the most important food source worldwide. Maize, sorghum, millet, wheat, rice, barley, oats, teff, and quinoa account for a large amount of the human diet. Maize (*Zea mays* L.) is the third crop by volume cultivated in the world and has a great economic importance, since it is used for animal and human foods as well as a raw material in the production of several industrial products [1]. There are over 59 maize races in Mexico [2] with different shapes and colors, such as white, yellow, blue, and red. In recent years, there has been a growing interest in pigmented varieties form the scientific point of view, since they are sources of phenolic compounds. Approximately 60% of polyphenols are flavonoids, which are regarded as the most important group of phenolic compounds. Among these are anthocyanins, the chemical compounds responsible for the color of red maize, which are located in the pericarp or aleurone layer of the grain [3].

Anthocyanins are generally regarded as the most important pigments in nature. These water-soluble compounds belong to the flavonoid family and can be found in the colors blue, red, purple, and dark violet [4]. From the chemical standpoint, anthocyanins in nature take the form of glycosides, whose aglycone is known as anthocyanidin. Anthocyanidins are made up of a benzopyril system (A–C) and a phenolic ring (B). Particular differences between anthocyanins depend upon the number of hydroxyl groups, the nature and number of glycosides attached to the molecule, the position of the point of attachment, and to the nature and number of aromatic acids joined to the glycoside in the molecule [5]. Since each anthocyanidin may be glycosylated and acylated by different acids at different points, there is a large number of chemical combinations [6]. Furthermore, the simplest or monomeric anthocyanins may react between themselves, producing more complex structures of higher molecular weights known as polymeric anthocyanins. They may also react with other compounds, and therefore, in nature more than 600 different anthocyanins are found.

On the other hand, there is a growing scientific interest on pigmented maize races such as red, due to their content of anthocyanins. Experts recommend the consumption of these because of their strong free-radical scavenging activity and chelation of divalent cations, which is given by their phenolic structure, double bonds of their rings, and hydroxyl groups. Anthocyanins may also modulate enzymes related to oxidative stress, and their preventive action in the development of disease has been reported [7].

Several epidemiological studies also reveal that the consumption of flavonoid-rich foods is associated with a lower risk of neurodegenerative diseases such as cancer [8–10]. Cancer is one of the main causes of death worldwide, causing 7.6 million deaths only in 2008. For 2030, this number of cancer victims is expected to increase to 11 million. Lung, breast, colorectal, stomach, and prostate cancers cause the majority of cancer deaths [11]. The treatment of this disease depends on the type of cancer as well as on the subtype or stage of the patient and is currently based on hormones, chemotherapy, radiotherapy, pharmaceuticals, and nanotherapy. However, these treatments are aggressive, highly expensive, and unattainable for many people [12]. Unfortunately, some cancer tumors are extremely resistant to current therapeutic agents; therefore, the development of new strategies in prevention, treatment, and control of cancer is urgently needed [13].

The scientific community is looking for alternatives, such as the use of natural compounds as chemopreventives or adjuvants. In this regard, red maize form Mixteco race may be an important source of flavonoids; however, the number of studies investigating the antiproliferative activity of this grain is scarce. Given the above, the aim of the present work was to evaluate the antiproliferative activity of purified anthocyanins from red maize in the prostate cancer cell line DU145.

2. Research methods

2.1. Plant material

Three samples of red maize from the Mixteco race (**Figure 1**), cultivated in the Mixteca region of the State of Oaxaca, Mexico, were used in the present study.

2.2. Preparation of red maize extract

To prepare the extracts, 200 g of red maize was ground using a coffee grinder (Krups model GX410011V) in order to reduce particle size and increase contact surface area. Then 500 mL of EtOH/ $C_6H_8O_7$ (85:15 1 M) were added, and the mix was thermosonicated in an ultrasonic homogenizer (Cole-Palmer, modelo VCX-750) for 30 min at 40% amplitude, pulse 05:05). Afterward, the extracts were left to stand for 24 h at 4°C and then centrifuged at 4000 rpm for 20 min at 5°C. Extracts were concentrated in a rotavapor (Büchi Rotavapor R-205, Büchi Vacuum Controller V-800, Büchi Heating Bath B-490) at 28°C under vacuum, and finally samples were lyophilized and stored at −20°C. This process was carried out in the dark [14].

Figure 1. Samples of red maize from Mixteco race.

2.3. Analysis of total polyphenols

Total polyphenols were analyzed by the method described by Folin and Ciocalteu and modified by Singleton and Rossi [15], which is based on an oxidation-reduction reaction and phenolic compounds react with the Folin-Ciocalteu reagent under alkaline conditions, resulting in a blue coloration that is evaluated by spectrophotometry. Previous to the analysis, a calibration curve with gallic acid was made (12 mg EAG/100 mL) and read at 730 nm. Results were expressed as mg equivalent of gallic acid/L.

2.4. Determination of monomeric anthocyanins

Quantification of anthocyanins was carried out by the pH differential method described by Giusti and Wrolstad [16]. The amount of anthocyanins was expressed as mg of cyanidin 3-glucoside/L of sample, and each sample was analyzed by triplicate.

2.5. Antioxidant activity by DPPH

The DPPH (1,1-diphenyl-2-picrylhydrazyl) method, as reported by Brand-Williams et al. [17], was used to determine antioxidant activity. A standard calibration curve (100–800 μmol) of trolox was used as reference. A measure of 2.9 mL of DPPH were vigorously mixed with 0.1 mL of each extract and then kept in the dark for 30 min at 25°C. Absorbance was read at 517 nm. Results were expressed as μmol equivalent of trolox/g of simple. All determinations were made by triplicate.

2.6. Antioxidant activity by FRAP

Analysis of antioxidant activity by FRAP (Ferric reducing antioxidant power) was carried out according to Benzie and Strain [18]. A standard trolox calibration curve (100–800 μmol) was used as reference. About 3 mL of FRAP reagent were mixed with 0.1 mL of each extract, then incubated for 5 min at 37°C, and the absorbance was read at 593 nm. Results were expressed in μmol equivalent of trolox/g of simple. All determinations were made by triplicate.

2.7. Isolation of anthocyanins

Purification of anthocyanins was made in the dark, using a chromatographic column packed with amberlite XAD-7.1 g of lyophilized red maize extract that was placed per column, and two solutions were added as mobile phases: solution A: H_2O/acetic acid (95:5 v:v) and solution B: EtOH/acetic acid (95:5 v:v). Fractions of anthocyanins were collected in ambar glass bottles and concentrated in a rotavapor (Büchi Rotavapor R-205, Büchi Vacuum Controller V-800, Büchi Heating Bath B-490) at 28°C under vacuum. Samples were lyophilized and stored at −20°C [14].

2.8. Analysis of anthocyanins by high-performance liquid chromatography with electrospray ionization mass spectrometry (HPLC-ESI-MS)

Anthocyanin analysis was carried out in a high-performance liquid chromatographer (Agilent model 1200) coupled with a mass spectrometer (Bruker model Esquire 6000), equipped with electrospray and ion trap with a nitrogen nebulizer 15 psi, nitrogen as drying gas at a rate

http://dx.doi.org/10.5772/67809

of 7 L/min, drying temperature of 300°C, 500 m/z target, and 50–1000 m/z scan. A Zorbax Eclipse plus C-18 column (2.1 mm × 100 mm × 3.5 μm) was used. The mobile phase consisted of acetonitrile as solvent A and 2% aqueous acetic acid solution as solvent B. The gradient was 7:93 (A:B) at 0 min, 35:65 (A:B) at 80 min, and 100% A in 35 min. Flow rate was 0.2 mL/min. Test time was of 40 min [14].

2.9. Evaluation of cell proliferation in vitro

DU-145 cells were seeded into 96-well plates at 5000 cells per well and incubated with different extract concentrations. The MTT ((3-(4, 5-dimethylthiazolyl-2)-2, 5-diphenyltetrazolium bromide) assay was performed after treatment time (24–48 h) [19]. Roswell Park Memorial Institute (RPMI) 1640 culture medium was replaced with 50 μL of MTT (5 mg/mL), and the cells were incubated for 4 h at 37°C in darkness. Formazan crystals were dissolved in 200 μL of DMSO (dimethyl sulfoxide). Absorbance was read at 570 nm on a micro plate reader (Awareness Technology, Stat Fax 4200 model).

2.10. Statistical analysis

Three independent repetitions were made for the cell culture experiments. In order to determine significant differences among data, ANOVA (analysis of variance) tests were performed followed by Tukey ($p < 0.05$) multiple range tests.

3. Results and discussion

3.1. Contents of total polyphenols, monomeric anthocyanins, and antioxidant activity

In the first part of the present work, the contents of total polyphenols, monomeric anthocyanins, and antioxidant activity by DPPH and FRAP were evaluated. Sample R-14 showed the highest values for total polyphenols (**Table 1**). Values of monomeric anthocyanins for red maize were higher than those reported for Chalco and Red Chihuahua maize [20] and lower than those reported for samples of red maize from Mexico City and Puebla, Mexico [21]. The values for antioxidant activity as evaluated by the DPPH method were between 10.5 and 12.7 μmol ET/g, whereas antioxidant activity by FRAP recorded 2.77 and 2.79 μmol ET/g. This suggests that red maize samples are a potential source of phenolic compounds with antioxidant properties. On the other hand, sample R-14 was selected to evaluate antioxidant profile and antiproliferative activity on prostate cancer cell line DU145, since this particular sample showed the highest values for total polyphenols and monomeric anthocyanins.

Samples	Total polyphenols (mg EAG/100 g)	Monomeric anthocyanins (mg C3G/100 g)	DPPH (μmol ET/g)	FRAP (μmol ET/g)
R-03	327	50.1	10.5	2.77
R-07	368	68.1	12.3	2.69
R-14	373	67.4	12.7	2.79

Table 1. Total polyphenols, monomeric anthocyanins, and antioxidant activity of red maize from Mixteco race.

3.2. Anthocyanin profile of red maize from Mixteco race

It is widely known that biological properties of anthocyanins depend upon their chemical structure, substitutions, conjugations, and polymerization. Therefore, in the present study, the anthocyanin profile of red maize was analyzed by HPLC-ESI-MS. **Figure 2** shows the total ion chromatogram (TIC) of red maize anthocyanins from Mixteco race. A total of 20 different compounds were detected, their retention times (tr) and main ions (m/z) are shown in **Table 2**. A total of 17 compounds were derived from cyanidin (287 m/z). On the other hand, 15 different compounds were identified in the anthocyanin profile of red maize from Mixteco race (**Table 2**), which includes monoglycosilated, acylated anthocyanins, and proanthocyanidins. Regarding the role of anthocyanins in the prevention of prostate cancer, a recent study shows that cyanidin-3-glucoside produces cancer cell apoptosis in line DU145 [22]. After the anthocyanin profile was completed on the selected sample, its antiproliferative activity was tested on cell line DU145.

3.3. Antiproliferative activity

Figure 3 shows the percentage of cell viability after 24 h of treatment with different concentrations of purified anthocyanins from red maize. As seen in the graph, no significant differences were detected between the different concentrations and the control ($p \leq 0.05$). This shows that red maize extract had no significant effect on prostate cancer cell line DU145 after 24 h.

Figure 4 shows the percentage of cell viability after 48 h of application of different concentrations of purified anthocyanins of red maize, where a significant reduction percentage on cell growth was observed at 1000 μg/mL as compared to the control, corresponding to 35% of cell viability; the same effect was observed in the micrograph (**Figure 5**), where a cytoplasmic vacuolization is present and the growth of live cells is haltered. Even though there are no reports on the effect of anthocyanins of red maize on cancer cell lines, recent studies show that cyanidin-3-glucoside produces an antiproliferative effect through the activation of caspase-3 on prostate cell lines LnCap and DU145 [22]. It has also been reported that flavonoids from blueberry inhibit the activity of metalloproteinase in DU145 [23]. These data suggest a potential for red maize flavonoids in the chemoprevention of prostate cancer, which is nowadays the first cause of death by cancer.

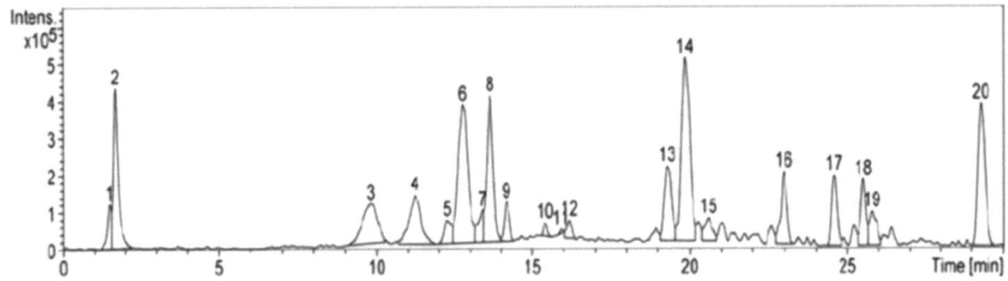

Figure 2. Anthocyanin profile of red maize from Mixteco race.

http://dx.doi.org/10.5772/67809

Peak	T_R (min)	Fragment ions (m/z)	Tentative identification
1	1.5	287,509	Nonidentified
2	1.7	287,449	Cyanidin-3-glucoside
3	9.8	287,491	Cyanidin-3-O-(6′-acetyl-glucoside)
4	11.2	287,493	Nonidentified
5	12.2	287,530	Cyanidin-3-O-sambubioside-5-rhamnoside
6	12.8	287,460	Nonidentified
7	13.4	287,599	Nonidentified
8	13.6	287,754	Cyanidin-3-O-(6′-acetyl-arabinoside)
9	14.2	287,491	Cyanidin 3-(6″-malonyl) glucoside
10	15.4	287,549	Cyanidin-3-succinylglucoside
11	15.9	287,501	Cyanidin-3-O-soforoside
12	16.2	287,451	Nonidentified
13	19.3	287,433	Cyanidin-3-O-(6″-p-coumaroil-glucoside)-5-O-(6″-malonyl-glucoside)
14	19.9	287,731	Cyanidin 3-p-hydroxy-benzoyl sophoroside-5-glucoside
15	20.6	287,773	Cyanidin 3-sophoroside-5-glucoside
16	23.0	287,901	Procyanidin dimer
17	24.6	339,843	Nonidentified
18	25.5	287,901	Procyanidin dimer
19	25.8	353,901	Procyanidin dimer
20	29.3	343,885	Procyanidin type B

Table 2. Identity of anthocyanins detected in red maize from Mixteco race.

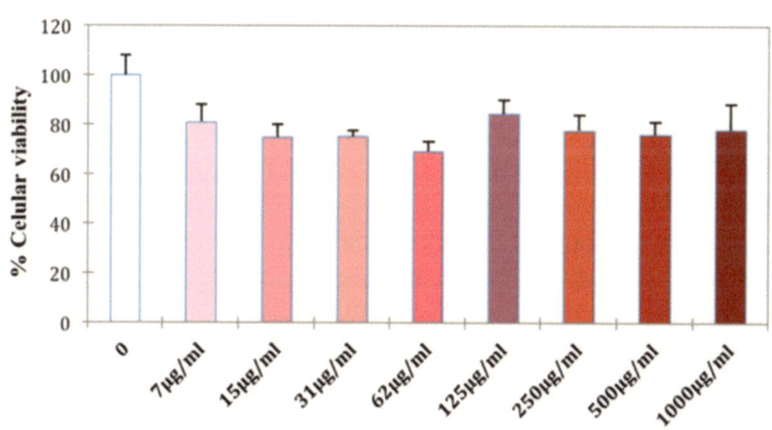

Figure 3. Percentage of cell viability after 24 h of treatment with red maize anthocyanins. Columns show the mean value and standard deviation of three independent experiments.

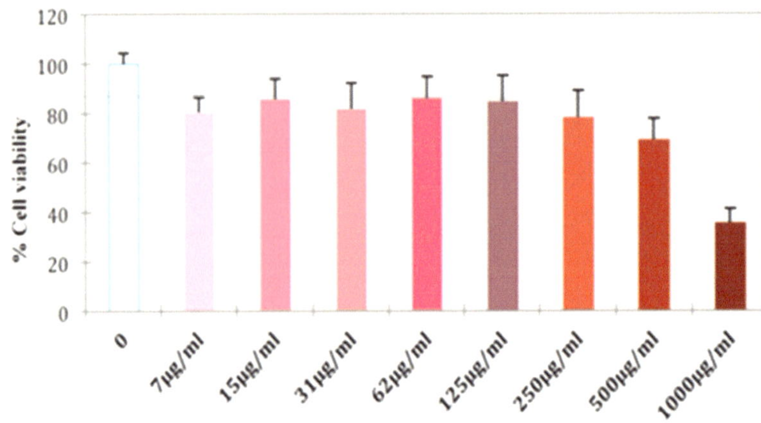

Figure 4. Percentage of cell viability after 48 h of treatment with red maize anthocyanins. Columns show the mean value and standard deviation of three independent experiments.

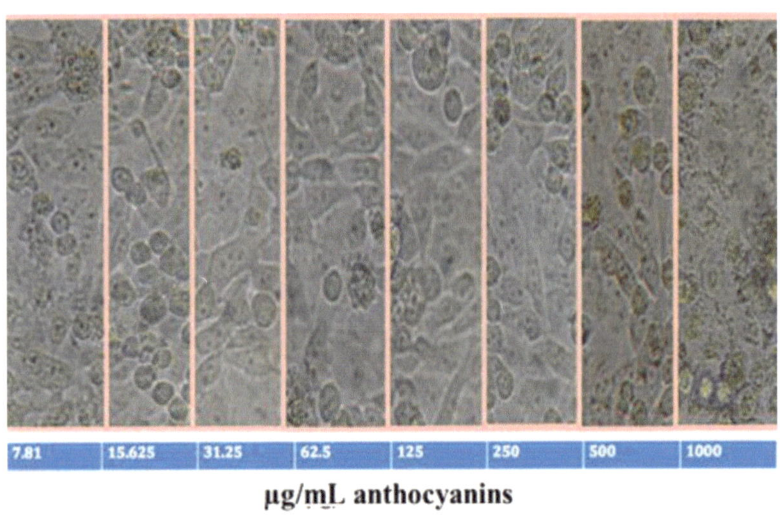

Figure 5. Microscopy of cancer cells DU145 after 48 h application of different concentrations of purified anthocyanin extract from red maize.

Figure 6 shows an image obtained by optical microscopy: a DU145 cell culture treated with the maximum concentration of red maize anthocyanins (1000 µg/mL) along with a culture in the absence of anthocyanins; after 48 h of incubation, cell vacuolization is observed in the experimental sample.

In summary, red maize from Mixteco race is a rich source of flavonoids such as anthocyanins, and their profile is mainly constituted by cyanidin-derived anthocyanins. These compounds have a potential application in the prevention of prostate cancer, showing antiproliferative activity on cell line DU145. Future research is needed.

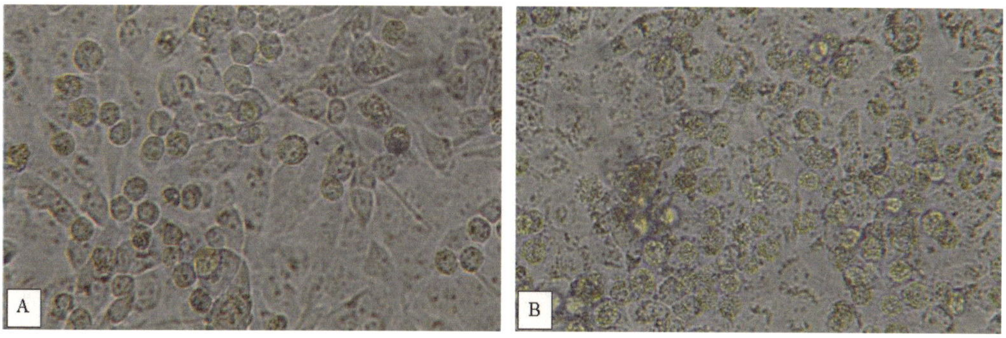

Figure 6. Microscopy of DU145 cells incubated with 1000 μg/ml of red maize anthocyanins for 48 h. (A) Control culture; (B) culture treated with anthocyanins (1000 μg/mL).

Acknowledgements

The authors like to thank the SINAREFI from México.

Author details

Mónica Herrera-Sotero[1], Francisco González-Cortés[2], Hugo García-Galindo[1], Enrique Juarez-Aguilar[3], Mauricio Rodríguez Dorantes[4], José Chávez-Servia[5], Rosa Oliart-Ros[1] and Rosa Guzmán-Gerónimo[2]*

*Address all correspondence to: roguzman@uv.mx

1 Food Research and Development Unit (UNIDA), Technological Institute of Veracruz, Veracruz, México

2 Basic Sciences Institute, University of Veracruz, Xalapa, Veracruz, México

3 Health Sciences Institute, University of Veracruz, Xalapa, Veracruz, Mexico

4 National Institute of Genomic Medicine, México City, México

5 The Interdisciplinary Research Center for Integrated Regional Development, Oaxaca Unit, National Polytechnic Institute, Oaxaca, México

References

[1] Paliwal RL, Granados G, Lafitte HR, Violic AD, editors. El maíz en los trópicos: Mejoramiento y producción 2001. 1st ed. Rome; Italia: FAO: 392 p.

[2] Ruiz CJ, Durán PN, Sánchez GJ, Ron PR, González ED, Medina GG, Holland BJ. Climatic adaptation and ecological descriptors of 42 maize races. Crop Sci Soc Am 2008;**48**:1502-1512. DOI: 10.2135/cropsci2007.09.0518.

[3] Trujillo EE, Mendoza CMC, Castillo GF, Ortiz CJ, Delgado DA, Carrillo SA. Acumulación de antocianinas en pericarpio y aleurona del grano y sus efectos genéticos en poblaciones criollas de maíz pigmentado. Rev Fitotec Mex 2009;**32**(4):303-309.

[4] Delgado-Vargas F, Jiménez AR, Paredes-López O. Natural pigments: Carotenoids, anthocyanins, and betalains—Characteristics, biosynthesis, processing, and stability 2003. 1st ed. Boca Raton, Florida: CRC press: 313 p. DOI: 10.1080/10408690091189257.

[5] Kong JM, Chiam SL, Goh NK, Chia TF, Brouillar C. Analysis and biological activities of anthocyanins. Phytochemistry 2003;**64**(5):923-933. DOI: 10.1016/S0031-9422(03)00438-2.

[6] Clifford MN. Anthocyanins: Nature, occurrence and dietary burden. J Sci Food Agric 2000;**80**(7):1063-1072. DOI: 10.1002/(SICI)1097-0010(20000515)80:7<1063::AID-JSFA605>3.0.CO;2-Q.

[7] Wang LS, Stoner GD. Anthocyanins and their role in cancer prevention. Cancer Lett 2008;**269**(281-290). DOI: 10.1016/j.canlet.2008.05.020.

[8] Kotesha R, Takami A, Espiniza JL. Dietary phytochemicals and cancer chemoprevention: a review. Oncotarget 2016;**7**:57517-52529. DOI: 10.18632/oncotarget.9593.

[9] Mehta M, Shike M. Diet and physical activity in the prevention of colorectal cancer. J Natl Compr Canc Netw 2014;**12**:1721-1726.

[10] Howes MJ, Simmonds MS. The role of phytochemicals as micronutrients in health and disease. Curr Opin Clin Nutr 2014;**17**:558-566. DOI: 10.1097/MCO.0000000000000115.

[11] WHO. http://www.who.int/gho/ncd/mortality_morbidity/cancer_text/en/ [Internet]. [Accessed: January 7].

[12] Wang CJ, Ying J, Kapur P, Wohlfeld B, Roehrborn C, Kim DW. Solitary recurrence of castration-resistant prostate cancer with low or undetectable levels of prostate specific antigen salvaged with local ablative radiation therapy: A case report. Oncol Lett 2016;**11**(1):713-716. DOI: 10.3892/ol.2015.3940.

[13] Priyadarsini RV, Nagini S. Cancer chemoprevention by dietary phytochemicals: promises and pitfalls. Curr Pharm Biotechnol 2012;**13**:125-136. DOI: 10.2174/138920112798868610.

[14] Alarcón AE. Polifenoles, antocianinas, actividad antioxidante y actividad biológica de tortilla y grano de maíz azul de la raza Mixteco. [Thesis] 2013. Xalapa, Veracruz. México: Universidad Veracruzana: 200 p.

[15] Singleton VL, Rossi, JA. Colorimetry of total phenolics with phosphomolybdic-phosphotungstic acid reagents. Am J Enol Vitic 1965;**16**:144-150.

[16] Giusti M, Wrolstad RE. Characterization and measurement of anthocyanins by UV-visible spectroscopy. Giusti M, Wrolstad RE. Characterization and measurement of anthocyanins by UV-visible spectroscopy. In: Current Protocols in Food Analytical Chemistry 2001. New York, N.Y.: John Wiley and Sons, Inc. p. F:F1:F1.2. DOI: 10.1002/0471142913.

[17] Brand-Williams W, Cuvelier ME, Berset C. Use of a free radical method to evaluate antioxidant activity. LWT—Food Sci Technol 1995;**28**:25-30. DOI: 10.1016/S0023-6438(95)80008-5.

[18] Benzie IF, Strain FJ. Ferric reducing ability of plasma (FRAP) as a measure of antioxidant power: The FRAP assay. Anal Biochem 1996;**239**(1):70-76. DOI: 10.1006/abio.1996.0292.

[19] Denizot F, Lang R. Rapid colorimetric assay for cell growth and survival. Modifications to the tetrazolium dye procedure giving improved sensitivity and reliability. J Immunol Methods 1986;**89**(2):271-277.

[20] Salinas MY, Martínez BF, Soto HM, Ortega PR, Arellano VJ. Efecto de a nixtamalización sobre las antocianinas del grano de maíces pigmentados. Agrociencia 2003;**37**(6):615-628.

[21] López-Martinez LX, Parkin KL, García HS. Phase II-Inducing, polyphenols content and antioxidant capacity of corn (*Zea mays* L.) from phenotypes of white, blue, red, and purple colors processed into masa and tortillas. Plant Food Hum Nutr 2011;**66**(1):41-47. DOI: 10.1007/s11130-011-0210-z.

[22] Sorrenti V, Vanella L, Acquaviva R, Cardile V, Giofre S, Giacomo DC. Cyanidin induces apoptosis and differentiation in prostate cancer cells. Int J Oncol 2015;**47**(4):1303-1310. DOI: 10.3892/ijo.2015.3130.

[23] Matchett MD, MacKinnon SL, Sweeney M, Gottschall-Pass KT, Hurta RA. Blueberry flavonoids inhibit matrix metalloproteinase activity in DU145 human prostate cancer cells. Biochem Cell Biol 2005;**83**(5):637-643.

Chapter 19

Flavonoid and Capsaicinoid Contents and Consumption of Mexican Chili Pepper (*Capsicum annuum* L.) Landraces

Araceli M. Vera-Guzmán, Elia N. Aquino-Bolaños, Elena Heredia-García, José C. Carrillo-Rodríguez, Sanjuana Hernández-Delgado and José L. Chávez-Servia

Additional information is available at the end of the chapter

http://dx.doi.org/10.5772/68076

Abstract

There is insufficient evidence to identify the precise health effects of chili pepper consumption. However, there is evidence of their topical use as an analgesic to decrease pain from rheumatoid arthritis, osteoarthritis, neuralgias, neuropathic diabetes, neuronal dysfunctions and inflammation, among others. In this work, the diversity and variety of consumed forms of chilis in Mexico, flavonoid and capsaicinoid content in fruits, and their potential health uses are documented, based on various research results and bibliographic information. In Mexico, more than 150 landraces of wild and cultivated origins are consumed and preserved and are distributed throughout the country; the greatest diversity is concentrated in the central and south-southeastern regions. Consumption per capita in urban households is from 8 to 9 kg, and in rural communities, it varies from 14 to 17 kg. Chili peppers contain up to 23 flavonoids and 20 capsaicinoids, differing among landraces because of crop management, maturation of fruits, postharvest management and ecological-environmental influences. Flavonoids and capsaicinoids confer antioxidant, anticarcinogenic properties on the fruit and have lipolytic and preventative effects on chronic degenerative diseases. However, in vitro and in vivo experimental trials of capsaicinoids and flavonoids with beneficial effects must be conducted with regard to human health.

Keywords: *Capsicum annuum*, pepper landraces diversity, antioxidants, benefits of chili pepper consumption, flavonoids, capsaicinoids on health

1. Introduction

Chili pepper (*Capsicum* spp.) has been used in Mexico as food and a condiment for over eight centuries [1] and is an integral component of the diet and cultural identity of Mexico. Although various chili species grow wild and are cultivated in the United States of America, Mexico is the center of origin, domestication and diversification of *Capsicum annuum* L. [2, 3]. Wild variants continue diversifying (*C. annuum* var. *glabriusculum* [Dunal] Heiser & Pickersgill) and are distributed throughout the country [4, 5]. In this Mesoamerican biocultural context and using traditional management, farmers preserve and utilize a great diversity of chili landraces.

A total of slightly more than 150 landraces, occupying extensive planted surfaces and providing for great consumption, have been described in Mexico. The central and south-southeastern regions are more diverse and relevant for the in situ preservation of chilis. In these regions, approximately 80 landraces have been preserved. In Mexico, 153,565 ha are cultivated annually, resulting in an average production of 2.8 million tons of dry and fresh fruits. Annually, approximately 2 million tons of fresh and dry chilis are consumed, and in some years, imports range between 20,757 and 41,000 tons [6]. Consumption per capita in urban areas is from 8 to 9 kg, and in rural communities, it varies between 13 and 17 kg.

The health benefits of chili consumption and its active ingredients continue to be investigated using in vitro and in vitro biological models as well as theoretical and experimental models. Capsaicinoids and flavonoids are the determinant compounds of color, flavor, texture and aroma of food prepared with chilis. A common question among health specialists is whether chili consumption improves health or contributes to disease development. The chili fruit contains up to 15 or more capsaicinoid compounds. Among the majority are capsaicin, dihydrocapsaicin and nordihydrocapsaicin; in flavonoid content, quercetin, apigenin and luteolin are notable as well as some catechins and cyanidins. Thus, one does not only ingest vitamins B_1, B_2 and C, minerals, carotenoids and phenols by chili consumption.

Flavonoids contain phenolic hydroxyl groups in their chemical structure and possess excellent chelating properties of iron and other transition metals—characteristics that confer on chilis a high antioxidant capacity and anti-free-radical properties, generating protection against oxidative damage [7]. Flavonoids have positive effects on health although their mechanisms of action continue to be investigated because it is difficult to quantify daily intake and their direct effects on health. Thus far, flavonoids are associated with antisclerotic, anti-inflammatory, antitumoral, antithromobogenic, antiviral and antiosteoporotic effects and may function as a preventive agent in cancer, among other effects [8–10].

Capsaicinoids are synthesized in the placenta of the fruit and are genetically determined (*Pun1* allele of pungency) by the presence of the *Pun1* or *pun1* gene with EST- or AT3-type cofactors that induce a quantitative effect of the gene and variations in the pungency of the fruit. Therefore, not all chili peppers are spicy, and various consumers consider chili varieties that carry the *pun1* recessive gene to be sweet fruits [11]. In addition, genetic or genotypic factors, crop management, cultivation environments, maturity of the fruit upon cutting, postharvest management of fruit and forms of processing or cooking the fruit for consumption influence the level of pungency and flavor at the time of consumption [12].

http://dx.doi.org/10.5772/68076

In the last decade, hundreds of articles have been published on the potential adverse and beneficial effects of capsaicin (8-methyl-N-vainillyl-6-nonenamide) on human health. Experimental results are controversial and require testing in humans, not only biological models. Consumption of capsaicin in chilis stimulates neuronal networks and gastric secretions and stimulation; in addition, capsaicin is considered to be a preventative agent for cancer, has lipolytic action by increasing the hydrolysis of triacylglycerol in adipocytes and is associated with antioxidant action. Direct effects as a carcinogenic agent have not been demonstrated, and recent studies have demonstrated that there is no conclusive evidence that confirms the association between consumption of extremely spicy red chili peppers and mortality in the European and North American populations [13–18].

The role of diet in health is increasingly important and warrants further investigation. Several investigators have focused their studies on the documentation of macro- and micro-nutrients, vitamins, minerals, proteins, gastronomic aspects and, recently, the functional and nutraceutical character of various bioactive compounds. Chili peppers and other species are consumed universally because these fruits confer color, flavor, aroma and texture and help preserve foods. In this work, the diversity and variety of consumed forms of chilis in Mexico, flavonoid and capsaicinoid content in the fruits and their potential health benefits, are documented based on various study results and bibliographic information. Additionally, elements of the culinary culture associated with chili in Mexico, high levels of consumption and repercussions on human health with an emphasis on flavonoids and capsaicinoids are provided.

2. Diversity of chili pepper landraces in Mexico

Depending on the archaeobotanic, genetic and pre-Columbian sociocultural diversity, it has been established that chili peppers (*Capsicum annum* L. and *C. frutescen* L.) have been collected, cultivated and consumed in Mexico for hundreds of years. For example, analysis of archaeological remains from the Tehuacán Valley in Mexico indicates that there were wild forms 8000 years ago, and different cultivation and domestication events of *C. annuum* occurred with evidence dating to 6000 years ago [19, 20]. Perry and Flannery [1], from excavations, collection and dating of archaeobotanic samples of the Silvia and Guila Naquitz´s caves in Oaxaca, indicated that cultivated forms of *C. annuum* and *C. frutescens* have existed for nearly 8000 years. In Chiapas, Mexico, from a sample of ceramic residue, extensive culinary use of chili peppers was demonstrated [2]. In general, archaeobotanic, genetic and cultural linguistic evidence identifies the Mesoamerican region as the center of origin, domestication and diversification of *C. annuum* and *C. frutescens*, and those species continued to evolve under domestication or as wild forms [3].

In Mexico, wild species of *Capsicum* have certain characteristics in common, such as being small spicy fruits of round, elongated or conical forms that regularly ripen in red or yellow colors. The regularly erect fruits are easily separated from the calyx. They have small seeds, which form a portion of the diet of birds and facilitate the chilis' dispersal over long distances. Today, wild forms have a wide distribution from the southern United States of America to South America (Argentina). Among these cases, *C. annuum* var. *annuum* and *C. annuum* var. *glabriusculum* (Dunal) Heiser & Pickersgill are notable [5, 21–23].

During the domestication process, chili fruits and plants have undergone various genetic transformations in response to human and natural selection. Thus, by the processes of natural environmental and artificial selection, the forms, sizes, colors and flavors of the fruit began to change. Obtaining a larger fruit and greater productivity per plant is, even today, one of the goals of breeders and growers. In the domesticated or cultivated forms of chili, the fruits are hanging and remain on the plant upon ripening, and in some cases, the fruit is covered and hidden from birds; the plant is interdependent with man to survive. That is, domesticated or cultivated species do not survive in their wild state because the plants have lost various defense mechanisms against natural enemies, including their seed dispersion mechanisms. In commercial varieties, the fruits and seeds are generally larger (e.g., Bell types) compared with wild forms [3, 24]. One of the evident changes during the domestication syndrome was the increase in the germination rate of cultivated forms [4].

The domestication of chili peppers generated heritable genotypic changes, which are expressed in several known phenotypic variants. Environmental and human selection modified several morphological characteristics of the fruit and plant and continues to generate significant changes to adapt to different cultivation systems (e.g., greenhouse and high input use). Current cultivation and genetic improvement impose strong selection pressures to the point of dividing cultivated forms into highly pungent, intermediate and sweet or non-pungent groups. In all cases, sources of resistance or tolerance to disease and pests are sought but with high productive efficiency [25, 26]. Chili consumers seek a great diversity of fruits with varied grades of pungency, flavor, aroma, color and sizes of ripe, immature, fresh or dry fruits. In rural households, a few plants grown in the garden or backyard produce enough fruit to satisfy the needs of a family. Traditionally, small producers keep the produced seed from a small number of cultivated plants year after year; the seeds tend to be homogeneous or homozygotic because of regular self-pollination or crossing one another. All of these factors generate high differentiation among cultivated populations [23, 24, 27].

The extensive diversity of landraces of chili in Mexico and Mesoamerica is the product of the geographic convergence of wild and cultivated species. Among the cultivated endemic species, *C. annuum* and *C. frutescens* are notable, and species introduced from South America to Mexico are *C. chinense* Jacq. and *C. pubescens* R. & P. among *C. annuum, C. frutescens* and *C. chinense*, there is a strong crosslinking capacity, called the white flower *annuum-chinense-frutescens* species complex; this complex has a high likelihood of crossing with wild variants of *C. annuum* var. *glabriusculum*, particularly when the latter acts as a male parent [28, 29]. In this biogeographical context, diversity of landraces in Mexico is generated by crossing related species, genetic flow generated by the dispersal of seeds and the selection of cultivators, including the effects of genetic drift products from the low number of plants and geographic isolation [4, 5, 27, 30–32].

In addition to the high diversity of landraces, in the Mesoamerican region of Mexico, there is a large ethnolinguistic diversity, with the presence of 28 indigenous groups including Otomí, Mazahua, Náhuatl, Popolucas, Zapoteco, Mixteco, Mixe, Amuzgo, Triqui, Mazateco, Chinanteco, Mayas, Chontales, Huaves, Chatino, Cuicateco, Chontal, Tzetzal, Tzotzil, Purépecha, Totonaco, Ocuilteco and Matlazinca [33]. Pre-Columbian cultures represented by current indigenous groups exerted a strong selection pressure on chili cultivation to satisfy all food requirements such as medicinal, cultural and ritual. Brown et al. [34] and Kraft et al. [3] studied paleobiolinguistic evidence of the domestication of *Capsicum* in Mexico. Among the primary findings was evidence of 17 words of proto-languages to designate the various cultivated and wild variants or populations of chili; among these terms are Uto-Aztecan, Otomanguean, Popolocan-Zapotecan, Chinantecan, Mixe-Zoquean, Zapotecan, Mayan, Chiapanec-Mangue, Sonoran, Tonacan, Otomapean, Mixtecan, Amuzgo-Mixtecan, Totozoquean and Popolocan, all related to the indigenous groups mentioned here. This renders it possible to distinguish relations between a high diversity of landraces and the extensive gastronomic diversity of dishes [35].

Based on the phenotypic and genetic diversity of chili peppers in Mexico, here, we use the concept of landraces as a combination of definitions proposed by Zeven [36] and Camacho-Villa et al. [37] and the dimensions of seed lots managed by growers proposed by Louette et al. [38]. Thus, a chili landrace is understood to be a dynamic population of cultivated plants in the backyard or crop plots with defined, highly variable or uniform evolutionary origins of the characteristics of the plant, strong similarity in the form of fruit, seed lots that are managed independently and high locally adapted genetic diversity, occasionally with a different name associated with traditional forms of consumption and cultivation. In various cases, the forms, flavors and aromas of the fruit are phenotypically distinguishable, and differentiation frequently obeys geographic-reproductive isolation and selection that farmers utilize. Nevertheless, from that diversity of landraces, numerous traditional and improved commercial varieties have been generated and continue to be generated.

2.1. Phenotypic and genetic diversity among landraces

The wild and cultivated chili of Mexico have been known since pre-Columbian times and are the result of genetic recombination between wild and cultivated forms, only cultivated forms or between primary and secondary genepools. Today we know these genepools as indigenous, traditional and regional varieties or landraces of high phenotypic variation in fruit and plant traits. However, these landraces have suffered genetic erosion as a result of habitat loss, changes in the use of soil from forests to cultivated or urban zones and displacement by improved varieties introduced and imposed by the demands of national or international markets [24].

There are marked phenotypic divergences among landraces, relative to sizes, forms, colors and pungency of the fruit. For example, Piquin has a fruit diameter of 2–20 mm compared with Chilaca, which has a cylindrical-elongated form that is 15–35 cm in length with a 2–6 cm diameter. Anchos are conical-triangular fruits that vary from 12 to 15 cm in length and from 8 to 10 cm in width [39]. Colors vary from coffee-reddish, yellow, light greens, dark greens, whitish and very light greens with purple anthocyaninic stains such as in the Pico Paloma and Tusta types of south-southeastern Mexican chili peppers (**Table 1**).

With regard to consumption, landraces can be classified as fresh or dry. The largest number of landraces are consumed fresh or dry; among the latter are Piquin, Guajillo or Mirasol, Costeño, Puya, Cascabel, Catarina, Canica, Chilhuacle, Ancho, Mulato, De árbol or Cola de Rata and Pasilla. A special case is the preparation of the Chipotle chili (a form of drying by smoking or by oven and pickled); Jalapeno variants are used before ripening, and the fruit is dehydrated and dried in an oven or smoked. In some cases, the fruit is dried in the sun or shade or by smoking [32, 39, 42].

Species and botanical varieties	Regionalization for cropping and on-farm conservation of landraces (local and regional names)
C. annuum var. *annuum* (Cultivated and wild forms)	Península de Yucatán: Pico de Paloma[s], Xcat'ik[m], Ya'ax ik[m] o Cha'hua[m], Dulce, Jalapeño, Maax[m], Canica, Sucurre, Kum[m]
	South-southeastern: Guajillo o Mirasol, Pasilla o Chilaca, Ancho o Poblano, Taviche, De Agua, Costeño, Miahuateco, Mulato, Gordo, Huacle o Chilhuacle, Jalapeño, Serrano, Loco, Coxle o Chicoxle[c], Gallo-Gallina, De Árbol o Cola de Rata o Yahualica, Tabaquero, Taviche, Apaxtleco, Achilito, Nanche, Tusta, Piquín, Piquín de Simojovel, Chocolate, De Onza, Paradito o Escuchito, Lajoyero o Joyeño o Chilaquita, Parado de Zitlala, De Monte, Tecpín de Zitlala, Morrón, Garbanzo, Ojo de Cangrejo, Güero, Bojo, Blanco de Chiapas y Tabasco, Bandeño de Guerrero, Bolita de la Frailesca, Siete Caldos, Morado, Serranito, De Gallo, Ardilla, Totic de Chiapas
	Central: Ancho o Poblano, Miahuateco, Mulato, De Chorro o Cristalino, Pasilla o Chilaca, Tecomatlán, Loco, Rayado, Jalapeño, Serrano, Criollo de Morelos, Soledad, Copi de Puebla, Guajillo o Mirasol, Tabaquero, Mirador o Chalinguero, Pico de Paloma, Piquín o Piquín Huasteco, Cascabel, Soltero, Morron, Morita, Güero, De Chorro, Comapeño de Veracruz, Catarino del Bajío, Chilacate de Jalisco, Pahueteco, Altamira Serrano de Guanajuato, Carricilo o Tornachile, Ozuluamero, Canica del Bajío
	Northwest: Ancho, Mulato, Pasilla o Chilaca, Guajillo o Mirasol, Puya, De Árbol o Cola de Rata o Yahualica, Cora, Piquín
	Northeast: Ancho, Mulato, De Árbol o Cola de Rata o Yahualica, Guajillo o Mirasol, Serrano, Puya, Cora, Piquín, Cascabel, De Chorro, Corazón de Durango, Caloro de Chihuahua, Guajón de Zacatecas, Sarta de Sonora, Negro de Chihuahua, Sinahuisa Serrano de Sonora, Vallero de Chihuahua
C. annuum var. *glabriusculum* (commonly as wild forms)	Península de Yucatán: Maax[m] o Mashito, Piquín
	South-southeastern: Mashito o Amashito, Amashito Grande, Ojo de Cangrejo, Garbanzo, Bolita, Chingolito Amarillo, Piquín de Tabasco, Piquín Amarillo de Chiapas, Güiña Shirunduu[z], Chilpete de Jalisco
	Central: Piquín o Chiltepín
	Northwest: Piquín o Chiltepín
	Northeast: Piquín o Chiltepín
C. frutescens (cultivated a wild forms)	Península de Yucatán: Mashito[m]
	South-southeastern: Chilpaya, Tabasco, Mashito, Pico de Paloma, Bolita, Güiña Shuladi[z], Mirasol de Oaxaca, Miraparriba, Tabasqueño de la Frailesca
	Central: Pico de Paloma, Zacapaleño
	Northwest: Chilpaya

[m]Maya, [s]Español, [c]Cuicateco and [z]Zapoteco (local languages).

Regions of distribution: (a) **Península de Yucatán** = Estados de Quintana Roo, Yucatán and Campeche, (b) **South-southern**: Include the south region from Veracruz, Tabasco, Chiapas, Oaxaca and Guerrero, (c) **Central**: Cover the states of Jalisco, Colima, Michoacán, Estado de México, Guanajuato, Querétaro, Hidalgo, Morelos, Tlaxcala, Puebla and Central region of Veracruz, (d) **Northwest**: Include Zacatecas Aguascalientes, Nayarit, Sinaloa, Durango Sonora, Chihuahua Baja California Norte and Baja California Sur, and (e) **Northeast**: From North of Zacatecas, San Luis Potosí, North of Veracruz, Tamaulipas, Nuevo León and Coahuila.

Sources: González-Jara et al. [4], Pozo-Campodonic et al. [39], Votava et al. [31], Aguilar-Meléndez et al. [32], Pérez-Castañeda et al. [29], Cazáres-Sánchez et al. [40], Castañón-Nájera et al. [41], Vela [42]; Aguilar-Rincón et al. [43], Narez-Jiménez et al. [44], Loaiza-Figueroa et al. [45].

Table 1. Regionalization of the diversity of *Capsicum annuum* landraces in Mexico.

With regard to the level of pungency or spiciness, chili peppers are classified as quite pungent, fairly pungent, sweet or without spiciness. Any consumer of chili in Mexico recognizes Habanero and Manzano as the most spicy chilis; both variants were introduced to the country and belong to the species *C. chinense* (150,000–325,000 Scoville units) and *C. pubescens* (30,000–60,000 Scoville units), respectively [42]. After these two, De Árbol, Serrano, Jalapeño and Piquín variants follow in order of pungency. Finally, there are the moderately pungent or sweet chilis such as Dulce (sweet) chili of the Yucatan Peninsula, De agua from Oaxaca, Poblano and Güero that are particularly notable. Pungent or spicy chilis are more often preferred for the preparation of numerous dishes (**Table 1**).

A total of slightly more than 150 landraces that are extensively planted and consumed have been described. The distribution of the diversity of chili landraces in Mexico can be divided into five regions. The Yucatan Peninsula includes the states of Quintana Roo, Yucatan and Campeche. Here, the Sucurre, Maax, Xcat'ik, Yaax ik and Dulce variants, among others, occur most frequently. Two regions are quite relevant to the in situ preservation of landraces: south-southeast and central, in which regionally, more than 80 local varieties are preserved (**Table 1**). In these regions, the greatest diversity of indigenous groups is also concentrated; therefore, current indigenous people preserve the great diversity of landraces. The landraces of northern Mexico (northeast and northwest) are less diverse than in central and south-southeastern areas.

Wild and cultivated forms of chilis of greater diversity, endemism and distribution in Mexico are *C. annuum* var. *annuum, C. annuum* var. *glabriusculum* and *C. frutescens*. Despite advances in the distribution, classification and quantification of the genetic diversity of *Capsicum*, interest-specific hybridizations in the *C. annuum-C. frutescens-C. chinense* complex that converges in Mexico remain unexplored, including the variability generated by hybridization between landraces and improved introduced varieties. Among landraces, strong genetic divergences have been documented by differences in regionally distributed populations. For example, by nine isoenzymatic systems, Loaiza-Figueroa et al. [45] determined high isoenzymatic diversity by the population description of 186 accessions.

The preservation of a large diversity of landraces is associated with a culture of exploitation of the species that originated in pre-Columbian times [1–3]. Among the aspects that continue to draw attention to gastronomic use are the variety of flavors, aromas and spicy characteristics, including variations in color in dishes and perceptions that are associated with compositions of the fruit in terms of capsaicinoids, flavonoids and polyphenols, among other compounds. In a traditional rural Mexican family, it is quite unlikely that the diet does not include chilis as a condiment, sauce or main dish.

Many authors have demonstrated that there is great genetic diversity among the populations of a landrace. With regard to diversity between and within landraces, Contreras-Toledo et al. [46] and Toledo-Aguilar et al. [47], using microsatellite markers, demonstrated great genetic diversity within the Ancho or Poblano, Piquín, Guajillo and Chilaca landraces. Kraft et al. [27] indicated that divergences among landraces are fundamentally a result of the form of selection of the seed and isolated management of the crop. This same pattern is observed among wild forms (e.g., Piquín), which grow spontaneously in backyards [4]. Pacheco-Olivera et al. [23], in evaluations of the genetic diversity of Mexican chili peppers by microsatellites, observed that the diversity and genetic differentiation of three landraces were statistically similar to the evaluated pattern in seven commercial hybrids.

2.2. Regionalization of production and consumption of chili landraces

In Mexico, the cultivation and consumption of chili peppers have great economic and social importance. In 2015, 153,565 ha were cultivated, with an average production of 2.8 million tons of dried and fresh fruit. Of the total production, 22,143 and 793,501 tons of dry and fresh fruit were exported, respectively. National consumption neared 2 million tons, and in some years, imports of approximately 20,757–41,000 tons have been reported [6]. Consumption per capita in urban areas is from 8 to 9 kg, and in rural communities, it varies from 14 to 17 kg. Landraces and varieties of greater production and consumption are Jalapeño, Ancho, Serrano, Guajillo, Poblano and Mirasol, with more than 9000 hectares planted. The varieties of highest consumption dried are Ancho, Guajillo and Mirasol. Serrano, Jalapeño and Ancho are grown throughout in nearly all of Mexico (**Table 2**).

State of production origin[1]	Landraces, improved varieties and local populations of fresh and dry chili peppers
Aguascalientes	Guajillo, Poblano o Ancho
Baja California Norte	Serrano, Jalapeño, Anaheim, Bell
Baja California Sur	Bell, Anaheim, Caloro, Jalapeño, Serrano, Poblano,
Campeche	Habanero[c], Jalapeño
Coahuila	Habanero[c], Bell, Anaheim, Serrano, Poblano, Jalapeño
Colima	Habanero[c], Jalapeño, Serrano
Chiapas	Habanero[c], Serrano, Costeño
Chihuahua	Habanero[c], Ancho, Colorado, De Árbol o Cola de Rata, Mirasol, Pasilla, Caloro, Cayenne, Chilaca, Jalapeño, Paprika, Poblano, Serrano,
Durango	Ancho, Puya, Jalapeño, Poblano, Anaheim, Bell
Guanajuato	Chilaca, Serrano, Jalapeño, Poblano, Bell, Anaheim
Guerrero	Ancho, Costeño, Puya, Chilaca, Guajillo, Mirasol, Serrano
Hidalgo	De Árbol o Cola de Rata, Jalapeño, Serrano, Bell, Anaheim
Jalisco	De Árbol o Cola de Rata, Tabaquero, Anaheim, Caloro, Chilaca, Jalapeño, Poblano, Serrano, Bell
México	Manzano[p], Jalapeño, Serrano
Michoacan	Habanero[c], Pasilla, Caloro, Chilaca, De Árbol o Cola de Rata, Jalapeño, Manzano o Peron[p], Piquín, Poblano, Serrano, Bell, Anaheim
Morelos	Anaheim, Serrano, Jalapeño
Nayarit	Habaneo[c], Cascabel, De Árbol o Cola de Rata, Caloro, Cristal, Jalapeño, Serrano, Anaheim, Bell

http://dx.doi.org/10.5772/68076

State of production origin[1]	Landraces, improved varieties and local populations of fresh and dry chili peppers
Nuevo Leon	Habanero[c], Serrano, Jalapeño
Oaxaca	Habanero[c], Ancho, Costeño, Pasilla, Tabaquero, De Agua, Jalapeño, Serrano, Soledad, Chilhuacle
Puebla	Ancho, Jalapeño, Serrano, Poblano, Miahuateco, Bell
Querétaro	Pasilla, Chilaca, Jalapeño, Serrano, Anaheim
Quintana Roo	Habanero[c], Jalapeño, Dulce
San Luis Potosi	Habanero[c], Ancho, Guajillo, Mirasol, Mulato, Pasilla, Puya, Chilaca, Jalapeño, Poblano, Serrano, Bell, Anaheim
Sinaloa	Habanero[c], Anaheim, Bell, Caloro, Chilaca, De Agua, De Árbol o Cola de Rata, Hungaro, Jalapeño, Poblano, Serrano
Sonora	Habanero[c], Anaheim, Bell, Jalapeño, Serrano
Tabasco	Habanero[c], Costeño, Tabaquero, Jalapeño, Dulce
Tamaulipas	Habanero[c], Jalapeño, Serrano, Piquín, Poblano, Bell, Anaheim
Tlaxcala	Serrano
Veracruz	Habanero[c], Jalapeño, Manzano[p], Serrano, Piquín, Soledad
Yucatan	Habanero[c], Mulato, De Árbol o Cola de Rata, Xcat'ik, Bell, Anaheim
Zacatecas	Ancho, De Árbol o Cola de Rata, Guajillo, Mirasol, Mulato, Pasilla, Puya, Caloro, Poblano

[1]SIAP [6]; Species: [p]*C. pubescens*, [c]*C. chinense* and without superscript are *C. annuum*.

Table 2. Regionalization of landraces and improved varieties cropping by production origin.

According to the geography, altitude and climates of Mexico, there are six primary patterns of cultivation: (a) Habanero variants (*C. chinense*) are sown preponderantly in Yucatán, Quintana Roo and Campeche, although recently, Habanero was cultivated in Coahuila, Chihuahua, Colima, Nayarit and Michoacán, (b) diverse Manzano (*C. pubescens*) chilis are cultivated in Central Mexico in Michoacán, Estado de México, Puebla and Veracruz, (c) the commercial Morrón (sweet bell) varieties are preponderantly distributed in the north such as in Bell, Anaheim, California and Victoria, which are exported to the United States of America, (d) Ancho, Guajillo, Poblano and Pasilla variants are grown from Guanajuato to Zacatecas and Tamaulipas and finally, (e) varieties with regional distribution include Chilhuacle in Oaxaca and Guerrero, Tabaquero in Tabasco and Soledad in Veracruz and Oaxaca, among other patterns (**Table 2**).

Several references contrast the beneficial and harmful effects of capsaicinoids consumed by chilis. Despite advances in the demythification of harmful effects, studies nevertheless refer to capsaicin as a risk factor for cancer [14, 48]. Recent studies indicate a lack of direct association between chili peppers and capsaicinoid consumption and cancer or several other diseases.

Capsaicin is used as a topically applied analgesic to decrease pain in rheumatoid arthritis, osteoarthritis, neuralgias, neuropathic diabetes, neuronal inflammations and dysfunctions, among others [15, 17]. The use of capsaicinoids is controversial, and it is necessary to further investigate their effects on human health because among other aspects, it is not possible to extrapolate from biological studies of animals (e.g., rats) to humans [13, 17].

2.3. Chili peppers in Mexican gastronomy

The varied forms of chili consumption are characteristic of traditional Mexican cuisine and are ancestral and living cultural expressions of the community, a part of the Mexican national identity [49]. In Mexico, there is a great preference for spicy chili, whose consumption is more common in people who live in hot, tropical or subtropical climates than in people who live in temperate-to-cold regions [50]. The sensation of heat or sting is the result of the presence of various types of capsaicinoids such as capsaicin, dihydrocapsaicin, homodihydrocapsaicin, nordihydrocapsaicin and homocapsaicin, among others, which have vanillin, 8-methyl-N-Vanillyl-6-nonemide, as a base compound. During ingestion, chemo-static stimuli are generated that affect chemical stimulation of thermoreceptors, nociceptors and somatosensory receptors [51].

The populations of Mexico, Central and South America show evolutionary adaptations according to the food these people consumed. Culinary culture in Mexico is as diverse as differences in ethnicity, geography and ecology; the current cuisine is a combination of indigenous pre-Columbian, Spanish and French cuisines and, in some regions, Lebanese. Recent times have seen a marked influence of American culture, increasing Mexicans' consumption of carbohydrates. In Mexican families, food consumption practices are taught from childhood; mothers give chili to children from 1 to 3 years of age. This evidence of cultural reinforcement is primarily focused on distinguishing the flavonoid principles of food consumption. Common foods are bean broth, soups prepared with tomato (*Solanum lycopersicum* L.) and dishes flavored with chili; as the child grows, chili consumption increases. Consuming such foods helps to transmit identity from parents to children and implicitly strengthens the food culture [52].

In practical terms, the continued consumption of chili pepper through several generations has engendered genetic modifications in the population of Mexican consumers. Capsaicin tolerance is closely associated with AVI (alanine, valine and isoleucine) allelic variants of the TAS2R38 type 2 gene in the tongue and palate. Consequently, homozygotic carriers of the AVI haplotype of the taste receptor perceive only a slight "burning" sensation. This genetic variant is also associated with a low perception of a bitter flavor. That is, a lower perception of bitterness in other foods and a lower perception of the burning of chili are because of TAS2R38 [50, 53, 54].

The Mexican consumer of chili has an extensive variety of landraces that the consumer produces himself or can buy in local and regional markets, from the "sweet" types or types without spice to moderately spicy and extremely spicy, and all are widely used in regional cuisine and family diets. Chilies are consumed in whole, ground, sliced, diced, dried, roasted forms or can be acquired pickled in cans from nearly all landraces. Chilis are essential to Mexican gastronomy and played a decisive role in UNESCO's designating Mexican cuisine the "intangible cultural heritage of mankind" (Decision 5.COM 6.30, 2010) [55].

The variety of chili landraces is directly associated with a wide variety of processing forms for consumption. Among chili and forms of consumption are the following groups of peppers and preparation styles:

(a) Group I. Sweet types such as Morrón (Anaheim, Bell and California), Ancho de Puebla, De Agua de Oaxaca or Dulce de Yucatán are eaten raw and used in salads, primarily the Morrón because that type ripens in red, green or yellow and adds color to dishes.

(b) Group II. Unripe or ripe spicy green peppers such as the Serrano, Jalapeño, Rayado and Loco. Ya'ax ik, Criollos de Morelos, De Árbol Verde, Xcat'ik, Copi, Soledad, Cora, Tabaquero, Costeño Verde and others are regularly used to prepare traditional green salsas and green moles, typical dishes of Mexican cuisine.

(c) Group III. Among chilis consumed, dried, powdered Piquín is notable, as are chilis used to prepare different types of regional moles: Ancho, Pasilla, Guajillo, Chilhuacle, Taviche, Costeño, Cascabel, De Onza, Mulato, Miahuateco, Cristalino, Gordo, Coxle, De Árbol Seco, Puya, Chawa, Tabaquero, Apaxtleco and others.

Mole is a typical Mexican national dish resulting from the fusion of Spanish and pre-Columbian indigenous cuisines. The word *mole* (in Spanish) comes from the indigenous language Nahuatl "*molli*", which means a mix of ingredients and flavors. Ingredients and forms of preparation vary from region to region or between states. For example, moles from Puebla, Oaxaca, Tlaxcala, Estado de México, Querétaro, Michoacán and Hidalgo are quite popular. The dish is a combination of different chili landraces (e.g., Ancho, Pasilla, Morita and others) and various fruits, seeds and bulbs, depending on the mole variant (**Table 3**). However, the essential ingredients, according to connoisseurs, are the chilis. In Oaxaca, seven types of moles are prepared, and the most popular are "negro (black mole) with Chilhuacle", "verde" (green mole) with Jalapeño or Serrano chilis, "coloradito" (reddish mole) with Ancho and Guajillo chilis, red mole with Ancho and Guajillo chilis and Amarillo (yellow mole) with Anchos, Guajillos and Costeños. Additionally, "chichilo" and "manchamantel" ("tablecloth-staining") moles are mentioned. For each type of mole, the complementary condiments vary [56].

Another common dish of Mexican cuisine is from spicy to extremely spicy salsas, which are prepared with fresh or dried chilis, cooked or uncooked, and are a traditional form of promoting spicy flavors (capsaicinoids), flavonoids, aromatic compounds and phenolic compounds. Occasionally, chilis are roasted and ground with green tomato (husk tomato, *Physalis ixocarpa* Brot.), red tomato (*S. lycopersicum*), garlic, onion, salt and water. In other cases, all of the ingredients are cooked and then ground (**Table 3**). From one region to another, the primary ingredients and forms of preparation vary. In this sense, Mexican guacamole is the addition of avocado to green salsa (e.g., Jalapeño or Serrano chilis cooked with green tomato, plus garlic and water) in which pieces of avocado are mashed slightly once the green salsa is ground, all of this in a "molcajete" (a rock with a hole) [57]. A group of Mexican dishes prepared with different local varieties of chilis is listed on **Table 3**. It is unlikely that the level of consumption of chilis in Mexico will decrease in the short term.

Typical dishes of the Mexican cuisine[1]	Landrace used or improved variety
Known molee: negro, coloradito, chichilo, rojo, verde, pipián, amarrillo, poblano (turkey in deep-brown sauce), "estofado", "manchamantel", de caderas, de hongos, de frijol, relleno negro, almendrado	**(a)** Blanck mole: Ancho, Chilhuacle **(b)** Red mole: Ancho and Guajillo **(c)** Green mole: Jalapeño, Rayado and Serrando
Salsas (sauces) prepared with fresh and dry chilis: Borracha ("drunk"), verde, roja, de ancho, de guajillo	**(a)** Freh: Commonly Jalapeño o Serrano buta ll high spiced chilis can be used. **(b)** Dry: Ancho, Pasilla, De árbol, Cascabel
Chilaquiles: verdes (green), rojos (red), de mole (made with mole)	Jalapeño, Serrano, Ancho, Guajillo, Pasilla, Cascabel, De Árbol
Tamales de verdes (greenish) o rojos (red), de mole, de rajas de chile (with slices of chili)	Jalapeño, Serrano, Ancho, Pasilla
Traditional guacamole	Serrano, Jalapeño
Chiles rellenos (stuffed chili peppers)	Poblano, Ancho, De Agua, Güeros, Jalapeños
Tortas ahogadas and Pambazos	Pasilla, Guajillo
Elotes (cob) o frutas (tropical fruits) with powdery dry chili pepper	Chili Piquín milled with sal
Chile en chipotle (smoked peppers)	Jalapeño and Rayado
Chili in encurtidos (pickled peppers)	Usually Jalapeño and Serrano but all chili peppers can be used.
Roasted chili to slice o eat directly ("morder") o en rajas:	Regularly Serrano and Jalapeño
Immature peppers or green to mature but fresh to slice and combine with onion and lemon, regionally known as "gatos" or "pico de gallo"	**(a)** Immature: Jalapeño, Serrano, De Agua **(b)** Muture: Habanero, Manzano, Bell, Anaheim o California
Chileatole	Serrano
Chiles en nogada (dish name)	Ancho, Poblano, Miahuateco
Flautas ahogadas (dish name)	Jalapeño, Serrano
Mixiote verdes de pollo (dish name)	Jalapeño, Serrano

[1]Sources: Urdaneta and Kanter [56] and Katz [57].

Table 3. Typical dishes of the Mexican cuisine where main ingredients are the diversity of landraces or improved varieties of chili pepper.

3. Flavonoid and capsaicinoid contents in fruits of chili pepper landraces

3.1. Flavonoids in *Capsicum*

In Mexico, there is a great diversity of phenotypic variants of chilis that are distinguished by color, flavor, spice, aroma, size and shape and are fundamental to and completely accepted in Mexican gastronomy. The species *C. annuum* originated in Mexico and is one of the most cultivated species in the world, with high genetic diversity [39, 43]. The organoleptic properties of each type of chilis are characterized by their chemical composition as a function of flavonoid content, capsaicinoids, phenolic acids, vitamins, minerals and various volatile compounds [40, 58–61]. Flavonoids and capsaicinoids are molecules that, in addition to influencing the sensory characteristics of the fruit, contribute to the prevention of chronic degenerative diseases [62, 63].

Flavonoids are secondary metabolites of low molecular weight that share a common skeleton with diphenylpropanes (C6-C3-C6) and comprise two phenyl rings (A and B) joined through a C ring of pyran (heterocyclic). Flavonoids are classified according to oxidation state and degree of unsaturation of the central heterocyclic ring. The primary families are flavones, isoflavones, flavanones, flavanols, antho-cyanidins and chalcones [64]. The structural diversity of flavonoids depends on the substitution of aromatic rings A and B by hydroxyl and methoxy groups and by extensive conjugations, including glucosides. In foods, flavonoids exist primarily as 3-O-glycoside and polymers [64, 65]. Flavonoids possess multiple properties for eliminating reactive oxygen species. Their activity as an antioxidant depends on the redox properties of their hydroxyphenolic groups and the structural relation between various components of the chemical structure [7]. Some of the characteristics that favor their antioxidant capacity are the ortho-hydroxylation of ring B, the number of free hydroxyl groups, a C2-C3 double bond at ring C and the presence of a 3-hydroxyl group [66].

In fruits and vegetables, it is estimated that more than 7000 flavonoids have been identified [64]. In chili fruit, the primary quantified flavonoids are quercetin, luteolin, kaempferol, catechin, epicatechin, rutin, luteolin, apigenin and myricetin [61, 67–76]. Variations in flavonoid concentrations are primarily the result of the diversity of genotypes, landraces, varieties and the ripening phase of the fruit; also implied are variations related to analytical laboratory parameters such as sample preparation, extraction method and quantification methods (**Table 4**).

In chili fruits, different flavonoids have been identified depending on morphotype, landrace, varietal group or variety. In sweet peppers (*C. annuum* L. cv. Vergasa), 23 flavonoids have been identified and quantified in the pericarp by high-performance liquid chromatography-diode array detection-electrospray ionization mass spectrometry (HPLC). These include O-glycosides of quercetin, luteolin and chrysoeriol and a large number of C-glycosyl flavones. The most abundant compounds were quercetin-3-O-rhamnoside and luteolin 7-O-(2-apiosyl-6-malonyl), which represented 41% of the total flavonoids [77]. Materska and Perucka [79] identified quercetin 3-O-α-L-rhamnopyranoside-7-*O*-*β*-D-glucopyranoside, luteolin 6-*C*-*β*-D-glucopyranoside-8-C-α-L-arabinopyranoside, apigenin 6-*C*-β-D-glucopyranoside-8-*C*-α-L-arabinopyranoside, lutoeolin 7-O-[2-(β-D-apiofuranosyl)-β-D-glucopyranoside], quercetin 3-*O*-α-L-rhamnopyranoside and luteolin 7-O-[2-(β-D-apiofuranosyl)-4-(β-D-glucopyranosyl)-6-malonyl]-β-D-glucopyranoside. These flavonoids are present in four varieties of *C. annuum*. In the "Italian sweet" (green), "Lamuyo" (yellow) and "California wonder" (red) varieties, 23 flavonoids and their glycoside derivatives were identified by high-performance liquid chromatography coupled with diode array and electrospray time-of-flight mass spectrometry detectors (HPLC), highlighting the group of flavonols known as glycosylated quercetin derivatives (rutin pentoside, quercetin 3,7-di-O-α-Lrhamnopyranoside and quercetin 3-O-α-L-rhamnoside) [79]. Various flavonoids continue to be identified in all species of *Capsicum*, and depending on advances, it is inferred that the task continues and the work is incipient in Mexican landraces.

Flavonoids[1]	Maturation stage	*C. annuum* landrace and variety groups							*C. frutescens*
		Jalapeño	Serrano	Ancho	Güero	Morron	Cayenne	Piquin	Tabasco
Quercetin	Immature	4.3.–151.2	9.30–159.8	276.0	42.4–210.2	4.76–276.0	6.0–22.9	ND	2.2
	Mature	ND	8.1	ND	23.9–64.4	3.29–448.0	3.7–24.7	ND	0.9
Kaempferol	Immature	5.9	2.1	ND	ND	ND	2.0–4.7	ND	ND
	Mature	1.4	ND	ND	ND	ND	6.3–6.4	ND	ND
Catechin	Immature	0.1	1.0	ND	ND	1.85–5.13	ND	ND	ND
	Mature	ND	ND	ND	ND	5.28–6.41	ND	ND	8.1
Epicatechin	Immature	0.1	1.2	ND	ND	3.70–7.35	ND	ND	ND
Rutin	Immature	0.2	2.0	ND	ND	0.38–1.90	ND	ND	ND
Luteolin	Immature	0.20–37.5	0.57–41.40	3.6	15.7–51.5	0.35–9.32	2.0–19.1	ND	43.6
	Mature	3.2	1.4	ND	5.96–16.8	0.36–11.0	7.1–17.3	ND	0.84–35.6
Myricetin	Immature	ND	ND	ND	ND	658.0	2.1.0–2.1	ND	ND
	Mature	ND	ND	ND	ND	171.0–244.0	5.9.0–7.2	ND	ND
Apigenin	Immature	ND	ND	ND	ND	272.0	ND	ND	ND
Total Flavonoids	Immature	10.2–332.0	11.4–441.0	309.6	58.1–309.0	5.4–31.7	10.0–48.8	97.40–544.6	45.8
	Mature	4.6	9.5	ND	29.9–81.3	3.6–892.0	42.1–44.3	50.1–425.0	36.5
References		[74, 76]	[74, 76]	[74]	[67, 74]	[67, 69, 71–73]	[67]	[61, 75]	[67, 73]

[1]Quercetin, catechin or quercetin + luteolin equivalents (μg/g or g/g), ND = Not determined.

Sources: Vera-Guzmán et al. [61], Howard et al. [67], Miean and Mohamed [69], Kim et al. [71], Blanco-Ríos et al. [72], Zhuang et al. [73], Lee et al. [74], Rochín-Wong et al. [75], Álvarez-Parrilla et al. [76].

Table 4. Flavonoid contents in fruits of *Capsicum annuum* and *C. frustences*.

The structure of each flavonoid influences antioxidant capacity, which varies between species and genotypes of chili. Materska [80] isolated three glycosylated flavonoids (luteolin 6-C-glucoside, luteolin 6,8-di-C-glucoside and apigenin 6-C-glucoside-8-C-arabinoside) from the chili fruit (*C. annuum* var. Capel Hot) and determined their antioxidant activity using in vitro methods to generate radicals in hydrophilic (superoxide radical) and lipophilic (2,2-diphenyl-1-picrylhydrazyl—DPPH—and peroxide radicals) media. This study demonstrates that luteolin 6-C-glucoside and luteolin 6,8-di-C-glucoside have a greater ability to eliminate superoxide radicals generated in enzymatic and non-enzymatic systems and thus engender high antioxidant activity [80]. The high and effective antioxidant capacity confers functional and nutraceutical properties to the fruit for health.

Metabolism, biochemical synthesis and concentration of flavonoids in chili fruits depend on species, genotype, landrace or commercial varieties [61, 67–69, 81] interacting with agroecological characteristics and crop management [82]. Evidence indicates that the concentration of flavonoids is related to the degree of maturity of the fruit. Bhandari et al. [83] observed that fruits had a higher flavonoid content in the initial-to-intermediate phases (breaker stage) than in the phase immediately after maturity (green mature phase) and red ripe phase. Similarly, Howard et al. [67] observed that total flavonoid content was reduced in the immature-to-mature phase of the fruit (**Table 4**). This fact is consistent with other studies; green fruits (immature) reach four to five times the flavonoid content of mature fruits [67, 72, 77, 84]. In C. *chinense* Jacq. (Habanero), flavonoid content also decreases with ripening [85]. However, the behavior of Morrón peppers is different because the peppers ripen in different colors; for example, quercetin and luteolin flavonoid content was higher in red fruits than in green [68]. Differences in the composition of flavonoids and other compounds define the organoleptic characteristics of each chili fruit and are directly related to usage and consumption preferences [86].

Chili fruits are considered a natural source of antioxidants for their bioactive compounds [61, 77, 87]. Phenolic compounds increase the antioxidant activity of chili fruits and are related to the structure of their molecules. In the case of Caribe and Bell varieties (*C. annuum*), the major antioxidant activity was the result of the content of the flavonoids catechin, epicatechin and rutin, among other compounds [81]. In addition, high activity has been reported when quercetin 3-Oα-L-rhamnopyranoside was identified; such activity was comparable with quercetin activity [78]. Quercetin is a highly antioxidative aglycone and is generated by enzymatic hydrolysis of the glycoside bond of rutin (quercetin 3-O-rhamnoglucoside) [66]. In evaluating the fruits of the landraces De Árbol, Chipotle, Guajillo and Morita (*C. annuum*), it was determined that there was greater bioaccessibility of polyphenols in the small intestine in a range of 72–77%. Therefore, these landraces are considered important sources of polyphenols and bioaccessible bioactive compounds in the intestine [88].

3.2. Capsaicinoids in chili pepper landraces

Capsaicinoids are compounds that confer pungency or spice to chili fruits and are synthesized by the condensation of vanillylamine with a branched short chain fatty acid. Their chemical structure comprises a phenolic nucleus joined by an amide bond to a fatty acid. The phenolic portion is vanillylamine and is synthesized from phenylalanine in the phenylpropanoid pathway. The fatty acid is generated from branched chain amino acids, valine or leucine [89]. Currently, more than 20 capsaicinoids are known [90], and differences in structures occur because of the nature of the side chain, which can vary between 8 and 10 carbons, and the number of double bonds. Capsaicinoids are classified into three groups of compounds: capsaicins possessing a methyl branched acyl residue with a carbon-carbon double bond, dihydrocapsaicins analogous to the previous class but being saturated compounds and N vanillyl-n-acylamides comprising saturated, unbranched alkyl chains [91]. Capsaicin [(E)-N(4-hydroxy-3-methoxybencil)-8-methyl-6-nonenamide)] and its analogue 6,7-dihydrocapsaicin represents more than 90% of total capsaicinoids in chili fruits (**Table 2**), primarily accumulating in the placenta of the fruit [58]. Schweiggert et al. [91] identified and characterized

15 capsaicinoids in the red fruits of *C. frutescens*, primarily capsaicin, dihydrocapsaicin and nordihydrocapsaicin, with nornorcapsaicin, norcapsaicin, homocapsaicin I and II, nornordihydrocapsaicin, homodihydrocapsaicin isomers I and II, N-Vanillyl-octanamide, N-vanillyl-nonanamide and N-anillyl-decanamide as minor compounds (**Figure 1**).

Capsaicinoids are synthesized in the placenta of the fruit and are genetically determined (*Pun1* allele of pungency) by the presence of the *Pun1* or *pun1* gene with EST- or AT3-type cofactors that induce a quantitative effect of the gene and variations in the pungency of the fruit. In consequence, not all chili peppers are spicy, and various consumers consider chili varieties that carry the *pun1* recessive gene to be sweet fruits [11]. In practical terms, it is difficult to determine the exact content of capsaicinoids in chili fruits, and the estimates or patterns that are obtained vary enormously depending on genotypes, ecological-environmental conditions, crop systems, the maturity of the fruit and harvest season, among other aspects [61, 87, 92–95]. **Table 5** presents the estimates of the capsaicinoid content in fruits of different landraces and varietal groups. Various studies have demonstrated that capsaicinoid content is greater in mature fruits than in immature one [58, 73, 78]. Some of the landraces with notable concentrations of capsaicinoids are Piquín, De Árbol and Serrano, and those with lower concentrations are the Morrón group (California, Bell, Anaheim), Pasilla, Ancho and Guajillo, among others [59, 95, 96].

Ecological-environmental conditions are factors that influence the accumulation of capsaicinoids in fruits; among the primary climatic elements are temperature and precipitation or irrigation interacting with crop management. For example, González-Zamora et al. [95] observed that temperatures of 40–48°C reduce the concentrations of capsaicinoids from 32.5 to 61.5% and

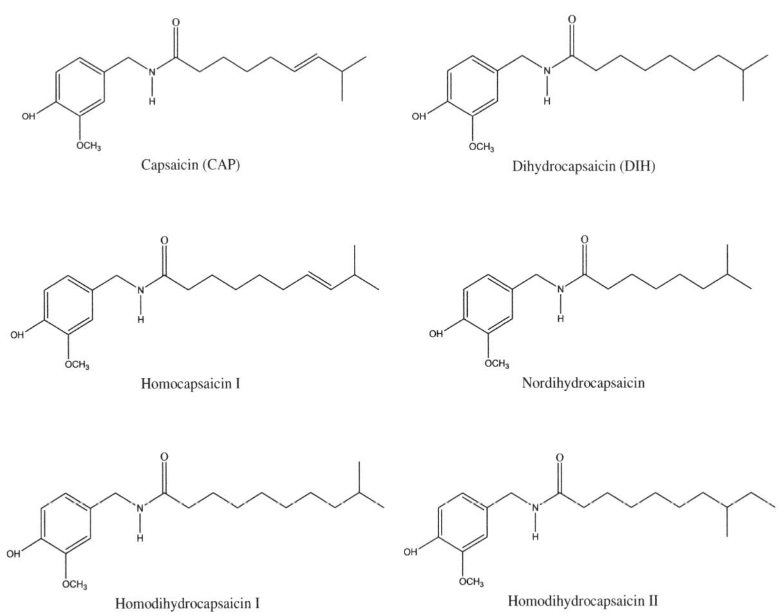

Figure 1. Structures of major capsaicinoids identified in *Capsicum* species.

http://dx.doi.org/10.5772/68076

Landrace or varietal groups	Capsaicin (CAP)[1]		Dihydrocapsaicin (DIH)		Nordihydro-capsaicin		Homo-CAP and Homo-DIH	References
	Immature	Mature	Immature	Mature	Immature	Mat.	Immature	
Jalapeño	115.8–8000	169.0–373.5	54.0–9390	113.5–237.8	13.3–2480	24.4	160–970	[75, 95]
Serrano	47.5–1606	81.3–627.48	57.2–3540	87.7–399.7	22.3–530	27.8	640.0	[75, 95, 96]
Mulato	ND	29.5	ND	49.5	ND	ND	ND	[59]
Pasilla	1.0	49.2	ND	68.8	1.0	ND	ND	[96]
Guajillo	170.0	22.9	610.0	36.9	120.0	ND	560.0	[95, 96]
Ancho	0.3	42.8	0.8	0.4	0.3	ND	0.6	[96]
Poblano	0.6	ND	ND	ND	ND	ND	ND	[96]
Pasado¤	ND	38.3	ND	58.4	ND	ND	ND	[96]
Puya	1180.0	53.9	2320.0	67.4	550.0	ND	68.0–80.0	[95, 96]
Tres Venas	ND	66.9	ND	77.1	ND	ND	ND	[96]
Mirasol	ND	353.8	ND	231.7	ND	ND	ND	[96]
Morita	ND	338.2	ND	334.9	ND	ND	ND	[96]
Chipotle	ND	163.8–883.0	ND	126.3–552.7	ND	ND	ND	[96]
De Árbol	138.5–4249	309.3–1293	146.4–6250	238.2–641.7	1070.0	ND	130.0–1290	[93, 95, 96]
Chiltepin	3790–15360	4170	13390	ND	2170.0	ND	220–690	[75, 86]
Piquin	ND	16.2–2657	ND	62.4–1031	ND	ND	ND	[61, 96]
Maax	640.0	750.0–3584	370.0	1707	ND	ND	ND	[40, 58]
Tusta	51.4	ND	33.5	ND	ND	ND	ND	[61]
Tabaquero	ND	6.7	ND	1.5	ND	ND	ND	[61]
Solterito	ND	142.0	ND	65.5	ND	ND	ND	[61]
Nanchita	ND	27.4	ND	13.2	ND	ND	ND	[61]

Landrace or varietal groups	Capsaicin (CAP)[1]		Dihydrocapsaicin (DIH)		Nordihydro-capsaicin		Homo-CAP and Homo-DIH	References
	Immature	Mature	Immature	Mature	Immature	Mat.	Immature	
Güero	44.5	ND	6.8	ND	ND	ND	ND	[61]
Costeño	ND	14.6	ND	4.0	ND	ND	ND	[61]
De agua	ND	4.9	ND	1.6	ND	ND	ND	[61]
Sucurre	26945	2930–18995	3652.0	4355–5043	ND	ND	ND	[40, 58]
Pico paloma	ND	2456.4	ND	1928	ND	ND	ND	[40]
Ya´x ik	ND	1777.6	ND	1811	ND	ND	ND	[40]
Chawa	12815	1415–11822	2175	1317–5555	ND	ND	ND	[40, 58]
Xcat ik	ND	748.4–3189	ND	831.3	ND	ND	ND	[40, 58]
Bobo	ND	204.7	ND	372.2	ND	ND	ND	[40]
Dulce	ND	42.4	ND	58.9	ND	ND	ND	[40]
Miahuateco	ND	63.6	ND	45.5	ND	ND	ND	[59]
Copi	ND	267.4	ND	167.7	ND	ND	ND	[59]
Tecomatlán	ND	54.6	ND	35.7	ND	ND	ND	[59]
Morrón	1.8–143.1	1.2–134.5	55.6–99.0	57.5–113.4	ND	ND	ND	[73, 84]
Cayenne	20.8–149.5	53.7–211.7	72.0	40.1–114.6	ND	ND	ND	[84]
Tabasco[2]	ND	746.8	ND	496.1	ND	ND	ND	[73]

[1]Content in μg/g of fresh or dry weight, [2]*Capsicum frutescens*, ND = Not determined.

Sources: Cazáres-Sánchez et al. [40], Cisneros-Pineda et al. [58], Morán-Bañuelos et al. [59], Vera-Guzmán et al. [61], Kim et al. [71], Zhuang et al. [73], Rochín-Wong et al. [75], Álvarez-Parrilla et al. [76], Bae et al. [84], Othman et al. [93], González-Zamora et al. [95], Orellana-Escobedo et al. [96], Ornelas-Paz et al. [77].

Table 5. Capsaicinoid contents in fruits of chili landrace and varietal groups.

32.5% in Jalapeño and De Árbol chilis, respectively. By contrast, in Guajillos and Serranos, the identical temperature generated up to a three-fold increase; a similar effect was identified in Puya and Ancho but only 21 and 8.6% more, respectively. In a trial in three regions of Peru, the change in temperature from 19.4 to 26.8°C generated a decrease in capsaicinoid content [92].

3.3. The effect of traditional processing on flavonoid and capsaicinoid contents

Chili fruits are consumed fresh, dried or processed. The chili's condition or processing experience changes their chemical composition because of the effects of temperature changes, pH, solar radiation, smoking or other treatment. For example, the composition of Chiltepín chili fruits changes because of the effect of the pickling process (cooked in 1:1 water and vinegar). In this case, total flavonoids, capsaicin and antioxidant activity are reduced to less than 25%. However, no significant changes are shown as a result of sun drying (temperatures between 34 and 40°C for 32 h) [75]. Similarly, Álvarez-Parrilla et al. [76] identified reductions in capsaicinoids in Serrano and Jalapeño chilis from the pickling process. During the preparation of salsas with Chiltepín, losses in capsaicinoids were detected from changes in pH and the milling process; with pH 2.7, the reduction of capsaicin and dihydrocapsaicin was 90% [98].

Regarding processing for consumption, in Mexico, the preparation of mole paste (a traditional dish) prepared with Pasilla (landrace) chili generates a decrease in the content of flavonoids, phenols and antioxidant activity as a result of cooking and changes in pH, which induce degradation of these compounds [99]. Ornelas-Paz et al. [97] indicated that cooking in water at only 96°C or on the grill at 210°C generates moderate losses of 1.1 and 28.1% of the initial content in Poblano, Bell, Chilaca, Caribe, Jalapeño and Serrano landraces of *C. annuum*, Habanero (*C. chinense*) and Manzano (*C. pubescens*). In the particular case of the Jalapeño landrace, the decrease was 10.6–52.2%. The roasting process increased capsaicin content from 6.1 to 924.9%, dihydrocapsaicin from 2.6 to 57% and nordihydrocapsaicin from 6.6 to 206.8%, depending on genotype. Notably, the compounds are highly volatile, and the increase from heat treatment is attributed to dehydration of the fruit, cell breakdown and inactivation of enzymes that degrade capsaicinoids, such as peroxidases [100]. In this sense, Turkmen et al. [101] observed that cooking by boiling, vapor and microwaves produces increases from 2 to 26% of phenolic compounds and up to 30% in antioxidant activity in spicy chilis. In *C. frutescens* cv. Sina and *C. annuum* cv. Coduion, Shaimaa et al. [102] identified increases in phenolic compounds, flavonoids and antioxidant activity by boiling.

In Mexico, a smoked-drying or oven-drying method (65% of humidity and 75°C) is used with Jalapeño landraces, producing what is known as "chipotle chilis" (special preparation). This process increases the content of flavonoids and antioxidant capacity up to 10 times as a result of the combined liberation of phenolic compounds and flavonoids, including flavonoids generated by wood combustion. By contrast, there is a reduction in capsaicinoid content because of increase in temperature [103].

3.4. Flavonoids and capsaicinoids in health

Chili fruits, in addition to conferring sensory characteristics on foods, also provide nutritional advantages from their chemical composition of vitamins, minerals, carotenoids, flavonoids and capsaicinoids, among others [40, 58, 59, 61]. Controversy remains regarding the beneficial and unfavorable health effects of chili consumption. Recently, Chopan and Littenberg [18] determined that there is no direct relation between high consumption of red chilis and mortality in various populations of North America and Europe. In these cases, adult consumers of red chilis showed a 13% lower risk of death than non-consumers. Lv et al. [104] also observed no direct relation between consumption of fresh or dried chilis and causes of mortality from cancer, diabetes, respiratory or cardiac diseases. Other reports argued that the bioactive or phytochemical compounds of *Capsicum* have anti-inflammatory, antidiabetic, antimicrobial, anticholesterolemic, anticoagulant and antioxidant properties [63].

Cancer is one of the primary causes of morbidity and mortality worldwide. Oxidative stress is implied in the etiology of this disease and results from imbalances in the production of reactive oxygen species (ROS) and the antioxidant defense system of cells. ROS deregulate redox homeostasis and promote formation of tumors by aberrant induction of signaling pathways that cause cancerous tumors [105]. ROS modulate different pathways of cell signaling, which are mediated by the transcription factors NF-κB and STAT3, hypoxia-inducible factor, growth factors, kinases and other proteins and enzymes involved in the development of cancer [106]. In this carcinogenic process, it is argued that capsaicinoids (capsaicin and hydrocapsaicin) help eliminate reactive oxygen species and consequently demonstrate anticarcinogenic, antimutagenic and preventative properties [107, 108]. These properties help prevent the proliferation of cells, migration and induction of apoptosis [109]. In in vivo and in vitro trials, capsaicin inhibited the growth and proliferation of prostate cancer [110]. In addition, capsaicin stimulates the cascade of MAP protein kinase signaling, extracellular signal-regulated protein kinase (ERK) and c-Jun N-terminal kinase (JNK), which have antiproliferative effects [111]. Capsaicin also induces apoptosis of both androgen receptors AR-positive (LNCaP) and -negative (PC-3, Du-145) prostate cancer cell lines by increasing p53, p21 and Bax [112]. Other studies determined that capsaicin can regulate the increase in IL-6 by secretion of TNF-α and the signaling responsible for activation of Akt, ERK and PKC-α [110].

Capsaicin can prevent the growth of colorectal cancer cells by suppression of pathways dependent on β-catenin/TCF by proteosomal degradation of β-catenin and breakdown of β-catenin/TCF-4 interactions [113]. In addition, capsaicin induces apoptosis in gastric cancer cells and can serve as an antitumor agent in gastric cancer [114]. In other studies, capsaicin generates ROS through mitochondria and the depletion of intracellular antioxidants and generates mitochondrial damage and apoptosis in pancreatic cancer cells [115].

Dihydrocapsaicin showed strong antibacterial activity against *Helicobacter pylori*, the bacteria associated with gastric cancer [116]. In biological trials in vivo, oral administration of quercetin generated decreases in infection from *H. pylori* in gastric mucosa and reduced the inflammatory response and lipid peroxidation [62].

Flavonoids are molecules that not only act as conventional antioxidant hydrogen donors but also exert a modulatory action on the signaling of protein and lipid kinases of cells [117]. The apigenin flavonoid is an inhibitor of protein kinases and has an antiproliferative effect on breast cancer cells [118]. The aglycone of quercetin has a protective effect on DNA, which is induced by mitomycin C and antiproliferative activities in human lymphocytes [119]. This molecule also contributes to the protective effect of nerve cells against neurotoxicity induced by oxidative stress, including in Alzheimer's [120].

Chili consumption can promote weight loss by the effect of capsaicin [121]. Capsaicin stimulates lipolysis, causing a reduction in intercellular lipids from the increase in the hydrolysis of triacylglycerol in adipocytes. This effect is mediated by the regulation of genes associated with the catabolic pathway of lipids, HSL and CPT-Ia and genes involved in thermogenesis such as UCP2 [16]. Capsaicin promotes the removal of visceral fat and prevents obesity induced by diets high in fats. Capsaicin activates transient receptor potential cation channel sub-family V member 1 (TRPV1) channels, increases levels of Ca^{2+} ions mediated by connexin 43 (Cx43) and promotes lipolysis in adipocytes and fat reduction [122]. Capsaicinoids demonstrate hypocholesterolemic activity from stimulation of the conversion of cholesterol to bile acids by expression of the cholesterol 7-hydrolase gene and increased secretion of bile acids in feces [123].

4. Final considerations and conclusions

Chili peppers are used worldwide as food and spices. In Mexico, chilis have great economic, social and culinary importance with variations in consumption per capita from 8 to 17 kg, and approximately 2.8×10^6 tons are produced and consumed annually. Mexico is the center of origin, domestication and diversification of *C. annuum*, which has engendered more than 150 cultivated and wild *landraces* of wide distribution. Nearly every region in Mexico cultivates or consumes chilis. Indigenous groups and traditional producers are the primary actors responsible for in situ preservation of the diversity of chili landraces, particularly in the central and south-southeastern regions. The wild forms of *C. annuum* var. *glabriusculum* and *C. annuum* var. *annuum* are distributed throughout the country.

Daily consumption ranges from two to four times a week, and it is estimated that between 70 and 80% of the total Mexican rural population regularly consume spicy chilis. This high frequency of consumption has generated an adaptation to consumption of spicy and extremely spicy chilis, and among consumers, the TAS2R38 type 2 gene has been identified as being associated with tongue and palate receptors, which confer upon the chili consumer a higher capacity to not perceive or barely perceive the "burning" sensation of chilis. Into the diversity of landraces and improved varieties, there are non-pungent landraces, for example, Morrón, Dulce, Ancho and Chilaca and types of Morrón varieties such as California, Victoria and others.

Chili fruit contains up to 23 flavonoid compounds and 20 capsaicinoids, differing between landraces because of the effect of crop management, the state of maturation of the fruits and postharvest management and ecological-environmental influences. Capsaicinoids (capsaicin, dihydrocapsaicin, nordihydrocapsaicin, homodihydrocapsaicin, homocapsaicin, etc.) exert multiple physiological effects on human health, including their potential value in pain relief, cancer prevention and decreased bone loss and as antioxidant activity. Experimentally, anticarcinogenic, antimutagenic and cancer preventative properties have been determined for capsaicin in addition to anti-lipid effects. However, these properties require direct evaluation with regard to human health. The flavonoids quercetin and apigenin, present in chili fruits, have antiproliferative effects on breast cancer cells and exert a protective effect on neuronal cells against the neurotoxicity induced by oxidative stress in Alzheimer's.

Acknowledgements

The authors are grateful for the financial support provided by the National Polytechnic Institute (projects SIP-20160113 and SIP-20170781) and COFAA-IPN and EDI-IPN fellows.

Author details

Araceli M. Vera-Guzmán[1], Elia N. Aquino-Bolaños[2], Elena Heredia-García[3], José C. Carrillo-Rodríguez[4], Sanjuana Hernández-Delgado[5] and José L. Chávez-Servia[1]*

*Address all correspondence to: jchavezs@ipn.mx

1 National Polytechnic Institute, CIIDIR-Oaxaca, Mexico

2 Basic Sciences Institute Universidad Veracruzana, Jalapa, Mexico

3 National Institute of Forest, Agricultural and Livestock Research (INIFAP), Experimental Station-El Bajio, Celaya, Mexico

4 Technologic Institute of the Oaxaca Valley, Oaxaca, Mexico

5 National Polytechnic Institute, Center of Genomic Biotechnology, Reynosa, Mexico

References

[1] Perry L, Flannery KV. Precolumbian use of chili peppers in the valley of Oaxaca, Mexico. Proceedings of National Academy of Science of the USA 2007. **104**: 11905-11909. DOI: 10.1073/pnas.0704936104.

[2] Powis TG, Gallaga-Murrieta E, Lesure R, Lopez-Bravo R, Grivetti L, Kucera H, Gaikwad NW. Prehispanic use of chili peppers in Chiapas, Mexico. PLoS One 2013. **8**(11): e79013. DOI: 10.1371/journal.pone.0079013.

[3] Kraft KH, Brown CH, Nabhan GP, Luedeling E, Luna-Ruiz JJ, d'Eeckenbrugge GC, Hijmans RJ, Gepts P. Multiple lines of evidence for the origin of domesticated chili pepper, *Capsicum annuum,* in Mexico. Proceedings of National Academy of Science of the USA 2014. **111**: 6165-6170. DOI: 10.1073/pnas.1308933111.

[4] González-Jara P, Moreno-Letelier A, Fraile A., Piñeiro D, García-Arenal F. Impact of human management of the genetic variation of wild pepper, *Capsicum annuum* var. *glabriusculum*. PLoS One 2011. **6**(12): e28715. DOI: 10.1371/journal.pone.0028715.

[5] Hayano-Kanashiro C, Gámez-Meza N, Medina-Juárez LM. Wild pepper *Capsicum annuum* L. var. *glabriusculum*: Taxonomy, plant morphology, distribution, genetic diversity, genome sequencing and phytochemical compounds. Crop Science 2016. **56**: 1-11. DOI: 10.2135/cropsci2014.11.0789.

[6] Servicio de Información Agroalimentaria y Pesquera (SIAP). Anuario Estadístico de la Producción Agrícola 2015. Servicio de Información Agroalimentaria y Pesquera, Secretaria de Agricultura, Ganadería, Desarrollo Rural, Pesca y Alimentación. México, DF. 2016. Available from: http://infosiap.siap.gob.mx/aagricola_siap_gb/ientidad/index.jsp [Accessed: January 7, 2017].

[7] Martínez-Flórez S, González-Gallegos J, Culebras JM, Tuñon MJ. Los flavonoides: Propiedades y acciones antioxidantes. Nutrición Hospitalaria. 2002. **7**: 271-278. http://www.bvs.sld.cu/revistas/ibi/vol22_1_03/ibi07103.pdf

[8] Birt DE, Hendrich S, Wang W. Dietary agents in cancer prevention: Flavonoids and isoflavonoids. Pharmacology and Therapeutics 2001. **90**: 157-177. DOI: 10.1016/S0163-7258(01)00137-1.

[9] Nijveldt RJ, van Nood E, van Hoorn EC, Boelens PG, van Norren K, van Leeuwen PAM. Flavonoids: A review of probable mechanism of action and potential applications. American Journal of Clinical Nutrition 2001. **74**: 418-425. http://ajcn.nutrition.org/content/74/4/418.full.pdf+html

[10] Dixon RA, Pasinetti GM. Flavonoids and isoflavonoids: From plant biology to agriculture and neuroscience. Plant Physiology 2010. **154**: 453-457. DOI: 10.1104/pp.110.161430.

[11] Stewart C, Kang B-C, Liu K, Mazourek M, Moore SL, Yoo EY, Kim B-D, Paran I, Jahn MM. The Pun1 gene for pungency in pepper encodes a putative acytltransferase. The Plant Journal 2005. **42**: 675-688. DOI: 10.1111/j.1365-313X.2005.02410.x.

[12] Ornelas-Paz JJ, Martínez-Burrola JM, Ruiz-Cruz S, Santana-Rodríguez V, Ibarra-Junquera V, Olivas GI, Pérez-Martínez JD. Effect of cooking on the capsaicinoids and phenolic contents of Mexican pepper. Food Chemistry. 2010 **119**: 1619-1625. DOI: 10.1016/j.foodchem.2009.09.054.

[13] Surh Y-J, Lee SS. Capsaicin in hot chili pepper: Carcinogen, co-carcinogen or anticarcinogen? Food and Chemical Toxicology 1996. **34**: 313-316. DOI: 10.1016/0278-6915(95)00108-5.

[14] Serra I, Yamamoto M, Calvo A, Cavada G, Báez S, Endoh K, Watanabe H, Tajima K. Association of chili pepper consumption, low socioeconomic status and longstanding gallstones with gallbladder cancer in a Chilean population. International Journal of Cancer 2002. **102**: 407-411. DOI: 10.1002/ijc.10716.

[15] Surh Y-J. Anti-tumor promoting of selected spice ingredients with antioxidant and anti-inflammatory activities: A short review. Food and Chemical Toxicology 2002. **40**: 1091-1097. DOI: 10.1016/S0278-6915(02)00037-6.

[16] Lee M-S, Kim C-T, Kim Y. Effect of capsaicin on lipid catabolism in 3T3-L1 adipocytes. Phytotherapy Research 2011. **25**: 935-939. DOI: 10.1002/ptr.3339.

[17] Bode AM, Dong Z. The two face of capsaicin. Cancer Research 2011. **71**: 2809-2814. DOI: 10.1158/0008-5472.CAN-10-3756.

[18] Chopan M, Littenberg B. The association of hot red chili pepper consumption and mortality: A large population-based cohort study. PLoS ONE 2017. **12**(1): e0169876. DOI: 10.1371/journal.pone.0169876.

[19] Smith CE. Plant Remains. In: Byers DS, editor. The Prehistory of the Tehuacan Valley, Vol. 1: Environment and Subsistence. Austin: University of Texas Press. 1967. pp. 220-255.

[20] Perry L, Dickau R, Zarrillo S, Holst I, Pearsall DM, Piperno DR, Berman MJ, Cooke RG, Rademaker K, Ranere AJ, Raymon JS, Sandweiss DH, Scaramelli F, Tarble K, Zeidler JA. Starch fossils and the domestication and dispersal of chili peppers (*Capsicum* spp. L.) in the Americas. Science 2007. **315**: 986-988. DOI: 10.1126/science.1136914.

[21] Pickersgill B. Migrations of chili pepper, *Capsicum* spp, in the Americas. In: Stone D, editor. Pre-Columbian Plant Migration. Cambridge: Harvard University Press. 1984. pp. 105-124.

[22] Pickersgill B. Domestication of plants in the Americas: Insights from Mendelian and molecular genetics. Annals of Botany 2007. **100**: 925-940. DOI: 10.1093/aob/mcm193.

[23] Pacheco-Olivera A, Hernández-Verdugo S, Rocha-Ramírez V, González-Rodríguez A, Oyama K. Genetic diversity and structure of pepper (*Capsicum annuum* L.) from Northwestern Mexico analyzed by microsatellite markers. Crop Science 2012. **52**: 231-241. DOI: 10.2135/cropsci2011.06.0319.

[24] Pickersgill B. Chile peppers (*Capsicum* spp.). In: Lira R, Casas A, Blancas J, editors. Ethnobotany of Mexico: Interactions of the People and Plants in Mesoamerica. New York: Springer Science+Business Media. 2016. pp. 417-437.

[25] Pickersgill B. Genetic resources and breeding of *Capsicum* spp. Euphytica 1997. **96**: 129-133. DOI: 10.1023/A:1002913228101.

[26] Paran I, Ben-Chaim A, Kang B-C, Jahn M. Capsicums. In: Kole C, editor. Genome Mapping and Molecular Breeding in Plants, Vol. 5: Vegetables. New York: Springer-Verlag Berlin Heidelberg; 2007. pp. 209-226.

[27] Kraft KH, Luna-Ruiz JJ, Gepts P. Different seed selection and conservation practices for fresh market and died chile farmers in Aguascalientes, Mexico. Economic Botany 2010. **64**: 318-328.

[28] Pickersgill B. Relationships between weedy wild and cultivated forms in some species of chili pepper (genus *Capsicum*). Evolution 1971. **25**: 683-691. DOI: 10.2307/2406949.

[29] Pérez-Castañeda LM, Castañón-Nájera G, Ramírez-Meraz M, Mayek-Pérez N. Avances y perspectivas sobre el estudio del origen y la diversidad genética de *Capsicum* spp. Ecosistemas y Recursos Agropecuarios 2015. **2**: 117-128. http://148.236.18.64/era/index.php/rera/article/view/721/591

[30] Hernández-Verdugo S, Luna-Reyes R, Oyama K. Genetic structure and differentiation of wild and domesticated populations of *Capsicum annuum* (Solanaceae) from Mexico. Plant Systematic and Evolution 2001. **226**: 129-142. DOI: 10.1007/s006060170061.

[31] Votava EJ, Baral JB, Bosland PW. Genetic diversity of chile (*Capsicum annuum* var. *annuum* L.) landraces from Northern New Mexico, Colorado and Mexico. Economic Botany 2005. **59**: 8-17. DOI: 10.1663/0013-0001(2005)059[0008:GDOCCA]2.0.CO.

[32] Aguilar-Meléndez A, Morrel PL, Roose ML, Kim S-C. Genetic diversity and structure in semiwild and domesticated chiles (*Capsicum annuum*; Solanaceae) from Mexico. American Journal of Botany 2009. **96**: 1190-1202. DOI: 10.3732/ajb.0800155.

[33] Boege E. El patrimonio biocultural de los pueblos indígenas de México. Hacia la conservación *in situ* de la biodiversidad y agrobiodiversidad en los territorios indígenas. Mexico, D.F: Instituto Nacional de Antropología e Historia y Comisión Nacional para el Desarrollo de los Pueblos Indígenas; 2010, 342 p.

[34] Brown CH, Clement CR, Epps P, Ludeling E, Wichmann. The paleobiolinguistics of domesticated chili pepper (*Capsicum* spp.). Ethnobiology Letters 2013. **4**: 1-11. DOI: 10.14237/ebl.4.2013.2.

[35] Long-Towell J. Los senderos prehispánicos del *Capsicum*. In: Long-Towell J, Attolini-Lecón A, editors. Caminos y Mercados en México. Mexico, D.F.: Universidad Nacional Autónoma de México e Instituto Nacional de Antropología e Historia; 2009. pp. 79-106.

[36] Zeven AC. Traditional maintenance breeding of landraces: 2. Practical and theoretical considerations on maintenance of variation of landraces by farmers and gardeners. Euphytica 2002. **123**: 147-158. DOI: 10.1023/A:1014940623838.

[37] Camacho-Villa TC, Maxted N, Scholten M, Ford-Lloyd B. Defining and identifying crop landraces. Plant Genetic Resources 2005. **3**: 373-384. DOI: 10.1079/PGR200591.

[38] Louette D, Charrier A, Berthaud J. *In Situ* conservation of maize in Mexico: Genetic diversity and maize seed management in a traditional community. Economic Botany 1997. **51**: 20-38. DOI: 10.1007/BF02910401.

[39] Pozo-Campodonic O, Montes-Hernández S, Redondo-Juárez E. Chile (*Capsicum* spp.). In: Ortega R, Palomino G, Castillo F, González VA, Livera M, editors. Avances en el estudio de los recursos fitogenéticos de México. Texcoco, México: SOMEFI, A. C., 1999. pp. 217-238.

[40] Cazáres-Sánchez E, Ramírez-Vallejo P, Castillo-González F, Soto-Hernández RM, Rodríguez-González MT, Chávez-Servia JL. Capsaicinoides y preferencias de usos en diferentes morfotipos de chile (*Capsicum annuum* L.) del centro-oriente de Yucatán. Agrociencia 2005. **39**: 627-638. http://www.colpos.mx/agrocien/Bimestral/2005/nov-dic/art-6.pdf

[41] Castañón-Nájera G, Latournerie-Moreno L, Mendoza-Elos M, Vargas-López A, Cárdenas-Morales H. Colección y caracterización de chile (*Capsicum* spp) en Tabasco, México. Phyton, International Journal of Experimental Botany 2008. **77**: 189-202. http://www.revistaphyton.fund-romuloraggio.org.ar/vol77/CASTANION.pdf

[42] Vela E. Los chiles de México: Catálogo visual. Arqueología Mexicana 2009. 32: 39-77.

[43] Aguilar-Rincón VH, Corona-Torres T, López-López P, Latournerie-Moreno L, Ramírez-Meraz M, Villalón-Mendoza H, Aguilar-Castillo JA. Los Chiles de México y su Distribución. Texcoco: SINAREFI, Colegio de Posgraduados, INIFAP, IT-Conkal, UANL y UAN; 2010, 114 p.

[44] Nárez-Jiménez CA, de-la-Cruz-Lázaro E, Gómez-Vázquez A, Castañeda-Nájera G, Cruz-Hernández A, Márquez-Quiroz. La diversidad morfológica *in situ* de chiles silvestres (*Capsicum* spp.) en Tabasco, México. Revista Fitotecnia Mexicana 2014. **37**: 209-215. http://www.revistafitotecniamexicana.org/documentos/37-3/3a.pdf

[45] Loaiza-Figueroa F, Ritland K, Laborde-Cancino JA, Tanksley SD. Patterns of genetic variation of the genus *Capsicum* (*Solanaceae*) in Mexico. Plant Systematic and Evolution 1989. **15**: 159-1988. DOI: 10.1007/BF00936000.

[46] Contreras-Toledo AR, López-Sánchez H, Santacruz-Varela A, Valadez-Moctezuma E, Aguilar-Rincón VH, Corona-Torres T, López PA. Diversidad genética en México de variedades nativas de chile 'Poblano' mediante microsatélites. Revista Fitotecnia Mexicana 2011. **34**: 225-232. http://www.revistafitotecniamexicana.org/documentos/34-4/1a.pdf

[47] Toledo-Aguilar R, López-Sánchez H, Santacruz-Varela A, Valadez-Moctezuma E, López PA, Aguilar-Rincón VH, González-Hernández VA, Vaquera-Huerta H. Characterization of genetic diversity of native 'Ancho' chili populations of Mexico using microsatellite markers. Chilean Journal of Agricultural Research 2016. **76**: 18-26. DOI: 10.4067/S0718-58392016000100003.

[48] López-Carrillo L, Fernández-Ortega MC, Costas-Dias R, Franco-Marina J, Alejandre-Badillo T. Creencias sobre el consumo de chile y la salud en la ciudad de México. Salud Pública de México 1995. **37**: 339-343. http://saludpublica.mx/index.php/spm/article/view/5854/6567

[49] Pilcher JM. Tamales or timbales: Cuisine and the formation of Mexican national identity, 1821-1911. The Americas 1996. **53**: 193-196. DOI: 10.2307/1007616.

[50] Nabhan GP. Why Some Like it Hot: Food, Genes, and Cultural Diversity. Washington DC: Island Press, 2004. 244 p.

[51] Abdel-Salam OME. Preference for hot pepper: A complex interplay of personal, cultural and pharmacological effects. Temperature 2016. **3**: 39-40. DOI: 10.1080/23328940.2015.1111289.

[52] Mennella JA, Turnbull B, Ziegler PJ, Martinez H. Infant feeding practices and early flavor experiences in Mexican infants: An intra-cultural study. Journal of the American Dietetic Association 2005. **105**: 908-915. DOI: 10.1016/j.jada.2005.03.008.

[53] Hayes JE, Baroshuk LM, Kidd JR, Duffy VB. Supertasting and PROP bitterness depends on more than the TAS2R38 gene. Chemical Senses 2008. **33**: 255-265. DOI: 10.1093/chemse/bjm084.

[54] Román S, Ojeda-Granados C, Panduro A. Genética y evolución de la alimentación de la población en México. Revista de Endocrinología y Nutrición 2013. **21**: 42-51. http://www.medigraphic.com/pdfs/endoc/er-2013/er131f.pdf

[55] United Nations Educational, Scientific and Cultural Organization (UNESCO). List of Intangible Cultural Heritage in Need of Urgent Safeguarding. Rome: United Nations Educational, Scientific and Cultural Organization; 2017. Available from: http://www.unesco.org/culture/ich/en/lists [Accessed: January 17, 2017].

[56] Urdaneta ML, Kanter DF. Deleites de la Cocina Mexicana: Healthy Mexican American Cooking. Austin: University of Texas Press. 1996. 256 p.

[57] Katz E. Chili pepper, from Mexico to Europe: Food, imaginary and cultural identity. In: Medina FX, Ávila R, De Garine I, editors. Food, Imaginary and Cultural Frontiers. Guadalajara: Universidad de Guadalajara; 2009, pp. 213-232.

[58] Cisneros-Pineda O, Torres-Tapia LW, Gutiérrez-Pacheco LC, Contreras-Martín F, González-Estrada T, Peraza-Sánchez SR. Capsaicinoids quantification in chili peppers cultivated in the state of Yucatan, Mexico. Food Chemistry 2007. **104**: 1755-1760. DOI: 10.1016/j.foodchem.2006.10.076.

[59] Morán-Bañuelos SH, Aguilar-Rincón VH, Corona-Torres T, Castillo-González F, Soto-Hernández RM, Miguel-Chávez S. Capsaicinoides en chiles nativos de Puebla, México. Agrociencia 2008. **42**: 807-816. Available from: http://www.colpos.mx/agrocien/Bimestral/2008/oct-nov/art-7.pdf

[60] Eggink PM, Maliepaard C, Tikunov Y, Haanstra JPW, Bovy AG, Visser RGF. A taste of sweet pepper: Volatile and non-volatile chemical composition of fresh sweet pepper (*Capsicum annuum*) in relation to sensory evaluation of taste. Food Chemistry 2012. 132: 301-310. DOI: 10.1016/j.foodchem.2011.10.081.

[61] Vera-Guzmán AM, Chávez-Servia JL, Carrillo-Rodríguez JC, López MG. Phytochemical evaluation of wild and cultivated pepper (*Capsicum annuum* L. and *C. pubescens* Ruiz & Pav.) from Oaxaca, Mexico. Chilean Journal of Agricultural Research 2011. **71:** 578-585. http://www.chileanjar.cl/files/V71_I4_2011_ENG_AraceliMinervaVera-Guzman.pdf

[62] González-Segovia R, Quintanar JL, Salinas E, Ceballos-Salazar R, Avilés-Jiménez F, Torres-López, J. Effect of the flavonoid quercetin on inflammation and lipid peroxidation induced by *Helicobacter pylori* in gastric mucosa of guinea pig. Journal of Gastroenterology 2008. **43:** 441-447. DOI: 10.1007/s00535-008-2184-7.

[63] Chamikara MDM, Dissanayake DRRP, Ishan M, Sooriyapathirana SDSS. Dietary, anticancer and medicinal properties of the phytochemicals in chili pepper (*Capsicum* spp.). Ceylon Journal of Science 2016. **45**: 5-20. DOI: 10.4038/cjs.v45i3.7396.

[64] de Villiers A, Venter P, Pasch H. Recent advances and trends in the liquid-chromatography–mass spectrometry analysis of flavonoids. Journal of Chromatography A 2016. **1430**: 16-78. DOI: 10.1016/j.chroma.2015.11.077.

[65] Heim KE, Tagliaferro AR, Bobilya DJ. Flavonoid antioxidants: Chemistry, metabolism and structure-activity relationships. The Journal of Nutritional Biochemistry 2002. **13**: 572-584. DOI: 10.1016/S0955-2863(02)00208-5.

[66] Burda S, Oleszek W. Antioxidant and antiradical activities of flavonoids. Journal of Agricultural and Food Chemistry 2001. **49**: 2774-2779. DOI: 10.1021/jf001413m.

[67] Howard LR, Talcott ST, Brenes CH, Villalon B. Changes in phytochemical and antioxidant activity of selected pepper cultivars (*Capsicum* species) as influenced by maturity. Journal of Agricultural and Food Chemistry 2000. **48**: 1713-1720. DOI: 10.1021/jf990916t.

[68] Sun T, Xu Z, Wu CT, Janes M, Prinyawiwatkul W, No HK. Antioxidant activities of different colored sweet bell peppers (*Capsicum annuum* L.). Journal of Food Science 2007. **72**: S98–S102. DOI: 10.1111/j.1750-3841.2006.00245.x.

[69] Miean KH, Mohamed S. Flavonoid (myricetin, quercetin, kaempferol, luteolin, and apigenin) content of edible tropical plants. Journal of Agricultural and Food Chemistry 2001. **49**: 3106-3112. DOI: 10.1021/jf000892m.

[70] Marinova D, Ribarova F, Atanassova M. Total phenolics and total flavonoids in Bulgarian fruits and vegetables. Journal of the University of Chemical Technology and Metallurgy 2005. **40**: 255-260. http://dl.uctm.edu/journal/node/j2005-3/Marinova.pdf

[71] Kim JS, Ahn J, Lee SJ, Moon B, Ha TY, Kim S. Phytochemicals and antioxidant activity of fruits and leaves of paprika (*Capsicum annuum* L., var. Special) cultivated in Korea. Journal of Food Science 2011. 76: C193–C198. DOI: 10.1111/j.1750-3841.2010.01891.x.

[72] Blanco-Ríos AK, Medina-Juárez LÁ, González-Aguilar GA, Gámez-Meza N. Antioxidant activity of the phenolic and oily fractions of different sweet bell peppers. Journal of the Mexican Chemical Society 2013. **57**: 137-143. http://www.jmcs.org.mx/PDFS/V57/2/11.-%20Blanco%20G3.pdf

[73] Zhuang Y, Chen L, Sun L, Cao J. Bioactive characteristics and antioxidant activities of nine peppers. Journal of Functional Foods 2012. **4**: 331-338. DOI: 10.1016/j.jff.2012.01.001

[74] Lee Y, Howard LR, Villalon B. Flavonoids and antioxidant activity of fresh pepper (*Capsicum annuum*) cultivars. Journal of Food Science 1995. **60**: 473-476. DOI: 10.1111/j.1365-2621.1995.tb09806.x.

[75] Rochín-Wong CS, Gámez-Meza N, Montoya-Ballesteros LC, Medina-Juárez LA. Efecto de los procesos de secado y encurtido sobre la capacidad antioxidante de los fitoquímicos del chiltepín (*Capsicum annuum* L. var. *glabriusculum).* Revista Mexicana de Ingeniería Química 2013. **12**: 227-239. http://www.rmiq.org/iqfvp/Pdfs/Vol.%2012,%20No.%202/Bio1/Bio1.pdf

[76] Álvarez-Parrilla E, de la Rosa LA, Amarowicz R, Shahidi F. Antioxidant activity of fresh and processed Jalapeno and Serrano peppers. Journal of Agricultural and Food Chemistry 2011. **59**: 163-173. DOI: 10.1021/jf103434u.

[77] Marín A, Ferreres F, Tomás-Barberán FA, Gil MI. Characterization and quantitation of antioxidant constituents of sweet pepper (*Capsicum annuum* L.). Journal of Agricultural and Food Chemistry 2004. **52**: 3861-3869. DOI: 10.1021/jf0497915.

[78] Materska M, Perucka I. Antioxidant activity of the main phenolic compounds isolated from hot pepper fruit (*Capsicum annuum* L.). Journal of Agricultural and Food Chemistry 2005. **53**: 1750-1756. DOI: 10.1021/jf035331k.

[79] Morales-Soto A, Gómez-Caravaca AM, García-Salas P, Segura-Carretero A, Fernández-Gutiérrez A. High-performance liquid chromatography coupled to diode array and electrospray time-of-flight mass spectrometry detectors for a comprehensive characterization of phenolic and other polar compounds in three pepper (*Capsicum annuum* L.) samples. Food Research International 2013. **51**: 977-984. DOI: 10.1016/j.foodres.2013.02.022.

[80] Materska M. Flavone C-glycosides from *Capsicum annuum* L.: Relationships between antioxidant activity and lipophilicity. European Food Research and Technology 2015. **240**: 549-557. DOI: 10.1007/s00217-014-2353-2.

[81] Medina-Juárez LÁ, Molina-Quijada DM, Del-Toro-Sánchez CL, González-Aguilar GA, Gámez-Meza N. Antioxidant activity of peppers (*Capsicum annuum* L.) extracts and characterization of their phenolic constituents. Interciencia 2012. **37**: 588-593. http://www.interciencia.org/v37_08/588.pdf

[82] Lee JJ, Crosby KM, Pike LM, Yoo KS, Leskovar DI. Impact of genetic and environmental variation on development of flavonoids and carotenoids in pepper (*Capsicum* spp.). Scientia Horticulturae 2005. **106**(3), 341-352. DOI: 10.1016/j.scienta.2005.04.008.

[83] Bhandari SR, Jung BD, Baek HY, Lee YS. Ripening-dependent changes in phytonutrients and antioxidant activity of red pepper (*Capsicum annuum* L.) fruits cultivated under open-field conditions. HortScience 2013. **48**: 1275-1282. http://hortsci.ashspublications.org/content/48/10/1275.full.pdf+html

[84] Bae H, Jayaprakasha GK, Crosby K, Yoo KS, Leskovar DI, Jifon J, Patil BS. Ascorbic acid, capsaicinoid, and flavonoid aglycone concentrations as a function of fruit maturity stage in greenhouse-grown peppers. Journal of Food Composition and Analysis 2014. **33**(2): 195-202. DOI: 10.1016/j.jfca.2013.11.009.

[85] Menichini F, Tundis R, Bonesi M, Loizzo MR, Conforti F, Statti G, De Cindio B, Houghton PJ, Menichini F. The influence of fruit ripening on the phytochemical content and biological activity of *Capsicum chinense* Jacq. cv Habanero. Food Chemistry 2009. **114**: 553-560. DOI: 10.1016/j.foodchem.2008.09.086.

[86] Castellón-Martínez E., Chávez-Servia JL, Carrillo-Rodríguez JC, Vera-Guzmán AM. Preferencias de consumo de chiles (*Capsicum annuum* L.) nativos en los valles centrales de Oaxaca, México. Revista Fitotecnia Mexicana 2012. **35**: 27-35. http://www.revistafitotecniamexicana.org/documentos/35-3_Especial_5/5a.pdf

[87] Deepa N, Kaur C, George B, Singh B, Kapoor HC. Antioxidant constituents in some sweet pepper (*Capsicum annuum* L.) genotypes during maturity. LWT-Food Science and Technology, 2007. **40**: 121-129. DOI: 10.1016/j.lwt.2005.09.016.

[88] Hervert-Hernández D, Sáyago-Ayerdi SG, Goñi I. Bioactive compounds of four hot pepper varieties (*Capsicum annuum* L.), antioxidant capacity, and intestinal bioaccessibility. Journal of Agricultural and Food Chemistry 2010. **58**: 3399-3406. DOI: 10.1021/jf904220w.

[89] Vázquez-Flota F, Miranda-Ham ML, Monforte-González M, Gutiérrez-Carbajal G, Velázquez-García C, Nieto-Pelayo Y. La biosíntesis de capsaicinoides, el principio picante del chile. Revista Fitotecnia Mexicana 2007. **30**: 353-360. http://www.revistafitotecnia-mexicana.org/documentos/30-4/2a.pdf

[90] Thompson RQ, Phinney KW, Sander LC, Welch MJ. Reversed-phase liquid chromatography and argentation chromatography of the minor capsaicinoids. Analytical and Bioanalytical Chemistry 2005. **381**: 1432-1440. DOI: 10.1007/s00216-005-3098-3.

[91] Schweiggert U, Carle R, Schieber A. Characterization of major and minor capsaicinoids and related compounds in chili pods (*Capsicum frutescens* L.) by high-performance liquid chromatography/atmospheric pressure chemical ionization mass spectrometry. Analytica Chimica Acta 2006. **557**: 236-244. DOI: 10.1016/j.aca.2005.10.032.

[92] Meckelmann SW, Riegel DW, van Zonneveld M, Ríos L, Peña K, Mueller-Seitz E, Petz, M. Capsaicinoids, flavonoids, tocopherols, antioxidant capacity and color attributes in 23 native Peruvian chili peppers (*Capsicum* spp.) grown in three different locations. European Food Research and Technology 2015. **240**: 273-283. DOI: 10.1007/s00217-014-2325-6.

[93] Othman ZAA, Ahmed YBH, Habila MA, Ghafar AA. Determination of capsaicin and dihydrocapsaicin in *Capsicum* fruit samples using high performance liquid chromatography. Molecules 2011. **16**: 8919-8929. DOI: 10.3390/molecules16108919.

[94] Zewdie Y, Bosland PW. Evaluation of genotype, environment, and genotype-by-environment interaction for capsaicinoids in *Capsicum annuum* L. Euphytica 2000. **111**: 185-190. DOI: 10.1023/A:1003837314929.

[95] González-Zamora A, Sierra-Campos E, Luna-Ortega JG, Pérez-Morales R, Ortiz JCR, García-Hernández JL. Characterization of different capsicum varieties by evaluation of their capsaicinoids content by high performance liquid chromatography, determination of pungency and effect of high temperature. Molecules 2013. **18**: 13471-13486. DOI: 10.3390/molecules181113471.

[96] Orellana-Escobedo L, Garcia-Amezquita LE, Olivas GI, Ornelas-Paz JJ, Sepulveda DR. Capsaicinoids content and proximate composition of Mexican chili peppers (*Capsicum* spp.) cultivated in the state of Chihuahua. CyTA-Journal of Food 2013. **11**: 179-184. DOI: 10.1080/19476337.2012.716082.

[97] Ornelas-Paz J, Martínez-Burrola JM, Ruiz-Cruz S, Santana-Rodríguez V, Ibarra-Junquera V, Olivas GI, Pérez-Martínez JD. Effect of cooking on the capsaicinoids and phenolics contents of Mexican peppers. Food Chemistry 2010. **119**: 1619-1625. DOI: 10.1016/j.foodchem.2009.09.054.

[98] Montoya-Ballesteros LC, Gardea-Béjar A, Ayala-Chávez GM, Martínez-Núñez YY, Robles-Ozuna LE. Capsaicinoides y color en chiltepín (*Capsicum annuum* var. *aviculare*): Efecto del proceso sobre salsas y encurtidos. Revista mexicana de Ingeniería Química 2010. **9**: 197-207. http://www.rmiq.org/iqfvp/Pdfs/Vol9%20no%202/RMIQVol9No2_8.pdf

[99] Álvarez-Parrilla E, Mercado-Mercado G, La Rosa LAD, Díaz JAL, Wall-Medrano A, González-Aguilar GA. Antioxidant activity and prevention of pork meat lipid oxidation using traditional Mexican condiments (pasilla dry pepper, achiote, and mole sauce). Food Science and Technology 2014. **34**(2): 371-378. DOI: 10.1590/fst.2014.0052.

[100] Schweiggert U, Schieber A, Carle R. Effects of blanching and storage on capsaicinoid stability and peroxidase activity of hot chili peppers (*Capsicum frutescens* L.). Innovative Food Science and Emerging Technologies 2006. **7**: 217-224. DOI: 10.1016/j.ifset.2006.03.003.

[101] Turkmen N, Sari F, Velioglu YS. The effect of cooking methods on total phenolics and antioxidant activity of selected green vegetables. Food chemistry 2005. **93**, 713-718. DOI: 10.1016/j.foodchem.2004.12.038.

[102] Shaimaa GA, Mahmoud MS, Mohamed MR, Emam AA. Effect of heat treatment on phenolic and flavonoid compounds and antioxidant activities of some Egyptian sweet and chilli pepper. Natural Products Chemistry and Research 2016. **4** (2). DOI: 10.4172/2329-6836.1000218.

[103] Moreno-Escamilla JO, Laura A, López-Díaz JA, Rodrigo-García J, Núñez-Gastélum JA, Alvarez-Parrilla E. Effect of the smoking process and firewood type in the phytochemical content and antioxidant capacity of red Jalapeño pepper during its transformation to chipotle pepper. Food Research International 2015. **76**: 654-660. DOI: 10.1016/j.foodres.2015.07.031.

[104] Lv J, Qi L, Yu C, Yang L, Guo Y, Chen Y, Bian Z, Aun D, Du J, Ge P, Tang Z, Hou W, Li Y, Chen J, Chen Z, Li Z. Consumption of spicy foods and total and cause specific mortality: Population based cohort study. BMJ 2015. 351: h3942. DOI: 10.1136/bmj.h3942.

[105] Acharya A, Das I, Chandhok D, Saha T. Redox regulation in cancer: A double-edged sword with therapeutic potential. Oxidative medicine and Cellular Longevity 2010. **3**: 23-34. DOI: 10.4161/oxim.3.1.10095.

[106] Prasad S, Gupta SC, Tyagi AK. Reactive oxygen species (ROS) and cancer: Role of antioxidative nutraceuticals. Cancer Letters 2017. **387**: 95-105. DOI: 10.1016/j.canlet.2016.03.042.

[107] Cao S, Chen H, Xiang S, Hong J, Weng L, Zhu H, Liu Q. Anti-cancer effects and mechanisms of capsaicin in chili peppers. American Journal of Plant Sciences 2015. **6**: 3075. DOI: 10.4236/ajps.2015.619300.

[108] Amruthraj NJ, Raj-Preetam JP, Saravanan S, Lebel-Antoine L. *In vitro* studies on anticancer activity of capsaicinoids from *Capsicum chinense* against human hepatocellular carcinoma cells. International Journal of Pharmacy and Pharmaceutical Sciences 2014. **6**: 254-558. http://www.ijppsjournal.com/Vol6Issue4/8998.pdf

[109] Luo XJ, Peng J, Li YJ. Recent advances in the study on capsaicinoids and capsinoids. European Journal of Pharmacology 2011. **650**(1):1-7. DOI: 10.1016/j.ejphar.2010.09.074.

[110] Venier, N.A., Yamamoto, T., Sugar, L.M., Adomat, H., Fleshner, N.E., Klotz, L.H. and Venkateswaran, V. Capsaicin reduces the metastatic burden in the transgenic adenocarcinoma of the mouse prostate model. Prostate 2015. 75: 1300-1311. DOI: 10.1002/pros.23013.

[111] Sánchez AM, Malagarie-Cazenave S, Olea N, Vara D, Chiloeches A, Díaz-Laviada I. Apoptosis induced by capsaicin in prostate PC-3 cells involves ceramide accumulation, neutral sphingomyelinase, and JNK activation. Apoptosis 2007. **12**: 2013-2024. DOI: 10.1007/s10495-007-0119-z.

[112] Mori A, Lehmann S, O'Kelly J, Kumagai T, Desmond JC, Pervan M, McBride WH, Kizaki M, Koeffler HP. Capsaicin, a component of red peppers, inhibits the growth of androgen-independent, p53 mutant prostate cancer cells. Cancer Research 2006. **66**: 3222-3229. DOI: 10.1158/0008-5472.CAN-05-0087.

[113] Lee SH, Richardson RL, Dashwood RH, Baek SJ. Capsaicin represses transcriptional activity of β-catenin in human colorectal cancer cells. The Journal of Nutritional Biochemistry 2012. **23**: 646-655. DOI: 10.1016/j.jnutbio.2011.03.009.

[114] Park S-Y, Kim J-Y, Lee S-M, Jun C-H, Cho S-B, Park C-H, Joo Y-E, Kim H-S, Choi SK, Rew JS. Capsaicin induces apoptosis and modulates MAPK signaling in human gastric cancer cells. Molecular Medicine Reports 2014. **9**: 499-502. DOI: 10.3892/mmr.2013.1849.

[115] Pramanik KC, Boreddy SR, & Srivastava SK (2011). Role of mitochondrial electron transport chain complexes in capsaicin mediated oxidative stress leading to apoptosis in pancreatic cancer cells. PloS ONE 2011. 6(5), e20151. DOI: 10.1371/journal.pone.0020151.

[116] Ochi T, Takaishi Y, Shibata H, Higuti T, Kataoka M. Chemical constituents of *Capsicum annuum* L var. *angulosum*, and anti *Helicobacter pylori* activity. Natural Medicines 2005. **59**: 76-84.

[117] Williams RJ, Spenser JP, Rice-Evans C. Flavonoids: Antioxidants or signalling molecules? Free Radical Biology and Medicine 2004. **36**(7): 838-849. 10.1016/j.freeradbiomed.2004.01.001.

[118] Way T, Kao M, Lin J. Apigenin induces apoptosis through proteosomal degradation of HER2/neu in HER2/neu-overexpressing breast cancer cells via the phosphotidylinositol 3-kinase/Akt-dependent pathway. The Journal of Biological Chemistry 2004. **6**: 4479-4489. DOI: 10.1074/jbc.M305529200.

[119] Ündeger U, Aydin S, Basaran AA, Basaran N. The modulating effect of quercetin and rutin on the mitomycin C induced DNA damage. Toxicology Letters 2004. **151**: 143-149. DOI: 10.1016/j.toxlet.2003.12.071.

[120] Heo HJ, Lee CY. Protective effects of quercetin and vitamin C against oxidative stress-induced neurodegeneration. Journal of Agricultural and Food Chemistry 2004. **52**: 7514-7517. DOI: 10.1021/jf049243r.

[121] Varghese S, Kubatka P, Rodrigo L, Gazdikova K, Caprnda M, Fedotova J, Zulli A, Kruzliak P, Büsselberg D. Chili pepper as a body weight-loss food. International Journal of Food Sciences and Nutrition 2016. 1-10. DOI: 10.1080/09637486.2016.1258044.

[122] Chen J, Li L, Li Y, Liang X, Sun Q, Yu H, Zhong J, Ni Y, Chen J, Zhao Z, Gao P, Wang B, Liu D, Zhu Z, Yan Z. Activation of TRPV1 channel by dietary capsaicin improves visceral fat remodeling through connexin43-mediated Ca 2+ influx. Cardiovascular Diabetology 2015. **14**: 1-14. DOI: I 10.1186/s12933-015-0183-6.

[123] Zhang L, Zhou M, Fang G, Tang Y, Chen Z, Liu X. Hypocholesterolemic effect of capsaicinoids by increased bile acids excretion in ovariectomized rats. Molecular Nutrition and Food Research 2013. **57**: 1080-1088. DOI: 10.1039/c3fo30321g.

Chapter 20

Onions: A Source of Flavonoids

Ana Sofia Rodrigues, Domingos P.F. Almeida,
Jesus Simal-Gándara and
Maria Rosa Pérez-Gregorio

Additional information is available at the end of the chapter

http://dx.doi.org/10.5772/intechopen.69896

Abstract

Flavonoids are a large and diverse group of polyphenolic compounds with antioxidant effects, and onion (*Allium cepa* L.) is one of the richest sources of dietary flavonoids. Flavonoid content is affected by endogenous factors—genotype and agro-environmental conditions. Considerable research has been directed toward understanding the nature of polyphenols in different products and the factors influencing their accumulation. This review examines the impacts of pre- and postharvest factors on onions' flavonoid content, highlighting how this knowledge may be used to modulate their composition and the potential use of onion by-products.

Keywords: polyphenols, plant foods, *Allium cepa*, preharvest factors, harvest handling, genotype

1. Introduction

Phenolic compounds are responsible for the major organoleptic characteristics of plant-derived foods and beverages, particularly color and taste properties, and they also contribute to the nutritional qualities of fruits and vegetables [1, 2].

Plants present diverse defense mechanisms, including physical and chemical barriers. Phenolic compounds are particularly abundant and play an important role in both strategies, as monomers for the synthesis of lignin and as chemical agents. Flavonoids are one of the most relevant secondary compounds in plants and currently more than 9000 being identified [3]. A most significant function of the flavonoids, especially the anthocyanins, together with flavones and flavonols as copigments, is their contribution to flower and fruit colors. This is important for attracting pollinators and seed-dispersing animals. Phenolics may influence the competition among plants "allelopathy." They act in plant defense mechanisms against herbivores or pathogens, contributing to the disease resistance mechanisms in plants, and act as supporting materials of cell walls as photoprotectors against UV radiation and plant-microbe symbiosis and involved in the repair of wounds and contribute to healing by lignifications of damaged areas. Stress conditions such as excessive UV light, wounding, or infection induce the biosynthesis of phenolic compounds [66, 67].

Plants composition can be affected by pre-harvest factors, including genotype (cultivar and variety), maturity at harvest and tissue distribution, and exogenous factors, including climate, soil micro-environment, and pest and disease attack [4, 5]. Environmental factors have a major effect on polyphenol content. These factors may be pedoclimatic (soil type, sun exposure, and rainfall) or agronomic (culture in greenhouses or fields, biological culture, hydroponic culture, fruit yield per tree, etc.). With the current state of knowledge, it is difficult to determine for each family of plant products the key variables that are responsible for the polyphenol variability. A huge amount of analysis would be required to obtain this information [10].

This paper reviews recent literature on the main factors affecting the flavonoid content in onion, as well as different approaches aiming to increase the accumulation of these compounds in onions, which provide an added functional value.

2. Occurrence and identity of flavonoids in onions

Onion has been reported as one of the major sources of dietary flavonoids in many countries [6–9], contributing to a large extent to the overall intake of flavonoids [10, 11]. Two flavonoid classes are mainly found in onion, the anthocyanins, which impart a red/purple color to some varieties, and flavonols such as quercetin and its derivatives, responsible for the yellow and brown skins of many other varieties (see **Table 1**).

Flavonols are the most ubiquitous flavonoids in onions. At least 25 different flavonols have been characterized in onion, being quercetin derivatives the most important ones in all onion cultivars [11]. Quercetin 4′-glucoside and quercetin 3,4′-diglucoside are reported as the main flavonols in onions, accounting for about 80–95% of total flavonols [12–26].

The quantitative content of anthocyanins in some red onion cultivars has been reported to be approximately 10% of the total flavonoid content or 39–240 mg kg^{-1} FW [11]. In red onions more than 50% of anthocyanins are cyanidin glucosides non-acylated or acylated with malonic acid. Delphinidin and petunidin do not have malonyl derivatives in detectable amounts, indicating that the presence of malonylated derivatives seems to occur only in the cyanidin derivatives [27]. Some of these pigments facilitate unique structural features like 4′-glycosylation and unusual substitution patterns of sugar moieties. Altogether at least 25 different anthocyanins have been reported from red onions, including 2 novel 5-carboxypyranocyanidin-derivatives [28].

Class	Flavonoid	mg/100 g FW	Reference
Total polyphenols			
		438.88 (Y) (DW)	[132]
		35.00	[133]
		443.20	[135]
		253.60–310.80 (R)	[136]
		216.70 (W)	[136]
		129.60	[137]
		73.33–180.84	[138]
		66.80	[139]
		76.10	[140]
		116.00 (R)	[141]
		70.00 (W)	[141]
		24.40 (W)	[142]
		154.10	[143]
		260.00–650.00 (W) (DW)	[144]
		21.60–58.30	[145]
Total flavonoids		12.21–52.43	[138]
		0.18 (W)	[15]
		69.20 (Y)	[29]
		76.58 (Y)	[15]
		61.05 (R)	[15]
		18.70	[143]
		56.00–1150 (W) (DW)	[144]
		35.3 (R)	[146]
Total flavonols		7.90–43.10 (R)	[148]
		9.85 (Y) (μmol/g DW)	[149]
		8.90–177.80	[22]
		55.40–62.10 (R)	[150]
		35.00–159.2	[50]
		28.55–51.64 (Y)	[14]
		58.09 (R)	[14]
	Quercetin aglicone	0.07 (W)	[15]
		111.70 (Y)	[22]
		5.00 (W)	[22]

Class	Flavonoid	mg/100 g FW	Reference
		105.2 (P)	[22]
		137.50 (R)	[22]
		0.50–9.90	[50]
	Total quercetin		
		8.11 (Y)	[15]
		23.95 (R)	[15]
		7.70–46.32	[138]
		28.40–48.60	[151]
		54.40	[152]
		6.17 (W)	[153]
		39.21 (R)	[153]
		19.20 (Y)	[154]
		30.70 (R)	[154]
		30.60 (R)	[155]
	Q. total (after hydrolysis)	22.00–48.00 (Y)	[156]
	"	237.03 (Y) (DW)	[132]
		83.00–330.00 (Y) μg/g	[26]
	Q. 3′-glucoside	1.70–2.30 (R)	[150]
		0.30	[16]
		0.76	[157]
		0.30–2.60	[50]
	Q. 4′-glucoside	20.80–23.00 (R)	[150]
		3.60 (W)	[22]
		36.00 (Y)	[22]
		30.20 (P)	[22]
		39.40 (R)	[22]
		9.70	[16]
		0.11 (W)	[15]
		57.18 (Y)	[15]
		13.77–26.75 (Y)	[14]
		30.01 (R)	[14]
		30.01 (R)	[27]
		29.89 (R)	[15]
		19.00–95.20	[50]
		33.08 (Y) (DW)	[132]

http://dx.doi.org/10.5772/intechopen.69896

Class	Flavonoid	mg/100 g FW	Reference
	Q. 3,4′-diglucoside	25.40–27.40 (R)	[150]
		20.22 (R)	[158]
		11.37–21.11 (Y)	[14]
		20.22 (R)	[14]
		4.92 (Y)	[15]
		2.48 (R)	[15]
		40.50	[16]
		111.70 (Y)	[22]
		5.00 (W)	[22]
		105.20 (P)	[22]
		137.50 (R)	[22]
		11.60–45.50	[50]
		241.04 (Y) (DW)	[132]
	Q. 7,4′-diglucoside	0.70–1.10 (R)	[150]
	Q. 3,7,4′-triglucoside	0.60–0.90 (R)	[150]
	Kaempferol	1.54–1.83	[138]
	Myricetin	2.77–4.13	[138]
	Isorhamnetin	5.19 (Y)	[15]
		0.63–5.04 (R)	[15]
	I. 4′-glucoside	4.10–4.90 (R)	[150]
		3.40 (R)	[15]
		1.20–710	[50]
		6.00 (R)	[159]
	I. 3,4′-diglucoside	2.10–2.50	[150]
		0.30–1.50	[50]
Anthocyanins	Total	7.00–21.00 (R)	[160]
	Cyanidin	3.19 (R)	[153]
	Cyanidin 3-O-(6″-malonyl-glucoside)	1.50 (R)	[27]
	Cyanidin 3-(6″-malonyl-3″-glucosyl-glucoside)	1.00 (R)	[27]
	Delphinidin 3-glucosyl-glucoside	6.50 (R)	[27]

DW, dry weight; FW, fresh weight.

Table 1. Phenolic compounds in onion (onion color: P, pink; R, red; Y, yellow; and W, white).

Flavonoids comprise a generous portion of the total antioxidant activity in onions [29]. Elhassaneen and Sanad [30], in a study with Egyptian onion varieties, concluded that phenolic compounds, particularly the flavonol quercetin, beside other factors including selenium and sulfur-containing amino acids, play the major role in the antioxidant activity of onion bulbs.

2.1. Approaches for the accumulation of antioxidant flavonoids in onions

Flavonoids play a lot of roles in plant physiology, mainly related to plant resistance [31, 32], in defense mechanisms against herbivore and pathogen attacks, UV radiation protection, plant-microbe symbiosis. They contribute, as copigments, to flower and fruit colors, especially the anthocyanins, flavones, and flavonols [33], important plant characteristic for attracting pollinators or seed-dispersing animals and allelopathy [34]. Flavonoids have also been shown to modulate transport of the phytohormone auxin [35] as well as the levels of reactive oxygen species (ROS) [36].

Thus, the strategies applied to obtain plant foods with higher level of flavonoids, increasing their functional value, must be based on the manipulation of interacting factors (genetic, environmental conditions, and agronomic practices) that are known to affect their content [4, 129].

The great challenge, due to the vast variables involved (intraspecific chemodiversity, genetic and ontogeny, postharvest, and biotic and abiotic factors) [129], is the implementation of a large-scale and low-cost suitable production systems to obtain onions rich in flavonoids with the maintenance of balance between phytochemical content and agro-production. An interdisciplinary overview and data collection, analysis, and evaluation of scattered data regarding the diverse factors involved in the optimization of plant production and postharvest and processing management are fundamental [37].

The plant needs to recognize the agro-environmental stimuli (see exogenous factors in **Table 3**), which is dependent on the sensitivity of organs and tissues and influences the metabolic response, depending on the gene expression transcribed to functional enzymes. Then, metabolic channeling may induce the accumulation of the target product. The agronomic activities, including climate modeling, modifying secondary metabolism, and the correspondent bioactive compound produced, may change plant physiological activity and affect their development and productivity [38].

2.1.1. Approaches based on endogenous factors

Research on genetics and plant metabolism has started, in the last years, to become interested in crop development with enhanced phytochemical concentrations. Although the genetic impact seems to be greater than the external factors, the synergistic effect of genetics with specific agronomic approaches could have a stronger capacity on improving certain phytochemicals. However, it is extremely complex to implement preharvest strategies to maximize the biosynthesis of specific phytochemicals and simultaneously maintain the level of productivity and other qualitative parameters of the products. Progress toward understanding the impact of key strategies will allow their integration into sustainable agricultural production systems aimed to alter the content and/or profile of phytochemicals in new crop varieties [39]. The main endogenous factors that affect onion flavonoids are summarized in **Table 2**.

2.1.1.1. *Cultivar selection*

Onion flavonoid content is highly explained by genetic factors [15], probably due to the diversity of onion cultivars, hybrids, and open pollinated. The genetic makeup of the onion varieties needs to be factored in when differences in flavonoids content and antioxidant activity are considered [29, 134].

Lee and Mitchell [40] studied six commercial onion varieties in which quercetin content ranged up to 18-fold between 93 and 1703 mg/100 g DW. The highest level has detected in a yellow, early, and long-day variety "milestone."

The flavonoid profile affects the color of onion bulbs. Red cultivars generally contain higher total flavonoid content [11, 20, 30, 41–43], because they are richer in flavonols but also contain anthocyanins, unlike white varieties. Dalamu et al. [44] evaluated 34 onion genotypes and verified great variation in total phenolic content between white (165.0), pink (702.0 mg kg^{-1}), and red varieties (867.8 mg kg^{-1}) which means an overall more than fivefold variation. Red onions, with highest levels of phenolics, also have about three times higher antioxidant activity than white onions. The quercetin content in these 34 genotypes ranges from 22.0 to 890.5 mg kg^{-1}. The largest variation occurred in yellow cultivars [11]. Patil et al. [45], in a study with 55 cultivars of yellow onions, verified a variation from 54 to 286 mg kg^{-1} FW in quercetin content. Grzelak et al. [12] also reported differences between three yellow onion varieties in flavonol glucosides and total flavonol content but no statistical differences between harvest seasons.

Quantitative data compilation presented in **Tables 1–3** indicates a great diversity in flavonoid content among the cultivars surveyed. Total phenolic content in onion genotype seems to present a definite hierarchy, highest in red and lowest in white. In contrast, Crozier et al. [46] reported the opposite but only for quercetin; they found only 201 mg kg^{-1} of quercetin in edible parts of red onion but much higher quercetin amount in white onions (185–634 mg kg^{-1}). Marotti and Piccaglia [15] also found higher levels of total flavonoids in a golden variety "Dorata Density," in relation to other different color varieties (including red onion).

There are many reports in the scientific literature on how resistant cultivars of different crops contained more phenolic compounds than susceptible ones suggesting that these compounds play an important role in the defense mechanism [47, 48]. Lachman et al. [49] found different profiles in polyphenol content between susceptible and resistant onion cultivars. Yang et al. [29] concluded that onion varieties, which have strong, bitter, and pungent flavors and high sugar contents, exhibited higher antioxidant and antiproliferative activities. Vågen and Slimestad [50], in 15 cultivars studied, also detected a positive correlation between pungency, amounts of fructooligosaccharides (FOS) and flavonols, and the highest Trolox equivalent antioxidant capacity (TEAC) values.

Okamoto et al. [51] reported differences in quercetin content between short-day and long-day onion cultivars. The long-day cultivars from Northern Europe and their close relatives contain higher concentrations of quercetin glucosides than those of Japanese and North American. In long-day cultivars, total quercetin content was higher than in short-day cultivars being independent of the growing origin [51, 52].

Factor	Evaluated parameters	Effect on flavonoids	References
Varietal differences	Different bulb colors	Red > yellow > gold *Exception	[46]* [49] [42] [15]* [11] [43] [41] [30] [44] [20]
	Yellow varieties		[45] [11] [12]
	Resistant and susceptible onion cultivars	Resistant > susceptible	[49]
	Long-day and short-day onion cultivars	Long-day > short-day	[51] [52]
Size and bulb weight		No differences *Small > large	[45] [64] [61]* [20]*
Bulb parts	Scales	Dry outer skins > outer edible > middle edible > inner edible *Exception: middle layers > outer scales > inner layers	[63] [55] [57] [27] [64, 65] [61] [56] [12] [60] [62]*
	Top to bottom	Top > bottom	[63] [57] [20]

Table 2. Endogenous factors affecting the accumulation of flavonoid in onions.

2.1.1.2. Tissue selection

Although flavonoids are derived from the same biosynthetic pathway, they accumulate differentially in plant tissues, depending on the developmental stage and the environmental conditions, since they fulfill different physiological functions [53]. The plant prioritization defense strategy to allocate defense compounds to the most valuable tissues can explain why young leaves have more phenolics than mature leaves. Tissues such as skin scales, with protective function, appear to have the same strategy. Similarly, ontogenetic changes in defensive allocation in seedling and juvenile plants may also be an evolutionary response to herbivore at this particularly susceptible stage of a plant's life cycle [54].

http://dx.doi.org/10.5772/intechopen.69896

Factor	Evaluated parameters	Effect on flavonoids	References
Soil type		Clay > sandy loam	[70]
Fertilization	N levels	No differences in quercetin	[80]
	NH_4^+:NO_3^- ratios	>Dominant nitrate supply	[72]
	Varieties Kamal and Robin (N and S)	Positive correlation with N and S fertilization	[146]
	White variety of Pueblo and yellow variety of Mundo were the most efficient when fertilized by nitrogen and sulfur in combination with iron	Positive correlation with N, S, and Fe fertilization	[147]
Mycorrhizal colonization/ inoculation		>Quercetin	[72]
		No effects	[64]
Organic versus conventional		Organic > conventional	[78] [81]
	Organic fertilizers, and no chemical herbicides or fungicides, or inorganic fertilizers	No differences	[64] [80]
Chemical treatments	Benzothiadiazole and K_2HPO_4 to control *Stemphylium*	>Phenolic	[85]
Yearly variation			[64] [80] [25]
Light	Global radiation in the end of production period	>Radiation > flavonoids	[64]
	Total global radiation during production period	>Radiation > flavonoids	[25]
	UV light lamps after harvest	>Quercetin	[77]
	Fluorescent light after harvest	>Flavonoid	[61]
CO_2	Elevated to 550 ppm in relation to atmospheric (365 ppm)	<Flavonoid >Anthocyanins and total phenols	[162]
Lifting	Lifting time	Late lifting > early lifting	[71]
Curing			[123]
	Evolution in relation of levels at harvest	After curing > at lifting	[70] [64] [23]
		Field > dark environment *Field curing: dark similar to light exposed	[64] [23]

Table 3. Exogenous factors affecting the accumulation of flavonoid in onions.

The outer onion skin-dry peel scales have more total flavonoids than the edible flesh scales [27, 55, 56]. Hirota et al. [57] found that outer scales and the upper portions of the edible scales had higher levels of 4′-Qmg and 3,4′-Qdg and Qag than lower (internal) scales.

The flavonol glucoside hydrolysis during the peel formation can explain why aglycones are the main flavonoids present in the peel [22, 58]. Quercetin is concentrated in the dry skin of most onions where its oxidation products, 3,4-dihydroxybenzoic acid, and 2,4,6-trihydroxyphenylglycosilic acid impart the brown color and provide the onion bulb protection from the soil microbial infection [58, 59]. Bilyk et al. [55] observed that as much as 53% of the total quercetin in onion skin was present as aglycon, occurring great differences between dry skin and edible scales. The dry skin of onion bulbs of red and pink varieties is richer in flavonols and anthocyanins, mainly in aglycon forms. In red onion, the dry skins contain ~63% of total anthocyanins present in bulb. It means that only 27% of the total anthocyanins will be consumed after bulb peeling [27].

Slimestad and Vågen [60], in edible scales, detected higher quantities of flavonols and fructose and simultaneously the highest antioxidant capacity in the outer fresh scales. An abrupt drop in flavonol quantity occurred from the first to the second scale, followed by a slight decrease further inward. Grzelak et al. [12] also reported that outer edible fresh scales of the bulb have threefold of mono- and diglucosides of quercetin and isorhamnetin than inner scales. Outer scales of triglycosides have ca. 1.5-fold greater in the middle scales. A graduated decrease in flavonoids, from the first to the seventh scale, was also observed in onion bulb [61]. Inversely to other authors, Beesk et al. [62] verified the following distribution order in scales of the flavonoid content: middle layers > outer scales > inner layers. Qdg was the major flavonoid in the inner layers, Qdg and Qmg were in equal amounts in the middle layers, and quercetin was the major flavonoid in outer scales followed by Qmg. Trammell and Peterson [63], considering vertical distribution, found that the flavonoid is presented in higher amounts (by twofold) in top than bottom (disk) of the bulbs. A two- to threefold increase in concentration from the center of the bulb outward is observed in a horizontal bulb distribution. The least pigmented line showed a 17-fold increase and had 56% of its total flavonols in the outer scale compared with about 30% for the other lines. Mogren et al. [64, 65] and Lee et al. [61] reported comparable gradient in total quercetin composition in the edible onion scales indicating that 90% of total flavonols are in epidermal tissue. Parenchymous storage tissue, the bulk of a bulb, only contains around 10% of the total pigment. It follows that any factors which modify the ratio of epidermal to storage tissue scale, including thickness, could indirectly change gross flavonol concentration.

In onion, quercetin concentration does not appear to be affected by bulb size or weight, and small bulbs contain the proportional quercetin concentration as larger bulbs [45]. Mogren et al. [64] obtained results that showed minor or no differences in quercetin glucoside content among small-, medium-, or large-sized onions, although Lee et al. [61] and Pérez-Gregorio et al. [20] detected higher flavonoid content in small onions than in large ones.

2.1.2. Approaches based on exogenous factors

In addition to genetics, other factors can affect the onion bulbs' flavonoid contents, mainly related with pedoclimatic conditions, agronomic practices, and postharvest handling and processing. Being secondary metabolism an integral part of the plant capacity to adaptation to the surrounding environment, it is not surprising that these factors can modulate its phytochemical profile.

As polyphenolic compounds are part of a complex defense mechanism of plants, environmental stress factors such as pests and diseases, ozone and UV light, cold, and nutritional stress can induce their biosynthesis [66, 67]. Therefore, regulating environmental stresses provides an opportunity to enhance the flavonoid content of plants. Nevertheless, because of their potential adverse effects on crop growth, yield, and even in commercial quality (sensorial attributes), such approach should be considered with caution.

Treutter [38] made a compilation of agricultural technologies influencing the biosynthesis and accumulation of phenolic compounds in plants, including remarks on the effects of temperature light, mineral and organic nutrition, water availability and moisture stress, grafting, atmospheric CO_2, growth and differentiation of the plant and application of stimulating agents, elicitors, and plant activators.

Table 3 compiles studies about the main exogenous factors affecting the flavonoid content in onions, as well as different strategies targeting to increase their content.

2.1.2.1. Soil nutrient status

Accumulation of phenolic compounds in plant can be influenced by mineral, being a limited nitrogen supply generally linked with higher levels of phenolics [10]. This reaction can be explained by the activity increase of phenylalanine ammonia lyase (PAL) enzyme to obtain ammonia from phenylalanine, as a source of nitrogen for amino acid metabolism. Cinnamic acid, as a result of the deamination process, is also released and further incorporated into the phenylpropanoid synthetic pathway, increasing the phenolic synthesis [68]. On the other hand, nitrogen limitation will affect photosynthesis, decreasing chlorophyll availability and disrupting photosynthetic membranes due to starch accumulation, which can explain the increased sensitivity to light intensity. Synthesis of photoprotective pigments such as anthocyanins and flavonols may give protection against light-induced oxidative damage [69].

Patil et al. [70] observed higher amounts of quercetin in onions growing under nitrogen limitation in both clay and sandy loam soils. Despite this, the location of growth, more than soil type or growth stage, is a key environmental factor for quercetin levels in onion.

Mogren et al. [64, 65, 71] compared diverse applications of organic fertilizers, and it was found that the nitrogen fertilization did not affect the yield or quercetin glucoside content in the onion. Additionally, it did not find significant differences between onions with or without nitrogen fertilization in quercetin glucoside content. High levels of nitrogen (80 kg ha^{-1}) do not improve yield or quercetin glucoside levels in the onions. Thus, it is preferable to fractionally apply small amounts of nitrogen fertilizers because it reduces the risk of leaching of mineral nutrients without reducing the crop yield or quercetin content of onion bulbs.

Perner et al. [72] studied the effect of mycorrhizal colonization and different ammonium/nitrate ratios as nitrogen fertilizer on onion yield and nutritional characteristics. It was concluded that the organosulfur compounds, quercetin glycosides, and antioxidant activity can be increased in suitably supplied onion plants if nitrate is dominant. Quercetin glycosides and antioxidant activity are also increased with mycorrhizal colonization. This was possibly due to amplified precursor production and induced defense mechanisms.

As these compounds are produced as part of plant defense mechanisms against stress factors, water availability and regulated deficit irrigation might also modulate metabolic pathway and considerably affect plant phenolic composition [4]. Mohamed and Aly [163] observed that seawater salt stress causes a reduction in the total phenolic compounds.

2.1.2.2. Light

The intensity, quality, and photoperiod of light (sunlight spectrum and proportion of ultraviolet and the red/far-red ratio) are the main environmental factors affecting the flavonoid synthesis. The regulation of expression of several genes that encode the activity of enzymes participating in the phenylpropanoid pathway such as cinnamate 4-hydroxylase (C4H) or PAL, is affected by light conditions during plant development and storage, playing an important role in the phenolic compounds [1].

Flavonoids protect against UV radiation and accumulate mainly in the epidermal cells of plant tissues [73]. However, the response to UV radiation, of various plant species, can vary substantially in terms of flavonoid synthesis [67]. The synthesis of specific flavonoids and other phenolics can be differently regulated in response to UV light depending of plant species, and the contribution to UV stress protection can vary between phenolics [74]. Light stimulates flavonoid synthesis, particularly anthocyanins and flavones, being PAL the major inducible enzyme [66, 75].

The levels of quercetin glucosides in the external dry skins, exposed to light, are less than 10% of the levels in fleshy and partly dried scales. The probable mechanism is that quercetin is formed by deglucosidation of quercetin glucosides on the border between drying and dried brown areas on individual scales [57, 76].

In the end of onion bulb growth, the global radiation seems to be one of the major determinants on quercetin glucoside content [17–20, 23, 25]. Mogren et al. [64] observed that the lower the global radiation in the last month of bulb growth, the lower the content of quercetin. Postharvest treatment of onion bulbs with UV light or fluorescent light lamps can induce quercetin production [77]. Exposure of onion bulbs to fluorescent light for 24 and 48 h induced time-dependent increases in the flavonoid content [61].

2.1.2.3. Organic versus conventional production

Manach et al. [10] verified that vegetables produced by organic or sustainable agriculture contain higher polyphenol content than vegetables grown in conventional production or hydroponic systems. Two main hypotheses have been proposed to explain the potential increases in polyphenol compounds in organic versus conventional production of vegetables. One hypothesis considers the influences of nutrient management and fertilizer application on plant metabolism. Synthetic fertilizers, used in conventional agriculture, normally present higher availability nitrogen that may accelerate plant growth more than organic fertilizers. Consequently, plant resources are allocated mainly for growth, and the plant tends to invest less in the production of secondary metabolites such as amino and organic acids and polyphenols. The second hypothesis considers the plant reactions to biotic stress such as pests and diseases and weed competition. Organic production methods, which limit the use of agrochemicals such as insecticides, herbicides, and fungicides, may induce greater stresses on plants that tend to allocate more resources toward the synthesis of their own chemical defense compounds [161].

Ren et al. [78] detected 1.3–10.4 times higher levels of flavonoids, quercitrin, caffeic acid, and baicalein and in various organic vegetables onion than conventional, suggesting the influence of cultivation techniques. All green vegetables tested also had greater antioxidant activity in organic production.

Grinder-Pedersen et al. [79] verified differences in quercetin levels between organic and conventional onions, but because different cultivars in the two different production systems were studied, it cannot be ruled out that the differences were due to cultivar (genetic factor).

Mogren et al. [80] did not find significant differences on quercetin glucoside levels between onions organically produced and onion treated with chemical fertilizers. The conclusion could be that the nitrogen source, organic or inorganic, and the absence of chemical fungicides seemed to have no effect on quercetin biosynthesis.

Faller and Fialho [81] suggest that the effect of organic practices results in different effect patterns according to the plant species analyzed, with fruits being more susceptible to the induction of polyphenol synthesis than vegetables. Organic onion pulp had higher antioxidant capacity than conventional [81].

Søltoft et al. [82] also did not find significant differences in the flavonoid level between organic and conventional onions.

In Lee et al. [83] study the organic onions usually start bulbing later than conventional onions because of black plastic film and delayed nitrogen mineralization. That might be an important cause of the lower level of phenolics in organic onions.

2.1.2.4. *Chemical treatments*

Herbicide and, to a lesser extent, insecticide and fungicide application can also affect the synthesis of phenolic compounds in plants. Diphenyl ethers (e.g., acifluorfen) act as herbicide mainly by oxidative damages (singlet oxygen of protoporphyrin). Plants, when treated with herbicides, as a possible defensive reaction to the oxidative damages, increase the PAL synthesis and produce more flavonoids. The risks of the combined natural and pesticide-induced modulating effects on human health and environmental protection should be evaluated [84]. Kamal et al. [85] observed that onion plants treated with di-potassium phosphate and benzothiadiazole (Bion) presented significantly higher PAL and PO activity and phenolic contents than the untreated plants. It was concluded that application of chemical solutions such as di-potassium phosphate and benzothiadiazole applied for pathogenic control can enhance phenolic compounds in onion plants [85]. But, the risks of the combined natural and pesticide-induced modulating effects on human health and environmental protection should be further evaluated [84].

2.1.2.5. *Harvest time and postharvest treatments*

Many phytochemicals are synthesized in parallel with the overall development and maturation of fruits and vegetables. Therefore, their content in plants can considerably vary with different stages of maturity [10].

Total flavonol content increased as spring onion plants matured (226–538 mg/100 g at 14 and 77 days, respectively) [86]. In bulbs, harvest date has been reported to have almost no effect to onion bulbs [70].

Mogren et al. [71] found that late lifting of onions (80% fallen leaves) leads up to 45% higher concentrations of quercetin glucosides compared with early lifting (50% fallen leaves).

Onions left in the field, to curing, after harvest accumulates more flavonols [70]. Mogren et al. [64] also detected a dramatic increase in quercetin glucoside content during field curing (between 100 and 300%, during the 10–14 days of curing). Price et al. [21] demonstrated a 50% loss in quercetin monoglucoside during the initial curing process. Flavonol and anthocyanin levels in onions cured in the dark were similar to those obtained in bulbs cured in the light [23]. Mogren et al. [64] observed that field curing onions presented an increase in quercetin content significantly higher compared to the onions stored in dark conditions. Removal of the foliage to the bulb, before the process of field curing, did not affect quercetin content, suggesting that no transport occurs, in mature bulbs, between the foliage and scales. During field curing an increase in quercetin content occurred, particularly when the flavonol concentrations were low at lifting [23].

Rodrigues et al. [24] observed that total flavonols increased during storage of onion bulbs, but when stored under traditional storage (without controlled temperature) showed higher increases of flavonoid levels than those stored under refrigeration. Bulbs stored in the field (at fluctuating ambient temperature) reached higher levels of flavonoids (64% maximum) than refrigerated onions (40% maximum). Regarding anthocyanins, after 7 months in both conditions (refrigeration and traditional treatment), the whole anthocyanin content was reduced to more than 40%. Gennaro et al. [27] also observed a decrease to 64–73% of total anthocyanins in onions stored under domestic conditions, which seems to indicate that flavonol glucosides are more resistant than anthocyanins during storage. Ethylene accumulated during onion storage can stimulate activity of phenylalanine ammonia lyase (PAL), a key enzyme in biosynthesis of phenolic compounds and accumulation of phenolic constituents [87, 88], and justify the significant increase in flavonols observed during storage [24]. Benkeblia [87] reported a positive relationship between PAL activity and total phenolic variations in long-term stored onion bulbs.

The effect of onion bulbs' storage conditions in the composition of flavonoids was studied by several authors. Price et al. [21], apart from a 50% loss of quercetin 4-monoglucoside during the initial drying process (after curing at 28°C), observed little change in composition over 6 months of bulbs storage. Benkeblia [87] evaluated total phenolics in onion bulb during storage at 4 and 20°C and observed a variation in phenolics relatively regular at both temperatures. Lachman et al. [42] observed an increase of total flavonoids, especially at higher temperatures, at the end of 36 weeks of storage, in red and yellow onion varieties. Gennaro et al. [27] concluded that home storage habits resulted in a decrease to 64–73% of total anthocyanins, but degradation is slower when onions are refrigerated. Rodrigues et al. [24] also observed that after 7 months of storage, total anthocyanin content was reduced between 40 and 60%.

2.1.3. Processing

Onion flavonoid effects of domestic treatments like slicing [89–91], cooking [23, 92], or frozen [19, 93] were also studied. Onion products could be processed before consuming, but processing may result in losses in those valuable flavonoids. As was already referred, some researches focused on the effect of domestic processing techniques such as chopping, shredding, peeling, roasting, cooking, or boiling on flavonoid content, and depending on the severity of heat treatment, losses were evident. Furthermore, onion could be also industrially processed. Thus, industrial processing not only includes all domestic treatments referred but also includes the effect of sanitizing technologies as well as freezing, freeze-drying, dehydration, packaging, and stored processes. Through this section how these applications and storage affect the flavonoid content and profile will be described.

2.1.3.1. Fresh-cut technology

Fresh-cut fruit and vegetable products hardly increase their presence in the marketplace due to demand by the consumer. In the coming years, it is commonly perceived that the fresh-cut food industry will have unprecedented growth. However, processors of fresh-cut fruit products face numerous challenges not commonly encountered during fresh-cut vegetable processing. The difficulties encountered with fresh-cut fruit, while not insurmountable, require a new and higher level of technical and operational sophistication. Physical changes resulting of minimally processed food production could induce physiological and therefore compositional changes that could affect the final food quality. The effect on flavonoid content of minimally processed onion will be discussed in each step of food processing.

2.1.3.1.1. Cutting

Wounding stress was largely studied as increasing the phenolic content and antioxidant activity of vegetables [94–96]. According with Cantos et al. [90], the three most important enzymes related to phenolic metabolism, polyphenol oxidase, peroxidase, and phenylalanine ammonia lyase, activity remain unaltered after wounding. Reyes et al. [97] further verified that the effect of this stress depends on the type of vegetable. In sliced onions, wounding was found to increase phenolic content and antioxidant activity [17].

Given the distribution of onion flavonoids in the bulb tissues, the wounding effect is also affected by the cutting technology. Hence, generally, the outer leafs contain the highest flavonoid levels, whereas inner layers have the lowest amount of flavonoids [17, 98, 99]. The greatest loss was during preprocessing steps such as peeling and trimming. Keeping in mind that onion human consumption is limited to edible part, the brown outer leafs are not actually being under consideration. As referred, flavonoid distribution was described as not homogeneous in edible onion bulb. Hence, the initial flavonoid content and evolution could depend on the cutting technique. Overall, trending to increasing the initial flavonoid content was generally observed in chopped onion [17] and sliced onion [100]. However onions could be also cut into half onion rings, onion rings, diced onions, and julienne strips. Recent studies evaluated the effect of the type of cutting in the flavonoid contents [89]. They found that slicing led to greater anthocyanin content in comparison to dicing. Another controversy could be extracted from the research about how cutting affects onion flavonoid content. Temperature, light presence or absence, and storage time have normally been studied in parallel to cut effect. Some authors attribute the only effect of storage time [89], whereas other authors verified differences promoted by temperature changes [17]. Further studies are needed to verify the differences in the flavonoid evolution and their mechanisms depending on the tissue analyzed.

2.1.3.1.2. Sanitizing technologies

Different sanitizing technologies emerged in food science to disinfect fresh-cut food prior to package. Fresh-cut or minimally processed food has been described by the USDA and FDA like fruits and vegetables cut, washed, packaged, and further maintained under refrigeration conditions. Fresh-cut products are therefore raw. Even though minimally processed food remains in a fresh state, it could be physically altered from the original form. Fresh-cut food is ready to eat or cook, without freezing, thermal processing, or treatments with additives or preservatives [101]. Given the nature of fresh-cut products which are not subjected to thermal processing, it is necessary to include some sanitizing technologies to maintain the hygienic quality of the raw food. Washing is one of the most important processing operations and uses physical and chemical treatments to eliminate, or at least reduce, the population of pathogenic and spoilage-inducing microorganisms. However, according with Perez-Gregorio et al. [102], the main effect contributing to the loss of flavonols in fresh-cut onion slices is their solubility in immersion in water leading to losses from 17 to 23% of flavonoids at 4 or 50°C. Despite that sodium hypochlorite is not allowed as sanitizer of fresh-cut vegetables in some European countries, it is still the most used for being inexpensive and easy to use and for having a broad spectrum of activity [102]. Chlorine can oxidize organic matter in foods or in water, and in the latter case, by-products such haloforms and haloacetic acids, which are potentially carcinogenic and mutagenic, can be formed [103]. Searching for organic chlorinated products (sodium dichloroisocyanurate, potassium dichloroisocyanurate, dichloroisocyanuric acid, and trichloroisocyanuric acid) as alternative sanitizing agents gained interest in recent years [104]; nevertheless the antimicrobial efficacy in onions of these sanitizers was lower than the others like hydrogen peroxide [105]. It was verified that onion flavonoid content experienced a significantly decrease for chlorine, organic chlorine, or hydrogen peroxide treatment [102]. Alternative treatment like nisin and citric acid in combination was also tested as sanitizer in fresh-cut onion manufacturer. Nisin and citric acid are generally recognized as safe (GRAS) for use as food ingredients [106] which is an advantage in the use of nisin and citric acid in the microbial cleaning of fresh-cut onions. Cheng et al. [100] verified an increase of total phenolics and antioxidant activity after using niacin and citric acid to wash fresh-cut onions. It was therefore highlighted that it might be used as a safe preservative for fresh-cut onions, whereas the phenolic content will be improved. Among the chemical methods for controlling postharvest diseases, other treatments as UV-C irradiation were assayed. UV radiation in the range of 250–260 nm is lethal to most microorganisms, including bacteria, viruses, protozoa, mycelial fungi, yeasts, and algae and also leads to increase the onion flavonoid content [102]. Other treatments like ozone [107] were also used as sanitizer agent; however, no scientific paper has been found to evaluate the effect of this treatment in onion flavonoid content.

2.1.3.1.3. Packaging: atmosphere and package material

As already referred, fresh-cut technology could promote several physiological changes that could induce microbial spoilage. Furthermore, color changes, softening, surface dehydration, water loss, translucency, and off-flavor and off-odor development are other frequent causes of quality loss in fresh-cut products. The use of innovative modified atmospheres as well as edible coatings is nowadays standing out against revolutionary techniques to control the food safety; likewise the fresh state was maintained. Even though some studies have already demonstrated the effectiveness of these proceedings, more studies are required to better keep the minimally processed organoleptic properties. Moreover, further studies about how packaging might affect onion flavonoids are still required. Little scientific information is available to better know the effect of "ready-to-eat" packaging onion in its flavonoid content. Flavonoid stability was evaluated during fresh-cut onion storage in perforated films [108] or polyethylene and polyethylene terephthalate cups [17]. Overall, the onions experienced changes in flavonoid content during storage time. Storage conditions like light presence or absence, temperature, and storage time marked the onion flavonoid evolution. Hence, anthocyanins increase under light but experienced a decrease under dark storage conditions [17]. Moreover, the individual flavonoid stability was very different, the malonated anthocyanins being much more stable than the corresponding non-acylated pigments [108]. In addition, the arabinosides were shown to be less stable than the corresponding glucosides [108].

There is still a gap in the knowledge as how the package material affects the onion flavonoid evolution during storage time. It is also necessary to deepen the study of package atmosphere influence or what is the best type of package in order to maintain the levels of onion flavonoids.

2.1.3.2. Cooking: frying, microwaving, baking, and boiling

The impact of common domestic and technological treatments on flavonoid composition in onions was studied [23, 91–93, 109–111]. During technological and culinary treatments, important chemical and biochemical reactions occur in onion tissue. Such reactions may have an impact on the flavonoid structure, resulting in changes of the bioavailability and activity of these compounds [112]. In general, papers report that cooking of onions led to a decrease in total flavonol content, but these losses vary depending on the culinary treatment (frying, boiling, roasting, etc.) and on the length of exposure to this treatment. Overall, slight conditions did not affect to the flavonol content, but intense treatments cause flavonol losses from 16 to 30% [23]. Boiling onions led higher losses of quercetin glycosides, which leached to the boiling water until 53% in intense treatments [23]. Quercetin degradation was higher for diglucosides than monoglycosilated quercetin derivatives, whereas anthocyanins experienced the greater losses under cooking temperature exposure [23].

2.1.3.3. Frozen onion

In addition to ready-to-use vegetables, the trend to find a higher number of preprocessing vegetables is increasing in the commercial areas. The modern lifestyle drives to a high consumption not only in minimally processed food but also in frozen vegetables that are ready to cook and cheaper than fresh vegetables. Frozen storage has also an economic advantage for manufacturers since the wastage of unused products is reduced and the shelf life increases. However, the freezing process could affect the food quality, and this is a worrisome point for consumers. It is well known that frozen vegetables may have a lower nutritional value than their respective commodities. Little knowledge is highlighted about how this technology could affect to onion flavonoid content. However few authors concluded that frozen onions lead to an increase of onion flavonoid content [23, 113]. This could be a potent strategy to increase the consumption of frozen vegetables.

2.1.3.4. Dehydrating and freeze-dried onion powder

Industry often carries out processes based in the food water extraction such us freezing and drying to achieve the objective of long-term storage. However, the health-promoting ability and nutritional attributes of fruits and vegetables depend on the type of processing employed. Onions can be marketed as powder for cooking purposes after drying processes [114]. Drying technological developments are driven by consumer who demands for healthy, fresh-like, and convenient food. The trending in consumer demand has increased for processed products that are ready to use, cook, and eat but keeping more of their original characteristics. The development of operations that minimize the adverse effects of processing is therefore required by an industrial point of view. The main concern in food drying is related to a loss of volatiles and flavors, changes in color and texture, and a decrease in nutritional value associated with the process. Hence, the effect of dehydration on onion quality was studied [115]. Mass production of dried foods is often accomplished through the use of convective dryers. This drying process suffers from quality losses regarding color, flavor (taste and aroma), and texture, while rehydration is often poor. Freeze-drying process produces the highest-quality dried food product since the food structure is not damaged during sublimation. Nevertheless the freeze-dried process has a strong disadvantage, is much more expensive than convective drying, and is therefore only used for the production of a minor volume of high-value products.

Regardless of the drying procedure used, dried food has residual enzyme and microbial activities, essential parameters to extend the food shelf life. On the other hand, the minimization of enzymatic activity given by the dehydration process might also influence quality factors like antioxidant activity and flavonoid content. Hence, it was verified that onion flavonoid content increases after freeze-drying process [19].

In recent years, there has been an increasing interest of the food industry in incorporating ingredients with health beneficial properties. Among these ingredients, spices are recognized by their flavoring and coloring potential. Spices may contain phenolic compounds and contribute to the intake of natural antioxidants. Therefore, the incorporation of purified extracts of bioactive compounds in many foods may represent an interesting alternative to increase consumption of these substances and allow the population to benefit from the positive effects attributed to them. Onion, therefore, would be used as freeze-dried powder to improve the antioxidant capacity of foods, and onion flavor could be added.

Overall, further studies are needed in order to improve the knowledge about how onion flavonoids are affected by domestic or industrial treatments. The scientific evidences about flavonoid content could be modulated by normal industry processes and could be also profited to offer food with high quality and high added value.

2.2. Valorization of onion by-products

The production of onion worldwide increased by a 25% over the past 10 years, with a production of about 83 million tons nowadays [116], which makes onions as the second most important world horticultural crop after tomatoes. This high level of production gives as a result more than 500,000 tons of onion skin waste (OSW) which are discarded within the European Union every year [117]. Therefore, the resulting wastes and by-products have become a major problem [118]. They include onion skins, the outer two fleshy scales, and roots generated during industrial peeling but also undersized, malformed, or damaged onion bulbs. They are not suitable as fodder because of their strong characteristic aroma and neither as an organic fertilizer due to the rapid development of phytopathogenic agents such as *Sclerotium cepivorum* [119]. Their disposal commonly involves landfill with high economical costs and important environmental impact [120].

The recovery of valuable phytochemicals with high potential for the pharmaceutical, food, and cosmetics manufacturing is of key importance [121]. The onion waste has been identified as a potential source of flavor compounds, dietary fiber components, nonstructural carbohydrates like fructans and fructooligosaccharides, and flavonoids particularly quercetin glycosides [117, 122]. Most of the studies have been performed at a laboratory scale, so further research is necessary in order to scale up these processes to the industry requirements, assessing their economical viability. Onion composition is variable and depends on cultivar, stage of maturation, environment, agronomic conditions, storage time, and bulb section.

It is key to know the composition of each industrial onion waste to know its potential health benefits. Quercetin 4′-glucoside and quercetin 3,4′-diglucoside are in most cases reported as the main flavonols of the flesh, whereas onion skins contain higher concentrations of quercetin aglycon [123, 124]. There is a big potential opportunity given the increasing demand of consumers for substituting synthetic compounds by natural substances [125]. The presence of these flavonoids in onion products confers them some healthy properties. Flavonoids are shown to have antioxidant activity, free radical scavenging capacity, coronary heart disease prevention, and anticancer activity. Some flavonoids exhibit potential for antihuman immunodeficiency virus functions. Quercetin is known for its anticancer, anti-inflammatory, and antiviral activity [126]. Future investigations on the bioactivity, bioavailability, and toxicology of onion product phytochemicals [127] and their stability and interactions with other food ingredients [128] should be performed and carefully assessed by in vitro and in vivo studies. Functional foods represent an important, innovative, and rapidly growing part of the overall food market.

3. Future challenges for plant scientists and growers

This chapter deals about the current state of knowledge on the main factors affecting the flavonoid content in onions, as well as different approaches that can be applied to increase the accumulation of these compounds. For example, red cultivars contain the highest flavonoid levels; in this sense, also resistant onions present higher flavonoid levels than those that are susceptible. The nonedible dry skin is richer in flavonoids than the flesh, promoting the nonedible portions as a source of natural antioxidants. Within the edible bulb, a decrease across the onion from the outer onion scales to the inners is also found. With regard to soil management factors, the nitrogen fertilizer levels should be minimized to favor flavonoid levels. It was also found that organically grown onions present higher levels of flavonoids and antioxidant activity than conventional. Late lifting of onions generally results in higher concentrations of quercetin glucosides than early lifting.

Phenolic compounds can affect sensory attributes such as color, flavor, bitterness, and texture affecting the consumer assessment. The identification of specific compounds in different onion cultivars and agronomic practices would lead to a better understanding of the physiological responses to onion consumption [17–20, 23–25]. This would aid the development of onion production systems that provide an increased health benefit [56] and the development of guidelines for the consumption of these compounds. An interesting and challenging aspect for future research is to clarify the interactions between genotype and agro-environmental factors on the flavonoid composition in onions [129].

The production of fresh "functional food" with defined health claims may be favorable for a premium market segment. In the future, the minimum quality of plant foods could be defined on the base of their content of bioactive components [130]. One of the projects that have been awarded over the years is given below as an example of the fruit and vegetables research community [131] to generate successful applications in the calls published by the EU Commission: FLAVO is the project for "flavonoids in fruits and vegetables: their impact on food quality, nutrition and human health." The project is centered on fruits widely available to Europeans—apple, grape, and strawberry—together with their derivatives. FLAVO aimed to monitor the flavonoids in fruits and vegetables and to optimize their beneficial effects. This action was promoted by the European Fruit Research Institutes Network (EUFRIN), and a similar project would be desirable for the vegetable sector with the support of the European Vegetable Research Institutes Network (EUVRIN) to cover areas such as (a) the study of consumer behavior about new products, (b) selection of improved plant foods by breeding, (c) the choice of agronomic techniques to maximize flavonoids, (d) knowledge on the appropriate dose of flavonoids for beneficial effects, and (e) the dissemination of the results to consumers and other stakeholders.

Author details

Ana Sofia Rodrigues[1,2]*, Domingos P.F. Almeida[3], Jesus Simal-Gándara[4] and Maria Rosa Pérez-Gregorio[5]

*Address all correspondence to: sofia@ipvc.pt

1 Instituto Politécnico de Viana do Castelo, Escola Superior Agrária, Ponte de Lima, Portugal

2 Centre for Research and Technology of Agro-Environmental and Biological Sciences—CITAB, Vila Real, Portugal

3 University of Lisbon, Instituto Superior de Agronomia, Lisbon, Portugal

4 Nutrition and Bromatology Group, Department of Analytical and Food Chemistry, Food Science and Technology Faculty, University of Vigo, Ourense, Spain

5 LAQV/REQUIMTE, Departamento de Química e Bioquímica, Faculdade de Ciências da Universidade do Porto, Porto, Portugal

References

[1] Cheynier V. Polyphenols in foods are more complex than often thought. American Journal of Clinical Nutrition. 2005;**81**:223S-229S

[2] Parr AJ, Bolwell GP. Phenols in the plant and in man. The potential for possible nutritional enhancement of the diet by modifying the phenols content or profile. Journal of Science and Food Agriculture. 2000;**80**:985-1012

[3] Williams CA, Grayer RJ. Anthocyanins and other flavonoids. Natural Product Reports. 2004;**21**:539-573

[4] Tomás-Barberán FA, Espín JC. Phenolic compounds and related enzymes as determinants of quality in fruits and vegetables. Journal of Science and Food Agriculture. 2001;**81**:853-876

[5] Wang SY. Effect of pre-harvest conditions on antioxidant capacity in fruits. Acta Horticulturae. 2006;**712**:299-305

[6] Hollman PCH, Katan MB. Dietary flavonoids: Intake, health effects and bioavailability. Food and Chemical Toxicology. 1999;**37**:937-942

[7] Knekt P, Järvinen R, Reunanen A, and Maatela J. Flavonoid intake and coronary mortality in Finland: A cohort study. British Medical Journal. 1996;**312**:478-481

[8] Nijveldt RJ, Van Nood E, Van Hoorn DEC, Boelens PG, Van Norren K, Van Leeuwen PAM. Flavonoids: A review of probable mechanisms of action and potential applications. American Journal of Clinical Nutrition. 2001;**74**:418-425

[9] Sampson L, Rimm E, Hollman PCH, De Vries JHM, Katan MB. Flavonol and flavone intakes in US health professionals. Journal of the American Dietetic Association. 2002;**102**:1414-1420

[10] Manach C, Scalbert A, Morand C, Rémésy C, Jiménez L. Polyphenols: Food sources and bioavailability. American Journal of Clinical Nutrition. 2004;**79**:727-747

[11] Slimestad R, Fossen T, Vågen IM. Onions: A source of unique dietary flavonoids. Journal of Agricultural and Food Chemistry. 2007;**55**:10067-10080

[12] Grzelak K, Milala J, Król B, Adamicki F, Badełek E. Content of quercetin glycosides and fructooligosaccharides in onion stored in a cold room. European Food Research and Technology. 2009;**228**:1001-1007

[13] Ioku K, Aoyama Y, Tokuno A, Terao J, Nakatani N, Takei Y. Various cooking methods and the flavonoid content in onion. Journal of Nutritional Science and Vitaminology. 2001;**47**:78-83

[14] Lombard K, Peffley E, Geoffriau E, Thompson L, Herring A. Quercetin in onion (*Allium cepa* L.) after heat-treatment simulating home preparations. Journal of Food Composition and Analysis. 2005;**18**:571-581

[15] Marotti M, Piccaglia R. Characterization of flavonoids in different cultivars of onion (*Allium cepa* L.). Journal of Food Science. 2002;**67**:1229-1232

[16] Moon JH, Nakata R, Oshima S, Inakuma T, Terao J. Accumulation of quercetin conjugates in blood plasma after the short-term ingestion of onion by women. American Journal of Physiology. 2000;**279**:R461-R467

[17] Pérez-Gregorio MR, García-Falcón MS, Simal-Gándara J. Flavonoids changes in fresh-cut onions during storage in different packaging systems. Food Chemistry. 2011;**124**:652-658

[18] Pérez-Gregorio MR, Regueiro J, Alonso-González E, Pastrana-Castro LM, Simal-Gándara J. Influence of alcoholic fermentation process on antioxidant activity and phenolic levels from mulberries (*Morus nigra* L.). LWT—Food Science and Technology. 2011;**44**:1793-1801

[19] Pérez-Gregorio MR, Regueiro J, González-Barreiro C, Rial-Otero R, Simal-Gándara J. Changes in antioxidant flavonoids during freeze-drying of red onions and subsequent storage. Food Control. 2011;**22**:1108-1113

[20] Pérez-Gregorio MR, García-Falcón MS, Simal-Gándara J, Rodrigues AS, Almeida DPF. Identification and quantification of flavonoids in traditional cultivars of red and white onions at harvest. Journal of Food Composition and Analysis. 2010;**23**:592-598

[21] Price KR, Bacon JR, Rhodes MJC. Effect of storage and domestic processing on the content and composition of flavonol glucosides in onion (*Allium cepa*). Journal of Agricultural and Food Chemistry. 1997;**45**:938-942

[22] Price KR, Rhodes MJC. Analysis of the major flavonol glycosides present in four varieties of onion (*Allium cepa*) and changes in composition resulting from autolysis. Journal of Science and Food Agriculture. 1997;**74**:331-339

[23] Rodrigues AS, Pérez-Gregorio MR, García-Falcón MS, Simal-Gándara J. Effect of curing and cooking on flavonols and anthocyanins in traditional varieties of onion bulbs. Food Research International. 2009;**42**:1331-1336

[24] Rodrigues AS, Pérez-Gregorio MR, García-Falcón MS, Simal-Gándara J, Almeida DPF. Effect of post-harvest practices on flavonoid content of red and white onion cultivars. Food Control. 2010;**21**:878-884

[25] Rodrigues AS, Pérez-Gregorio MR, García-Falcón MS, Simal-Gándara J, Almeida DPF. Effect of meteorological conditions on antioxidant flavonoids in Portuguese cultivars of white and red onions. Food Chemistry. 2011;**124**:303-308

[26] Yoo KS, Lee EJ, Patil BS. Quantification of quercetin glycosides in 6 onion cultivars and comparisons of hydrolysis-HPLC and spectrophotometric methods in measuring total quercetin concentrations. Journal of Food Science. 2010;**75**:C160-C165

[27] Gennaro L, Leonardi C, Esposito F, Salucci M, Maiani G, Quaglia G, Fogliano V. Flavonoid and carbohydrate contents in tropea red onions: Effects of homelike peeling and storage. Journal of Agricultural and Food Chemistry. 2002;**50**:1904-1910

[28] Fossen T, Andersen ØM. Anthocyanins from red onion, *Allium cepa,* with novel aglycone. Phytochemistry. 2003;**62**:1217-1220

[29] Yang J, Meyers KJ, Van Der Heide J, Lui RH. Varietal differences in phenolic content and antioxidant and antiproliferative activities of onions. Journal of Agricultural and Food Chemistry. 2004;**52**:6787-6793

[30] Elhassaneen YA, Sanad MI. Phenolics, selenium, vitamin C, amino acids and pungency levels and antioxidant activities of two Egyptian onion varieties. American Journal of Food Technology. 2009;**4**:241-254

[31] Dixon RA, Achnine L, Kota P, Liu CJ, Reddy MSS, Wang L. The phenylpropanoid pathway and plant defence—A genomics perspective. Molecular Plant Pathology. 2002; **3**:371-390

[32] Treutter D. Significance of flavonoids in plant resistance: A review. Environmental Chemistry Letters. 2006;**4**:147-157

[33] Alén-Ruiz F, Pérez-Gregorio MR, Martínez-Carballo E, García-Falcón MS, Simal-Gándara J. Influence of polyphenols on colour and antioxidant value in plant foods. Electronic Journal of Environmental, Agricultural and Food Chemistry. 2008;**7**:3171-3176

[34] Peer W, Murphy A. Flavonoids as signal molecules: Targets of flavonoid action. In: Grotewold E, editor. The Science of Flavonoids. New York: Springer; 2006. pp. 239-268

[35] Peer WA, Murphy AS. Flavonoids and auxin transport: Modulators or regulators? Trends in Plant Science. 2007;**12**:556-563

[36] Taylor LP, Grotewold E. Flavonoids as developmental regulators. Current Opinion in Plant Biology. 2005;**8**:317-323

[37] Bruni R, Sacchetti G. Factors affecting polyphenol biosynthesis in wild and field grown St. John's Wort (*Hypericum perforatum* L. Hypericaceae/Guttiferae). Molecules. 2009;**14**:682-725

[38] Treutter D. Managing phenol contents in crop plants by phytochemical farming and breeding-visions and constraints. International Journal of Molecular Sciences. 2010; **11**:807-857

[39] Leskovar DI, Crosby K, Jifon JL. Impact of agronomic practices on phytochemicals and quality of vegetable crops. Acta Horticulturae. 2009;**841**:317-322

[40] Lee J, Mitchell AE. Quercetin and isorhamnetin glycosides in onion (*Allium cepa* L.): Varietal comparison, physical distribution, coproduct evaluation, and long-term storage stability. Journal of Agricultural and Food Chemistry. 2011;**59**:857-863

[41] Kaur C, Joshi S, Kapoor HC. Antioxidants in onion (*Allium cepa* L.) cultivars grown in India. Journal of Food Biochemistry. 2009;**33**:184-200

[42] Lachman J, Pronek D, Hejtmankovà A, Dudjak J, Pivec V, Faitovà K. Total polyphenol and main flavonoid antioxidants in different onion (*Allium cepa* L.) varieties. Horticultural Science. 2003;**30**:142-147

[43] Lako J, Trenerry VC, Rochfort S. Routine analytical methods for use in South Pacific regional laboratories for determining naturally occurring antioxidants in food. International Food Research Journal 2008;**15**:313-323

[44] Dalamu, Kaur C, Singh M, Walia S, Joshi S, Munshi AD. Variations in phenolics and antioxidants in Indian onions (*Allium cepa* L.): Genotype selection for breeding. Nutrition & Food Science. 2010;**40**:6-19

[45] Patil BS, Pike LM, Yoo KS. Variation in the quercetin content in different colored onions (*Allium cepa* L.). Journal of the American Society for Horticultural Science. 1995;**120**:909-913

[46] Crozier A, Lean MEJ, McDonald MS, Black C. Quantitative analysis of the flavonoid content of commercial tomatoes, onions, lettuce, and celery. Journal of Agricultural and Food Chemistry. 1997;**45**:590-595

[47] Cherif M, Arfaoui A, Rhaiem A. Phenolic compounds and their role in bio-control and resistance of chickpea to fungal pathogenic attacks. Tunisian Journal of Plant Protection. 2007;**2**:7-21

[48] Kavousi HR, Marashi H, Mozafari J, Bagheri AR. Expression of phenylpropanoid pathway genes in chickpea defense against race 3 of ascochyta rabiei. Plant Pathology Journal. 2009;**8**:127-132

[49] Lachman J, Orsak M, Pivec V. Flavonoid antioxidants and ascorbic acid in onion (*Allium cepa*). Zahradnictvi-UZPI. 1999;**26**:125-134

[50] Vågen IM, Slimestad R. Amount of characteristic compounds in 15 cultivars of onion (*Allium cepa* L.) in controlled field trials. Journal of Science and Food Agriculture. 2008;**88**:404-411

[51] Okamoto D, Noguchi Y, Muro T, Morishita M. Genetic variation of quercetin glucoside content in onion (*Allium cepa* L.). Journal of the Japanese Society for Horticultural Science. 2006;**75**:100-108

[52] Lombard KA, Geoffriau E, Peffley EB. Total quercetin content in onion: Survey of cultivars grown at various locations. HortTechnology. 2004;**14**:628-630

[53] Hichri I, Barrieu F, Bogs J, Kappel C, Delrot S, Lauvergeat V. Recent advances in the transcriptional regulation of the flavonoid biosynthetic pathway. Journal of Experimental Botany. 2011;**62**(8):2465-2483

[54] Pizarro CL, Bisigato AJ. Allocation of biomass and photoassimilates in juvenile plants of six Patagonian species in response to five water supply regimes. Annals of Botany (London). 2010;**106**:297-307

[55] Bilyk A, Cooper PL, Sapers GM. Varietal differences in distribution of quercetin and kaempferol in onion (*Allium cepa* L.) tissue. Journal of Agricultural and Food Chemistry. 1984;**32**:274-276

[56] Németh K, Piskula MK. Food content, processing, absorption and metabolism of onion flavonoids. CRC Critical Reviews in Food Science and Nutrition. 2007;**47**:397-409

[57] Hirota S, Shimoda T, Takahama U. Tissue and spatial distribution of flavonol and peroxidase in onion bulbs and stability of flavonol glucosides during boiling of the scales. Journal of Agricultural and Food Chemistry. 1998;**46**:3497-3502

[58] Takahama U, Hirota S. Deglucosidation of quercetin glucosides to the aglycone and formation of antifungal agents by peroxidase-dependent oxidation of quercetin on browning of onion scales. Plant and Cell Physiology. 2000;**41**:1021-1029

[59] Takahama U, Oniki T, Hirota S. Phenolic components of brown scales of onion bulbs produce hydrogen peroxide by autooxidation. Journal of Plant Research. 2001;**114**:395-402

[60] Slimestad R, Vågen IM. Distribution of non-structural carbohydrates, sugars, flavonols and pyruvate in scales of onions, *Allium cepa* L. Journal of Food, Agriculture and Environment. 2009;**7**:289-294

[61] Lee SU, Lee JH, Choi SH, Lee JS, Ohnisi-Kameyama M, Kozukue N, Levin CE, Friedman M. Flavonoid content in fresh, home-processed, and light-exposed onions and in dehydrated commercial onion products. Journal of Agricultural and Food Chemistry. 2008;**56**:8541-8548

[62] Beesk N, Perner H, Schwarz D, George E, Kroh LW, Rohn S. Distribution of quercetin-3,4′-O-diglucoside, quercetin-4′-O-monoglucoside, and quercetin in different parts of the onion bulb (*Allium cepa* L.) influenced by genotype. Food Chemistry. 2010;**122**:566-571

[63] Trammell KW, Peterson CE. Quantitative differences in the flavonol content of yellow onion, *Allium cepa* L. Journal of the American Society for Horticultural Science. 1976;**101**:205-207

[64] Mogren LM, Olsson ME, Gertsson UE. Quercetin content in field-cured onions (*Allium cepa* L.): Effects of cultivar, lifting time, and nitrogen fertilizer level. Journal of Agricultural and Food Chemistry. 2006;**54**:6185-6191

[65] Mogren LM, Olsson ME, Gertsson UE. Quercetin content in stored onions (*Allium cepa* L.): Effects of storage conditions, cultivar, lifting time and nitrogen fertiliser level. Journal of Science and Food Agriculture. 2007;**87**:595-1602

[66] Dixon RA, Paiva NL. Stress-induced phenylpropanoid metabolism. Plant Cell. 1995;**7**: 1085-1097

[67] Harborne JB, Williams CA. Advances in flavonoid research since 1992. Phytochemistry. 2000;**55**:481-504

[68] Stewart AJ, Chapman W, Jenkins GI, Graham I, Martin T, Crozier A. The effect of nitrogen and phosphorus deficiency on flavonol accumulation in plant tissues. Plant, Cell & Environment. 2001;**24**:1189-1197

[69] Guidi L, Lorefice G, Pardossi A, Malorgio F, Tognoni F, Soldatini GF. Growth and photosynthesis of *Lycopersicon esculentum* (L.) plants as affected by nitrogen deficiency. Biologia Plantarum. 1997;**40**:235-244

[70] Patil BS, Pike LM, Hamilton BK. Changes in quercetin concentration in onion (*Allium cepa* L.) owing to location, growth stage and soil type. New Phytologist. 1995;**130**:349-355

[71] Mogren LM, Olsson ME, Gertsson UE. Effects of cultivar, lifting time and nitrogen fertiliser level on quercetin content in onion (*Allium cepa* L.) at lifting. Journal of Science and Food Agriculture. 2007;**87**:470-476

[72] Perner H, Rohn S, Driemel G, Batt N, Schwarz D, Kroh LW, George E. Effect of nitrogen species supply and mycorrhizal colonization on organosulfur and phenolic compounds in onions. Journal of Agricultural and Food Chemistry. 2008;**56**:3538-3545

[73] Jaakola L, Määttä-Riihinen K, Kärenlampi S, Hohtola A. Activation of flavonoid biosynthesis by solar radiation in bilberry (*Vaccinium myrtillus* L.) leaves. Planta. 2004;**218**:721-728

[74] Winkel-Shirley B. Biosynthesis of flavonoids and effects of stress. Current Opinion in Plant Biology. 2002;**5**:218-223

[75] Macheix JJ, Fleuriet A, Billot J. Fruits Phenolics. Boca Raton, FL: CRC Press; 1990

[76] Patil BS, Pike LM. Distribution of quercetin content in different rings of various coloured onion (*Allium cepa* L.) cultivars. Journal of Horticultural Sciences. 1995;**70**:643-650

[77] Higashio H, Hirokane H, Sato F, Tokuda S, Uragami A. Effect of UV irradiation after the harvest on the content of flavonoid in vegetables. Acta Horticulturae. 2005;**682**:1007-1012

[78] Ren H, Endo H, Hayashi T. Antioxidative and antimutagenic activities and polyphenol content of pesticide-free and organically cultivated green vegetables using water-soluble chitosan as a soil modifier and leaf surface spray. Journal of Science and Food Agriculture. 2001;**81**:1426-1432

[79] Grinder-Pedersen L, Rasmussen SE, Bügel S, Jørgensen LV, Dragsted LO, Gundersen V, Sandström B. Effect of diets based on foods from conventional versus organic production on intake and excretion of flavonoids and markers of antioxidative defense in humans. Journal of Agricultural and Food Chemistry. 2003;**51**:5671-5676

[80] Mogren LM, Caspersen S, Olsson ME, Gertsson UE. Organically fertilized onions (*Allium cepa* L.): Effects of the fertilizer placement method on quercetin content and soil nitrogen dynamics. Journal of Agricultural and Food Chemistry. 2008;**56**:361-367

[81] Faller ALK, Fialho E. Polyphenol content and antioxidant capacity in organic and conventional plant foods. Journal of Food Composition and Analysis. 2010;**23**:561-568

[82] Søltoft M, Nielsen J, Holst Laursen K, Husted S, Halekoh U, Knuthsen P. Effects of organic and conventional growth systems on the content of flavonoids in onions and phenolic acids in carrots and potatoes. Journal of Agricultural and Food Chemistry. 2010;**58**:10323-10329

[83] Lee J, Sunkyoung H, Ha I, Min B, Hwang H, Lee S. Comparison of bulb and leaf quality, and antioxidant compounds of intermediate-day onion from organic and conventional systems. Horticulture, Environment, and Biotechnology. 2015;**56**(4):427-436

[84] Daniel O, Meier MS, Schlatter J, Frischknecht P. Selected phenolic compounds in cultivated plants: Ecologic functions, health implications, and modulation by pesticides. Environmental Health Perspectives. 1999;**107**:109-114

[85] Kamal AAM, Mohamed HMA, Aly AAD, Mohamed HAH. Enhanced onion resistance against stemphylium leaf blight disease, caused by Stemphylium vesicarium, by di-potassium phosphate and benzothiadiazole treatments. Plant Pathology Journal. 2008;**24**:171-177

[86] Thompson L, Morris J, Peffley E, Green C, Paré P, Tissue D, Jasoni R, Hutson J, Wehner B, Kane C. Flavonol content and composition of spring onions grown hydroponically or in potting soil. Journal of Food Composition and Analysis. 2005;**18**:635-645

[87] Benkeblia N. Phenylalanine ammonia-lyase, peroxidase, pyruvic acid and total phenolics variations in onion bulbs during long-term storage. Lebensmittel-Wissenschaft & Technologie. 2000;**33**:112-116

[88] Leja M, Mareczeka A, Benb J. Antioxidant properties of two apple cultivars during long-term storage. Food Chemistry. 2003;**80**:303-307

[89] Berno ND, et al. Storage temperature and type of cut affect the biochemical and physiological characteristics of fresh-cut purple onions. Postharvest Biology and Technology. 2014;**93**:91-96

[90] Cantos E, Espín JC, Tomás-Barberán FA. Effect of wounding on phenolic enzymes in six minimally processed lettuce cultivars upon storage. Journal of Agricultural and Food Chemistry. 2001;**49**(1):322-330

[91] Islek M, Nilufer-Erdil D, Knuthsen P. Changes in flavonoids of sliced and fried yellow onions (*Allium cepa* L. var. zittauer) during storage at different atmospheric, temperature and light conditions. Journal of Food Processing and Preservation. 2015;**39**(4):357-368

[92] Harris S, et al. Human exposure modelling of quercetin in onions (*Allium cepa* L.) following thermal processing. Food Chemistry. 2015;**187**:135-139

[93] Ewald C, et al. Effect of processing on major flavonoids in processed onions, green beans, and peas. Food Chemistry. 1999;**64**(2):231-235

[94] Cisneros Zevallos L. The use of controlled postharvest abiotic stresses as a tool for enhancing the nutraceutical content and adding value of fresh fruits and vegetables. Journal of Food Science. 2003;**68**(5):1560-1565

[95] Reyes LF, Cisneros-Zevallos L. Wounding stress increases the phenolic content and antioxidant capacity of purple-flesh potatoes (*Solanum tuberosum* L.). Journal of Agricultural and Food Chemistry. 2003;**51**(18):5296-5300

[96] Ke D, Saltveit ME. Wound induced ethylene production, phenolic metabolism and susceptibility to russet spotting in iceberg lettuce. Physiologia Plantarum. 1989;**76**(3):412-418

[97] Reyes LF, Villarreal JE, Cisneros-Zevallos L. The increase in antioxidant capacity after wounding depends on the type of fruit or vegetable tissue. Food Chemistry. 2007;**101**(3):1254-1262

[98] Feng X, Liu W. Variation of quercetin content in different tissues of welsh onion (*Allium fistulosum* L.). African Journal of Agricultural Research. 2011;**6**(26):5675-5679

[99] Qureshi AA, et al. Colour and tissue differences in distribution of quercetin in Indian onions (*Allium cepa*). Indian Journal of Agricultural Sciences. 2012;**82**(7):629-631

[100] Chen C, et al. Levels of phenolic compounds, antioxidant capacity, and microbial counts of fresh-cut onions after treatment with a combination of nisin and citric acid. Horticulture, Environment and Biotechnology. 2016;**57**(3):266-273

[101] AMS. U.S. Department of Agriculture, A.M.S. Quality Through Verification Program for the Fresh-cut Produce Industry. Federal Register, **63**; 1998. pp. 47220-47224

[102] Pérez-Gregorio MR, et al. Comparison of sanitizing technologies on the quality appearance and antioxidant levels in onion slices. Food Control. 2011;**22**(12):2052-2058

[103] Ölmez H, Kretzschmar U. Potential alternative disinfection methods for organic fresh-cut industry for minimizing water consumption and environmental impact. LWT—Food Science and Technology. 2009;**42**(3):686-693

[104] Zhang J, Yang H. Effects of potential organic compatible sanitisers on organic and conventional fresh-cut lettuce (*Lactuca sativa* Var. Crispa L). Food Control. 2017;**72**:20-26

[105] Beerli KMC, et al. Effect of sanitizers on the microbial, physical and physical-chemical characteristics of fresh-cut onions (*Allium cepa* L). Ciencia e Agrotecnologia. 2004;**28**(1):107-112

[106] Stevens K, et al. Nisin treatment for inactivation of Salmonella species and other gram-negative bacteria. Applied and Environmental Microbiology. 1991;**57**(12):3613-3615

[107] Carletti L, et al. Use of ozone in sanitation and storage of fresh fruits and vegetables. Journal of Food, Agriculture and Environment. 2013;**11**(3-4):585-589

[108] Ferreres F, Gil MI, Tomas-Barberan FA. Anthocyanins and flavonoids from shredded red onion and changes during storage in perforated films. Food Research International. 1996;**29**(3):389-395

[109] Ozyurt D, et al. Effect of oven and microwave heating on the total antioxidant capacity of dietary onions grown in Turkey. International Journal of Food Properties. 2013;**16**(3):536-548

[110] Juániz I, et al. Influence of heat treatment on antioxidant capacity and (poly)phenolic compounds of selected vegetables. Food Chemistry. 2016;**197**:466-473

[111] Makris DP, Rossiter JT. Domestic processing of onion bulbs (*Allium cepa*) and asparagus spears (*Asparagus officinalis*): Effect on flavonol content and antioxidant status. Journal of Agricultural and Food Chemistry. 2001;**49**(7):3216-3222

[112] Ross JA, Kasum CM. Dietary flavonoids: Bioavailability, metabolic effects, and safety. Annual Review of Nutrition. 2002;**22**(1):19-34

[113] Pinho C, et al. Impact of freezing on flavonoids/radical-scavenging activity of two onion varieties. Czech Journal of Food Sciences. 2015;**33**(4):340-345

[114] Alezandro MR, et al. Commercial spices and industrial ingredients: Evaluation of antioxidant capacity and flavonoids content for functional foods development. Ciência e Tecnologia de Alimentos. 2011;**31**(2):527-533

[115] Sahoo NR, et al. Impact of pretreatment and drying methods on quality attributes of onion shreds. Food Technology and Biotechnology. 2015;**53**(1):57-65

[116] FAO—Food and Agriculture Organization of the United Nations. 2013. FAOSTAT URL: http://faostat.fao.org/site/567/DesktopDefault.aspx?PageID=567#ancor

[117] Benítez V, Mollá E, Martín-Cabrejas MA, Aguilera Y, López-Andréu FJ, Cools K, Terry LA, Esteban RM. Characterization of industrial onion wastes (*Allium cepa* L.): Dietary fiber and bioactive compounds. Plant Food for Human Nutrition. 2011;**66**:48-57

[118] Schieber A, Stintzing FC, Carle R. By-products of plant food processing as a source of functional compounds—Recent developments. Trends in Food Science & Technology. 2001;**12**:401-413

[119] Coventry E, Noble R, Mead A, Whipps JM. Control of Allium white rot (*Sclerotium cepivorum*) with composted onion waste. Soil Biology and Biochemistry. 2002;**34**:1037-1045

[120] González-Sáiz JM, Esteban-Díez I, Rodríguez-Tecedor S, Pizarro C. Valorization of onion waste and by-products: MCR-ALS applied to reveal the compositional profiles of alcoholic fermentations of onion juice monitored by near-infrared spectroscopy. Biotechnology and Bioengineering. 2008;**101**:776-787

[121] Laufenberg G, Kunz B, Nystroem M. Transformation of vegetable waste into value added products: (A) the upgrading concept; (B) practical implementations. Bioresource Technology. 2003;**87**:167-198

[122] Roldán E., Sánchez-Moreno C, de Ancos B, Cano MP. Characterisation of onion (*Allium cepa* L.) by-products as food ingredients with antioxidant and antibrowning properties. Food Chemistry. 2008;**108**:907-916

[123] Downes K, Chope GA, Terry LA. Effect of curing at different temperatures on biochemical composition of onion (*Allium cepa* L.) skin from three freshly cured and cold stores UK-grown onion cultivars. Postharvest Biology and Technology. 2009;**54**:80-86

[124] Downes K, Chope GA, Terry LA. Postharvest application of ethylene and 1-methylcyclopropene either before or after curing affects aonio (*Allium cepa* L.) bulb quality during long term cold storage. Postharvest Biology and Technology. 2010;**55**:36-44

[125] Benítez V, Mollá E, Martín-Cabrejas MA, Aguilera Y, López-Andréu FJ, Esteban RM. Onion Products: Source of Healthy Compounds. Hauppauge, N.Y. (USA): Nova Sic. Pub., Inc.; 2012

[126] Yao LH, Jiang YM, Shi J, Tomás-Barberán FA, Datta N, Singanusong R, Chen SS. Flavonoids in food and their health benefits. Plant Foods for Human Nutrition. 2004;**59**:113-122

[127] Pérez-Gregorio R, Simal-Gandara J. Critical review on bioactive food components and their functional mechanisms, biological effects and health outcomes. Current Pharmaceutical Design. 2017; **23** (E-pub Ahead of Print) DOI: 10.2174/138161282366617 0317122913 (Curr Pharm Des. 2017 Mar 17). In press

[128] Pérez-Gregorio R, Simal-Gandara J. A critical review of the characterization of polyphenol–protein interactions and of their potential use for improving food quality. Current Pharmaceutical Design. 2017; **23** (E-pub Ahead of Print) DOI: 10.2174/138161282366617 0202112530 (Curr Pharm Des. 2017 Feb 2). In press

[129] Pérez-Gregorio MR, Regueiro J, Simal-Gándara J, Rodrigues AS, Almeida DPF. Increasing the added-value of onions as a source of antioxidant flavonoids: A critical review. Critical Reviews in Food Science and Nutrition. 2014;**55**(2):202-218

[130] Choi IS, Cho EJ, Moon JH, Bae HJ. Onion skin waste as a valorization resource for the by-products quercetin and biosugar. Food Chemistry. 2015;**188**:537-542

[131] EUFRIN—European Fruit Research Institutes Network. 2016. Strategic innovation and research agenda for the fruit and vegetable sector. http://freshfel.org/wp-content/uploads/2016/11/Final-STRATEGIC-RESEARCH-AND-INNOVATION-AGENDA-FOR-THE-FRUIT-AND-VEGETABLE-SECTOR-v2016-21-10-2016.pdf

[132] Roldán-Marin E, Sanchez-Moreno C, Lloria R, De Ancos B, Cano MP. Onion high-pressure processing: Flavonol content and antioxidant activity. LWT—Food Science and Technology. 2009;**42**(4):835-841

[133] Scalbert A, and Williamson G. Dietary intake and bioavailability of polyphenols. Journal of Nutrition. 2000;**130**:2073S-2085S

[134] Pobłocka-Olech L, et al. TLC determination of flavonoids from different cultivars of *Allium cepa* and *Allium ascalonicum*. Acta Pharmaceutica. 2016;**66**(4):543-554

[135] Morris J. Composition and flavonoid levels in onions (*A. cepa*) grown in hydroponics in greenhouses and growth chambers [MSc thesis]. Lubbock, TX: Texas Tech University. Available from: University Library, Lubbock, TX; AC805.T3 no 138.; 2001. 145pp

[136] Lin JY, Tang CY. Determination of total phenolic and flavonoid contents in selected fruits and vegetables, as well as their stimulatory effects on mouse splenocyte proliferation. Food Chemistry. 2007;**101**:140-147

[137] Escarpa A, González MC. Approach to the content of total extractable phenolic compounds from different food samples by comparison of chromatographic and spectrophotometric methods. Analytica Chimica Acta. 2001;**427**:119-127

[138] Sellappan S, Akoh C. Flavonoids and antioxidant capacity of Georgia-grown Vidalia onions. Journal of Agricultural and Food Chemistry. 2002;**50**:5338-5342

[139] Lister CE, Podivinsky EP. Antioxidants in New Zealand grown fruit and vegetables. Polyphenols Communications. 1998;**98**:273

[140] Brat P, Georgé S, Bellamy A, Du Chaffaut L, Scalbert A, Mennen L, Arnault N, Amiot MJ. Daily polyphenol intake in France from fruits and vegetables. Journal of Nutrition. 2006;**136**:1-6

[141] Vinson JA, Hao Y, Su X, Zubik L. Phenol antioxidant quantity and quality in foods: Vegetables. Journal of Agricultural and Food Chemistry. 1998;**46**(9):3630-3634

[142] Ninfali P, Bacchiocca M. Polyphenols and antioxidant capacity of vegetables under fresh and frozen conditions. Journal of Agricultural and Food Chemistry. 2003;**51**:2222-2226

[143] Marinova D, Ribarova F, Atanassova M. Total phenolics and total flavonoids in Bulgaria fruits and vegetables. Journal of the University of Chemical Technology and Metallurgy. 2005;**40**:255-260

[144] Santas J, Carbó R, Gordon MH, Almajano MP. Comparison of the antioxidant activity of two Spanish onion varieties. Food Chemistry. 2008;**107**:1210-1216

[145] Zill-E-Huma, Vian MA, Mangonnat JF, Chemat F. Clean recovery of antioxidant flavonoids from onions: Optimising solvent free microwave extraction method. Journal of Chromatography A. 2009;**1216**:7700-7707

[146] Golisová A, Slamka P, Kóňa J. Content of flavonoids in onion (*Allium cepa* L.) under various fertilization. Acta Horticulturae et Regiotecturae. 2008;**11**(2):54-56

[147] Golisová A, Slamka P, Loek O, Hanáková E. Content of phenols in onion (*Allium cepa* L.) fertilized with nitrogen, sulphur and iron. Agrochémia. 2009;**13**(3):14-18

[148] Smith C, Lombard KA, Peffley EB, Liu W. Genetic analysis of quercetin in onion (*Allium cepa* L.) 'Laddy Raider'. Texas Journal of Agriculture and Natural Resources. 2003;**16**:24-28

[149] Németh K, Takàcsova M, Piskua MK. Effect of cooking on yellow onion quercetin. Polish Journal of Food and Nutrition Sciences. 2003;**12**(53):170-174

[150] Bonaccorsi P, Caristi C, Gargiulli C, Leuzzi U. Flavonol glucoside profile of southern Italian red onion (*Allium cepa* L.). Journal of Agricultural and Food Chemistry. 2005;**53**: 2733-2740

[151] Hertog MGL, Hollman PCH, Katan MB. Content of potentially anticarcinogenic flavonoids of 28 vegetables and 9 fruits commonly consumed in The Netherlands. Journal of Agricultural and Food Chemistry 1992;**40**:2379-2383

[152] Hertog MGL, Hollman PCH, Venema DP. Optimization of quantitative HPLC determination of potentially anticarcinogenic flavonoids in fruits and vegetables. Journal of Agricultural and Food Chemistry. 1992;**40**:1591-1598

[153] Bhagwat S, Haytowitz DB. USDA Database for the Flavonoid Content of Selected Foods, Release 3.2. U.S. Department of Agriculture, Agricultural Research Service. Nutrient Data Laboratory Home Page; 2015. https://www.ars.usda.gov/ARSUserFiles/80400525/Data/Flav/Flav3.2.pdf

[154] Mattila P, Astola J, Kumpulainen J. Determination of flavonoids in plant material by HPLC with diode-array and electro-array detection. Journal of Agricultural and Food Chemistry. 2000;**48**(12):5834-5841

[155] Nuutila AM, Kammiovirta K, Oksman-Caldentey KM. Comparison of methods for the hydrolysis of flavonoids and phenolic acids from onion and spinach for HPLC analysis. Food Chemistry. 2002;**76**:519-525

[156] Ostrowska E, Gabler NK, Sterling SJ, Tatham BG, Jones RB, Eagling DR, Jois M, Dunshea FR. Consumption of brown onions (*Allium cepa* var. cavalier and var. destiny) moderately modulates blood lipids, haematological and haemostatic variables in healthy pigs. British Journal of Nutrition. 2004;**91**:211-218

[157] Proteggente AR, Pannala AS, Pagana G, Van Buren L, Wagner E, Wiseman S. The antioxidant activity of regularly consumed fruit and vegetables reflects their phenolic and vitamin C composition. Free Radical Research. 2002:**36**:217-233

[158] Lombard KV, Geoffriau E, Peffley E. Flavonoid quantification in onion by spectrophotometric and high performance liquid analysis. Horticultural Science. 2002;**37**:682-685

[159] Tsushida T, Suzuki M. Content of flavonol glucosides and some properties of enzymes metabolizing the glucosides in onion. Flavonoids in fruits and vegetables, part II. Nippon Shokuhin Kagaku Kogaku Kaishi. 1996;**43**:642-649

[160] Mazza G, Maniati E. In: Mazza G, Miniati E, editors. Anthocyanins in Fruits, Vegetables, and Grains. FL: CRC Press; 1993

[161] Winter CK, Davis SF. Organic foods. Journal of Food Science. 2006;**71**:R117-R124

[162] Srinivasa Rao NK. Onion. In: Srinivasa Rao NK, Shivashankara KS, Laxman RH, editors. Abiotic Stress Physiology of Horticultural Crops. Part II-Abiotic Stress Tolerance in Horticultural Crops: Vegetables, Springer India; 2016. 133-149 pp Copyright Holder: Springer India (ISBN: 978-81-322-2723-6 (Print) 978-81-322-2725-0 (Online))

[163] Mohamed AA, Aly AA. Alteration of some secondary metabolites and enzymes activity by using exogenous antioxidant compound in onion plants grown under seawater salt stress. American-Eurasian Journal of Scientific Research. 2008;**3**(2):139-146